THE BEGINNINGS OF WESTERN SCIENCE

The Beginnings of Western Science

The European Scientific Tradition in
Philosophical, Religious, and Institutional
Context, Prehistory to A.D. 1450

SECOND EDITION

David C. Lindberg

The University of Chicago Press CHICAGO AND LONDON

DAVID C. LINDBERG is the Hilldale Professor Emeritus of the History of Science at the University of Wisconsin–Madison and past president of the History of Science Society. He is the author or editor of many books, including, with coeditor Ronald L. Numbers, *When Science and Christianity Meet*, also published by the University of Chicago Press.

The University of Chicago Press, Chicago 60637
The University of Chicago Press, Ltd., London
© 1992, 2007 by The University of Chicago
All rights reserved. Published 2007
Printed in the United States of America

16 15 14 13 12 11 10 09 08 07   1 2 3 4 5

ISBN-13: 978-0-226-48205-7 (paper)
ISBN-10: 0-226-48205-7 (paper)

Library of Congress Cataloging-in-Publication Data

Lindberg, David C.
   The beginnings of western science : the European scientific tradition in philosophical, religious, and institutional context, prehistory to A.D. 1450 / David C. Lindberg. —2nd ed.
      p. cm.
   Includes bibliographical references and index.
   ISBN-13: 978-0-226-48205-7 (pbk. : alk. paper)
   ISBN-10: 0-226-48205-7 (pbk. : alk. paper)   1. Science, Ancient—History.   2. Science, Medieval—History.   I. Title.
   Q124.95.L55   2007
   509.4—dc22

                                                          2007029485

♾ The paper used in this publication meets the minimum requirements of the American National Standard for Information Sciences—Permanence of Paper for Printed Library Materials, ANSI Z39.48-1992.

To Greta, Chris, John, Erik, Liana, Annie, and Davey,
who have brought so much joy to my life.

Contents

List of Illustrations • xi
Preface • xv

1   **SCIENCE BEFORE THE GREEKS** • 1
What Is Science? • 1
Prehistoric Attitudes toward Nature • 3
The Beginnings of Science in Egypt and Mesopotamia • 12

2   **THE GREEKS AND THE COSMOS** • 21
The World of Homer and Hesiod • 21
The First Greek Philosophers • 25
The Milesians and the Question of Underlying Reality • 27
The Question of Change • 32
The Problem of Knowledge • 33
Plato's World of Forms • 34
Plato's Cosmology • 38
The Achievement of Early Greek Philosophy • 43

3   **ARISTOTLE'S PHILOSOPHY OF NATURE** • 45
Life and Works • 45
Metaphysics and Epistemology • 46
Nature and Change • 49
Cosmology • 52
Motion, Terrestrial and Celestial • 56
Aristotle as a Biologist • 60
Aristotle's Achievement • 65

4   **HELLENISTIC NATURAL PHILOSOPHY** • 67
Schools and Education • 67
The Lyceum after Aristotle • 73
Epicureans and Stoics • 76

5 THE MATHEMATICAL SCIENCES IN ANTIQUITY · 82
The Application of Mathematics to Nature · 82
Greek Mathematics · 83
Early Greek Astronomy · 86
Cosmological Developments · 95
Hellenistic Planetary Astronomy · 98
The Science of Optics · 105
The Science of Weights · 109

6 GREEK AND ROMAN MEDICINE · 111
Early Greek Medicine · 111
Hippocratic Medicine · 113
Hellenistic Anatomy and Physiology · 119
Hellenistic Medical Sects · 122
Galen and the Culmination of Hellenistic Medicine · 124

7 ROMAN AND EARLY MEDIEVAL SCIENCE · 132
Greeks and Romans · 132
Popularizers and Encyclopedists · 136
Translations · 146
The Role of Christianity · 148
Roman and Early Medieval Education · 150
Two Early Medieval Natural Philosophers · 157
Learning and Science in the Greek East · 158

8 ISLAMIC SCIENCE · 163
Eastward Diffusion of Greek Science · 163
The Birth, Expansion, and Hellenization of Islam · 166
Translation of Greek Science into Arabic · 169
Islamic Reception and Appropriation of Greek Science · 173
The Islamic Scientific Achievement · 176
The Fate of Islamic Science · 189

9 THE REVIVAL OF LEARNING IN THE WEST · 193
The Middle Ages · 193
Carolingian Reforms · 194
The Schools of the Eleventh and Twelfth Centuries · 203
Natural Philosophy in the Twelfth-Century Schools · 209
The Translation Movement · 215
The Rise of Universities · 218

10 THE RECOVERY AND ASSIMILATION OF GREEK
AND ISLAMIC SCIENCE · 225
The New Learning · 225
Aristotle in the University Curriculum · 226

Points of Conflict · 228
Resolution: Science as Handmaiden · 233
Radical Aristotelianism and the Condemnations of 1270 and 1277 · 243
The Relations of Philosophy and Theology After 1277 · 249

11 THE MEDIEVAL COSMOS · 254
The Structure of the Cosmos · 254
Mathematical Astronomy · 261
Astrology · 270
The Surface of the Earth · 277

12 THE PHYSICS OF THE SUBLUNAR REGION · 286
Matter, Form, and Substance · 286
Combination and Mixture · 288
Alchemy · 290
Change and Motion · 295
The Nature of Motion · 297
Mathematical Description of Motion · 299
The Dynamics of Local Motion · 306
Quantification of Dynamics · 309
The Science of Optics · 313

13 MEDIEVAL MEDICINE AND NATURAL HISTORY · 321
The Medical Tradition of the Early Middle Ages · 321
The Transformation of Western Medicine · 329
Medical Practitioners · 330
Medicine in the Universities · 333
Disease, Diagnosis, Prognosis, and Therapy · 335
Anatomy and Surgery · 343
Development of the Hospital · 348
Natural History · 351

14 THE LEGACY OF ANCIENT AND MEDIEVAL SCIENCE · 357
The Continuity Question · 357
Candidates for Revolutionary Status · 359
The Scientific Revolution · 364

Notes · 369
Bibliography · 413
Index · 463

Illustrations

FIGURES

1.1   A Babylonian mathematical problem · 14
1.2   A Babylonian zig-zag function, representing arithmetic series · 17
1.3   A column from the Edwin Smith surgical papyrus (ca. 1600 B.C.) · 19
2.1   A shrine to the earth goddess Gaia at Delphi (4th century B.C.) · 23
2.2   A bronze statue of Zeus · 24
2.3   The ruins of ancient Ephesus · 30
2.4   Plato · 35
2.5   The five Platonic solids: tetrahedron, octahedron, icosahedron, cube, and dodecahedron · 40
2.6   The celestial sphere according to Plato · 42
3.1   Aristotle · 46
3.2   Square of opposition of the Aristotelian elements and qualities · 54
3.3   The Aristotelian cosmos · 56
4.1   The schools of Hellenistic Athens · 71
4.2   The Parthenon · 72
4.3   Epicurus · 76
5.1   The incommensurability of the side and the diagonal of a square · 84
5.2   Determining the area of a circle by the method of "exhaustion" · 85
5.3   Two-sphere model of the cosmos · 87
5.4   The observed retrograde motion of Mars · 88
5.5   The Eudoxan spheres for one of the planets · 89
5.6   The Eudoxan spheres and the hippopede · 91
5.7   Aristotelian nested spheres · 93
5.8   Aristarchus's method for determining the ratio between the solar and lunar distances from the earth · 96
5.9   Eratosthenes' calculation of the earth's circumference · 97
5.10   Ptolemy's eccentric model · 100
5.11   Ptolemy's epicycle-on-deferent model · 101
5.12   Ptolemy's epicycle-on-deferent model with the planet on the inward side of the epicycle · 101

5.13   Retrograde motion of a planet explained by the epicycle-on-deferent model · 101

5.14   Ptolemy's equant model · 103

5. 15   Ptolemy's model for the superior planets · 104

5.16   The geometry of vision according to Euclid · 106

5.17   Vision by reflected rays according to Ptolemy · 107

5.18   Ptolemy's theory of refraction · 108

5.19   Ptolemy's apparatus for measuring angles of incidence and refraction · 108

5.20   The balance beam in a state of balance · 109

5.21   The dynamic explanation of the balance beam · 109

5.22   Archimedes' static proof of the law of the lever · 110

6.1   Asclepius · 112

6.2   The theater at Epidaurus (4th c. B.C.) · 114

6.3   Hippocrates · 114

6.4   Greek physician, grave relief, 480 B.C. · 123

7.1   The ancient forum in Rome · 134

7.2   Cicero · 138

7.3   Pliny's monstrous races · 141

7.4   Macrobius on rainfall · 143

7.5   Attempts to capture Martianus Capella's theory of the motions of Venus and Mercury in relation to the sun · 146

7.6   A monk in the monastery library · 153

7.7   A medieval scribe at work · 156

8.1   The ibn Tūlūn Mosque (9th c.), Cairo · 175

8.2   The motion of Mercury according to Ibn al-Shāṭir · 180

8.3   Underground sextant in Samarqand · 182

8.4   The eyes and visual system according to Ibn al-Haytham · 183

8.5   Ḥunayn ibn Isḥāq on the anatomy of the eye · 186

8.6   Interior of the Great Mosque of Cordoba • 190

9.1   Planetary apsides • 198

9.2   Personification of the quadrivium • 200

9.3   A sixteenth-century armillary sphere • 202

9.4   A grammar school scene • 204

9.5   The chained library of Hereford Cathedral (England) • 207

9.6   Hugh of St. Victor teaching in Paris • 208

9.7   God as architect of the universe • 211

9.8   Mob Quad, Merton College, Oxford • 222

9.9   Doorway to a medieval school • 222

10.1   The beginning of Avicenna's Physics • 229

10.2   The Basilica of St. Francis, Assisi • 232

10.3   The skeleton of Robert Grosseteste • 235

10.4   Albert the Great • 238

10.5   Cathedral of Notre Dame, Paris • 245

11.1   The simplified Aristotelian cosmology • 258

11.2   Astrolabe, Italian, ca. 1500 • 262

11.3   An "exploded" view of the astrolabe • 263

11.4   Stereographic projection of the almucantars • 264

11.5   The "new quadrant" of Profatius Judaeus • 265

11.6   The model for one of the superior planets • 266

11.7   The *Alfonsine Tables* • 268

11.8   An astronomer observing with an astrolabe • 269

11.9   Ibn al-Haytham's physical-sphere model of the Ptolemaic deferent and epicycle • 270

11.10   The Arabic astrologer Albumasar or Abū Maʿshar • 276

11.11   Theodoric of Freiberg's theory of the rainbow • 278

11.12   A T–O map • 280

11.13   A modified T–O map, the Beatus map (1109 A.D.) • 281

11.14   A portolan chart by Fernão Vaz Dourado (ca.1570) • 282

11.15   Nicole Oresme • 284

12.1   Alchemical apparatus, including furnaces and stills • 293

12.2   The use of a line segment to represent the intensity of a quality • 302

12.3   The distribution of temperatures in a rod • 302

12.4   Oresme's system for representing the distribution of any quality in a subject • 302

12.5   The distribution of velocities in a rod rotating about one end • 303

12.6   Velocity as a function of time • 303

12.7   The representation of various motions • 304

12.8   Oresme's geometrical proof of the Merton rule • 305

12.9   Incoherent radiation from two points of a luminous body • 315

12.10   Rays issuing from the end points of the visible object mixing within the eye • 316

12.11   The visual cone and the eye in Alhacen's intromission theory of vision • 317

12.12   A page from the *Perspectiva communis* of John Pecham • 319

13.1   A page from a Greek manuscript of Dioscorides' *Materia medica* • 323

13.2   The miraculous healing of a leg • 326

13.3   Arabic surgical instruments • 328

13.4   Constantine the African practicing uroscopy • 330

13.5   Fetuses in the womb • 332

13.6   Trotula • 332

13.7   Medical instruction • 334

13.8   An apothecary shop • 337

13.9   A urine color chart • 340

13.10   Diagnosis by pulse • 341
13.11   A physician's girdle book • 342
13.12   Operation for cataract and nasal polyps • 344
13.13   Operation for scrotal hernia • 344
13.14   Human dissection • 346
13.15   Human anatomy • 348
13.16   A medieval hospital • 350
13.17   A page from the *Herbal* of Pseudo-Apuleius • 352
13.18   A page from a medieval bestiary • 355

MAPS

2.1   The Greek world about 450 B.C. • 26
4.1   Alexander the Great's empire • 68
7.1   The Roman Empire • 133
8.1   Islamic expansion • 167
9.1   Carolingian Empire about 814 • 195
9.2   Medieval universities • 220

Preface

The original edition of this book drew on two decades of experience teaching the history of ancient and medieval science to university undergraduates. Now, with another two decades of teaching experience under my belt and an array of recent scholarship on my bookshelves, I have been given the privilege of producing a revised version. In many ways this is the same book: same chapter titles, mostly the same illustrations, and basically the same story—but with many improvements, large and small. The chapter on Islamic science has been entirely rewritten—altered in both substance and presentation, to reveal the magnitude and sophistication of the medieval Islamic scientific achievement. The concluding chapter, which assesses the medieval contribution to scientific developments of the sixteenth and seventeenth centuries, has also been entirely rewritten. The section on Byzantine science has been enlarged. In the last couple of decades I have acquired a much sharper awareness of the importance of the Mesopotamian contribution to astronomy and have added material accordingly. Medieval alchemy and medieval astrology, generally viewed by the public as pseudoscience, have been given a larger place in the story, thanks to illuminating research by John North and William Newman, which yields some surprises about the relationship of medieval astrology and alchemy to the broader scientific enterprise.

These revisions are just a few of the many improvements. I think it unlikely that a single page of the book has managed to escape the process of revision unaltered. I've had the pleasure of copyediting my own prose, attempting to breathe life into a dead sentence, retracting a claim, softening a judgment, clarifying an explanation, correcting an error. My hope and expectation is that this book, in its second incarnation, will continue to reach a general audience, including students, with the startling news that the ancient and medieval periods were the scene of impressive scientific achievements, which provided a solid foundation for scientific developments of the sixteenth and seventeenth centuries and beyond.

Although I have written with a general audience in mind, I have not shrunk from opportunities to resolve contemporary scholarly disputes when the occasions present themselves. Passages in which I lecture the reader on the proper ways of doing history and warn against a variety of perils will be immediately recognized as the products of long classroom experience; and it is my hope that this book will continue to prove itself suitable for classroom use. But I believe that it will also interest the general educated reader and scholars who do not specialize in the history of ancient and medieval science. No other book of which I am aware covers the same breadth of material, over the same chronological span, at the same level of presentation. In this edition, as in the first, I have more persistently attempted to place ancient and medieval science in philosophical, religious, and institutional (largely educational) context than have the authors of other surveys. And I am quite certain that no other survey of this material has paid as much serious attention to the religious context, without embarrassment and without an apologetic or polemical agenda.

Two remarks about endnotes and bibliography: First, I have used the notes not only for purposes of documentation and acknowledgment of scholarly debt, but also as an opportunity for a running bibliographical commentary, in which I suggest sources where the subject at hand may be fruitfully pursued. Second, I have enlarged the bibliography of this second edition to take into account recent scholarship, increasing the number of entries by about two hundred. In both the notes and bibliography, I have (with the student audience and the general reader in mind) emphasized English-language literature. Sources in foreign languages are included where it seems to me that there is nothing comparable in English.

Finally, nobody covers a subject as large as this without a great deal of help, and I am profoundly indebted to friends and colleagues who have done their best to instruct me in the intricacies of their various specialties and rescue me from confusion and error. I have not always been a compliant pupil, and some will still find in this book interpretations that they do not like. The preface to the first edition of this book contains a long list of scholars to whom I owe a continuing debt of gratitude for their contributions. For advice on portions of this revised version, it is a pleasure to thank Emilie Savage-Smith, A. Mark Smith, and especially my colleague and comrade-in-arms, Michael Shank. My wife, Greta, has been her usual loving, patient, supportive self, and I hope that completion of this revised edition convinces her that I'll finally bring order to my study and become available for yard work.

David Lindberg, October 2007

1 ✸ *Science before the Greeks*

WHAT IS SCIENCE?

The opinion that there was *no science* in the two thousand years covered by this book continues to be stated with considerable regularity and dogmatic fervor. If the claim is true, I have written a book about a nonexistent subject—no mean feat, but not my goal. This book proclaims in its title that it will portray the beginnings of Western science over the approximately three millennia ending about the year A.D. 1450. Was there truly such a thing as science in those times? And if the answer is affirmative, was there enough of it to merit book-length coverage?

Before we can answer these questions, we need a definition of "science"—something that turns out to be surprisingly difficult to come by. There is, of course, the dictionary definition, according to which "science" is organized, systematic knowledge of the material world. But this proves to be so general as to be of little help. For example, do craft traditions and technology count for science, or are science and technology to be distinguished from one another—the former dedicated to theoretical knowledge, the latter to its application? If only theoretical knowledge counts as genuine science, we then need to decide which theories (or which kinds of theory) pass the test. Do astrology and parapsychology, both of which are chock full of theories, count as sciences?

Perceiving that the "theoretical knowledge" criterion is heading toward a dead end, some participants in the debate argue that true science can be recognized by its methodology—specifically, the experimental method, according to which a theory, if it is to be truly scientific, must be built on and tested against the results of observation and experiment. (In the minds of many of its advocates, a series of rigorously defined steps must be employed.) Theories that meet this test are often credited with superior epistemological status or warrant and thus are representative of a privileged way of knowing.

I

Finally, for many people—scientists and general public alike—true science is defined simply by its content—the current teachings of physics, chemistry, biology, geology, anthropology, psychology, and so forth.

This brief foray into lexicography ought to remind us that many words, especially the most interesting ones, have multiple meanings that shift with the contexts of usage or the practices of specific linguistic communities. Every meaning of the term "science" discussed above is a convention accepted by a sizable group of people, who are unlikely to relinquish their favored usage without a fight. From which it follows that we have no choice but to accept a diverse set of meanings as legitimate and do our best to determine from the context of usage what the term "science" means on any specific occasion.

But where does that leave us? Was there anything in Europe or the Near East in the twenty centuries covered by this book that merits the name "science"? No doubt! Many of the ingredients of what we now regard as science were certainly present. I have in mind languages for describing nature, methods for exploring or investigating it (including the performance of experiments), factual and theoretical claims (stated mathematically wherever possible) that emerged from such explorations, and criteria for judging the truth or validity of the claims thus made. Moreover, it is clear that pieces of the resulting ancient and medieval knowledge were, for all practical purposes, identical to what all parties would now judge to be genuine science. Planetary astronomy, geometrical optics, field biology or natural history, and certain branches of medicine are excellent examples.

This is not to deny significant differences—in motivation, instrumentation, institutional support, methodological preferences, mechanisms for the dissemination of theoretical results, and social function. Despite these differences, I believe that we can comfortably employ the expression "science" or "natural science" in the context of antiquity and the Middle Ages. In so doing, we declare that the ancient and medieval activities that we are investigating are the ancestors of modern scientific disciplines and therefore an integral part of their history. It is like my relationship to my paternal grandfather. The differences between us may outweigh the similarities; but I am his descendant, bearing to some extent both his genetic and his cultural stamp. And both of us may honorably claim the family name.

There is a danger that must be avoided. If historians of science were to investigate past practices and beliefs only insofar as those practices and beliefs resemble modern science, the result would be serious distortion. We would not be responding to the past as it existed, but examining it through a modern grid. If we wish to do justice to the historical enterprise, we must take the past for what it was. And that means that we must resist the temptation to scour the past for examples or precursors of modern science. We must respect the way earlier generations approached nature, acknowledging that although it may differ from the modern way, it is nonetheless of interest because it is part of our intellectual ancestry. This is the only suitable way of understanding how we became what we are. The historian, then, requires a very broad definition of "science"—one that will permit investigation of the vast range of practices and beliefs that lie behind, and help us to understand, the modern scientific enterprise. We need to be broad and inclusive, rather than narrow and exclusive; and we should expect that the farther back we go, the broader we will need to be.[1]

I will do my best to heed my own advice, adopting a definition of "science" as broad as that of the historical actors whose intellectual efforts we are attempting to understand. This does not mean, of course, that all distinctions are forbidden. I will distinguish between the craft and theoretical sides of science—a distinction that many ancient and medieval scholars would themselves have insisted upon—and I will focus my attention on the latter.[2] The exclusion of technology and the crafts from this narrative is not meant as a commentary on their importance, but rather as an acknowledgment of the magnitude of the problems confronting the history of technology and its status as a distinct historical specialty having its own skilled practitioners. My concern will be with the beginnings of scientific *theories*, the methods by which they were formulated, and the uses to which they were put; and that will prove a sufficient challenge.

A final word about terminology. Until now, I have consistently employed the word "science" to denote the object of our historical study. The time has come, however, to introduce the alternative expressions "natural philosophy" and "philosophy of nature," which will also appear frequently in this book. These are expressions that ancient and medieval scholars themselves applied to investigations of the natural world that concentrated on questions of material causation, as opposed to mathematical analysis. For the latter, the term "mathematics" did service. And finally, a vocabulary developed for identifying subdisciplines such as astronomy, optics, meteorology, metallurgy, the science of motion, the science of weights, geography, natural history (including both plants and animals), and medicine. Close attention by the reader to context should make the meaning clear in every case.

PREHISTORIC ATTITUDES TOWARD NATURE

From the beginning, the survival of the human race has depended on its ability to cope with the natural environment. Prehistoric people developed impressive technologies for obtaining the necessities of life. They learned how to make tools, start fires, obtain shelter, hunt, fish, and gather fruits and vegetables. Successful hunting and food gathering (and, after about 7000 or 8000 B.C., settled agriculture) required a substantial knowledge of animal behavior and the characteristics of plants. At a more advanced level, prehistoric people learned to distinguish between poisonous and therapeutic herbs. They developed a variety of crafts, including pottery, weaving, and metalworking. By 3500 they had invented the wheel. They were aware of the seasons and perceived the connection between the seasons and various celestial phenomena. In short, they knew a great deal about their environment.

But the word "know," seemingly so clear and simple, is almost as tricky as the term "science"; indeed, it brings us back to the distinction between technology and theoretical science. It is one thing to know *how* to do things, another to know *why* they behave as they do. One can engage in successful and sophisticated carpentry, for example, without any theoretical knowledge of stresses in the timbers one employs. An electrician with only the most rudimentary knowledge of electrical theory can successfully wire a house. It is possible to differentiate between poisonous and therapeutic herbs without possessing any biochemical knowledge that would explain poisonous or therapeutic properties. The point is simply that practical rules of thumb can be effectively employed even in the face of total ignorance of the theoretical principles that lie behind them. You can have "know-how" without theoretical knowledge.

It should be clear, then, that in practical or technological terms, the knowledge of prehistoric humans was great and growing. But what about theoretical knowledge? What did prehistoric people "know" or believe about the origins of the world in which they lived, its nature, and the causes of its numerous and diverse phenomena? Did they have any awareness of general laws or principles that governed the particular case? Did they even ask such questions? We have very little evidence on the subject. Prehistoric culture is by definition oral culture; and oral cultures, as long as they remain exclusively oral, leave no written remains. However, an examination of the findings of anthropologists studying preliterate tribes in the nineteenth and twentieth centuries, along with careful attention to remnants of prehistoric thought carried over into the earliest written records, will allow us to formulate a few tentative generalizations.

Critical to the investigation of intellectual culture in a preliterate society is an understanding of the process of communication. In the absence of writing, the only form of verbal communication is the spoken word; and the only storehouses of knowledge are the memories of individual members of the community. The transmission of ideas and beliefs in such a culture occurs only in face-to-face encounter, through a process that has been characterized as "a long chain of interlocking conversations" between its members. The portion of these conversations considered important enough to remember and pass on to succeeding generations forms the basis of an oral tradition, which serves as the principal repository for the collective experience and the general beliefs, attitudes, and values of the community.[3]

There is an important feature of oral tradition that demands our attention—namely, its fluidity. Oral tradition is typically in a continuous state of evolution, as it absorbs new experiences and adjusts to new conditions and needs within the community. Now, this fluidity of oral tradition would be extremely frustrating if the function of oral tradition were conceived as the communication of abstract historical or scientific data—the oral equivalent of a historical archive or a scientific report. But an oral culture, lacking the ability to write, certainly cannot create archives or reports; indeed, an oral culture lacks even the idea of writing and must therefore lack even the idea of a historical archive or a scientific report.[4] The primary function of oral tradition is the very practical one of explaining, and thereby justifying, the present state and structure of the community, supplying the community with a continuously evolving "social charter." For example, an account of past events may be employed to legitimate current leadership roles, property rights, or distribution of privileges and obligations. And in order to serve this function effectively, oral tradition must be capable of adjusting itself fairly rapidly to changes in social structure.[5]

But here we are principally interested in the content of oral traditions, especially those portions of the content that deal with the nature of the universe—the portions, that is, that might be thought of as the ingredients of a worldview or a cosmology. Such ingredients exist within every oral tradition, but often beneath the surface, seldom articulated, and almost never assembled into a coherent whole. It follows that we must be extremely reluctant to articulate the worldview of preliterate people on their behalf, for this cannot be done without our supplying the elements of coherence and system, thereby distorting the very conceptions we are attempting to portray. But we may, if we are careful, formulate certain conclusions about the ingredients or elements of worldview within preliterate oral traditions.

It is clear that preliterate people, no less than those of us who live in a modern scientific culture, need explanatory principles capable of bringing order, unity, and especially meaning to the apparently random and chaotic flow of events. But we should not expect the explanatory principles accepted by preliterate people to resemble ours: lacking any conception of "laws of nature" or deterministic causal mechanisms, their ideas of causation extend well beyond the sort of mechanical or physical action acknowledged by modern science. It is natural that in the search for meaning they should proceed within the framework of their own experience, projecting human or biological traits onto objects and events that seem to us devoid not only of humanity but also of life. Thus, the beginning of the universe is typically described in terms of birth, and cosmic events may be interpreted as the outcome of struggle between opposing forces, one good and the other evil. There is an inclination in preliterate cultures not only to personalize but also to individualize causes, to suppose that things happen as they do because they have been willed to do so. This tendency has been described by H. and H. A. Frankfort:

> Our view of causality . . . would not satisfy primitive man because of the impersonal character of its explanations. It would not satisfy him, moreover, because of its generality. We understand phenomena, not by what makes them peculiar, but by what makes them manifestations of general laws. But a general law cannot do justice to the individual character of each event. And the individual character of the event is precisely what early man experiences most strongly. We may explain that certain physiological processes cause a man's death. Primitive man asks: Why should this man die thus at this moment? We can only say that, given these conditions, death will always occur. He wants to find a cause as specific and individual as the event which it must explain. The event . . . is experienced in its complexity and individuality, and these are matched by equally individual causes.[6]

Oral traditions typically portray the universe as consisting of sky and earth, and perhaps also an underworld. An African myth describes the earth as a mat that has been unrolled but remains tilted, thereby explaining upstream and downstream—an illustration of the general tendency to describe the universe in terms of familiar objects and processes. Deity is an omnipresent reality within the world of oral traditions, though in general no clear distinction is drawn between the natural, the supernatural, and the human; the gods do not transcend the universe but are rooted in it and subject to its principles. Belief in the existence of ghosts of the dead, spirits, and a variety of invisible powers, which magical ritual allows one to control, is another universal feature of oral tradition. Reincarnation (the idea that after death the soul returns in another body, either human or animal) is widely believed in. Conceptions of space and time are not (like those of modern physics) abstract and mathematical, but are invested with meaning and value drawn from the experience of the community. For example, the cardinal directions for a community whose existence is closely connected to a river might be "upstream" and "downstream," rather than north, south, east, and west. Some oral cultures have difficulty conceiving of more than a very shallow past: an African tribe, the Tio, for example, cannot situate anybody farther back in time than two generations.[7]

There is a strong tendency within oral traditions to identify causes with beginnings, so that to explain something is to identify its historical origins. Within such a conceptual framework, the distinction that we make between scientific and historical understanding cannot be sharply drawn and may be nonexistent. Thus, when we look for the features of oral tradition that count for worldview or cosmology, they will almost always include an account of origins—the beginning of the world, the appearance of the first humans, the origin of animals, plants, and other important objects, and finally the formation of the community. Related to the account of origins is often a genealogy of gods, kings, or other heroic figures in the community's past, accompanied by stories about their heroic deeds. It is important to note that in such historical accounts the past is portrayed not as a chain of causes and effects that produce gradual change, but as a series of decisive, isolated events by which the present order came into existence.[8]

These tendencies can be illustrated with examples from both ancient and contemporary oral cultures. According to the twentieth-century Kuba of equatorial Africa,

> Mboom or the original water had nine children, all called Woot, who in turn created the world. They were, apparently in order of appearance: Woot the ocean; Woot the digger, who dug riverbeds and trenches and threw up hills; Woot the flowing, who made rivers flow; Woot who created woods and savannas; Woot who created leaves; Woot who created stones; Woot the sculptor, who made people out of wooden balls; Woot the inventor of prickly things such as fish, thorns, and paddles; and Woot the sharpener, who first gave an edge to pointed things. Death came to the world when a quarrel between the last two Woots led to the demise of one of them by the use of a sharpened point.[9]

Notice how this tale not only accounts for the origin of the human race and the major topographical features of the Kuba world, but also explains the invention of what the Kuba clearly considered a critically important tool—the sharpened object.

Similar themes abound in early Egyptian and Babylonian creation myths. According to one Egyptian account, in the beginning the sun-god, Atum, spat out Shu, the god of air, and Tefnut, the goddess of moisture. Thereafter,

> Shu and Tefnut, air and moisture, gave birth to earth and sky, the earth-god Geb and the sky-goddess Nut. . . . Then in their turn Geb and Nut, earth and sky, mated and produced two couples, the god Osiris and his consort Isis, the god Seth and his consort Nephthys. These represent the creatures of this world, whether human, divine, or cosmic.[10]

A Babylonian myth attributes the origin of the world to the sexual activity of Enki, god of the waters. Enki impregnated the goddess of the earth or soil, Ninhursag. This union of water and earth gave rise to vegetation, represented by the birth of the goddess of plants, Ninsar. Enki subsequently mated first with his daughter, then with his granddaughter, to produce various specific plants and plant products. Ninhursag, angered when Enki devoured eight of the new plants before she had the opportunity to name them, pronounced a curse on him. Fearing the consequences of Enki's demise (apparently a drying up of the waters), the other gods prevailed on Ninhursag to withdraw the curse and heal Enki of the various ailments induced by the curse, which she did by giving birth to eight healing deities, each associated with a part of the body—thus accounting for the origin of the healing arts.[11]

It will be convenient to pause for a moment on the healing arts, which can serve to illustrate some important characteristics of oral cultures. There can be no doubt that healing practices were extremely important in ancient oral cultures, where primitive conditions made disease and injury everyday realities.[12] Minor medical problems, such as wounds and lesions, were no doubt treated by family members. More dramatic ailments—major wounds, broken bones, severe and unexpected illness—might require assistance from somebody with more advanced knowledge and skill. A certain amount of medical specialization thus came into existence: some members of the tribe or the village became known for herb-gathering ability, proficiency in the setting of bones or the treatment of wounds, or experience in assisting at childbirth.

But so described, the primitive medicine practiced in preliterate societies sounds remarkably like a rudimentary version of modern medicine. A more careful look reveals the healing arts within oral cultures to be inseparable and indistinguishable from religion and magic. The "wise woman" or the "medicine man" was valued not simply for pharmaceutical or surgical skill, but also for knowledge of the divine and demonic causes of disease and the magical and religious rituals by which it could be treated. If the problem was a splinter, a wound, a familiar rash, a digestive complaint, or a broken bone, the healer responded in the obvious way—by removing the splinter, binding the wound, applying a substance (if one were known) that would counteract the rash, issuing dietary prohibitions, and setting and splinting the broken limb. But if a family member became mysteriously and gravely ill, one might suspect sorcery or invasion of the body by an alien spirit. In such cases, more dramatic remedies would be called for—exorcism, divination, purification, songs, incantations, and other ritualistic activities.

One last feature of belief in oral cultures (both ancient and contemporary) demands our attention—namely, the simultaneous acceptance of what seem to us incompatible alternatives, without any apparent awareness that such behavior could present a problem. Examples are innumerable, but it may suffice to note that the story of the nine Woots related above is one of seven (or more) myths of origin that circulate among the Kuba, while the Egyptians had a variety of alternatives to the story of Atum, Shu, Tefnut, and their off-spring; and nobody seemed to notice, or else to care, that all of them could not be true. Add to this the seemingly "fanciful" nature of many of the beliefs described above, and the question of "primitive mentality" is inevitably raised: did the members of preliterate societies possess a mentality that was prelogical or mystical or in some other way different from our own; and, if so, exactly how is this mentality to be described and explained?[13]

This is an extremely complex and difficult problem that has been hotly debated by anthropologists and others for the better part of the past century, and I am not likely to resolve it here. But I can at least offer a word of meth-odological advice: namely, that it is wasted effort, contributing absolutely nothing to the cause of understanding, to spend time wishing that preliterate people had employed a conception and criteria of knowledge that they had never encountered—a conception, in the case of prehistoric people, that was not invented until centuries later. We make no progress by assuming that preliterate people were trying, but failing, to live up to our conceptions of knowledge and truth. It requires only a moment of reflection to realize that they were operating within quite a different linguistic and conceptual world, and with different purposes; and it is in the light of these that their achieve-ments must be judged.

The stories embodied in oral traditions are intended to convey and reinforce the values and attitudes of the community, to offer satisfying explanations of the major features of the world as experienced by the community, and to legitimate the current social structure; stories enter the oral tradition (the collective memory) because of their effectiveness in achieving those ends, and as long as they continue to do so there is no reason to question them. There are no rewards for skepticism in such a social setting and few resources to facilitate challenge. Indeed, our highly developed conceptions of truth and the criteria that a claim must satisfy in order to be judged true (internal coherence, for example, or correspondence with an external reality) do not generally exist in oral cultures and, if explained to a member of an oral culture, would be greeted with incomprehension. Rather, the operative principle among preliterates is that of sanctioned belief—the sanction in question emerging from community consensus.[14]

Finally, if we are to understand the development of science in antiquity and the Middle Ages, we must ask how the preliterate patterns of belief that we have been examining yielded to, or were supplemented by, a new conception of knowledge and truth (represented most clearly in the principles of Aristotelian logic and the philosophical tradition it spawned). A necessary condition, if not the full explanation, was the invention of writing, which occurred in a series of steps. First there were pictographs, in which the written sign stood for the object itself. Around 3000 B.C. a system of word signs (or logograms) appeared, in which signs were created for the important words, as in Egyptian hieroglyphics. But in hieroglyphic writing, signs could also stand for sounds or syllables—the beginnings of syllabic writing. The development of fully syllabic systems about 1500 B.C. (that is, systems in which all nonsyllabic signs were discarded) made it possible and, indeed, reasonably easy for people to write down everything they could say. And finally, fully alphabetic writing, which has a sign for each sound (both consonants and vowels), made its appearance in Greece about 800 B.C. and became widely disseminated in Greek culture in the sixth and fifth centuries.[15]

One of the critical contributions of writing, especially alphabetic writing, was to provide a means for the recording of oral traditions, thereby freezing what had hitherto been fluid, translating fleeting audible signals into enduring visible objects.[16] Writing thus served a storage function, replacing memory as the principal repository of knowledge. This had the revolutionary effect of opening knowledge claims to the possibility of inspection, comparison, and criticism. Presented with a written account of events, we can compare it with other (including older) written accounts of the same events, to a degree unthinkable within an exclusively oral culture. Such comparison encourages skepticism and, in antiquity, helped to create the distinction between truth, on the one hand, and myth or legend, on the other; that distinction, in turn, called for the formulation of criteria by which truthfulness could be ascertained; and out of the effort to formulate suitable criteria emerged rules of reasoning, which offered a foundation for serious philosophical activity.[17]

But giving permanent form to the spoken word does not merely encourage inspection and criticism. It also makes possible new kinds of intellectual activity that have no counterparts (or only weak ones) in an oral culture. Jack Goody has argued convincingly that early literate cultures produced large quantities of written inventories and other kinds of lists (mostly for administrative purposes), far more elaborate than anything an oral culture could conceivably produce; and, moreover, that these lists made possible new kinds of inspection and called for new thought processes or new ways of organizing thoughts. For one thing, the items in a list are removed from the context that gives them meaning in the world of oral discourse, and in that sense they have become abstractions. And in this abstract form they can be separated, sorted, and classified according to a variety of criteria, thereby giving rise to innumerable questions not likely to be raised in an oral culture. To give a single example, the lists of precise celestial observations assembled by early Babylonians could never have been collected and transmitted in oral form; their existence in writing, which allowed them to be minutely examined and compared, made possible the discovery of intricate patterns in the motions of the celestial bodies, which we associate with the beginnings of mathematical astronomy and astrology.[18]

Two conclusions may be drawn from this argument. First, the invention of writing was a prerequisite for the development of philosophy and science in the ancient world. Second, the degree to which philosophy and science flourished in the ancient world was, to a very significant degree, a function of the efficiency of the system of writing (alphabetic writing having a great advantage over all of the alternatives) and the breadth of its diffusion among the people. We see the earliest benefits of the use of word signs or logograms in Egypt and Mesopotamia, beginning about 3000 B.C. However, the difficulty and inefficiency of logographic writing inevitably limited its diffusion and made it the property of a small scholarly elite. In sixth- and fifth-century Greece, by contrast, the wide dissemination of alphabetic writing contributed to the spectacular development of philosophy and science. We must not imagine that literacy was sufficient of itself to produce the "Greek miracle" of the sixth and fifth centuries; other factors no doubt contributed, including prosperity, new principles of social and political organization, contact with Eastern cultures, and the introduction of a competitive style into Greek intellectual life. But surely a fundamental element in the mix was the emergence of Greece as the world's first widely literate culture.[19]

THE BEGINNINGS OF SCIENCE IN EGYPT AND MESOPOTAMIA

The earliest roots of what would become Western science are to be found in ancient Mesopotamia (the region between the Tigris and Euphrates Rivers, site of ancient Babylonia and Assyria) and Egypt (the Nile River and its environs)—see map 4.1. I have said enough about creation myths in the preceding section to reveal key features of Egyptian and Mesopotamian cosmogony (concerned with the origins of the universe) and cosmology (concerned with the structure of the universe). Here I will restrict myself to the Egyptian and Mesopotamian contribution to several other disciplines that subsequently found a place within Greek and medieval European science: mathematics, astronomy, and medicine. The evidence is scanty by comparison with materials available on Greek science, but sufficient to convey a general picture.

The Greeks themselves believed that mathematics originated in Egypt and Mesopotamia. Herodotus (fifth century B.C.) reported that Pythagoras traveled to Egypt, where he was introduced by priests to the mysteries of Egyptian mathematics. From there, according to ancient tradition, he was carried captive to Babylon, where he came into contact with Babylonian mathematics. Eventually he made his way home to the island of Samos, bearing gifts of Egyptian and Babylonian mathematical treasure to the Greeks. Whether this and similar tales regarding other mathematicians are historically accurate or legendary is less important than the larger truth they convey—namely, that the Greeks were (and knew they were) the beneficiaries of Egyptian and Babylonian mathematical knowledge.

By about 3000 B.C., the Egyptians developed a number system that was decimal in character, employing a different symbol for each power of 10 (1, 10, 100, and so forth). These symbols could be lined up, as in Roman numerals, to form any desired number. Thus if | represented 1, and ∩ represented 10, then the number 34 could be expressed as ||||∩∩∩. By about 1800 B.C. additional symbols had been devised for other numbers, so that, for example, 7 could be represented by a sickle ʔ rather than by seven vertical strokes. Addition and subtraction were simple operations in Egyptian arithmetic, performed as with Roman numerals, but multiplication and division were extremely clumsy; and the generalized concept of a fraction was unknown, the general rule allowing only unit fractions (fractions with a numerator of 1). Elementary problems of the following type could be solved: if one-seventh of a quantity is added to the quantity, and the sum equals 16, how large is the quantity?[20]

Egyptian geometrical knowledge appears to have been oriented toward practical problems, including those of surveyors and builders. Egyptians were able to calculate the areas of simple plane figures, such as the triangle and the rectangle, and the volumes of simple solids, such as the pyramid. For example, to find the area of a triangle they took one-half the length of the base times the altitude; and, to find the volume of a pyramid, one-third the area of the base times the altitude. For calculating the area of a circle, the Egyptians worked out rules that correspond to a value for π of about 3.17. Finally, in one of the most obvious areas of applied mathematics, the Egyptians devised an official calendar consisting of twelve months of thirty days each, plus an additional five days at the end of the year—a calendar substantially simpler, because of its fixed character, than contemporary Babylonian calendars and those of the early Greek city-states, which attempted to take into account the lunar, as well as the solar, cycle.[21]

The contemporary mathematical achievement in Mesopotamia was an order of magnitude superior to that of the Egyptians. Clay tablets (see fig. 1.1) recovered in large quantities reveal a Babylonian number system, fully developed by about 2000 B.C., that was simultaneously decimal (based on the number 10) and sexagesimal (based on the number 60). We retain sexagesimal numbers today in our system for measuring time (60 minutes to an hour) and angles (60 minutes in a degree and 360 degrees in a circle). The Babylonians had separate symbols for 1 (▼) and 10 (◀); these could be combined like Roman numerals to form numbers up to 59. The number 32, for example, would be expressed by three of the tens symbol plus two of the units symbol, as in table 1.1.

But beyond 59 an important difference appears. Instead of forming the number 60 by lining up six symbols for 10, the Babylonians used a place system similar to our own. In our number 234, the numeral 4, situated in the units column, signifies simply the number 4; the numeral 3, situated in the tens column, represents the number 30; while the numeral 2, situated in the hundreds column, stands for the number 200. Thus 234 is 200 + 30 + 4. The Babylonian place system worked similarly, except that successive columns represent powers of 60 rather than powers of 10. Thus in the second example in table 1.1, the two unit symbols in the 60 column represent not 2, but $2 \times 60 = 120$; and in the third example the unit symbol in the 60^2 column represents not 1 but $1 \times 60^2 = 3600$. There was no equivalent of the decimal point by which to locate the units column, and this information would therefore have to be inferred from context. Multiplication tables, tables of reciprocals, and

Fig. 1.1. A Babylonian clay tablet (ca. 1900–1600 B.C.), containing a mathematical problem text dealing with bricks, their volumes, and their coverage. Yale Babylonian Collection, YBC 4607. The text is translated and discussed in O. Neugebauer and A. Sachs, eds., *Mathematical Cuneiform Texts*, pp. 91–97.

Table 1.1. Five Babylonian Sexagesimal Numbers and Their Hindu-Arabic equivalents.

	60^3	60^2	60	1	$1/60$	$1/60^2$	Modern Hindu-Arabic Equivalent
(1)				◄◄◄▼▼			32
(2)			▼▼	◄▼▼▼			$2 \times 60 + 16 = 136$
(3)		▼	◄▼▼	◄◄▼▼▼▼			$1 \times 3600 + 12 \times 60 + 23 = 4{,}343$
(4)	▼▼	◄◄▼▼					$2 \times 216000 + 22 \times 3600 = 511{,}200$
(5)					◄◄	◄▼▼	$2 \times 1/60 + 12 \times 1/3600 = 1/30 + 1/300 = 11/300$

▼ = 1   ◄ = 10

tables of powers and roots were used to facilitate calculation. One of the great advantages of the sexagesimal system was the ease with which calculations could be performed using fractions.[22]

The full superiority of Babylonian mathematics over its Egyptian counterpart is evident when we turn to more difficult problems, which we would solve algebraically. Historians of mathematics sometimes refer to these problems as "algebra"—useful shorthand for this aspect of the Babylonian mathematical enterprise, perhaps, but misleading if it is taken to mean that they practiced genuine algebra—that is, that they had a generalized algebraic notation or an understanding of what we consider algebraic rules. What we can safely say is that Babylonian mathematicians used arithmetical operations to solve problems for which we would employ a quadratic equation. For example, we find many Babylonian tablets, including teaching texts, demonstrating how to solve problems such as the following: given the product of two numbers and their sum or difference, find the two numbers.[23]

The heavens have been objects of observation and speculation since the dawn of human existence. But our first evidence of close, systematic observation, measurement, and cataloging of the stars and planets is found among Babylonians during the second millennium B.C. Astronomical activity in other ancient cultures (Greece, India, and Egypt) not only emerged later, but also apparently owed its existence to Babylonian influence.[24] It would be reasonable to suppose that Babylonian astronomy grew naturally out of Babylonian mathematics, as the riches of the latter were applied to celestial phenomena. But reality has a way of violating our reasonable expectations. It is true that Babylonians eventually developed a predictive mathematical astronomy, but not until centuries of celestial divination (the art of reading the heavenly signs as predictors of future events) had paved the way.

So the story begins with Babylonian divination. It was universally believed within ancient Near Eastern cultures that a wide range of natural phenomena, including those associated with the heavens, concealed messages from the gods—signs or omens—that might be deciphered and interpreted by the adept. The goal of the scholar-scribes who claimed this as their task was to learn the language of the gods, the meaning of those signs, in order to advise their clients how to take appropriate measures to evade, mitigate, or otherwise prepare for the promised event—the defeat of an army, a flood, a stillborn child, a time of peace, a promise of wealth or longevity, or some other event, favorable or unfavorable, personal or public. The gods were believed to speak through all sorts of terrestrial objects or events, including animal entrails, dreams, malformed births, the "color of a dog that urinates on a man," but also (important for our purposes) through celestial phenomena. Astronomical phenomena probably drew special attention because of their apparent regularity, their celestial location, and identification of the planets with the gods. In any case, by the middle of the first millennium B.C. clay tablets containing nonmathematical diaries, almanacs, and numerical planetary tables (ephemerides) were readily available. These documents provided the resources required for calculation of time and place of a variety of planetary and lunar phenomena, including eclipses, conjunctions (two or more planets meeting in their celestial rounds), and first and last visibilities of planets in the night sky. This was the beginning of computational astronomy.[25]

By the end of the fifth century B.C., Babylonian celestial divination had expanded to embrace horoscopic astrology, which used planetary positions at the moment of birth (or near the date of birth for such exceptional phenomena as lunar eclipses) to predict individual fortunes. The exact relationship between this new form of divination and preexisting Babylonian computational astronomy is obscure and a matter of dispute among experts.[26] What is clear is that Babylonian horoscopic astrology at least rubbed shoulders with Babylonian computational astronomy and absorbed its aims and methodology. When this astrology was subsequently communicated to Hellenistic Greeks in the third and second centuries B.C., it was an astrology/astronomy (the two were inseparable) that had acquired both computational aims and numerical methods. And what is of critical importance, the Greek astronomy of the Hellenistic period took its shape—its commitment to computational aims and quantitative methods—from this Babylonian endowment.[27] As a result, the astronomical enterprise set out on a course that would culminate millennia later in the achievements of Nicolaus Copernicus, Johannes Kepler, and others.

Zig-zag Functions: A Problem in Babylonian Computational Astronomy

A surviving Babylonian tablet for the year 133/32 B.C. calculates where in the zodiac the moon will be at the beginning of consecutive months. The aim is to predict the first appearance of the new moon (important for the calendar, since the new moon signified the beginning of a new month). What makes this a mathematically difficult calculation is the fact that the moon's speed through the zodiac is variable, gradually increasing and decreasing by turns over the course of a year. Unable to deal mathematically with continuously changing variables, the author of the tablet employs an arithmetic progression consisting of three discontinuous arithmetic series to approximate variations in lunar speed and, consequently, predict the approximate location of the moon at the beginning of consecutive months. On the clay tablet in question (see just below), the speed assigned to the lunar motion (measured in degrees of the zodiac traversed per month) is assumed to *decrease* by a fixed amount per month for the first three months, to *increase* by a fixed amount per month for the next six months, and finally to *decrease* by a fixed amount per month for the remainder of the year. If graphed, as in fig. 1.2 (something that we can do, but they could not), the lunar speed over the course of the year appears as a "zig-zag function."[28]

Months of 133/132 B.C.	Distance (in degrees) traversed by the moon in a given month at the end of the month in degrees, minutes, seconds, and thirds[29]	Position in the zodiac
I	28, 37, 57, 58	20, 46, 16, 14 Taurus
II	28, 19, 57, 58	19, 6, 14, 12 Gemini
III	28, 19, 21, 22	17, 25, 35, 34 Cancer
IV	28, 37, 21, 22	16, 2, 56, 56 Leo
V	28, 55, 21, 22	14, 58, 18, 18 Virgo
VI	29, 13, 21, 22	14, 11, 39, 40 Libra
VII	29, 31, 21, 22	13, 43, 1, 2 Scorpio
VIII	29, 49, 21, 22	13, 32, 22, 24 Sagittarius
IX	29, 56, 36, 38	13, 28, 59, 2 Capricorn
X	29, 38, 36, 38	13, 7, 35, 40 Aquarius
XI	29, 20, 36, 38	12, 28, 12, 18 Pisces
XII	29, 2, 36, 38	11, 30, 48, 56 Aries

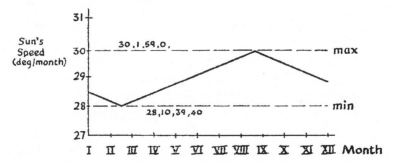

Fig. 1.2. A Babylonian zig-zag function, representing arithmetic series. From Stephen Toulmin and June Goodfield, *The Fabric of the Heavens*, p. 50.

The final area of Egyptian and Mesopotamian achievement to be considered is medical. A number of Egyptian medical papyri (written in the period 2500–1200 B.C.) have survived, and these offer us a fragmentary picture of the healing arts in ancient Egypt. From several of the papyri it becomes clear that a principal cause of disease was thought to be invasion of the body by evil forces or spirits. Relief was to be gained through rituals designed to appease or frighten the spirits—exorcism, incantation, purification, or the wearing of an appropriate amulet. The gods could be appealed to for protection: a prayer to the god Horus, found in the Leyden papyrus, reads in part, "Hail to thee, Horus. . . . I come to thee, I praise thy beauty: destroy thou the evil that is in my limbs."[30] Certain gods came especially to be associated with healing functions or healing cults: Thoth, Horus, Isis, and Imhotep. The view that each bodily organ was ruled by a specific god, who could be invoked for healing of that organ, seems to have been widespread. And all of this ritual, of course, required the assistance of an expert who was of acknowledged purity, who knew the required incantations, and who could assure that the ritual was properly performed; this was the priest-healer.

Healing therapies in ancient Egypt were not limited to prayer, incantation, and ritual. Pharmacological remedies, prepared from animal, vegetable, or mineral substances, were also widespread—though their effectiveness was believed to depend on preparation and administration under appropriate ritual conditions. The Ebers papyrus (written about 1600 B.C., but containing material copied from much older texts) contains medical recipes for dealing with diseases of the skin, eyes, mouth, extremities, digestive and reproductive systems, and other internal organs; for treating wounds, burns, abscesses, ulcers, tumors, headaches, swollen glands, and bad breath.[31]

Surgery is dealt with in another papyrus, known as the Edwin Smith papyrus (written about the same time as the Ebers papyrus), which contains a surgical manual that systematically describes the treatment of wounds, fractures, and dislocations (see fig. 1.3).[32] One of the notable features of the Ebers and Edwin Smith papyri is the careful arrangement of case studies, beginning with a description of the problem and proceeding to diagnosis, verdict (as to whether or not the ailment is treatable), and treatment.

Mesopotamian medicine displays many of the same characteristics as Egyptian healing practices. Babylonian clay tablets, like Egyptian papyri, contain case studies, systematically organized by type, many of them revealing careful observation of symptoms and intelligent prognosis. Mesopotamian healers displayed equal skill in surgery and the preparation of pharmaceutical remedies. As in Egypt, a certain amount of medical specialization

Fig. 1.3. A column from the Edwin Smith surgical papyrus (ca. 1600 B.C.). New York
Academy of Medicine.

developed—different categories of healers coming to have somewhat different specialties and functions. And again we find healing intimately mingled with religion and with practices that we would now view as magical. Disease was regarded as the result of invasion of the body by evil spirits (owing to fate, carelessness, sin, or sorcery). Therapy was directed toward elimination of the invading spirit through divination (including the interpretation of astrological omens), sacrifice, prayer, and magical ritual.[33]

This brief sketch of Egyptian and Mesopotamian contributions to mathematics, astronomy, and the healing arts offers us a glimpse of the beginnings of the Western scientific tradition, as well as a background against which to view the Greek achievement. There is no doubt that the Greeks were aware of the work of their Egyptian and Mesopotamian predecessors, and benefited from it. In the chapters that follow, we will see how these products of Egyptian and Mesopotamian thought entered and helped shape Greek natural philosophy.

2 ❄ *The Greeks and the Cosmos*

THE WORLD OF HOMER AND HESIOD

Sing to me of the man [Odysseus], Muse, the man of twists and turns
    driven time and again off course, once he had plundered
    the hallowed heights of Troy.
Many cities of men he saw and learned their minds,
    many pains he suffered, heartsick on the open sea,
    fighting to save his life and bring his comrades home.
But he could not save them from disaster, hard as he strove—
    the recklessness of their own ways destroyed them all,
    the blind fools, they devoured the cattle of the Sun
    and the Sungod blotted out the day of their return.
Launch out on his story, Muse, daughter of Zeus,
    start from where you will—sing for our time too.
                                 By now,
all the survivors, all who avoided headlong death
    were safe at home, escaped the wars and waves.
But one man alone . . .
    his heart set on his wife and his return—Calypso
    the bewitching nymph, the lustrous goddess, held him back,
    deep in her arching caverns, craving him for a husband.
But then, when the wheeling seasons brought the year around,
    that year spun out by the gods when he should reach his home,
    Ithaca—though not even there would he be free of trials,
    even among his loved ones—then every god took pity,
    all except Poseidon. He raged on, seething against
    the great Odysseus till he reached his native land.[1]

So begins Homer's *Odyssey*, recounting the return of Odysseus to Ithaca at the
conclusion of the Trojan War—alternately thwarted and aided by the gods.[2]

Committed to writing, probably, in the seventh century B.C., the *Odyssey* is a tale of heroic deeds performed in the face of divine intervention and interference. Along with the *Iliad* (also attributed to Homer), it is the closest thing we have to a history of the Greek people before the sixth century. Not history as we moderns would write it, of course, but the two poems do offer an account of historical events and heroic deeds of the past. And they are the best window we've got on the intellectual furniture—the language, learning, and culture—of the ancient Greek mind.

But Homer is not our only source for Greek mythological thought. Hesiod, Homer's rough contemporary, provided a mythological cosmogony in his *Theogony*:

> First came Chasm; then broad-breasted Earth, secure seat for ever of all the immortals who occupy the peak of snowy Olympus. . . . Out of Chasm came Erebos and dark Night, and from Night in turn came Bright Air and Day, whom she bore in shared intimacy with Erebos. Earth bore first of all one equal to herself, starry Heaven . . . and she bore the long Mountains, pleasant haunts of the goddesses, the Nymphs who dwell in mountain glens; and she bore also the undraining Sea and its furious swell.[3]

Gaia (mother earth; fig. 2.1) proceeded to mate with her offspring, Ouranos (father heaven), and from that union issued Oceanus (the river that encircles the world, father of all other rivers), the twelve Titans, and a collection of monsters. Eventually Kronos, one of the Titans, castrated and overthrew his father, Ouranos; Kronos, in turn, was deposed by his son Zeus (fig. 2.2). Zeus obtained the thunderbolt from the Cyclopes and used it to defeat the Titans and establish his own Olympian rule.[4]

We also have short fragments from other early mythographers and many later collections from the Hellenistic period (after 335 B.C.). What strikes one about the world defined by this mythological literature is that it is drenched with the divine. The gods and humans shared a common history. This was a world of anthropomorphic deities interfering in human affairs, using humans as pawns in their own plots and intrigues—acting out of spite, anger, love, lust, benevolence, pleasure, or simple caprice. The gods were also implicated in natural phenomena. Sun and moon were conceived as deities, offspring of Theia and Hyperion. Storms, lightning bolts, winds, and earthquakes were not regarded as inevitable outcomes of impersonal, natural forces, but mighty feats willed by the gods. The result was a capricious world, in which nothing

Fig. 2.1. A shrine to the earth goddess Gaia at Delphi (4th c. B.C.).

could be safely predicted because of the boundless possibilities of divine intervention.

What are we to make of this? Did the ancient Greeks take the stories constituting what we now call "Greek mythology" to be literally true? Did they really believe in divine beings, lodged on Mount Olympus or in some other mysterious place, seducing one another and bedeviling humans who crossed their path? Was there nobody who doubted that storms and earthquakes were a result of divine caprice? We have seen in the previous chapter, in the discussion of preliterate thought, how difficult these questions are.[5] What is clear is that any attempt to measure such beliefs by modern criteria of scientific truth is a sure road to misunderstanding. But perhaps we can learn something by comparing Homeric mythology with modern beliefs outside the scientific realm. When a professional athlete thanks God for victory, does he or she really believe that victory was obtained through supernatural intervention? Or

Fig. 2.2. A bronze statue of
Zeus. Museo Archeologico,
Florence. Alinari/Art Resource
N.Y.

is this simply a case of conventional athlete-talk? At some level, most of the athletes in question would probably expect their claim to be taken seriously. But it has probably never occurred to them that such claims might be asked to survive philosophical or scientific scrutiny. They are not casting about for defensible philosophical or scientific truth, but celebrating victory by expressing conventional, unconsciously assimilated beliefs, widely held in the culture to which they belong. By the same token, although the writings of Homer and Hesiod appear to address questions of causation, we must understand that they were not intended as scientific or philosophical treatises—the very idea of which did not yet exist. Homer and Hesiod—and the bards whose epic poems lie behind theirs—were recording heroic deeds in conventional terms, in order to instruct and entertain; if we treat them as failed philosophers or scientists, we will inevitably misunderstand their achievement.

Yet we must not dismiss these ancient sources too quickly. Homer and Hesiod, after all, are among the few sources at our disposal that reveal anything of archaic Greek thought; and if they do not represent primitive Greek philosophy, they were nonetheless central to Greek education and culture for centuries and cannot have been without influence on the Greek mind. It is abundantly clear that the language and the images people employ affect the reality they perceive. If the content of Homer's and Hesiod's poems was not "believed" in the same way as we believe the content of modern physics or chemistry, the mythology of the Olympian gods, as well as local deities, was nonetheless a central feature of early Greek culture, affecting the way Greeks thought, talked, and behaved.

THE FIRST GREEK PHILOSOPHERS

However, a fresh wind was about to blow from another direction. Early in the sixth century, Greek culture experienced a burst of a radically new kind of discourse—speculation unprecedented in its rationality (*nous* in Greek), its concern for evidence, and its acknowledgment that claims were open to dispute and needed to be defended.[6] Speculations ranged over a broad subject matter, including the cosmos and its origins, the earth and its inhabitants, celestial bodies, striking phenomena such as earthquakes, thunder, and lightning, disease and death, and the nature of human knowledge.

The Greek-speaking people who produced this burst of intellectual activity were distributed geographically over an area that extended well beyond the boundaries of the modern Greek state. Colonization, conquest, and the absorption of invading tribes had created a territory of Greek-speaking people

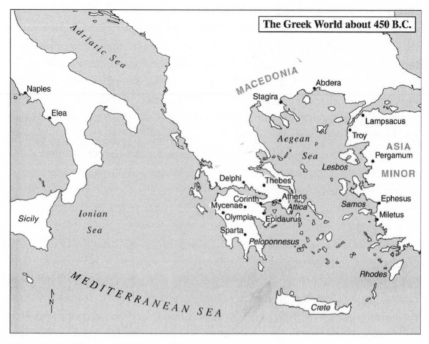

Map 2.1. The Greek world about 450 B.C.

that extended as far north as Macedonia; east to Asia Minor (modern Turkey), especially the region of Ionia along its Aegean coastline (see map 2.1); and west, across the Adriatic Sea to southern Italy and Sicily. The mingling of peoples and cultures in these territories may help to explain the appearance of philosophical and cosmological thinking in the sixth and fifth centuries.[7]

What were these new modes of thought that we identify as "philosophy"? A small band of thinkers in the sixth century embarked on a serious, critical inquiry into the nature of the world in which they lived—an inquiry that has stretched from their day to ours. They asked about its ingredients, its composition, its operation, and its shape. They inquired whether it is composed of one kind of thing or many. They sought to understand the causes of change, by which things come into being or change character. They contemplated extraordinary natural phenomena, such as earthquakes and eclipses, and sought universal explanations applicable not only to a particular earthquake or eclipse but to earthquakes and eclipses in general. And they began to reflect on the rules of argumentation and proof.

These early philosophers did not merely pose a new set of questions; they also sought new kinds of answers. They did not personify Nature, and the gods disappeared from their explanations of natural phenomena. Whereas Hesiod regarded earth and sky as divine offspring, for the philosophers Leucippus (fl. 435) and Democritus (fl. 410) the world and its various parts result from mechanical sorting of lifeless atoms in a primeval vortex or whirlpool. To be sure, these philosophical developments did not signal the end of Greek mythology. As late as the fifth century, the historian Herodotus retained much of the old mythology, sprinkling tales of divine intervention through his *Histories*. Poseidon, by his account, used a high tide to flood a swamp the Persians were crossing. And Herodotus regarded an eclipse that coincided with the departure of the Persian army for Greece as a supernatural omen. But the philosophers offered a new and (judging from its subsequent growth) powerful alternative, containing no hint of supernatural intervention. Anaximander (fl. 555) judged eclipses to be the result of blockage of the apertures in rings of celestial fire. According to Heraclitus (fl. 500), the heavenly bodies are bowls filled with fire, and an eclipse occurs when the open side of a bowl turns away from us. These theories of Anaximander and Heraclitus do not seem particularly sophisticated (fifty years after Heraclitus the philosophers Empedocles and Anaxagoras understood that eclipses were simply a case of cosmic shadows), but what is of critical importance is that they exclude the gods. The explanations are entirely naturalistic; eclipses do not reflect personal whim or the arbitrary fancies of the gods, but simply the nature of fiery rings or of celestial bowls and their fiery contents.

The world of the philosophers, in short, was an orderly, predictable world in which things behave according to their natures. The Greek term used to denote this ordered world was *kosmos*, from which we draw our word "cosmology." The capricious world of divine intervention was being pushed aside, making room for order and regularity; *chaos* was yielding to *kosmos*. A clear distinction between the natural and the supernatural was emerging; and there was wide agreement that causes (if they are to be dealt with philosophically) must be sought only in the natures of things. The philosophers who introduced these new ways of thinking were called by Aristotle *physikoi* or *physiologoi*, from their concern with *physis* or nature.

THE MILESIANS AND THE QUESTION OF UNDERLYING REALITY

The philosophical developments described above appear to have emerged first in Ionia. There Greek colonists had established thriving cities, including Ephesus, Miletus, and Pergamum—cities whose prosperity was built on trade and the exploitation of local natural resources. Ionia may, like many frontier societies, have encouraged hard work and self-sufficiency and, in return, offered prosperity and opportunity. It also brought Greeks into contact with the art, religion, and learning of the Near East, with which Ionia had cultural, commercial, diplomatic, and military relations. This, along with the mingling of cultures, growth of literacy within the ruling class, and other causes now beyond our reach led to a burst of creativity in lyric poetry and philosophy.

Later authors, including Aristotle, identify Thales (fl. 585) of Miletus as the earliest of the Ionian philosophers. The surviving fragments (which do not contain any original writings by Thales himself) portray him as a geometer, astronomer, and engineer. He has acquired modern fame for allegedly predicting a solar eclipse in 585, but it is unlikely that Greek astronomical knowledge had developed to the point, in Thales' lifetime, where such a prediction was possible. Other fragments assign him the theory that the earth emerged out of water, on which it now floats like a log—a notion that may be a truer measure of his astronomical and cosmological sophistication. But what is more interesting to us in the present context is the claim (attributed to him by Aristotle two-and-a-half centuries later) that there must be some underlying matter in the universe—water, he believed—out of which everything else is composed and which persists through apparent change: Most of the early philosophers, Aristotle says, conceived that the fundamental, originating stuff that underlies all things is material:

> That of which all things consist, from which they first come and into which, on their destruction, they are ultimately resolved, the substance persisting but changing its qualities—this they say is the element and the origin of all things. ... Thales, the founder of this school of philosophy, claims that this fundamental stuff is water, which explains why he claimed that the earth floats on water.[8]

This theme of underlying matter or stuff was developed by two younger Milesians, Anaximander (fl. 550) and Anaximenes (fl. 535). The former identified the origin or fundamental stuff of the universe as the *apeiron*, a spacially unlimited and undefined something (and therefore unlike any known substance), out of which the material cosmos emerged.[9] Finally, Anaximenes is reported by Aristotle and Theophrastus to have maintained that the underlying stuff is air, which can be rarefied or condensed to produce the variety of substances found in the world as we know it. This air "differs in its substantial nature by rarity and density. Being made finer it becomes fire, being made thicker it becomes wind, then cloud, then (when thickened still more) water, then earth, then stones; and the rest come into being from these."[10] (This same idea was explored by Newton in the seventeenth century.) It is clear that the Milesian philosophers were monists (for they identify a *single* underlying reality) and materialists.

All of this may seem primitive. And judged by twenty-first-century criteria, it surely is. But comparing the past with the present is a sure recipe for distorting the achievements of the past. It is when we compare the Milesians with their predecessors that their importance becomes apparent. In the first place, our three Milesian philosophers posed a new sort of question, never before (as far as we know) asked in Greek or Middle Eastern culture: what is the material origin of things—the single and simple underlying reality that can take on a variety of forms to produce the diversity of substances that we perceive? This is a search for unity behind diversity and order behind chaos. Second, in the answers offered by these Milesians we find no personification or deification of nature; a conceptual chasm separates their worldview from the mythological world of Homer and Hesiod. The Milesians left the gods out of the story. What they may have thought about the Olympian gods we do not (in most cases) know; but they did not invoke the gods to explain the origin, nature, or cause of things. Third, the Milesians seem to have been aware of the need not simply to report their theories, but also to defend them against critics and competitors. This was the beginning of a tradition of critical assessment, which also continues to the present day.[11]

Milesian speculations about the underlying stuff were the beginning of a quest that has continued from their day to ours. In antiquity, the Milesians were succeeded by various schools of thought. Fifty years later Heraclitus (fl. 500) of Ephesus (an Ionian city not far from Miletus; fig. 2.3) argued for a world without beginning or end, composed ultimately of fire—an "ever-living fire," in a state of continuous transformation between fire in its "kindled" form (which we call "fire" or "flame") and its two other forms: water (fire liquified) and earth (fire soldified). A dynamic balance between these three forms, according to Heraclitus, assures an eternal, stable universe.[12] Heraclitus thus postulates a world of simultaneous stability and change. It was he, according to Plato, who compared this world with the flow of a river and authored the famous maxim that we can never step twice into the same river.[13]

The materialism of the sixth century was extended in the second half of the fifth century by the atomists Leucippus of Miletus (fl. 440) and Democritus of Abdera (fl. 410), who argued that the world consists of an infinity of tiny atoms moving randomly in an infinite void. These atoms, solid corpuscles too small to be seen, come in an infinitude of shapes; by their motions, collisions,

Fig. 2.3. The ruins of ancient Ephesus. SEF/Art Resource N.Y.

and transient configurations, they account for the great diversity of substances and the complex phenomena that we experience. At a cosmic level, the atoms move in huge vortices or whirlpools, out of which worlds (including ours) emerge and into which they again disappear.[14]

The atomists offered ingenious accounts of many other natural phenomena, but we must not allow ourselves to be diverted from the main point. What is important about the atomists is their vision of reality as a lifeless piece of machinery, in which everything that occurs is the necessary outcome of inert, material atoms moving according to their nature. No mind and no divinity intrude into this world. Life itself is reduced to the motions of inert corpuscles. No room exists for purpose or freedom; iron necessity alone rules. This mechanistic worldview would fall out of favor with Plato and Aristotle and their followers. And atomism survived during the Middle Ages principally as an object of abuse, but occasionally as the subject of serious interest.[15] It returned with a vengeance (and a few novel twists) in the seventeenth century and has been a powerful force in scientific discussions ever since.

Not all who investigated the underlying stuff were monists or materialists. Nor were the gods altogether absent from their explanations.[16] Empedocles of Acragas (fl. 450), a rough contemporary of Leucippus in the second half of the fifth century, identified four elements or "roots" (as he called them) of all material things: fire, air, earth, and water (introduced in mythological garb as Zeus, Hera, Aidoneus, and Nestis). From these four roots, Empedocles wrote, "sprang all things that were and are and shall be, trees and men and women, beasts and birds and water-bred fishes, and the long-lived gods too, most mighty in their prerogatives. For there are these things alone, and running through one another they assume many a shape."[17] But material ingredients alone cannot explain motion and change. Empedocles therefore introduced two additional, *immaterial* principles: love and strife, which induce the four roots to congregate and separate.

Empedocles was not the only ancient philosopher to include immaterial principles among the most fundamental things. The Pythagoreans of the sixth and fifth centuries (concentrated especially in the Greek colonies of southern Italy and known to us not as individuals but as a "school" of thought) seem to have argued, if we interpret them literally, that the ultimate reality is numerical rather than material—not matter, but number. Aristotle reports that in the course of their mathematical studies the Pythagoreans were struck by the power of numbers to account for phenomena such as the musical scale. According to Aristotle, "Since . . . all other things seemed in their whole nature to be modeled after numbers, and numbers seemed to be the first things in the whole of nature, they [the Pythagoreans] supposed the elements of numbers to be the elements of all things, and the whole heaven to be a musical scale and a number."[18] Now this is an obscure passage, and our uncertainty is compounded by the likelihood that Aristotle did not fully understand the Pythagorean teaching and the possibility that, with his own axes to grind, he was not altogether fair to it. Did the Pythagoreans literally believe that material things were constructed out of numbers? Or did they mean only to claim that material things have fundamental numerical properties, which determine the nature of those things? We will never know for certain. A sensible reading of the Pythagorean position is that in some sense numbers came first, and everything else is their offspring; number is in that sense the fundamental reality, and material things derive their properties, and possibly their existence, from number. If we wish to be more cautious, we can affirm at the very least that the Pythagoreans regarded number as a fundamental aspect of reality and mathematics as a basic tool for investigating this reality.

THE QUESTION OF CHANGE

If the most prominent philosophical problem of the sixth century was this question of the origins and fundamental ingredients of the world, a related issue came to dominate the philosophical enterprise in the fifth century. Is change possible, and if so, how? This may seem a ludicrous question to twenty-first-century readers, but with a little effort we may be able to understand its saliency for fifth-century philosophers. In the first place, we need to understand that the question was not addressed to laymen or in the context of the daily activities of laborers, craftsmen, merchants, and the like. It was a logical conundrum, thrown out as a challenge to philosophers: can there be change, motion, and activity in the material world as we experience it if the ingredients of that world are absolutely unchangeable, totally passive stuff? If the fundamental building blocks of the universe simply sit passively in their place, how (as a question of either logic or metaphysics) are motion and other forms of change possible?

The metaphysical approach (which probes the nature and structure of reality at its deepest level) was taken by Heraclitus, who (as we saw in the preceding section) offered a ringing declaration of the reality (indeed, universality) of change—the struggle of opposites—within an overall state of equilibrium or stability.[19] What Heraclitus affirmed, his younger contemporary Parmenides (fl. 480, from the Greek city-state at Elea in southern Italy) denied. Parmenides wrote a long philosophical poem (philosophy had not yet settled on prose as its preferred form of presentation), large sections of which have survived. In it, he adopted the radical position that change—all change—is a logical impossibility. He began by denying, on various logical grounds, the possibility that a thing should pass from nonexistence to existence: for example, if a thing were to come into being, why at one moment rather than another, and by what means? Moreover, this would be getting something from nothing—a logical impossibility. On analogous grounds it is impossible for a thing to undergo change. If A becomes B, either it was already B (that is, it possessed some B-ness) or it was not already B. If A was already B, then no change occurred; if A was not already B, then change would require the acquisition of B-ness from something that did not possess that quality—which brings us back to the impossibility of getting something from nothing. In either case, then, no change occurred.[20]

Parmenides' pupil Zeno (fl. 450) extended and defended this Parmenidean doctrine with a set of proofs against the possibility of one kind of change—motion, or change of place, but presumably applicable to other forms of change as well. One of these proofs, the "stadium paradox," will illustrate Zeno's approach. It is impossible, Zeno argued, ever to traverse a stadium, because before you cover the whole you must cover the half; and before you cover the half, you must cover the quarter; before the quarter, the eighth; and so on to infinity. To traverse a stadium is therefore to traverse an infinite sequence of halves, and it is impossible to traverse, or even "to come into contact with" (as Aristotle put it in his discussion of the paradox), an infinity of intervals in a finite time. The same argument can be applied to any spatial interval whatsoever—from which it follows that all motion is impossible.[21]

We have no way of knowing whether (or how) Parmenides and Zeno attempted to carry these logical conclusions over into the real world. There is little doubt that they got up in the morning, enjoyed a good breakfast, and made their way to the agora (the public square) for a hard day's philosophizing. But when they reached the agora, did they spend the rest of the day arguing that they were still at home in bed? I doubt it. They knew full well where they were and how long it had taken to get there; but as long as they were wearing their logicians' hats, they were (we may presume) prepared to ponder the logical consequences and range of applicability of what they took to be secure logical premises concerning the possibility of change.

Parmenides' denial of the possibility of change was enormously influential, offering a challenge that generations of philosophers felt compelled to address. Empedocles answered with his theory of four material "roots" or elements, plus love and strife. The elements do not come into being or pass away, and so the fundamental Parmenidean requirement is met. But they do congregate and separate and mix in various proportions, from which it follows that change is also genuine. The atomists Leucippus and Democritus granted that the individual atom is absolutely immutable, so that at the atomic level there is no generation, corruption, or alteration of any kind. However, these immutable atoms are perpetually moving, colliding, and congregating; and through the various motions and configurations of the atoms the endless variety in the world of sense experience is produced. According to the atomists, therefore, stability of the underlying reality (the atoms) underlies change at the sensory level; both are genuine.[22]

THE PROBLEM OF KNOWLEDGE

Poking through these discussions of the underlying reality and the problem of change and stability has been a third basic issue, which early Greek philosophers also addressed—namely, the problem of knowledge (more technically known as epistemology). It is implicit in the quest for the fundamental reality underlying the variety of substances revealed by the senses: if the senses do not reveal what the intellect attests—fundamental stability, for example—then we must abandon the senses as a guide to the truth. Parmenides' radical stance on the question of change had clear-cut epistemological implications: if the senses reveal change, their unreliability would seem to be demonstrated; it follows that truth is to be gained only by the exercise of reason. The atomists, too, had reason to denigrate sense experience. After all, the senses revealed the "secondary" qualities—color, taste, odor, and the tactile qualities—whereas reason taught that only atoms and the void truly exist. In a surviving fragment, Democritus identifies "two forms of knowledge, one genuine, one obscure. To the obscure belong all the following: sight, hearing, smell, taste, touch."[23] The fragment breaks off before the idea is completed, but we may assume that in Democritus's judgment genuine knowledge is rational knowledge.

If the early philosophers were inclined to favor reason over sense, this tendency was neither universal nor without qualification. Empedocles defended the senses against the attack of Parmenides. The senses may not be perfect, he argued, but they are useful guides if employed with discrimination. "But come, consider with all thy powers how each thing is manifest," he wrote, "neither holding sight in greater trust as compared with hearing, nor loud-sounding hearing above the clear evidence of thy tongue, nor withholding thy trust from any of the other limbs, wheresoever there is a path for understanding." And Anaxagoras (fl. 450) of Clazomenae (another Ionian coastal city) argued in a brief fragment that the senses offer "a glimpse of the obscure."[24]

One of the benefits gained from Greek epistemological concerns (from Greek rationalism in particular) was that they directed attention to the rules of reasoning, argumentation, and theory assessment. Formal logic would be the creation of Aristotle; but his sixth- and fifth-century predecessors became increasingly aware of the need to test the soundness of an argument and to assess the grounds on which a theory rested. The sophistication with which Parmenides and Zeno could argue—their sensitivity, for example, to the rules of inference and the criteria of proof—demonstrates how far Greek philosophy had come in a century and a half.

PLATO'S WORLD OF FORMS

The death of Socrates in 399 B.C., coming as it did around the turn of the century (not on their calendar, of course, but on ours), has made it a convenient point of demarcation in the history of Greek philosophy. Thus Socrates' predecessors

Fig. 2.4. Plato (1st c. A.D. copy).
Museo Vaticano, Vatican City.
Alinari/Art Resource N.Y.

of the sixth and fifth centuries (the philosophers who have occupied us until now in this chapter) are commonly called the "pre-Socratic philosophers." But Socrates' prominence is more than an accident of the calendar, for Socrates represents a shift in emphasis within Greek philosophy, away from the cosmological concerns of the sixth and fifth centuries toward political and ethical matters. Nonetheless, the shift was not so dramatic as to preclude continuing attention to the major problems of pre-Socratic philosophy. We find both the new and the old in the work of Socrates' younger friend and disciple, Plato (fig. 2.4).

Plato (427–348/47) was born into a distinguished Athenian family, active in affairs of state; he was undoubtedly a close observer of the political events that led up to Socrates' execution. After Socrates' death, Plato left Athens and visited Italy and Sicily, where he seems to have come into contact with Pythagorean philosophers. In 388 Plato returned to Athens and founded a school of his own, the Academy, where young men could pursue advanced studies (see fig. 4.1). Plato's literary output appears to have consisted almost entirely of dialogues, the majority of which have survived. We will find it necessary to be highly selective in our examination of Plato's philosophy; let us begin with his quest for the underlying reality.[25]

In a passage in one of his dialogues, the *Republic*, Plato reflected on the relationship between the actual tables constructed by a carpenter and the idea or definition of a table in the carpenter's mind. The carpenter replicates the mental idea as closely as possible in each table he makes, but always imperfectly. No two manufactured tables are alike down to the smallest detail, and limitations in the material (a knot here, a warped board there) ensure that none will fully measure up to the ideal.

Now, Plato argued, there is a divine craftsman who bears the same relationship to the cosmos as the carpenter bears to his tables. The divine craftsman (the Demiurge) constructed the cosmos according to an idea or plan, so that the cosmos and everything in it are replicas of eternal ideas or forms—but always imperfect replicas because of limitations inherent in the materials available to the Demiurge. In short, there are two realms: a realm of forms or ideas, containing the perfect form of everything; and the material realm in which these forms or ideas are imperfectly replicated.

Plato's notion of two distinct realms will seem strange to many people, and we must therefore stress several points of importance. The forms are incorporeal, intangible, and insensible; they have always existed, sharing the property of eternality with the Demiurge; and they are absolutely changeless. They include the form, the perfect idea, of everything in the material world. One does not speak of their location, since they are incorporeal and therefore not spatial. Although incorporeal and imperceptible by the senses, they objectively exist; indeed, true reality (reality in its fullness) is located only in the world of forms. The sensible, corporeal world, by contrast, is imperfect and transitory. It is less real in the sense that the corporeal object is a replica of, and therefore dependent for its existence upon, the form. The form has primary existence, its corporeal replica secondary existence.

Plato illustrated this conception of reality in his famous "allegory of the cave," found in book VII of the *Republic*. Men are imprisoned within a deep cave, chained so as to be incapable of moving their heads. Behind them is a wall, and beyond that a fire. People walk back and forth behind the wall, holding above it various objects, including statues of humans and animals; the objects cast shadows on the wall that is visible to the prisoners. The prisoners see only the shadows cast by these objects; and, having lived in the cave from childhood, they no longer recall any other reality. They do not suspect that these shadows are but imperfect images of objects that they cannot see; and consequently they mistake the shadows for the real.

So it is with all of us, says Plato. We are souls imprisoned in bodies. The shadows of the allegory represent the world of sense experience. The soul, peering out from its prison, is able to perceive only these flickering shadows, and the ignorant claim that this is all there is to reality. However, there do exist the statues and other objects of which the shadows are feeble representations and also the humans and animals of which the statues are imperfect replicas. To gain access to these higher realities, we must escape the bondage of sense experience and climb out of the cave, until we find ourselves able, finally, to gaze on the eternal realities, thereby entering the realm of true knowledge.[26]

What are the implications of these views for the concerns of the pre-Socratic philosophers? First, Plato equated his forms with the underlying reality, while assigning derivative or secondary existence to the corporeal world of sensible things. Second, Plato has made room for both change and stability _sense_ by assigning each to a different level of reality: the corporeal realm is the scene of imperfection and change, while the realm of forms is characterized by eter- _universals_ nal, changeless perfection. Both change and stability are therefore genuine; each characterizes something; but changelessness belongs to the forms and thus shares their fuller reality.

Third, as we have seen, Plato addressed epistemological questions, placing observation and true knowledge (or understanding) in opposition. Far from leading upward to knowledge or understanding, the senses are chains that tie us down; the route to knowledge is through philosophical reflection. This is explicit in the *Phaedo*, where Plato maintains the uselessness of the senses for the acquisition of truth and points out that when the soul attempts to employ them it is inevitably deceived.

Now the short account of Plato's epistemology frequently ends here; but there are important qualifications that it would be a serious mistake to omit. Plato did not, in fact, dismiss the senses altogether, as Parmenides had done and as the passage from the *Phaedo* might suggest Plato did. Sense experience, in Plato's view, served various useful functions. First, sense experience may provide wholesome recreation. Second, observation of certain sensible objects (especially those with geometrical properties) may serve to direct the soul toward nobler objects in the realm of forms. Plato used this argument as justification for the pursuit of astronomy. Third, Plato argued (in his theory of reminiscence) that sense experience may actually stir the memory and remind the soul of forms that it knew in a prior existence, thus stimulating a process of recollection that will lead to actual knowledge of the forms.

Finally, although Plato firmly believed that knowledge of the eternal forms (the highest, and perhaps the only true, form of knowledge) is obtainable only through the exercise of reason, the changeable realm of matter is also an acceptable object of study. Such studies serve the purpose of supplying examples of the operation of reason in the cosmos. If this is what interests us (as it sometimes did Plato), the best method of exploring it is surely to observe it. The legitimacy and utility of sense experience are clearly implied in the *Republic*, where Plato acknowledged that a prisoner emerging from the cave first employs his sense of sight to apprehend living creatures, the stars, and finally the most noble of visible (material) things, the sun. But if he aspires to apprehend "the essential reality," he must proceed "through the discourse of reason unaided by any of the senses." Both reason and sense are thus instruments worth having; which one we employ on a particular occasion will depend on the object of study.[27]

There is another way of expressing all of this, which may shed light on Plato's achievement. When Plato assigned reality to the forms, he was, in fact, identifying reality with the properties that classes of things have in common. The bearer of true reality is not (for example) this dog with the droopy left ear or that one with the menacing bark, but the idealized form of a dog shared (imperfectly, to be sure) by every individual dog—those characteristics by virtue of which we are able to classify all of them as dogs. Therefore, to gain true knowledge, we must set aside all characteristics peculiar to things as individuals and seek the shared characteristics that define them into classes. Now stated in this modest fashion, Plato's view has a distinctly modern ring. Idealization is a prominent feature of a great deal of modern science; we develop models or laws that overlook the incidental in favor of the essential. However, Plato went beyond this, maintaining not merely that true reality is to be found in the common properties of classes of things, but also that this common property (the idea or form) has objective, independent, and indeed prior existence.

PLATO'S COSMOLOGY

The doctrines that we have been considering—Plato's response to the pre-Socratics, found in his *Republic*, *Phaedo*, and various other dialogues—represent only a small portion of his total philosophy. Plato also wrote a dialogue, the *Timaeus*, that reveals his interest in the world of nature. Here we find his views on astronomy, cosmology, light and color, the elements, and human physiology. Since the *Timaeus* was the only Platonic dialogue to survive through the Middle Ages in more than fragmentary form, it represents one of the principal channels of continuing Platonic influence. It is important for our purposes because it provided the early Middle Ages (before the twelfth century) with its most coherent natural philosophy.

Plato referred to the contents of the *Timaeus* as a "likely story," and this has misled some readers to view it as a myth in which Plato himself placed no stock. In fact, Plato stated quite clearly that this was the best account possible, that anything better than a likely account was precluded by the subject matter. Certainty is attainable only when we give an account of the eternal and unchanging forms; when we describe the imperfect and changeable objects in the material world, our description will inevitably share in the imperfection and changeability of its subject and will therefore be no more than "likely."

What do we find in the *Timaeus*? One of its most striking characteristics is Plato's vehement opposition to certain features of pre-Socratic thought. The *physikoi* had deprived the world of divinity; in the process, they had also deprived it of plan and purpose. According to these philosophers, things behave according to their inherent natures, and this alone accounts for the order and regularity of the cosmos. Order, then, is intrinsic, rather than extrinsic; it is not imposed by an outside agent but arises from within.

Now Plato found such an opinion not only foolish but dangerous. He had no intention of restoring the gods of Mount Olympus, who interfered in the day-to-day operation of the universe, but he was convinced that the order and rationality of the cosmos could be explained only as the imposition of an outside mind. If the *physikoi* found the source of order in *physis* (nature), he would locate it in *psyche* (mind).[28]

Plato depicted the cosmos as the handiwork of a divine craftsman, the Demiurge. According to Plato, the Demiurge is a benevolent craftsman, a rational god (indeed, the very personification of reason) who struggled against the limitations inherent in the materials at his disposal in order to produce a cosmos as good, beautiful, and intellectually satisfying as possible. The Demiurge took a primitive chaos filled with the unformed material out of which the cosmos would be constructed and imposed order according to a rational plan. This was not creation of the cosmos from nothing, as in the Judeo-Christian account of creation, for the raw materials were already present and contained properties over which the Demiurge had no control; nor was the Demiurge omnipotent, for he was constrained and limited by the available materials. Nevertheless, Plato clearly intended to portray the Demiurge as a supernatural being, distinct from, and outside of, the cosmos that he constructed. Whether Plato meant his readers to take the Demiurge literally is another matter, much debated and perhaps incapable of ever being resolved. What is not open to dispute is Plato's wish to declare that the cosmos is the product of reason and planning, that the order in the cosmos is rational order, imposed on recalcitrant materials from outside.

Besides being a rational craftsman, the Demiurge is a mathematician, for he constructed the cosmos on geometrical principles. Plato's account borrowed

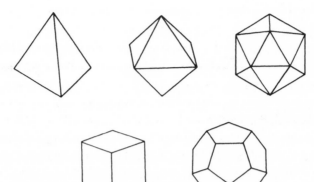

Fig. 2.5. The five Platonic solids: tetrahedron, octahedron, icosahedron, cube, and dodecahedron. Courtesy of J. V. Field.

the four roots or elements of Empedocles: earth, water, air, and fire. But (probably under Pythagorean influence) he reduced them to mathematical ingredients or components. It was already known in Plato's day that there are five,
and only five, regular geometrical solids (symmetrical solid figures formed
of plane surfaces, all identical); these are the tetrahedron (four equilateral
triangles), the cube (six squares), the octahedron (eight equilateral triangles),
the dodecahedron (twelve pentagons), and the icosahedron (twenty equilateral triangles). (See fig. 2.5.) Plato made these the basis of a "geometrical
atomism"—associating each of the elements with one of the geometrical solids. Fire is the tetrahedron (the smallest, sharpest, and most mobile of the
regular solids), air is the octahedron, water the icosahedron, and earth the
cube (the most stable of the regular solids). Plato also found a function for
the docedahedron (the regular solid closest to the sphere) by identifying
it with the cosmos as a whole. This was not the end of Plato's geometrical
analysis, for he reduced each of the three-dimensional geometrical figures
representing the elements into its two-dimensional components, as we will
see just below.[29]

Three features of this scheme deserve discussion. First, it accounts for change
and diversity in the same way as does Empedocles' theory: the elements can
mix in various proportions to produce variety in the material world. Second,
it allows for transmutation of the three elements that are composed of equilateral triangles (the tetrahedron, octahedron, and icosahedron), one to another, thus further accounting for change. For example, a single corpuscle of
water (the icosahedron) can be dissolved into its twenty constituent equilateral

triangles, which can then recombine into, say, two corpuscles of air (the octahedron) and one of fire (the tetrahedron). Only earth, which is composed of squares (and the square divided diagonally does not yield an equilateral triangle), is excluded from this process of transmutation. Third, Plato's geometrical corpuscles represent a significant step toward the mathematization of nature. Indeed, it is important for us to see just how large a step it is. Plato's elements are not material substance shaped as a square, tetrahedron, and so forth; in such a scheme matter would still be acknowledged as the fundamental stuff. For Plato, the shape—the geometrical figure—is all there is. The geometrical atoms are nothing more than the regular solids, which are reducible without residue to plane geometrical figures. Water, air, and fire are not *triangular*, they are (in the final analysis) nothing more than *triangles*, appropriately arranged. The Pythagorean program of reducing everything to mathematical first principles has been fulfilled.

Plato proceeded to describe many features of the cosmos; let us glance at a few of them. He demonstrated a sophisticated command of cosmology and astronomy. He proposed a spherical earth, surrounded by the spherical envelope of the heavens. He defined various circles on the celestial sphere, marking the paths of the sun, moon, and other planets. He understood that the sun moves around the celestial sphere once a year on a circle (which we call the ecliptic) tilted in relation to the celestial equator (see fig. 2.6). He knew that the moon makes a monthly circuit of approximately the same path. He knew that Mercury, Venus, Mars, Jupiter, and Saturn do the same, each at its own pace and with occasional reversals, and that Mercury and Venus never stray far from the sun. He even knew that the overall motion of the planetary bodies (if we combine their slow motion around the ecliptic with the daily rotation of the celestial sphere) is a spiral. And what is perhaps most important of all, Plato seems to have understood that the irregularities of planetary motion can be explained by the compounding of uniform circular motions.[30]

When Plato descended from the cosmos to the human frame, he offered an account of respiration, digestion, emotion, and sensation. He had a theory of sight, for example, that supposed that visual fire issues from the eye, interacting with external light to create a visual pathway that could transmit motions from the visible object to the observer's soul. The *Timaeus* even offered a theory of disease and outlined a regimen that was to ensure health.

It was an admirable cosmos that Plato portrayed. What were its most prominent features? From triangles and regular solids the Demiurge fashioned a final product of the utmost beauty and rationality. And the cosmos, if rational, is necessarily a living creature. The Demiurge, we read in the *Timaeus*, "wishing

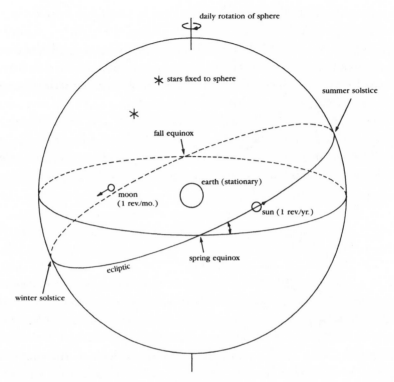

Fig. 2.6. The celestial sphere according to Plato.

to make the world most nearly like that intelligible thing which is best and in every way complete, fashioned it as a single visible living creature." But a living creature must possess a soul; and so in the center of the cosmos the Demiurge "set a soul and caused it to extend throughout the whole and further wrapped its body round with soul on the outside; and so he established one world alone, round and revolving in a circle, solitary but able by reason of its excellence to bear itself company, needing no other acquaintance or friend but sufficient to itself." The world soul is ultimately responsible for all motions in the cosmos, just as the human soul is responsible for the motions of the human body. We see here the origins of the strong animistic strain that was to remain an important feature of the Platonic tradition. Repelled by the lifeless necessity of the atomistic world, Plato has described an animated cosmos, permeated by rationality, replete with purpose and design.[31]

Nor is deity absent. We have the Demiurge, of course; but in addition Plato assigned divinity to the world soul and considered the planets and the fixed stars to be a host of celestial gods. However, unlike the gods of traditional

Greek religion, Plato's deities never interrupt the course of nature. Quite the contrary, it is the very steadfastness of the gods that, in Plato's view, guarantees the regularity of nature; the sun, moon, and other planets must move with some combination of uniform circular motions precisely because such motion is most perfect and rational, and consequently the only kind of motion conceivable for a divine being. Thus Plato's reintroduction of divinity does not represent a return to the unpredictability of the Homeric world. Quite the contrary, the function of divinity for Plato was to undergird and account for the order and rationality of the cosmos. Plato restored the gods in order to account for precisely those features of the cosmos that, in the view of the *physikoi*, required the banishing of the gods.[32]

THE ACHIEVEMENT OF EARLY GREEK PHILOSOPHY

If we survey early Greek philosophy with a modern scientific eye, certain pieces of it look familiar. The pre-Socratic inquiry into the shape and arrangement of the cosmos, its origin, and its fundamental ingredients reminds us of questions still investigated in modern astrophysics, cosmology, and particle physics. However, other pieces of early philosophy look considerably more foreign. Working scientists today do not inquire whether change is logically possible or where true reality is to be found; and it would be a considerable feat to turn up, say, a physicist or chemist who worries with any regularity about how to balance the respective claims of reason and observation. These matters are no longer talked about by scientists. Does it follow that such questions were a waste of time and that the early philosophers who devoted their lives to them were misguided or intellectually deficient?

This question needs to be handled with some delicacy. Surely the fact that the *physikoi* were concerned about some matters no longer of interest is no indictment of their enterprise; in the course of any intellectual endeavor some problems get resolved, while others go out of fashion. But the objection may go deeper than that: Are there issues that are intrinsically inappropriate or illegitimate, questions that were futile (and ought to have been recognized as such) from the beginning? And did Plato and the *physikoi* waste time and energy on any of these? Perhaps we can answer in this way. Themes such as the identity of the ultimate reality, the distinction between natural and supernatural, the source of order in the universe, the nature of change, and the foundations of knowledge are quite different from the explanation of arcane experimental data (say, a chemical reaction, a physiological process, or a meteorological event) that now occupy most scientists; but to be different is not to be useless or insignificant. At least until Isaac Newton, these larger themes demanded as much attention from the student of nature as did the problems that now fill up a university course in one of the sciences. Such questions were interesting and essential precisely because they were part of the effort to create conceptual foundations and vocabulary for investigating the world; and it is often the fate of foundational questions to seem pointless to later generations who take the foundations for granted. Today, for example, we may find the distinction between the natural and supernatural obvious; but until the distinction was carefully drawn, the investigation of nature could not properly begin.

Thus the early philosophers began at the only possible place: the beginning. They created a conception of nature that has served as the foundation of scientific belief and investigation in the intervening centuries—the conception of nature presupposed, more or less, by modern science. In the meantime many of the questions they asked have been resolved—often with rough-and-ready solutions, rather than definitive answers, but resolved sufficiently to slip from the forefront of scientific attention. As they have sunk from view, their place has been taken by a collection of much narrower investigations. If we would understand the scientific enterprise in all of its richness and complexity, we must see that its two parts—the foundation and the superstructure—are complementary and reciprocal. Modern laboratory investigation occurs within a broad conceptual framework and cannot even begin without expectations (created by predecessors) about nature or the underlying reality; in turn, the conclusions of laboratory research reflect back on these most fundamental notions, forcing refinement and (occasionally) revision. The historian's task is to appreciate the enterprise in all of its diversity. If the garden of the *physikoi* is situated at the beginning of the road to modern science, then the historian of science may profitably dally in its shady corners before embarking on his journey.

3 ❄ Aristotle's Philosophy of Nature

LIFE AND WORKS

Aristotle (fig. 3.1) was born in 384 B.C. in the northern Greek town of Stagira, into a privileged family. His father was personal physician to the Macedonian king, Amyntas II (grandfather of Alexander the Great). Aristotle had the advantage of an exceptional education: at age seventeen, he was sent to Athens to study with Plato. He remained in Athens as a member of Plato's Academy for twenty years, until Plato's death about 347. Aristotle then spent several years in travel and study, crossing the Aegean Sea to Asia Minor and its coastal islands. During this period he undertook biological studies, and he encountered Theophrastus (from the island of Lesbos), who was to become his pupil and lifetime colleague. He returned to Macedonia in 342 to become the tutor of the young Alexander (later "the Great"). In 335, when Athens fell under Macedonian rule, Aristotle returned to the city and began to teach in the Lyceum, a public garden frequented by teachers. He remained there, establishing an informal school, until shortly before his death in 322.[1]

In the course of his long career as student and teacher, Aristotle systematically and comprehensively addressed the major philosophical issues of his day. He is credited with more than 150 treatises, approximately 30 of which have come down to us. The surviving works appear to consist mainly of lecture notes or unfinished treatises not intended for wide circulation; whatever their exact origin, they were obviously directed to other philosophers, including advanced students. In modern translation, they occupy well over a foot of bookshelf, and they present a philosophical system overwhelming in power and scope. It is out of the question for us to survey the whole of Aristotle's philosophy, and we must be content with examining the fundamentals of his philosophy of nature—beginning with his response to positions taken by the pre-Socratics and Plato.[2]

Fig. 3.1. Aristotle. Museo Nazionale, Rome. Alinari/Art Resource N.Y.

METAPHYSICS AND EPISTEMOLOGY

Through his long association with Plato, Aristotle had, of course, become thoroughly versed in Plato's theory of forms. Plato had drastically diminished (without totally rejecting) the reality of the material world observed by the senses. Reality in its perfect fullness, Plato argued, is found only in the eternal forms, which are dependent on nothing else for their existence. The objects that make up the sensible world, by contrast, derive their characteristics and their very being from the forms; it follows that sensible objects exist only derivatively or dependently.

Aristotle refused to accept this diminished, dependent status that Plato assigned to sensible objects. They must exist fully and independently, for in Aristotle's view they were what make up the real world. Moreover, the traits that give an individual object its character do not, Aristotle argued, have a prior and separate existence in a world of forms, but belong to the object itself. There is no perfect form of a dog, for example, existing independently in the world of forms and replicated imperfectly in individual dogs, imparting to them their attributes. For Aristotle, there were just individual dogs. These dogs certainly shared a set of attributes—for otherwise we would not be entitled to call them "dogs"—but these attributes exist in, and belong to, individual dogs.

Perhaps this way of viewing the world has a familiar ring. Making individual sensible objects the primary realities ("substances," Aristotle called them) will seem like good common sense to most readers of this book, and probably struck Aristotle's contemporaries the same way. But if it makes good common sense, can it also be good philosophy? That is, can it deal successfully, or at least plausibly, with the difficult philosophical issues raised by the pre-Socratics and Plato—the nature of the fundamental reality, epistemological concerns, and the problem of change and stability? Let us take up these problems one by one.[3]

The decision to locate reality in sensible, corporeal objects does not yet tell us very much about reality—only that we should look for it in the sensible world. Already in Aristotle's day, any philosopher would demand to know more: one thing he would demand to know was whether the corporeal materials of daily experience (wood, water, air, stone, metal, flesh, etc.) are themselves the fundamental, irreducible constituents of things, or whether they are composites of still more fundamental stuff. Aristotle addressed this question by drawing a distinction between properties and their subjects. He maintained (as most of us would) that a property has to be the property *of* something; we call that something its "subject." To be a property is to belong to a subject; properties cannot exist independently.

Individual corporeal objects, then, have both properties (color, weight, texture, and the like) and something other than properties to serve as their subject. These two roles are played by "form" and "matter," respectively. Corporeal objects are "composites" of form and matter—form consisting of the properties that make the thing what it is, matter serving as the subject or substratum for the form. A white rock, for example, is white, hard, heavy, and so forth, by virtue of its form; but matter must also be present, to serve as subject for the form, and this matter brings no properties of its own to its union with form.[4] (Aristotle's doctrine will be further discussed in chap. 12, below, in connection with medieval attempts to clarify and extend it.)

We can never, in actuality, separate form and matter; they are presented to us only as a unitary composite. If they were separable, we should be able to put the properties (no longer the properties *of* anything) in one pile, the matter (absolutely propertyless) in another—an obvious impossibility. But if form and matter can never be separated, is it not meaningless to speak of them as the real constituents of things? Isn't this a purely logical distinction, existing in our minds, but not in the external world? Surely not for Aristotle, and perhaps not for us; most of us would think twice before denying the real existence of cold or red, although we can never collect a bucket of either one.

In short, Aristotle once again surprises us by using commonsense notions to build a persuasive philosophical edifice.

Aristotle's claim that the primary realities are concrete individuals surely has epistemological implications, since true knowledge must be knowledge of truly real things. By this criterion, Plato's attention was naturally directed toward the eternal forms, knowable through reason or philosophical reflection. Aristotle's metaphysics of concrete individuals, by contrast, directed his quest for knowledge toward the material world of individuals, of nature, and of change—a world encountered through the senses.

Aristotle's epistemology is complex and sophisticated. It must suffice here to indicate that the process of acquiring knowledge begins with sense experience. From repeated sense experience follows memory; and from memory, by a process of "intuition" or insight, the experienced investigator is able to discern the universal features of things. By the repeated observation of dogs, for example, an experienced dog breeder comes to know what a dog really is; that is, he comes to understand the form or definition of a dog, the crucial traits without which an animal cannot be a dog. Note that Aristotle, no less than Plato, was determined to grasp the universal traits or properties of things; but, unlike his teacher, Aristotle argued that one must start with the individual material thing. Once we grasp the universal properties or definition, we can put it to use as the premise of deductive demonstrations.[5]

Knowledge is thus gained by a process that begins with experience (a term broad enough, in some contexts, to include common opinion or the reports of distant observers). In that sense knowledge is empirical; nothing can be known apart from such experience. But what we learn by this "inductive" process does not acquire the status of true knowledge until put into deductive form; the end product is a deductive demonstration (nicely illustrated in a Euclidean proof) beginning from universal definitions as premises. Although Aristotle discussed both the inductive and deductive phases (the latter far more than the former) in the acquisition of knowledge, he stopped considerably short of later methodologists, especially in the analysis of induction.

This is the theory of knowledge outlined by Aristotle in the abstract. Is it also the method actually employed in Aristotle's own scientific investigations? Probably not—with perhaps an occasional exception. Like modern scientists, Aristotle did not proceed by following a methodological recipe book, but rather by rough and ready methods, familiar procedures that had proved themselves in practice. Somebody has defined science as "doing your damnedest, no holds barred"; when it came (for example) to his extensive biological researches, this is exactly what Aristotle did. It is not a surprise, and certainly no character defect, that Aristotle should, in the course of thinking about the nature and the foundations of knowledge, formulate a theoretical scheme (an epistemology) not perfectly consistent with his own scientific practice.[6]

NATURE AND CHANGE

The problem of change had become a celebrated philosophical issue (within the quite small community of philosophers) in the fifth century B.C. In the fourth century, Plato had dealt with it by restricting change to the imperfect material replica of the changeless world of forms. For Aristotle, a distinguished naturalist who was philosophically committed to the full reality of the changeable individuals that make up the sensible world, the problem of change was a most pressing one.[7]

Aristotle's starting point was the commonsense assumption that change is genuine. But this does not, by itself, get us very far; it remains to be demonstrated that the idea of change can withstand philosophical scrutiny; it must also be shown how change can be explained. Aristotle had various weapons in his arsenal by which to achieve these ends. The first was his doctrine of form and matter. If every object is constituted of form and matter, then Aristotle could make room for both change and stability by arguing that when an object undergoes change, its form changes (by a process of replacement, the new form replacing the old one) while its matter remains unchanged. Aristotle went on to argue that change in form takes place between a pair of opposites or contraries, one of which is the form to be achieved, the other its privation or absence. When the dry becomes wet, or the cold becomes hot, this is change from privation (dry or cold) to the intended form (wet or hot). Change, for Aristotle, is thus never random, but confined to the narrow corridor connecting pairs of contrary qualities; order is thus discernible even in the midst of change.

A determined Parmenidean might protest that to this point the analysis does nothing to escape Parmenides' objection to all change on the ground that inevitably it calls for the emergence of something out of nothing. Aristotle's reply is found in his doctrine of potentiality and actuality. Aristotle would undoubtedly have granted that if the only two possibilities are being and nonbeing—that is, if things either exist or do not exist—then the transition from non-hot to hot would indeed involve passage from nonbeing to being (the nonexistence of hot to the existence of hot) and would thus be vulnerable to Parmenides' objection. But Aristotle believed that the objection could be successfully circumvented by supposing that there are three categories associated with being instead of two: not just being and nonbeing, but (1) nonbeing, (2) potential being, and (3) actual being. If such is the state of things, then change can occur between potential being and actual being without nonbeing ever entering the picture. What Aristotle has in mind is perhaps most easily illustrated by examples from the biological realm. An acorn is potentially, but not actually, an oak tree. In becoming an oak tree, it becomes actually what it originally was only potentially. The change thus involves passage from potentiality to actuality—not from nonbeing to being, but from one kind or degree of being to another. Or for a pair of nonbiological examples, a heavy body held above the earth falls in order to fulfill its potential of being situated with other heavy things about the center of the universe. And a sculptor, with mallet and hammer, reveals in actuality a shape that existed potentially within the original block of marble.

If these arguments allow us to escape the logical dilemmas associated with the idea of change, and therefore to believe in its possibility, they do not yet tell us anything about the cause of change. Why should an acorn move from the status of potential oak tree to that of actual oak tree, or an object change from black to white, rather than remaining in its original state? Aristotle answered with an intricate, subtle, and not always consistent, theory of nature and causation. Given these difficulties, we will spare ourselves the pain of an exhaustive account and treat ourselves to the short version.

The world we inhabit is an orderly one, in which things generally behave in predictable ways, Aristotle argued, because every natural object has a "nature"—an attribute (associated primarily with form) that makes the object behave in its customary fashion, provided no insurmountable obstacle intervenes—or, as a modern commentator has put it, "that within a thing which determines basically what that thing does when it is being itself." For Aristotle, a brilliant zoologist, the growth and development of biological organisms were easily explained by the activity of such an inner driving force. An acorn becomes an oak tree because its nature is to do so. But the theory was applicable beyond biological growth and, indeed, beyond the biological realm altogether. Dogs bark, rocks fall, and marble yields to the hammer and chisel of the sculptor because of their respective natures. Ultimately, Aristotle argued, all change and motion in the universe can be traced back to the natures of things. For the natural philosopher, who by definition is interested in change and things capable of undergoing change, these natures are the central object of study.[8]

To this general statement of Aristotle's theory of "nature," we need to add a qualification—namely, that an artificially produced object is a special case, for such an object possesses no nature other than the natures of its ingredients. If a chariot is constructed of wood and iron, the nature of wood and nature of iron do not yield to a composite "nature of a chariot." By contrast, in the organic world the natures of the organs and tissues that make up an organism yield to the nature of the organism as a whole. The nature of the human body is not the sum of the natures of its various tissues and organs, but a unique nature characteristic of that living human as an organic whole.

With this theory of nature in mind, we can understand a feature of Aristotle's scientific practice that has puzzled and distressed modern commentators and critics—namely, the absence from his work of anything resembling controlled experimentation. Unfortunately, such criticism overlooks Aristotle's aims, which drastically limited his methodological options. If, as Aristotle believed, the nature of a thing is to be discovered through the behavior of that thing in its natural, unfettered state, then artificial constraints will merely interfere and corrupt.[9] If, despite interference, the object behaves in its customary fashion, we have troubled ourselves for no purpose. If we set up conditions that prevent the nature of an object from revealing itself, all we have learned is that it can be interfered with to the point of remaining concealed. Contrived experimentation violates, rather than reveals, the natures of things. Aristotle's scientific practice is not to be explained, therefore, as a result of stupidity or deficiency on his part—failure to perceive an obvious procedural improvement—but as a method compatible with the world as he perceived it and suited to the questions that interested him. Experimental science emerged not when, at long last, the human race produced somebody clever enough to perceive that artificial conditions would assist in the exploration of nature, but when a rich variety of conditions were fulfilled—including the emergence of questions to which such a procedure promised to provide answers.[10]

To complete our analysis of Aristotle's theory of change, we must briefly consider the celebrated four Aristotelian causes. To understand a change or the production of an artifact is to know its causes (perhaps best translated "explanatory conditions and factors"). There are four of these: the form of a thing; the matter underlying that form, which persists through the change; the agency that brings about the change; and the purpose served by the change. These are called, respectively, formal cause, material cause, efficient cause, and final cause. To take an extremely simple example—the production of a statue—the formal cause is the shape given the marble, the material cause is the marble that receives this shape, the efficient cause is the sculptor, and the final cause is the purpose for which the statue is produced (perhaps the beautification of Athens or the celebration of one of its heroes). There are cases in which identifying one or another of the causes is difficult, or in which one or more causes merge, but Aristotle was convinced that his four causes provided an analytical scheme of general applicability.

We have said enough about the form-matter distinction to make clear what was meant by "formal" and "material" causes, and "efficient" cause is close enough to modern notions of causation to require no further comment; but "final" cause requires explanation. In the first place, the expression "final cause" is an English cognate derived from the Latin word *finis*, meaning "goal," "purpose," or "end," and it has nothing to do with the fact that it often appears last in the list of Aristotelian causes. Aristotle argued, quite rightly, that many things cannot be understood without knowledge of purpose or function. To explain the arrangement of teeth in the mouth, for example, we must understand their functions (sharp teeth in front for tearing, molars in back for grinding). Or to take an example from the inorganic realm, it is not possible to grasp why a saw is made as it is without knowing the function the saw is meant to serve. Aristotle went so far as to give final cause priority over material cause, noting that the purpose of the saw determines the material (iron) of which it must be made, whereas the fact that we possess a piece of iron does nothing to determine that we will make it into a saw.[11]

Perhaps the most important point to be made about final cause is its clear illustration of the role of purpose (the more technical term is "teleology") in Aristotle's universe. The world of Aristotle is not the inert, mechanistic world of the atomists, in which the individual atom pursues its own course mindless of all others. Aristotle's world is not a world of chance and coincidence, but an orderly, organized world, a world of purpose, in which things develop toward ends determined by their natures. It would be unfair and pointless to judge Aristotle's success by the degree to which he anticipated modern science (as though his goal was to answer our questions, rather than his own); it is nonetheless worth noting that the emphasis on functional explanation to which Aristotle's teleology leads would prove to be of profound significance for all of the sciences and remains to this day a dominant mode of explanation within the biological sciences.

COSMOLOGY

Aristotle not only devised methods and principles by which to investigate and understand the world: form and matter, nature, potentiality and actuality, and the four causes. In the process, he also developed detailed and influential theories regarding an enormous range of natural phenomena, from the heavens above to the earth and its inhabitants below.[12]

Let us start with the question of origins. Aristotle adamantly denied the possibility of a beginning, insisting that the universe must be eternal. The alternative—that the universe came into being at some point in time—he regarded as unthinkable, violating (among other things) Parmenidean strictures about something coming from nothing. Aristotle's position on this question would prove troublesome for medieval Christian Aristotelians.

Aristotle considered this eternal universe to be a great sphere, divided into an upper and a lower region by the spherical shell in which the moon is situated. Above the moon is the celestial region; below is the terrestrial region; the moon, spatially intermediate, is also of intermediate nature. The terrestrial or sublunar region is characterized by birth, death, and transient change of all kinds; the celestial or supralunar region, by contrast, is a region of eternally unchanging cycles. That this scheme had its origin in observation would seem clear enough; in his *On the Heavens*, Aristotle noted that "in the whole range of time past, so far as our inherited records reach, no change appears to have taken place either in the whole scheme of the outermost heaven or in any of its proper parts."[13] If in the heavens we observe eternally unvarying circular motion, he continued, we can infer that the heavens are not made of the terrestrial elements, the nature of which (observation reveals) is to rise or fall in transient rectilinear motions. The heavens must consist of an incorruptible fifth element (there are four terrestrial elements): the quintessence (literally, the fifth essence) or aether. The celestial region is completely filled with this quintessence (no void space) and divided, as we shall see, into concentric spherical shells bearing the planets. It had, for Aristotle, a superior, quasi-divine status.[14]

The sublunar region is the scene of generation, corruption, and impermanence. Aristotle, like his predecessors, inquired into the basic element or elements to which the multitude of substances found in the terrestrial region can be reduced. He accepted the four elements originally proposed by Empedocles and subsequently adopted by Plato—earth, water, air, and fire. He agreed with Plato that these elements are in fact reducible to something even more fundamental; but he did not share Plato's mathematical inclination and therefore refused to accept Plato's regular solids and their constituent triangles. Instead, he expressed his own commitment to the reality of the world of sense experience by choosing sensible qualities as the ultimate building blocks. Two pairs of qualities are crucial: hot-cold and wet-dry. These combine in four pairs, each of which yields one of the elements (see fig. 3.2). Notice the use made once again of contraries. There is nothing to forbid any of the four

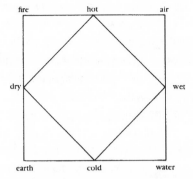

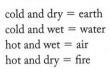

cold and dry = earth
cold and wet = water
hot and wet = air
hot and dry = fire

Fig. 3.2. Square of opposition of the Aristotelian elements and qualities. For a medieval (9th c.) version of this diagram, see John E. Murdoch, *Album of Science: Antiquity and the Middle Ages*, p. 352.

qualities being replaced by its contrary, as the result of outside influence. If water is heated, so that the cold of water yields to hot, the water is transformed into air. Such a process easily explains changes of state (from solid to liquid to vapor, and conversely), but also more general transmutation of one substance into another. On such a theory as this, alchemists could easily build.[15]

The various substances that make up the cosmos totally fill it, leaving no empty space. To appreciate Aristotle's view, we must lay aside our almost automatic inclination to think atomistically; we must conceive material things not as aggregates of tiny particles but as continuous wholes. If it is obvious that, say, a loaf of bread is composed of crumbs separated by small spaces, there is no reason not to suppose that those spaces are filled by some finer substance, such as air or water. And there is certainly no simple way of demonstrating, nor indeed any obvious reason for believing, that water and air are anything but continuous. Similar reasoning, applied to the whole of the universe, led Aristotle to the conclusion that the universe is full, a *plenum*, containing no void space. This claim would be attacked by medieval scholars.

Aristotle defended this conclusion with a variety of arguments, such as the following. The speed of a falling body is dependent on the density of the medium through which it falls—the less the density, the swifter the motion of the falling body. It follows that in a void space (density zero), there is nothing to slow the descent of the body, from which we would be forced to conclude that the body would fall with infinite speed—a nonsensical notion, since it implies that the body could be at two places at the same time. Critics have frequently noted that this argument can just as well be taken to prove that the absence of resistance does not entail infinite speed as to prove that void does not exist. The point is, of course, well taken. However, we need to understand that Aristotle's denial of the void did not rest on this single piece of reasoning. In fact, this was but one small part of a lengthy campaign against the atomists, in which Aristotle battled the notion of void space (or void place) with a variety of arguments, some more and some less persuasive.[16]

In addition to being hot or cold and wet or dry, each of the elements is also heavy or light. Earth and water are heavy, but earth is the heavier of the two. Air and fire are light, fire being the lighter of the two. In assigning levity to two of the elements, Aristotle did not mean (as we might, if we were making the claim) simply that they are less heavy, but that they are light in an absolute sense; levity is not a weaker version of gravity, but its contrary. Because earth and water are heavy, it is their nature to descend toward the center of the universe; because air and fire are light, it is their nature to ascend toward the periphery (that is, the periphery of the terrestrial region, the spherical shell that contains the moon). If there were no hindrances, therefore, earth and water would collect at the center; because of its greater heaviness, earth would achieve a lower position, forming a sphere at the very center of the universe; water would collect in a concentric spherical shell just outside it. Air and fire naturally ascend, but fire, owing to its greater levity, occupies the outermost region, with air as a concentric sphere just inside it. In the ideal case (in which there are no mixed bodies and nothing prevents the natures of the four elements from fulfilling themselves), the elements would thus form a set of concentric spheres: fire on the outside, followed by air and water, and finally earth at the center (see fig. 3.3). But in reality, the world is composed largely of mixed bodies, one always interfering with another, and the ideal is never attained. Nonetheless, the ideal arrangement defines the natural place of each of the elements; the natural place of earth is at the center of the universe, of fire just inside the sphere of the moon, and so forth.[17]

It must be emphasized that the arrangement of the elements is spherical. Earth collects at the center to form *the earth*, and it too is spherical. Aristotle defended this belief with a variety of arguments. Arguing from his natural philosophy, he pointed out that since the natural tendency of earth is to move toward the center of the universe, it must arrange itself symmetrically about that point. But he also called attention to observational evidence, including the circular shadow cast by the earth during a lunar eclipse and the fact that north-south motion by an observer on the surface of the earth alters the apparent position of the stars. Aristotle even reported an estimate by mathematicians of the earth's circumference (400,000 stades = about 45,000 miles, roughly

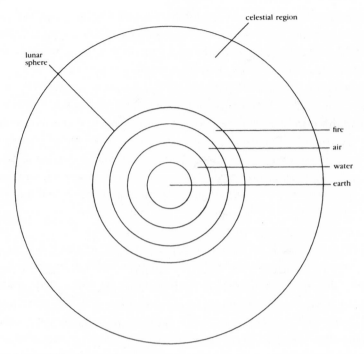

Fig. 3.3. The Aristotelian cosmos.

1.8 times the modern value). The sphericity of the earth, thus defended by Aristotle, would never be forgotten or seriously questioned. The widespread myth that medieval people believed in a flat earth is of modern origin.[18]

Finally, we must note one of the implications of this cosmology, namely that space, instead of being a neutral, homogeneous backdrop (analogous to our modern notion of geometrical space) against which events occur, has properties. Or to express the point more precisely, ours is a world of space, whereas Aristotle's was a world of place. Heavy bodies move toward their place at the center of the universe not because of a tendency to unite with other heavy bodies located there, but simply because it is their nature to seek that central place; if by some miracle the center happened to be vacant (a physical impossibility in an Aristotelian universe, but an interesting imaginary state of affairs), it would remain the destination of every heavy body.[19]

MOTION, TERRESTRIAL AND CELESTIAL

We can best understand Aristotle's theory of motion by grasping its two most fundamental claims. The first is that motion is never spontaneous; there is no motion without a mover. The second is the distinction between two types of motion: motion toward the natural place of the moving body is "natural" motion; motion in any other direction occurs only under coercion from an outside force and is therefore a "forced" or "violent" motion.

The mover in the case of natural motion is the nature of the body, which is responsible for its tendency to move toward its natural place as defined by the ideal spherical arrangement of the elements. Mixed bodies have a directional tendency that depends on the proportion of the various elements in their composition. When a body undergoing natural motion reaches its natural place, its motion ceases. The mover in the case of forced motion is an external force, which compels the body to violate its natural tendency and move in a direction or manner other than straight-line motion toward its natural place. Such motion ceases when the external force is withdrawn.[20]

So far, this seems sensible. One obvious difficulty, however, is to explain why a projectile hurled horizontally, and therefore undergoing forced motion, does not come to an immediate halt when it loses contact with whatever propelled it. Aristotle's answer was that the medium takes over as mover. When we project an object, we also act on the surrounding medium (air, for instance), imparting to it the power to move objects; this power is communicated from part to part, in such a way that the projectile is always in contact with a portion of the medium capable of keeping it in motion. If this seems implausible, consider the greater implausibility (from Aristotle's standpoint) of the alternative—that a projectile, which is inclined by nature to move toward the center of the universe, moves horizontally or upward despite the fact that there is no longer anything causing it to do so.

Force is not the only determinant of motion. In all real cases of motion in the terrestrial realm, there will also be a resistance or opposing force. And it seemed clear to Aristotle that the quickness of motion must depend on these two determining factors—the motive force and the resistance. The question arose: what is the relationship between force, resistance, and speed? Although it probably did not occur to Aristotle that there might be a quantitative law of universal applicability, he was not without interest in the question and did make several forays into quantitative territory. In reference to natural motion in his *On the Heavens* and again in his *Physics*, Aristotle claimed that when two bodies of differing weight descend, the times required to cover a given distance will be inversely proportional to the weights. (A body twice as heavy will require half the time). In the same chapter of the *Physics*, Aristotle introduced resistance into the analysis of natural motion, arguing that if bodies of equal weight move through media of different densities, the times required to traverse a given distance are proportional to the densities of the respective media; that is, the greater the resistance the slower the body moves. Finally, Aristotle also dealt with forced motion in his *Physics*, claiming that if a given force moves a given weight (against its nature) for a given distance in a given time, the same force will move half that weight twice the distance in that same time (or the same distance in half that time); alternatively, half the force will move half the weight the same distance in the same time.[21]

From such statements, some of Aristotle's successors have made a determined effort to extract a general law. This law is customarily stated as:

$$v \propto F/R.$$

That is, velocity (v) is proportional to the motive force (F) and inversely proportional to the resistance (R). For the special case of the natural descent of a heavy body, the motive force is the weight (W) of the body, and the relationship then becomes:

$$v \propto W/R.$$

Such relationships probably do no great violence to Aristotle's intent for most cases of motion; however, giving them mathematical form, as we have done, suggests that they hold for all values of v, F (or W), and R—a claim that Aristotle would certainly have denied. He stated explicitly, for example, that a resistance equal to the motive force will prevent motion altogether, whereas the formula above offers no such result. Moreover, the appearance of velocity in these relationships seriously misrepresents Aristotle's conceptual framework, which contained no concept of velocity as a quantifiable measure of motion, but described motion only in terms of distances and times. Velocity as a technical scientific term to which numerical values might be assigned was a contribution of the Middle Ages (see below, chap. 12).

Aristotle has been severely criticized for this theory of motion, on the assumption that any sensible person should have recognized its fatal flaws. Is such criticism justified? In the first place, our goal is to understand the behavior, beliefs, and achievements of historical actors against the background of the culture in which they lived, rather than to assess credit or blame according to the degree to which those historical actors resemble us. In short, historians must always contextualize their subjects. Second, some of the criticisms of Aristotle's theories of motion apply only to theories foisted onto Aristotle by followers and critics, rather than to his own. Third, the theory in its genuinely Aristotelian (and properly contextualized) version makes quite good sense today and would surely have made good sense in the fourth century B.C. For example, various surveys have shown that the majority of modern, university-educated people are prepared to assent to many of the basics of Aristotle's theory of motion. Fourth, the relatively modest level of quantitative content in Aristotle's theory is easily explained as the outcome of his larger philosophy of nature. His primary goal was to understand essential natures, not to explore quantitative relationships between such incidental factors as the space-time (or place-time) coordinates applicable to a moving body; even an exhaustive investigation of the latter gives us no useful information about the former. You may criticize Aristotle, if you like, for not being interested in whatever interests modern scientists, but we do not thereby learn anything significant about Aristotle.

Motion in the celestial sphere is an altogether different sort of phenomenon. The heavens, composed of the incorruptible quintessence, possess no contraries and are therefore incapable of qualitative change. It might seem fitting for such a region to be absolutely motionless, but this hypothesis is defeated by the most casual observation of the heavens. Aristotle therefore assigned to the heavens the most perfect of motions—continuous uniform circular motion. Besides being the most perfect of motions, uniform circular motion appears to have the capability of explaining the observed celestial cycles.

By Aristotle's day, these cycles had been an object of study for centuries in the Greek world and for millennia in its predecessor civilizations. It was understood that the "fixed" stars move with perfect uniformity, as though fixed to a uniformly rotating sphere, with a period of rotation of approximately one day. But there were seven stars, the wandering stars or planets, that displayed a more intricate motion, apparently crawling around on the stellar sphere as it went through its daily rotation. These seven were the Sun, Moon, Mercury, Venus, Mars, Jupiter, and Saturn. The sun crawls slowly (about 1°/day), west to east with small variations in speed, through the sphere of fixed stars along a path called the ecliptic, which passes through the center of the zodiac (see fig. 2.6). The moon follows approximately the same course, but at the more rapid rate of about 12°/day. The remaining planets also move along the ecliptic (or in its vicinity) with variable speed and with an occasional reversal of direction.

Are such complex motions compatible with the requirement of uniform circular motion in the heavens? Eudoxus, a generation before Aristotle, had already shown that they are. I will return to this subject in chap. 5; for the moment, it will be sufficient to point out that Eudoxus treated each complex planetary motion as a composite of a series of simple uniform circular movements.

He did this by assigning to each planet a set of concentric spheres, and to each sphere one component of the complex planetary motion. Aristotle took over this scheme, with various modifications. When he was finished, he had produced an intricate piece of celestial machinery, consisting of fifty-five planetary spheres plus the sphere of the fixed stars.

What is the cause of movement in the heavens? Aristotle's natural philosophy would not allow such a question to go unasked. The celestial spheres are composed, of course, of the quintessence; their motion, being eternal, must be natural rather than forced. The cause of this eternal motion must itself be unmoved, for if we do not postulate an unmoved mover, we quickly find ourselves trapped in an infinite regress: a moving mover must have acquired its motion from yet another moving mover, and so on. Aristotle identified the unmoved mover for the planetary spheres as the "Prime Mover," a living deity representing the highest good, wholly actualized, totally absorbed in self-contemplation, nonspatial, separated from the spheres it (or he or she) moves, and not at all like the traditional anthropomorphic Greek gods. How, then, does the Prime Mover or Unmoved Mover cause motion in the heavens? Not as efficient cause, for that would require contact between the mover and the moved, but as final cause. That is, the Prime Mover is the object of desire for the celestial spheres, which endeavor to imitate its changeless perfection by assuming eternal, uniform circular motions. Any reader who has followed this much of Aristotle's discussion would be justified in assuming that there is a single Unmoved Mover for the entire cosmos; it comes as something of a surprise, therefore, when Aristotle announces that, in fact, each of the celestial spheres has its own Unmoved Mover, the object of its affection and the final cause of its motion.[22]

ARISTOTLE AS A BIOLOGIST

There is no way of determining when or how Aristotle became interested in the biological sciences. That his father was a physician is a factor that we must surely take into account. Aristotle's biological studies no doubt occurred over an extended period, but several years on the island of Lesbos (off the coast of Asia Minor) offered him an exceptional opportunity for the observation of marine life. He was probably assisted in the gathering of biological data by his students, and he certainly relied on the reports of other observers, including physicians, fishermen, and farmers. The product of this research effort was a series of large zoological treatises and short works on human physiology and psychology that occupy well over four hundred pages in a modern translation;

these works laid the foundations of systematic zoology and profoundly shaped thought on human biology for some two thousand years.[23]

In Aristotle's day human anatomy and physiology had long attracted attention for their medical import and presumably required no further justification, but Aristotle felt obliged to defend zoological research. In *On the Parts of Animals* he admitted that animals are ignoble by comparison with the heavens and acknowledged that zoological studies are distasteful to many people. However, he considered this distaste to be childish, and he argued that in zoological studies the quantity and richness of the available data compensate for the absence of nobility in the object of study. He argued, moreover, that zoological studies contribute to knowledge of the human frame owing to the close resemblance between animal and human nature; he noted the pleasure of discovering causes in the zoological realm; and he pointed out that order and purpose are displayed with particular clarity in the animal kingdom, providing us with a golden opportunity to refute the atomists' notion that the "works of nature" are products of chance alone.[24]

Aristotle saw that biology has both a descriptive and an explanatory side. He considered the explanation of biological phenomena the ultimate goal, but acknowledged the gathering of biological data as the first order of business. His *History of Animals*, which was intended to meet this first need, is a vast storehouse of biological information. Aristotle began with the human body, as a standard to which other animals could be compared. He subdivided the human body into head, neck, thorax, arms, and legs; and he proceeded to discuss both internal and external features, including brain, digestive system, sexual organs, lungs, heart, and blood vessels.

However, Aristotle made his greatest contribution not in the area of human anatomy but in descriptive zoology. More than five hundred species of animals are mentioned in his *History of Animals*; the structure and behavior of many are described in considerable detail, often on the basis of skillful dissection. Although he devoted considerable attention to the theoretical problems of classification, in practice Aristotle adopted "natural" or popular groupings based on multiple attributes. He divided animals into two major categories—"blooded" (that is, red-blooded) and "bloodless." The former category he subdivided into viviparous quadrupeds (four-footed mammals that bring forth living young), oviparous (or egg-laying) quadrupeds, marine mammals, birds, and fish; the latter into mollusks (such as the octopus and cuttlefish), crustacea (including crabs and crayfish), testacea (including the snail and oyster), and insects. These major categories Aristotle arranged hierarchically in a scale of being according to what he judged to be their degree of vital heat.[25]

Although he ranged over the whole of the animal kingdom, Aristotle was no doubt most at home when it came to marine life, of which he exhibited intimate firsthand knowledge. It has often been noted, for example, that he described the placenta of the dogfish (*Mustelus laevis*) in terms that were not confirmed until the nineteenth century. But Aristotle displayed impressive skill in other parts of the animal kingdom as well. His description of the incubation of birds' eggs is an excellent example of meticulous observation:

> Generation from the egg proceeds in an identical manner with all birds, but the full periods from conception to birth differ. . . . With the common hen after three days and three nights there is the first indication of the embryo. . . . Meanwhile the yolk comes into being, rising towards the sharp end, where the primal element of the egg is situated, and where the egg gets hatched; and the heart appears, like a speck of blood, in the white of the egg. This point beats and moves as though endowed with life, and from it . . . two vein ducts with blood in them trend in a convoluted course . . . , and a membrane carrying bloody fibres now envelops the white, leading off from the vein-ducts. A little afterwards the body is differentiated, at first very small and white. The head is clearly distinguished, and in it the eyes, swollen out to a great extent. . . .[26]

Natural history, which enumerates and describes the population of the universe, is no doubt an appealing occupation and may be regarded by some as an end in itself. But for Aristotle it was a means to a higher end—the source of factual data that would lead to physiological understanding and causal explanations. And for him, true knowledge was always causal knowledge.

Aristotle's understanding of physiology was based on the same principles that functioned in other realms of his natural philosophy. (Whether they were first developed in the biological realm and then applied to metaphysics, physics, and cosmology, or vice versa, is a matter of dispute among scholars.)[27] Thus form and matter, actuality and potentiality, the four causes, and especially the element of purpose or function associated with final cause are central to his biology. Aristotle summarized the ingredients of a proper biological explanation in his *On the Generation of Animals*: "Everything that comes into being or is made must (1) be made out of something, (2) be made by the agency of something, and (3) must become something."[28] That out of which an organism is made is, of course, its material cause; the agency by which it is made is its formal or efficient cause (two causes that are often merged in Aristotle's biology); and that which it becomes, the goal of its full development, is its final cause.

Each living organism, then, is constituted of matter and form: the matter consists of the various organs that make up the body; the form is the organizing principle that molds these organs into a unified organic whole. Aristotle identified form with organisms' soul and assigned it responsibility for the vital characteristics of the organism—nutrition, reproduction, growth, sensation, movement, and so forth. Indeed, Aristotle arranged living things in a hierarchy on the basis of their participation in several kinds of form or soul, each of which performs certain functions. Plants possess a nutritive soul, which enables them to obtain nourishment, grow, and reproduce. Animals possess, in addition, a sensitive soul, which accounts for sensation and (indirectly) for movement. Finally, humans add to these two lower kinds of soul a rational soul, which supplies the higher capacities of reason. If, as Aristotle maintained, soul is but the form of the organism, then it is clear that soul (including the human soul) is not immortal; at death the organism disintegrates, and its form evaporates into nonbeing. This doctrine would become a bone of contention when Aristotle's works entered the Christian culture of the Middle Ages.[29]

How do Aristotle's metaphysics of form and matter and the four causes explain biological reproduction—one of the central questions of Aristotle's biology? First, Aristotle argued that the existence of two genders—male and female—reflects the distinction between (1) formal or efficient cause (here taken to be the same thing) and (2) the matter on which this cause works. In humans and higher animals the female supplies the matter—namely, menstrual blood. Male semen bears the form and impresses this on the menstrual blood to produce a new organism. The young in higher animals, which have a large measure of vital heat, are brought forth live, as fully developed members of the species; in animals somewhat deficient in vital heat, the offspring are eggs hatched internally; as we descend the scale of perfection, we come to animals that produce eggs hatched externally, the eggs being more or less perfect depending on the exact degree of heat; at the bottom of the scale, bloodless animals produce a grub or maggot:

> We must observe how rightly Nature orders generation in regular gradation. The more perfect and hotter animals produce their young perfect in respect of quality..., and these generate living animals within themselves from the first. The second class do not generate perfect animals within themselves from the first (for they are only viviparous after first

laying eggs)....The third class do not produce a perfect animal, but an egg, and this egg is perfect. Those whose nature is still colder than these produce an egg, but an imperfect one, which is perfected outside the body....The fifth and coldest class does not even lay an egg from itself; but so far as the young ever attain to this condition at all, it is outside the body of the parent....For insects produce a grub first; the grub after developing becomes egg-like.[30]

The idea of perfection, so prominent in Aristotle's theory of generation, brings us to the third and last element of biological explanation—final cause or that which a biological organism is in the process of becoming. The biologist, in Aristotle's view, always needs to know the complete, mature form or nature of an organism. Only such knowledge will enable him to understand the structure of the organism and the existence and interrelations of its parts. For example, Aristotle explained the presence of lungs in land animals by reference to the needs of the organism as a whole. Blooded animals, he argued, require an external cooling agent because of their warmth. In fish, this agent is water, and consequently fish have gills instead of lungs. Animals that breathe, however, are cooled by air and consequently come equipped with lungs.[31] Knowledge of the mature form is also part of the explanation of the organism's development, for there is an upward movement in the organic realm, as organisms strive to actualize the potentialities that exist within them. We cannot understand the changes that occur within an acorn, for example, if we do not understand the oak tree that is its final destination. Finally, purpose and function enter Aristotle's biology not merely as an explanation of the form or development of the individual or species, but on a universal or cosmic level, to explain the interdependence and interrelationships of species in the order of nature.

There is, of course, much more to Aristotle's biological system. He explained nutrition, growth, locomotion, and sensation. He considered the functions of the principal organs, including brain, heart, lungs, liver, and reproductive organs. It is important to note that he made the heart the central organ of the body, the seat of emotions and sensation as well as of vital heat. He developed the notion of hierarchy in the biological realm: form, he believed, is superior to matter, living to nonliving, male to female, blooded to bloodless, mature to immature. Indeed, he arranged living things in a single, hierarchical scale of being, beginning with the Prime Mover at the top and descending through the human race to viviparous, oviparous, and vermiparous animals, and finally to plants.

Let us conclude this discussion with a brief analysis of method in Aristotle's biological works. If there is any branch of the scientific enterprise that demands observation, surely it is biology (and especially natural history). It is inconceivable that Aristotle would have attempted to describe the structure and habits of animals on any other basis. The observation in question was frequently his own, and we find in his works plentiful evidence of empirical method, including dissection. However, no naturalist working alone could amass the quantity of data contained in Aristotle's biological works, and it is apparent that he relied on the reports of travelers, farmers, and fishermen, the help of assistants, and the writings of his predecessors. Aristotle was generally critical of his sources, and displayed a healthy skepticism even about his own observations. However, he was not always skeptical enough, and there are many examples of descriptive error in his biological works. When it came to biological theory, Aristotle (like any theorist) was obliged to make inferences from the observational data; if his inferences were not always the ones we would make, they nonetheless display the insight of one of the most brilliant biologists ever to live. They also, of course, display the powerful influence of Aristotle's larger philosophical system, which continually influenced the questions he asked, the details he noticed, and the theoretical interpretation he placed upon them.[32]

ARISTOTLE'S ACHIEVEMENT

My frequent mini-lectures on proper historical methodology may seem to be the flogging of a dead horse, but conscience requires me to continue until I am certain that this particular horse has succumbed. The proper measure of a philosophical system or a scientific theory is not the degree to which it anticipated modern thought, but its degree of success in treating the philosophical and scientific problems of its own day. If a comparison is to be made, it must be between Aristotle and his predecessors, not Aristotle and the present. Judged by such criteria, Aristotle's philosophy is an astonishing achievement. In natural philosophy, he offered a subtle and sophisticated treatment of the major problems posed by the pre-Socratics and Plato: the nature of the fundamental stuff, the proper means of knowing it, the problems of change and causation, the basic structure of the cosmos, and the nature of deity and its relationship to material things.

But Aristotle also went far beyond any predecessor in the analysis of specific natural phenomena. It is no exaggeration to claim that, almost single-handedly, he created entirely new disciplines. His *Physics* contains a detailed discussion of terrestrial dynamics. He devoted the better part of his *Meteorology* to phenomena of the upper atmosphere, including comets, shooting stars, rain and the rainbow, thunder, and lightning. His *On the Heavens* developed the work of certain predecessors into an influential account of planetary astronomy. He touched upon geological phenomena, including earthquakes and mineralogy. He undertook a thorough analysis of sensation and the sense organs, particularly vision and the eye, developing a theory of light and vision that would remain influential until the seventeenth century. He concerned himself with what we might regard as the basic chemical processes—mixtures and combinations of substances. He wrote a book on the soul and its faculties. And, as we have seen, he contributed monumentally to developments in the biological sciences.

We will consider Aristotle's influence in subsequent chapters. I will conclude here simply by stating that his powerful influence in late antiquity and his dominance from the thirteenth century through the Renaissance resulted not from intellectual subservience on the part of scholars during those periods or from interference on the part of the church, but from the overwhelming explanatory power of Aristotle's philosophical and scientific system. Aristotle prevailed through persuasion, not coercion.

4 ❀ *Hellenistic Natural Philosophy*

Aristotle's death in 322 B.C. nearly coincided with the end of the military campaigns of Alexander the Great (334–323 B.C.), which established a far-flung Greek empire and sounded the death knell for the autonomous Greek city-states. Alexander dramatically enlarged Greek territory, carrying Greek language and culture as far east as the Hindu Kush (now northeastern Afghanistan), across the Indus River into India, and as far south as Egypt (see map 4.1). However, Alexander and his successors also borrowed from the conquered peoples, creating a synthesis of Greek and foreign elements designated by the adjective "Hellenistic"—meaning "Greekish." Although Greek elements were overwhelmingly dominant, the historians who coined this term wished to distinguish the Hellenistic period from what they regarded as the unadulterated Greek culture of ancient "Hellenic" times. The phrase "Hellenistic natural philosophy," then, denotes thought about nature among scholars and educated people throughout this Greek empire. In the short run, the center of gravity remained in the traditional Greek territories; in the long run, leadership shifted southward to Alexandria in Egypt and westward to Rome.

SCHOOLS AND EDUCATION

Before we examine the content of Hellenistic natural philosophy, we need to examine its social underpinnings—the social and institutional mechanisms by which learning in general and natural philosophy in particular were cultivated and transmitted. Knowledge can, of course, be transmitted individually, from parent to child, friend to friend, or master to apprentice. But as that knowledge increases in complexity and sophistication, pressure for a more formalized, collective educational system is likely to grow. Did this occur in ancient Greece? If so, what was the nature of the educational system that resulted?[1]

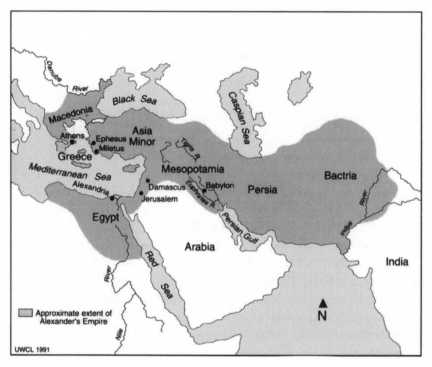

Map 4.1. Alexander the Great's empire.

No formal education was required in any ancient society, but several years of instruction at the elementary level became an ideal among the early Greek ruling class. Because it was intended for preadolescent children (*paides*), this education was referred to as *paideia*. Traditionally, *paideia* consisted of two parts: *gymnastike* for the body and *mousike* for the mind or spirit. *Gymnastike* included physical culture and athletics. *Mousike* covered all of the arts presided over by the Muses, especially music and poetry. However, social needs eventually outran this bipartite system, and by the beginning of the fifth century B.C. schools for reading and writing had emerged.

Instruction in *gymnastike* took place most frequently in a sports ground or wrestling school, or possibly in a public gymnasium. *Mousike* and literary education could be carried on almost anywhere, including a public building or the teacher's home. It must be understood that there was nothing resembling modern, compulsory mass education. Teachers entered the educational enterprise privately, on their own initiative; and the aristocracy took advantage of teachers' services according to individual need and inclination.

A major change in this educational pattern occurred in the fifth century B.C., with the coming of the itinerant teachers known as "sophists." Education to this point had been strictly elementary, largely athletic and artistic in its orientation. About the middle of the fifth century, the sophists made their appearance in Athens, offering something new. First, they offered education at a more advanced level. Second, their goal was the training of citizens and statesmen, and this called for a shift in the content of education toward intellectual, and especially political, matters. The sophists offered what we would regard as group tutorials, with no fixed curriculum or universal pattern, and certainly no shared philosophical system, for a duration negotiated by the parties concerned. (A figure of three or four years is often suggested by historians, but it has recently been argued that in some cases the duration of instruction may have been "as little as a week or an hour.")[2] In order to attract business, sophist teachers needed to be visible, and it therefore became customary for them to teach in a public place, such as the agora (public marketplace) or a large public gymnasium (of which Athens at the time had three). When business dried up or the teacher wore out his welcome, he moved on.

Against this background, we can begin to understand the teaching of Socrates and Plato. Socrates and Plato doubtless differed from the sophists in various respects—they were not itinerant but remained in Athens, and they departed from sophistic methods of instruction—but these distinctions were probably lost on contemporary Athenians, who must have viewed both men as typical representatives of the sophistic movement. When Plato returned to Athens in 388 after his travels in Italy, he established a school in the Academy, a monumental public gymnasium just outside the city walls, which had long been used for educational purposes. If there was anything unusual about this venture, it was that Plato's school acquired sufficient permanency to endure long beyond his death.[3]

Plato's school was a philosophical community, consisting of scholars who had reached various levels of maturity and attainment and who interacted as equals. Plato was no doubt the dominant force, inspiring his colleagues by example and assisting less advanced scholars by his critical powers; but he was not above criticism, and (like the teacher in a modern graduate seminar) he may have learned as much as he taught.[4] There was undoubtedly a religious undercurrent to the enterprise; the Academy was devoted to the service of the Muses, and there may have been what we would take to be religious ceremonies. However, doctrinal orthodoxy was certainly not required, and the school was (at least in principle) open to students of any persuasion. No fees were charged, and a scholar could participate in the activities of the Academy until he tired of them or his means of support ran out. At some point, Plato purchased a plot of land close to the Academy, which could be used for some of its activities. The possession of private property, along with Plato's provision for the selection of a successor, doubtless contributed to the longevity of the school.

Aristotle was a member of Plato's school for twenty years, until Plato's death in 348 or 347. When he returned to Athens in 335, following the imposition of Macedonian rule, Aristotle did not resume membership in the Academy, as he could perfectly well have done, but founded a rival school in another Athenian gymnasium, the Lyceum. The Lyceum, like the Academy, had long been the scene of educational activity. Aristotle and his followers were accustomed to gather in a colonnaded walk (*peripatos*) in the Lyceum and thereby acquired (or assumed) the designation "peripatetics," by which they have since been known. Aristotle's Lyceum and Plato's Academy were alike in many respects, but they differed in method and emphasis. Methodologically, Aristotle inaugurated the practice of cooperative research, exhibited in his natural history and also in the systematic collection of earlier philosophical literature. As for emphasis, Aristotle's biological interests stand in rather strong contrast to Plato's mathematical ones; and there is the obvious divergence between Platonic and Aristotelian metaphysics.[5]

Athens had by this time acquired educational leadership within the Greek world, and other teachers soon arrived to take advantage of the opportunities. Zeno of Citium arrived in Athens about 312 and subsequently began to teach in the *stoa poikile* (painted colonnade) in a corner of the Athenian agora, thus founding a school of what came to be called "Stoic" philosophy. Epicurus, an Athenian citizen born on the island of Samos, returned to Athens about 307, purchased a house and garden, and there founded a school of "Epicurean" philosophy that survived into the Christian era.

The Academy, the Lyceum, the Stoa, and the Garden of Epicurus—the four most prominent schools in Athens (fig. 4.1)—all developed institutional identities that enabled them to survive their founders. The Academy and the Lyceum seem to have had continuous existence until the beginning of the first century B.C. (perhaps until the sack of Athens by the Roman general Sulla in 86 B.C.). It is often claimed that the Academy survived until it was closed by the Emperor Justinian in A.D. 529. The truth seems to be that Neoplatonists (so-called because of their departure from or reinterpretation of various Platonic doctrines) *refounded* the Academy in the fifth century A.D. and managed to keep it alive until about 560 or later; however, there was no institutional

Fig. 4.1. The schools of Hellenistic Athens. © Candace H. Smith.

continuity between this and Plato's school. The Stoa survived into the second century A.D., and the Epicurean school into the following century.[6]

Meanwhile the Athenian model had been exported to other parts of the Greek world, particularly Alexandria (in Egypt). At the death of Alexander the Great, his generals divided his empire, Egypt and Palestine falling to Ptolemy. Alexandria became Ptolemy's capital, and through his patronage

Fig. 4.2. The Parthenon
(a temple to Athena)
on the Acropolis, Ath-
ens. Built 5th c. B.C.

and that of his successors it grew in size and magnificence and soon achieved a position of educational superiority. When Demetrius Phaleron, formerly a member of Aristotle's Lyceum, was overthrown as dictator of Athens in 307, Ptolemy invited him to Alexandria, where he probably influenced his patron's decision to found the Museum—not a building where artifacts could be displayed, but a temple to the Muses and therefore simultaneously a religious shrine and a place of learning. Connections between the Museum and the Lyceum are further illustrated by the fact that Strato, the third head of the Lyceum, spent a period at the Ptolemaic court, tutoring the royal offspring. The Museum seems to have consisted of certain buildings in the royal quarter and (since it was a temple) to have been presided over by a priest. Its associated library, generously patronized by the Ptolemaic kings, contained, by one ancient estimate, almost half a million rolls.[7] With the eventual decline of Athenian schools, it became the major research institution of the Hellenistic period—one of the primary links between early Greek thought and the Roman and medieval periods.[8]

The establishment of the Museum in Alexandria is important not only because of the significant research carried out there, but also because it is the first instance of the support of advanced learning through public or royal patronage. This pattern was extended in the period A.D. 140–80 by the Roman emperors Antoninus Pius and Marcus Aurelius, who endowed imperial chairs for teachers of rhetoric and philosophy in Athens and elsewhere. Marcus Aurelius (reign, 218–222) saw to the establishment in Athens of chairs for each of the major philosophical traditions—Platonic, Peripatetic, Stoic, and Epicurean—a model quickly imitated elsewhere in the Greek world. In the long run, this pattern exerted a strong influence on Roman and Christian educational practice.

THE LYCEUM AFTER ARISTOTLE

Aristotle made the acquaintance of Theophrastus (ca. 371–ca. 286) during his travels in Asia Minor, probably during his stay on the island of Lesbos (Theophrastus's birthplace) in the 340s. They became close associates, and when Aristotle returned to Athens in 335, Theophrastus joined him and participated for the next thirteen years in the activities of the Lyceum. Upon Aristotle's death, Theophrastus assumed the headship of the Lyceum, a position that he held for thirty-six years.

Theophrastus appears to have shared Aristotle's general philosophical outlook, his methodological commitments, and his range of interests. He continued to teach and to carry out the collaborative research projects in natural history and the history of philosophy begun during Aristotle's lifetime. He collected the opinions of the pre-Socratic philosophers in a book, which gave rise to what we now call the "doxographic" (or "opinion") tradition—a series of handbooks in which philosophical opinion on a variety of topics was collected and preserved. Most of the writings of Theophrastus are now lost, but among the surviving treatises are two botanical works and a treatise on minerals that reveal a high level of commitment to the Aristotelian research program. Like Aristotle's zoological works, the botanical works contain meticulous descriptions of plant life (more than five hundred varieties are mentioned), thoughtful attempts at classification, and intelligent physiological theorizing. Theophrastus accepted many of Aristotle's explanatory principles (association of life with vital heat, for example) and stressed the necessity of employing a rigorous empirical methodology. In his work *On Stones*, he followed Aristotle in dividing minerals into metals (in which the element water predominates) and "earths" (in which the element earth predominates). He proceeded to a systematic description of a wide variety of rocks and minerals.

While carrying out Aristotle's research program, Theophrastus was also prepared to question and disagree with aspects of Aristotelian natural philosophy. Three examples will serve to illustrate. Theophrastus expressed reservations about Aristotle's teleology, pointing out that not all features of the universe serve any identifiable purpose and that the world exhibits substantial random behavior. He reconsidered Aristotle's theory of the four elements and called into question the status of fire as an element. And he disagreed with Aristotle on light and vision, questioning Aristotle's opinion that light is the actualization of the transparency of the medium and expressing the view that the eyes of animals contain a kind of fire, the emission of which explains nocturnal vision.[9]

An achievement of quite a different sort was Theophrastus's acquisition of property for the Lyceum. Though not an Athenian citizen, Theophrastus received special dispensation to purchase a parcel of land close to the gymnasium; there, in several buildings, the school's library was presumably housed and work space provided. This real estate Theophrastus bequeathed to his scholarly colleagues in his will: "I give the garden, the *peripatos*, and all the houses along the garden to those of my friends, named herein, who wish continually to practice education and philosophy together in them . . . ; my condition is that no one alienate the property or devote it to private use but that all should hold it in common as if it were a sanctuary."[10]

The library of the peripatetic school experienced a more complicated fate. In his will Theophrastus bestowed the library (which contained not only his own books but also Aristotle's) on Neleus, whom he may have intended to succeed him. When the senior members of the community chose Strato instead, Neleus returned home to Skepsis in Asia Minor, taking the books (or many of them, at least) with him, thereby depriving the Lyceum of vital resources. This library survived more or less intact until early in the first century B.C., when (according to the historian Strabo, not to be confused with Strato) it was purchased from Neleus's heirs and restored to the peripatetic school in Athens. Shortly thereafter Athens was sacked by the Roman general Lucius Cornelius Sulla, who shipped the books to Rome. There they came into the hands of Andronicus of Rhodes about 40 B.C., who arranged and edited them, bringing them into prominence and wider circulation.[11]

Meanwhile Strato (from Lampsacus in Asia Minor) had taken over leadership of the Lyceum, a position that he held for eighteen years (286–268). Strato appears to have had interests almost as broad as those of Aristotle and Theophrastus. However, none of his works has survived intact, and we must be content with a fragmentary picture of his philosophical and scientific activity, reconstructed from scattered quotations and paraphrases in the works of later writers. It appears that Strato endeavored to correct and extend the work of Aristotle and Theophrastus on a variety of subjects. He certainly did not hesitate to question their views or to borrow from other philosophical traditions when there seemed good reason for doing so.

Strato's most notable contributions (as transmitted to us) were related to motion and the underlying structure of the physical world. Strato proposed a fundamental revision in Aristotle's theory of motion when he denied the distinction between heavy and light bodies, arguing that all bodies have weight in greater or lesser degree. Air and fire rise, then, not because they are absolutely light, but because they are less heavy and consequently are displaced by heavier bodies. Strato also opposed Aristotle's theory of place and space. And he set forth observational evidence to demonstrate that heavy bodies accelerate as they descend (a feature of falling bodies not treated by Aristotle). Strato pointed to the fact that a stream of water falling from a height is continuous at the top, but discontinuous near the bottom—a fact to be explained by steadily increasing speed. In support of the same conclusion, he noted that the impact made by a falling body is a function not simply of its weight, but also of the height of its fall.[12]

Although there is no question that Strato remained fundamentally Aristotelian in his view of the underlying structure of the corporeal world, it is also clear that he imported corpuscular notions into Aristotelian natural philosophy—possibly through the influence of Epicurus, who taught for a time in Strato's home town, Lampsacus, and also overlapped with Strato in Athens. Corpuscular ideas are most obvious in Strato's belief that light is a material emanation and that bodies are not continuous but contain interparticulate void spaces. Strato used the notion of void spaces to explain various properties of matter, including condensation, rarefaction, and elasticity. While acknowledging the existence of tiny void spaces distributed through matter, Strato denied the natural existence of continuous void space. We must be careful not to make Strato into a full-fledged atomist, for it appears that he retained a belief in the infinite divisibility of corporeal substance, thus rejecting the one absolutely essential feature of any atomistic philosophy—namely, belief in the existence of irreducible atoms.

Some of Strato's successors as head of the Lyceum are known by name, down to the end of the second century B.C. There can be no doubt that the school was the scene of regular lectures on peripatetic philosophy and continuing attempts to clarify Aristotle's philosophy and organize the materials that he had left behind. However, we have no record of fresh contributions to natural philosophy, nor of particularly sharp and telling criticism of traditional peripatetic philosophy, until after the Lyceum had ceased to function. Nonetheless, Aristotle's works continued to be known and commented upon, especially after Andronicus of Rhodes produced his new edition of the Aristotelian texts. In the middle of the first century B.C. we find commentaries by Boethius of Sidon (a student of Andronicus) and Nicholas of Damascus (historian at the court of Herod the Great). About A.D. 200 Alexander of Aphrodisias was lecturing on peripatetic philosophy in Athens and writing important and influential commentaries on a variety of Aristotelian works. Finally, the Aristotelian commentaries of Simplicius and John Philoponus (both Neoplatonists) testify to the persistence of the Aristotelian tradition as late as the sixth century A.D. Renewed attention to this tradition in Islam and medieval Christendom would once again restore Aristotle's philosophy to a position of leadership.[13]

EPICUREANS AND STOICS

During the Hellenistic period, followers of Plato and Aristotle continued to discuss, clarify, and modify Platonic and Aristotelian philosophy. At the same time, alternative philosophical systems were developing, two of which became serious rivals. Both contained familiar elements but were new in the prominence they gave to ethical questions. Indeed, what is striking about both is their determination to subordinate all other aspects of philosophy to ethical concerns.

The aim of philosophy, according to Epicurus (341–270 B.C.; fig. 4.3), is to secure happiness. "To say that the season for studying philosophy has not yet come, or that it is past and gone," Epicurus wrote to Menoeceus, "is like saying that the season for happiness is not yet or that it is now no more." The way to achieve happiness, Epicurus believed, was to eliminate fear of the unknown and the supernatural, and for this purpose, it appeared to Epicurus,

Fig. 4.3. Epicurus. Museo Vaticano, Vatican City. Alinari/Art Resource N.Y.

natural philosophy was ideally suited. A maxim attributed to Epicurus reads as follows: "If we had never been molested by alarms over celestial and atmospheric phenomena, nor by the misgiving that death somehow affects us, nor by neglect of the proper limits of pains and desires, we should have had no need to study natural philosophy." If we take Epicurus seriously on this point, the achievement of happiness is natural philosophy's sole purpose.[14]

Epicurus's natural philosophy borrowed many elements from ancient atomism. It conceived the universe to be eternal, consisting of an infinite void within which an infinity of atoms engage in perpetual motion, tossed about "as if in everlasting combat," like particles of dust in a beam of bright light. All things and all phenomena in our world (and in the infinity of other worlds that exist) are reducible to atoms and the void; the gods themselves must be of atomic composition. The sensible qualities of things (we now call them "secondary qualities"), such as flavor, color, and warmth, have no existence in the individual atom, the only genuine properties of which are shape, size, and weight. This is a mechanistic world of passive atoms and mechanical causation (with one exception, to be mentioned below); there is no ruling mind, no divine providence, no destiny, no life after death. And there are no final causes: as Lucretius (d. ca. 55 B.C.) was to put it, in his account of Epicurean philosophy, "all the members [of the body] . . . existed before their use; they could not then have grown up for the sake of use."[15]

But Epicurus and his followers did more than propagate the philosophical system of the ancient atomists. They also had to adapt the atomic philosophy to the service of ethical functions. And they modified its content to resolve difficulties, meet objections, and generally increase its explanatory power. For example, Epicurus opposed the rationalism of Democritus, arguing that all sensation is fundamentally trustworthy.[16] From this it seemed to follow that sensible or secondary qualities truly exist at the macroscopic level, even though (as Democritus had argued) they do not exist in the atoms.

A more significant modification of the content of atomistic natural philosophy was Epicurus's doctrine of the swerve, designed not only to save atomistic cosmology from fatal objections, but also to eliminate the threat of determinism from Epicurean ethics. According to Epicurus, atoms possess shape and size (as Leucippus and Democritus had argued), but also weight. Their weight causes them to fall in the infinite void, producing what might be regarded as a primeval cosmic rain. Because none of the atoms encounter resistance, all descend at the same speed, and none is ever overtaken by another. Now this is a totally unsatisfactory cosmology, because it seems to rule out the very collisions that give atomism its explanatory power. Epicurus dealt with the difficulty by postulating an infinitesimal swerve: an atom shifts its line of descent the least possible amount, setting off a chain reaction of collisions. The most troublesome feature of this theory is that this primeval swerve must have been an uncaused event, since in Epicurus's world atomic collisions are the only kind of cause there is—and no such collision caused that original swerve. And the absence of such collisions in that primeval rain was precisely the state of affairs that Epicurus was trying to escape.[17]

If we are tempted to judge Epicurus harshly for the invention of uncaused events (which are still a philosophical embarrassment, even if they do appear in some interpretations of modern quantum mechanics), we need to notice that the swerve not only accounts for the origin of the atomic maelstrom, which in turn explains the world in which we live; it also breaks the deterministic chain that would eliminate human responsibility and destroy Epicurus's ethical system. If the world is totally subject to rigid mechanical causation, then human action is a random result of atomic motions and collisions and, therefore, cannot be free; and if humans do not choose freely, they are without responsibility. But the swerve introduces an element of indeterminism into the universe; and even if this does not explain how free choice is actually exercised (a question to which we still do not know the answer), by revealing a break in the chain of rigid causal necessity it makes room for the possibility of free human volition. This is doubtless not an entirely satisfactory solution, but to have perceived the problem of free will in a mechanical universe (Epicurus was the first to do so) is of itself a significant achievement.

The founder of Stoic philosophy was Zeno (ca. 333–262 B.C.) from Citium on the island of Cyprus. This Zeno, who is not to be confused with Parmenides' disciple of the same name, came to Athens and spent about a decade studying in a variety of Athenian schools, including Plato's Academy, before setting up his own school in the *stoa poikile* about 300. Zeno was succeeded by Cleanthes of Assos (331–232 B.C.) and Chrysippus of Soli (ca. 280–207 B.C.), powerful thinkers in their own right, who contributed as much as Zeno to the development of Stoicism as a systematic philosophy. Stoic philosophy, as an active scholarly tradition, survived into the second century A.D.; its influence, however, can be traced as late as the seventeenth century.[18]

Stoics and Epicureans were in radical opposition on most subjects, but they agreed on a few things. They agreed, in the first place, on the subordination of natural philosophy to ethics; in both philosophical schools, the pursuit of happiness was regarded as the goal of human existence. Stoics believed that happiness could be obtained only through living in harmony with nature and natural law; and living in harmony with nature required a knowledge of natural philosophy. In the second place, members of both philosophical schools were committed materialists, arguing strenuously that nothing exists except material stuff.

This shared materialism was important common ground; it meant that Stoics and Epicureans were allies in the battle against proponents of any non-materialist philosophy, such as Plato and his followers. However, once we move beyond this basic proposition, we find that Stoics and Epicureans had fundamentally different visions of the universe. Epicureans believed that matter was discontinuous and passive—consisting of discrete, unbreakable, lifeless atoms, which moved mindlessly in infinite void space. Theirs was a mechanistic universe. Stoics, by contrast, created a model of an organic universe, characterized by continuity and activity. These contrasts (continuity-discontinuity and activity-passivity) will serve as useful points of entry into Stoic natural philosophy.[19]

Matter, Stoics believed, does not present itself in the form of atoms, each with a permanent identity, but as an infinitely divisible continuum, containing no natural breaks and no void spaces. Size and shape, therefore, are not permanent attributes of matter, for matter can be mentally chopped up into pieces of whatever size and shape we please. While allowing no void within the world, the Stoics did acknowledge an extracosmic void, viewing the cosmos as an island of continuous matter surrounded by an infinite void space.

Stoics followed the Epicureans in acknowledging a passive side to material things, but they were convinced that it could not be the whole story. The Epicurean position was vulnerable to the following objection. If an individual object derives all of its properties from the chance configuration of tiny lifeless pieces of matter, there can be no convincing explanation of many of the properties of the whole. The only properties possessed by Epicurean atoms are size, shape, and weight. How, then, could an Epicurean explain as simple and basic a property as cohesiveness—the fact that a rock remains a rock, resisting disintegration into its constituent particles? Where does the coldness of a block of ice come from, since coldness is not a property of its constituents? And how do we explain color, flavor, and texture? Or to turn to a far more difficult case, where do the characteristics of living things come from—the life cycle of a plant, the reproductive behavior of an insect, or the personality of a human being? If the family dog is merely a chance configuration of inert matter, how do we explain its obsession with chasing mail carriers? It would seem that besides passive matter there must be what came to be called "active principles," having the capacity to organize passive matter into an organic unity and account for its characteristic behavior. There must be something that is acted upon; but there must also be something that acts, and in a materialistic world that something must be material.

The Stoics identified this active principle with breath or pneuma, the subtlest of substances, which totally interpenetrates everything, binding the recipient passive matter into unified objects and endowing these objects with their characteristic properties. But it is important to keep in mind that pneuma is more than a subtle, all-penetrating substance; it is also an active and rational substance, the source of vitality and rationality in the cosmos. Indeed, Stoics identified pneuma with divine rationality and with deity itself. The equation pneuma = reason = god may seem odd from a modern point of view, and certainly wrong from a Judeo-Christian point of view, but it was basic to Stoic cosmology. Deity had been brought down from the heavens, materialized, and made to account for activity and order in the universe.

Let us scrutinize this pneuma more closely, inquiring into its structure (if any), the source of its organizing capabilities, and its relationship to passive matter. Stoics accepted the existence of the four Aristotelian elements, but divided them into two groups on the basis of activity. They regarded earth and water, the principal ingredients of tangible objects, as passive elements, and air and fire as active elements. Air and fire mix in various proportions (the Stoics had in mind a total, homogeneous blending) to produce a variety of pneumas. Thus air and fire act, while water and earth are acted upon.

Pneumas come in various grades. At the lowest level, the pneuma that accounts for the cohesion of what we would regard as inorganic bodies—rocks and minerals, for example—is called *hexis*. The pneuma of plants and animals, which gives them their vital properties, is *physis*. And the highest grade of pneuma, possessed by humans and accounting for their rationality, is *psyche*. Now Stoics identified the pneuma of an object with soul. It follows that every individual thing is permeated with soul, and this soul functions as its organizing principle. There must even be a cosmic pneuma, a world soul, since the cosmos too is an organic unity, having characteristics that require active principles for their explanation. The profoundly vitalistic character of Stoic natural philosophy should be evident.

Pneumas exist in a state of tension or elasticity. This tension accounts for that most basic property of all objects, cohesion. At higher levels, different tensions account for the variety of properties and personalities observable in the world. And finally, it may be well to reiterate that the relationship of pneuma to its host body is one of total mixture or interpenetration, both substances occupying the same space.

Stoic cosmology, like Plato's and Aristotle's, was geocentric. However, the Stoics followed the atomists and departed decisively from Aristotle, refusing to make any kind of radical distinction between the terrestrial and celestial regions; when it came to such fundamental matters as the composition and laws of nature, the Stoic cosmos was homogeneous. The Stoics agreed with Aristotle on the eternity of the universe, but in place of his belief in cosmic stability they substituted a cyclic theory inspired by pre-Socratic thought. According to a variety of Stoic thinkers, there is an eternal cosmic cycle of expansion and contraction, conflagration and regeneration. In the expansive phase, the world dissolves into fire; in the contractive phase, fire yields again to the other elements, and the world as we know it is regenerated. This cycle is repeated eternally, producing an everlasting sequence of identical worlds.[20]

Finally, we must note that the Stoic universe was conceived to be both purposeful and deterministic. Permeated as it was with mind and deity, the Stoic cosmos was inevitably suffused with purpose, rationality, and providence. At the same time, its course was rigidly determined. Stoic philosophy maintained that there are causal chains (themselves the products of divine rationality) that cannot be violated and that totally determine the sequence of events. As Cicero put it in *On Divination*, "nothing has happened which was not going to be, and likewise nothing is going to be of which nature does not contain causes working to bring that very thing about. This makes it intelligible that fate should be, not the 'fate' of superstition, but that of physics."[21]

We have seen that the natural philosophies of the Stoics and the Epicureans were, in many ways, opposites. Whereas Epicurean philosophy had as one of its principal aims the combating of Platonic and Aristotelian teleology, Stoic philosophy was directed toward the discovery of purpose and the defense of teleology. Whereas Epicureans portrayed a mechanistic universe, Stoics discovered an organic one. Whereas Epicurus struggled to introduce an element of indeterminacy into his otherwise mechanistic universe, the Stoics were content with an organic universe where rigid determinacy reigned. In the short run, the Stoic vision of the cosmos seemed the more plausible of the two and became a prominent philosophical option in late antiquity. In the long run, both Stoic and Epicurean philosophy were resurrected in the early modern period and presented as alternatives to the Platonic and Aristotelian world pictures; and each played a part in shaping the new philosophy of the seventeenth century.

5 ✣ The Mathematical Sciences in Antiquity

THE APPLICATION OF
MATHEMATICS TO NATURE

The applicability of mathematics to nature has been the subject of a long debate within the Western scientific tradition. The question is whether the world is fundamentally mathematical, in which case mathematical analysis offers a secure route to deeper understanding, or whether mathematics is applicable only to the superficial quantifiable aspects of things, leaving the underlying realities untouched. There can be no doubt that natural scientists (especially those representing certain scientific specialties) seem increasingly inclined to resolve the question in favor of the mathematical approach. However, the alternative also has its defenders, and the dispute lives on.

The ancient Pythagoreans appear to have maintained that nature is mathematical through-and-through. If Aristotle can be trusted, the Pythagoreans went to the extreme of claiming that the ultimate reality is number (see chap. 2 for additional discussion). Plato took up with a vengeance the Pythagorean program in his theory of matter, arguing that the four elements are reducible to regular geometrical solids, which are reducible in turn to triangles. Thus for Plato the fundamental building blocks of the visible world were not material, but geometrical; moreover, what binds everything together into a unified cosmos, Plato argued, is not a physical or mechanical force, but simply geometrical proportion.[1]

There can be no question that Aristotle was mathematically informed. He modeled his theory of knowledge on mathematical demonstration, utilized geometry in his theory of the rainbow (if the theory attributed to him is indeed his), and employed theory of proportion in his analysis of motion. But Aristotle was convinced that there was a difference between mathematics and natural philosophy or physics. The latter, by his definition, considers natural things in their entirety, as sensible, changeable bodies. The mathematician, by contrast, strips away all of the sensible qualities of bodies and concentrates on the mathematical remainder. So the mathematician is concerned only with the geometrical properties of things, but these by no means exhaust reality. Reintroduce weight, hardness, warmth, color, and the other qualities that exist in the real world, and you have moved out of the mathematical realm and back into the subject matter of physics, which has its own practitioners. And finally, some disciplines, such as astronomy (one of the mixed sciences), combine both mathematics and physics.[2] So Aristotle took a middle road on the question of the applicability of mathematics to nature. He was convinced that mathematics and physics are both useful, but it was clear to him that they are not the same thing; the mathematician and the physicist may study the same object, but they concentrate on different aspects of it. Finally, practitioners of the "middle" or "mixed" sciences concentrate on both the physical and mathematical side of things.

Plato and Aristotle thus provided two theories of the relationship between mathematics and nature, and these became the poles between which natural scientists have vacillated from antiquity to the present. However, our interest is not only in theories about the applicability of mathematics to nature, but also in the ways in which these theories were put into practice. To observe the Greeks actually at work, applying mathematics to nature, we will examine the subjects of astronomy, optics, and the balance or lever. In order to prepare ourselves for that investigation, we must first glance at the Greek achievement in pure mathematics.

GREEK MATHEMATICS

We know little about the origins of Greek mathematics. Unquestionably, early Greek mathematicians had access to the Egyptian, and especially the Babylonian, mathematical achievement (see chap. 1). But from the beginning Greek mathematics was different—the difference lying chiefly in its tendency to favor Greek geometry, with its orientation toward abstract geometrical knowledge and its formal methods of inference and proof. One reason for the Greek emphasis on geometry, it has been suggested, may have been the discovery that no matter how small the unit of measure, the side and the diagonal of a square cannot both be whole numbers. The side and the diagonal are thus said to be "incommensurable." (See fig. 5.1, where the diagonal is the irrational number $\sqrt{2}$.) This irrationality, the hypothesis continues, may have persuaded Greek mathematicians of the unsuitability of numbers (positive integers, on the Greek view) for the representation of reality and thus encouraged the development of geometry.[3]

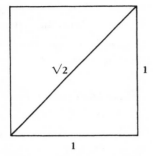

Fig. 5.1. The incommensurability of the side and the diagonal of a square.

We have only fragmentary evidence for specific mathematical developments in the period before Euclid (who flourished about 300 B.C.), but it is universally acknowledged that those developments were codified in Euclid's own mathematical textbook, the *Elements*.[4] Here we find mathematics highly developed as an axiomatic, deductive system. The *Elements* begins with a set of definitions: of a point ("that which has no part"); a line ("length without breadth"); a straight line; a surface; a plane surface; a plane angle, right, acute, and obtuse angles; various plane figures; parallel lines; and so forth. The definitions are followed by five postulates: that a line can be drawn from a point to any other point, that a straight line can be extended continuously from either end, that a circle of any radius can be drawn about any point, and that all right angles are equal; also a statement of the conditions under which straight lines will intersect. The postulates are followed by five "common notions," or axioms—self-evident truths needed for the practice of correct thinking in general and mathematics in particular. These include the claim that things equal to the same thing are equal to each other, that equals added to equals yield equal sums, and that the whole is greater than the part. These preparatory claims lay the groundwork for the propositions that fill the thirteen books that follow. A typical proposition begins with an enunciation, followed by an example, a further definition or specification of the proposition, and a construction; it ends with a proof and a conclusion. What is important to note is that the conclusion of a proper Euclidean demonstration follows *necessarily* from definitions, postulates, axioms, and previously proved propositions. So effectively did Euclid wield this method that, through his influence—and that of Aristotle, whose method resembled Euclid's in certain critical respects—it became the standard for scientific demonstration down to the end of the seventeenth century.

We need not pause long over the content of Euclid's *Elements*, for it bears a close resemblance to the geometry taught in modern secondary schools.

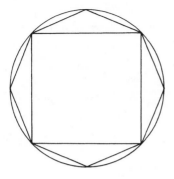

Fig. 5.2. Determining the area
of a circle by the method of
"exhaustion."

Books I-VI develop the elements of plane geometry; book X is devoted to a classification of incommensurable magnitudes; and books XI-XIII treat solid geometry. In books VII-IX Euclid takes up arithmetical topics, including the theory of numbers and of numerical proportion. Of the many achievements in the *Elements*, one that must be mentioned because of its future significance is the development of the method of "exhaustion"—probably borrowed by Euclid from his predecessor Eudoxus and destined to influence a variety of followers, including Archimedes. Euclid shows (XII, 2) how to "exhaust" the area of a circle by means of an inscribed polygon; if we successively double the number of sides in the polygon, we will eventually reduce the difference between the area of the polygon (known) and the area of the circle (unknown) to the point where it is smaller than any magnitude we choose (see fig. 5.2). This method made it possible to calculate the area of a circle to any desired degree of accuracy; with a little further development it could be used to calculate the area within (or under) other curves as well. Another important feature of the *Elements* is its exploration of the properties of the five regular geometrical solids, sometimes known as the "Platonic solids," and its demonstration (XIII, 18) that there are no regular geometrical solids beyond these five.[5]

Euclid was followed by a series of brilliant Hellenistic mathematicians, the greatest of whom was undoubtedly Archimedes (ca. 287–212 B.C.). Archimedes contributed to both theoretical and applied mathematics, but he is especially esteemed for the elegance of his mathematical proofs. In some of his most important work, Archimedes developed the method of exhaustion and applied it to the calculation of areas and volumes, including the area enclosed within a segment of a parabola, the area bounded by certain spirals, and the surface area and volume of a sphere. He calculated an improved value for π (the ratio of the circumference to the diameter of a circle), showing that it must fall between 3 10/71 and 3 1/7. Archimedes had a profound influence on the subsequent development of mathematics and mathematical physics, particularly after his works were rediscovered and reissued in the Renaissance.[6]

A final Greek mathematical achievement that must be mentioned is the work of Apollonius of Perga (fl. 210 B.C.) on conic sections. Apollonius studied the ellipse, parabola, and hyperbola—the plane figures formed when a circular cone is cut at various angles by a plane surface—and proposed a new approach to their definition and methods of generation. His book on conic sections, like the works of Archimedes, was destined to have a major influence in the early modern period.[7]

EARLY GREEK ASTRONOMY

Early Greek astronomy appears to have been concerned primarily with observation and mapping of the stars, with the calendar, and with the solar and lunar motions that had to be plotted before a satisfactory calendar could be constructed. The major difficulty in the calendar arose from the fact that the solar year is not an integral multiple of the lunar month. That is, in the time required for the sun to complete one circuit of the zodiac, the moon completes twelve circuits plus a fraction. Thus a calendar based on twelve lunar months, each of thirty days, is too short by about five days, and the calendar and the seasons will not be synchronized. Various schemes were developed for inserting an additional month as needed, in order to bring the calendar back into phase with the seasons. These calendric efforts culminated in the Metonic cycle, proposed by an Athenian astronomer, Meton (fl. 425 B.C.), based on the understanding that to a very close approximation nineteen years contain 235 months. Therefore, in a nineteen-year cycle, there will be twelve years of twelve months and seven years of thirteen months. It seems that Meton intended this as an astronomical, rather than a civil, calendar; and it was put to astronomical use for several centuries.[8]

Greek astronomy took a decisive turn in the fourth century with Plato (427-348/47) and his younger contemporary Eudoxus of Cnidus (ca. 390-ca. 337 B.C.). In their work we find (1) a shift from stellar to planetary concerns, (2) the creation of a geometrical model, the "two-sphere model," for the representation of stellar and planetary phenomena, and (3) the establishment of criteria governing geometrical theories designed to account for planetary observations. Let us consider these achievements in some detail.

The two-sphere model devised by Plato and Eudoxus conceives of the heavens and the earth as a pair of concentric spheres. To the celestial sphere are affixed the stars, and along its surface move the sun, the moon, and the remaining

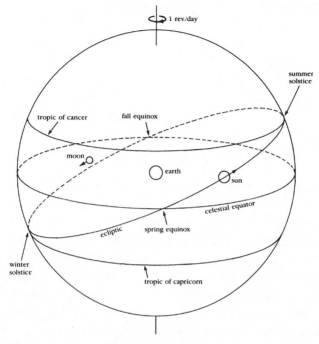

Fig. 5.3. Two-sphere model of the cosmos.

five planets. The daily rotation of the celestial sphere accounts for the observed daily rising and setting of all of the celestial bodies. Corresponding circles on the two spheres divide them into zones and mark out the motions of the wandering stars. Figure 5.3 reveals roughly what Plato and Eudoxus had in mind. The terrestrial sphere is fixed in the center, while the celestial sphere rotates daily about a vertical axis. The equator of the earth projected onto the celestial sphere defines the celestial equator. The annual path of the sun around the celestial sphere is the "ecliptic," a circle tilted at an angle to the equator of approximately 23°. This line, the center of the zodiac, is the line along which the sun and moon must move if they are to produce eclipses. The ecliptic and celestial equator intersect at the equinoxes; when the sun, in its annual trip around the ecliptic, reaches the fall equinox (approximately September 21 on our calendar), autumn begins; when it reaches the vernal equinox, spring begins. The points at which the ecliptic is most distant from the equator are the solstices; when the sun reaches the summer solstice (approximately June 21 on our calendar), summer begins. The circles drawn parallel to the equator through the summer and winter solstice, respectively, are the Tropics of Cancer and Capricorn.[9]

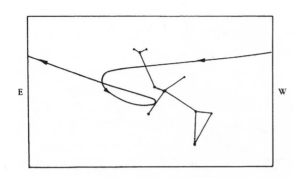

Fig. 5.4. The observed retrograde motion of Mars in the vicinity of the constellation Sagittarius, 1986. Data supplied by Jeffrey W. Percival.

The movements of the sun, moon, and remaining planets had been carefully observed and well charted by the fourth century. In the model of Plato and Eudoxus, the sun circles the ecliptic once a year, while the moon completes its circuit in a month, both moving in a west-to-east direction and with nearly uniform speed. The other planets—Mercury, Venus, Mars, Jupiter, and Saturn—also follow the ecliptic (deviating from it by no more than a few degrees), moving in the same direction as the sun and moon but with considerable variations in speed. Mars, for example, circles the ecliptic once in about 22 months (687 days); about once every 26 months it slows to a halt, reverses itself (moving now in an east-to-west direction), halts again, and then resumes its usual west-to-east motion. This reversal of direction is called "retrograde motion," and it is exhibited by all of the planets except the sun and moon. Figure 5.4 illustrates the observed retrograde motion of Mars.

One further striking feature of planetary motion known to Plato and Eudoxus was the fact that Mercury and Venus never stray far from the sun (the maximum elongation is 23° for Mercury, 44° for Venus). Like dogs on a leash, they can run ahead of the sun or lag behind it, but they can never be more distant from it than the fixed length of the leash allows. Finally, to appreciate the achievement embodied in the two-sphere model, we must understand that all of the planetary motions (including those of the sun and moon) discussed in this and the preceding paragraph occur on the surface of the celestial sphere while that sphere is going through its daily rotation around the earth. The resulting motion, observed from the fixed earth, will be a combination of the irregular motion of the planet around the ecliptic and the uniform daily rotation of the celestial sphere; it brings geometrical meaning to the (otherwise) bewildering complexity of observed planetary motions. The two-sphere model is thus a geometrical way of conceiving and talking about planetary phenomena.

It is a significant achievement to create a geometrical language for talking about planetary motion and to put that language to use in the presentation of a rough description of planetary motion around the ecliptic. But there is something more to which an ancient astronomer might aspire: if he wishes to bring a high level of order and intelligibility to this "bewildering complexity" in the heavens, there is no better way than to take the intricate, variable motion of each planet and reduce it to some combination of uniform motions. That is, he must assume that complexity conceals simplicity, that beneath irregularity lies regularity, and that this underlying order or regularity is discoverable. A late (and possibly unreliable) account gives Plato the credit for laying out this assumption as a research program, challenging astronomers or mathematicians to determine what combination of uniform circular motions would account for the apparent, irregular motions of the planets.[10]

Whether it was indeed Plato who first posed the question, it is clear that Eudoxus first proposed an answer to it. Eudoxus's idea was ingenious, but fundamentally simple. The goal was to treat each irregular planetary motion as a composite of a series of simple uniform circular movements. To achieve this goal, Eudoxus assigned to each planet a set of nested concentric spheres, and to each sphere one component of the complex planetary motion (see fig. 5.5). Thus the outermost sphere for, say, the planet Mars rotates uniformly once a day,

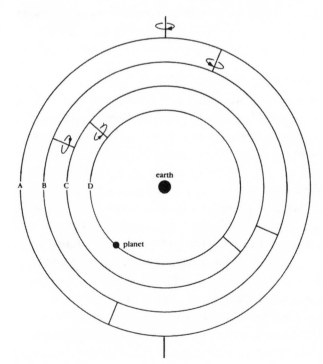

Fig. 5.5. The Eudoxan spheres for one of the planets.

accounting for Mars' daily rising and setting; the second sphere in the series also rotates uniformly about its axis (tilted in relation to the axis of the outermost sphere), but in the opposite direction, once in 687 days, thus accounting for Mars' slow west-to-east motion around the ecliptic. The two innermost spheres account for changes in speed and latitude and for retrograde motion. Mars, finally, is situated on the equator of the innermost sphere, participating not only in the motion proper to that sphere, but also in the motions transmitted downward from the three spheres above it. A similar system works for Mercury, Venus, Jupiter, and Saturn. The sun and moon, which do not exhibit retrograde motion, require only three spheres apiece.

Eudoxus thus created the first serious geometrical model of planetary motion. Three questions naturally present themselves. First, did Eudoxus attach physical reality to the model? That is, did he conceive the spheres as physical objects? The answer seems unambiguously in the negative. There is every reason to believe that the concentric spheres of Eudoxus were intended as a purely mathematical model, with no claim to represent physical reality. Eudoxus, as far as we can tell, did not imagine that the cosmos consists of physically distinct spheres, mechanically linked to one another; rather, he was endeavoring, by means of a geometrical model, to identify and put to use the separate components of *uniform* motion that underlie and make sense

The Inner Eudoxan Spheres and Retrograde Motion

If, for the sake of simplicity, we treat the interaction of spheres C and D (fig. 5.5) in isolation from the rest of the system, we find that assigning them equal and opposite rotations around axes tilted with respect to each other determines that the planet (on the equator of D) will move along a path resembling a hippopede (horse-fetter) or figure eight. We can thus visualize the motion that results from the four Eudoxan spheres by replacing C and D with a hippopede affixed to the equator of sphere B (fig. 5.6). Sphere A completes a uniform rotation once a day, carrying with it the axis of B; meanwhile B rotates uniformly around this axis in the sidereal period of the given planet (the time required for that planet to make one complete circuit of the celestial sphere), carrying the hippopede around the ecliptic; and all the while the planet is moving around the hippopede in the direction indicated by the arrows.[11] Note that the vantage point of the observer is not the same for the two figures, owing to the difficulties of reducing three-dimensional motions to two-dimensional drawings.

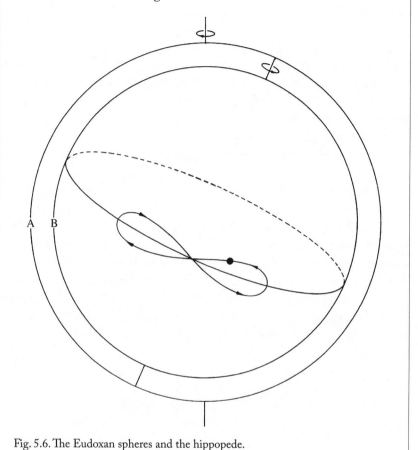

Fig. 5.6. The Eudoxan spheres and the hippopede.

of the *complicated* motions of the planets. He was searching not for physical structure, but for mathematical order.

Second, did Eudoxus combine the separate planetary models (the requisite collection of spheres for each planet) into a unified, composite cosmological system? Nothing in the historical record suggests that he did. We should rather think of the Eudoxan achievement as separate models, one for each planet, existing in the mind of Eudoxus (and astronomically literate colleagues)— perhaps also as physical drawings, sketched independently on papyrus, one set of planetary spheres at a time.

Third, did the model succeed? Because no treatise by Eudoxus has survived, we have no knowledge of the geometrical details of his system. Several things can be said, however. Although Eudoxus's model is clearly mathematical, it was not designed to yield quantitatively exact results. Indeed, it appears certain that the notion of exact, quantitative prediction had not yet entered Greek astronomy or any other Greek scientific subject; nobody aspired to more than rough qualitative agreement between theory and observations. We can, if we wish, speak of the *potential* of the Eudoxan model, under the assumption that the best values were chosen in every case. Under those circumstances, the system would (with one or two exceptions) have given results that correspond in a rough qualitative way, but without quantitative precision, to the astronomical observations as we know them today. Given the more limited state of astronomical knowledge and the modest aims of astronomical theory in the fourth century, this was a considerable achievement.

A generation after Eudoxus, his system was improved by Callippus of Cyzicus (b. ca. 370), who added a fourth sphere for the sun and the moon and a fifth sphere each for Mercury, Venus, and Mars. The function of the additional spheres for the sun and moon was to take into account their variations in speed as they circle the ecliptic—the fact, for example, that the times required for the sun to pass from the summer solstice to the fall equinox and from the fall equinox to the winter solstice (see fig. 5.3, above) may differ by several days.[12]

Further development of the system of concentric spheres was undertaken by Aristotle (384-322) in a short passage of a few hundred words in his *Metaphysica*[13]—important not so much for its ingenuity or originality as for its powerful future influence. Aristotle took the Eudoxan model as modified by Callippus, but with two important differences. First, he took the Eudoxan/Callippan spheres, one set for each planet, and put them all together, merging what had been individual models into a single, grand astronomical model running from the stellar sphere on the outside, through the multiple spheres required for each planet, to the spherical earth at the center of it all.

Second, whereas the concentric spheres appear to have been conceived merely as geometrical constructions by Eudoxus and Callippus, Aristotle endowed them with physical reality. The celestial region, he believed, was composed of an incorruptible aether or quintessence—subdivided, for astronomical purposes, into separate spheres, one within another, tightly fitted in order to leave no unoccupied space between spheres—all concentric with the earth at the center. Thus Aristotle specifies the *material* cause (the aether) of the stellar and planetary motions.

But what causes the spheres to rotate, each with its own speed and orientation? Aristotle did not supply much detail, and what he did provide contains contradictions, leaving us to reconstruct his thought with as much care as possible. The really short answer is that each sphere has a double motion. The first, a natural rotation, unique in speed about its axis, motivated by a *final* cause: love or desire for the perfection of the eternally unchanging and unchanged Unmoved Mover (see above, chap. 3).[14] Second, the sphere also follows the motion of its axis, imposed on it from the sphere immediately above.

There is also a longer answer, which readers who appreciate a geometrical or spatial challenge may find interesting. With an eye on fig. 5.7, consider the

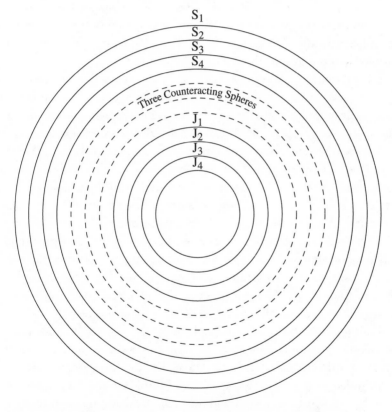

Fig. 5.7. Aristotelian nested spheres. The primary spheres for Saturn and Jupiter (four spheres apiece) are represented by solid lines (Saturn on the outside, Jupiter on the inside). Between these two sets of spheres are three counteracting spheres (dotted lines), which counteract or "unroll" the motion of Saturn's four spheres, above, in order to transmit a simple daily rotation to the outermost sphere of Jupiter, below.

four spheres for the outermost planet, Saturn. The outermost of these spheres (call it S_1) rotates uniformly, east-to-west, one complete revolution per day, in order to account for the daily motion of the planet. Just inside it, S_2 rotates uniformly, but very slowly about its own axis in the opposite direction (west-to-east), accounting for the slow motion of Saturn (its "sidereal motion") around the ecliptic. However, while rotating uniformly about its own axis, S_2 is also a recipient of the motion of S_1. It is *as though* S_2 has poles (of the long, cylindrical kind) that connect it to S_1 and transmit motion downward mechanically from the one to the other. As a result, S_2 achieves a wobbly motion, a composite of its own natural motion and the motion acquired from S_1. The same thing happens twice more, as the motions of S_2 and S_3 are passed downward, so that in the end the planet Saturn, situated on the equator of S_4 has a very complex, composite motion—the sum of the motions of its four spheres. If this is done right it will account at least roughly for Saturn's observed motions against the stellar background.

Now the same thing happens to each of the planets, situated on the equator of the innermost of its set of spheres. Each, if all turns out well, will move with a composite motion consistent with the observational data for that planet. But there is another problem—a consequence of Aristotle's decision to transform the Eudoxan spheres from geometrical ideas to physical objects. Aristotle never tells us how motion is actually transmitted mechanically downward from one sphere to the next below it. Whatever the identity of that cause, it should also act between the lowest of Saturn's spheres and the highest sphere of the next planet in line, Jupiter. That would mean that Jupiter's outermost sphere J_1 would *not* begin as did Saturn's outermost sphere S_1, with a simple, daily rotation about its axis, but rather with a complex motion compounded from its own daily rotation and the motions of all four of Saturn's spheres above it. And the problem gets progressively more challenging, as we work our way down through all seven planets. To account successfully for the astronomical observations using the spheres in the manner of Eudoxus, the outermost sphere for *each* planet must begin with a simple, uniform, twenty-four-hour rotation around its vertical axis. How can this be achieved?

Aristotle's solution was to insert a set of three counteracting spheres between S_4 and J_1, which unwind, and thus cancel, the motions contributed by S_2, S_3, and S_4, allowing J_1 to begin as S_1 did, with only its natural, daily, east-to-west rotation (see fig. 5.7). This unwinding is achieved by giving the three counteracting spheres the same speed and axial orientations as their counterparts S_2, S_3, and S_4, but opposite directions of rotation. Aristotle inserts similar counteracting spheres between each pair of planets, thus solving his problem but bequeathing to his successors an enormously complicated piece of celestial machinery, consisting of a grand total of fifty-five planetary and counteracting spheres, plus the sphere of the fixed stars. I quote the heart of Aristotle's brief account:

> It is necessary, if all the spheres combined are to explain the phenomena, that for each of the planets there should be other spheres (one fewer than those hitherto assigned) which counteract those already mentioned and bring back to the same position the first sphere of the [planet] which in each case is situated below the [planet] in question; for only thus can all the forces at work produce the motion of the planets.[15]

He also bequeathed to future generations an important question: In astronomy, where does the balance lie between the mathematical and the physical? Is astronomy primarily a mathematical art, as Eudoxus seems to have conceived it? Or must the astronomer also concern himself with the real structure of things, as Aristotle's astronomical scheme suggests? Astronomers would ponder this question for the next two thousand years.[16]

COSMOLOGICAL DEVELOPMENTS

Contemporary with Aristotle and in the next century, we find several cosmological developments of interest to astronomers. One was the proposal by Heraclides of Pontus (ca. 390-after 339), a member of Plato's Academy under Plato and Plato's successor, that the earth rotates on its axis once in twenty-four hours. This idea, which came to be fairly widely known (though it was rarely accepted as true), explains the daily rising and setting of all of the celestial bodies. Heraclides has been credited also with the claim that the motions of Mercury and Venus are centered on the sun, but modern scholarship has revealed this interpretation to be groundless.[17]

A generation or two after Heraclides, Aristarchus of Samos (ca. 310-230 B.C.) proposed a heliocentric system, in which the sun is fixed in the center of the cosmos, while the earth circles the sun as a planet. It is usually assumed that Aristarchus also gave the other planets sun-centered orbits, although the historical evidence does not address this point. In all likelihood, Aristarchus's idea was a development of Pythagorean cosmology, which had already removed the earth from the center of the universe and put it in motion around the "central fire."[18] Aristarchus has been celebrated for his anticipation of Copernicus, and his successors have been vilified for their failure to adopt his

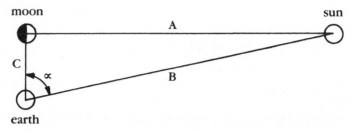

Fig. 5.8. Aristarchus's method for determining the ratio between the solar and lunar distances from the earth. The angle ∝ separating the two
lines of sight is measured when the moon is at quarter-phase (A and C
are then known to intersect at right angles). From this the ratio of B to
C can be calculated. But the method has several drawbacks. In the first
place, the exact moment when the moon's disk is half illuminated cannot be accurately determined. In the second place, a small error in the
measurement of the angle ∝ (Aristarchus put it at 87°, whereas the true
value is 89° 52′) leads to a very large error in the ratio of B to C.

proposal. It takes only a moment's reflection, however, to realize that we do
no justice to the situation in the third century B.C. if we judge Aristarchus's
hypothesis by twenty-first-century evidence. The question is not whether *we*
have persuasive reasons for being heliocentrists, but whether *they* had any
such reasons; and the answer, of course, is that they did not. Putting the earth
in motion and giving it planetary status violated ancient authority, common
sense, religious belief, and Aristotelian physics; it also predicted stellar parallax (variation in the geometrical relationship between pairs of stars as the observer approaches, recedes, or otherwise changes position), which could not be
observed. Meanwhile, whatever observational advantages it might have (for
example, its ability to explain variations in the brightness of the planets) were
available in other systems that did no violence to traditional cosmology.

There were attempts, early in the Hellenistic period, to calculate various cosmological constants. Aristarchus himself compared the earth-to-sun distance
with the earth-to-moon distance, figuring the former to be about twenty times
the latter (the correct ratio is about 400:1). Aristarchus's method is revealed in
figure 5.8. Hipparchus (d. after 127 B.C.) calculated absolute values for the solar
and lunar distances from the absence of solar parallax[19] and from data on solar
eclipses. The assumption that solar parallax is just below the threshold of visibility
gave him a value for the solar distance of 490 earth radii. From eclipse data he
then obtained a value of between 59 and 67 earth radii for the lunar distance.

As for the size of the earth, Eratosthenes (fl. 235 B.C.), a geographer and
mathematician who headed the library in Alexandria, had calculated the earth's

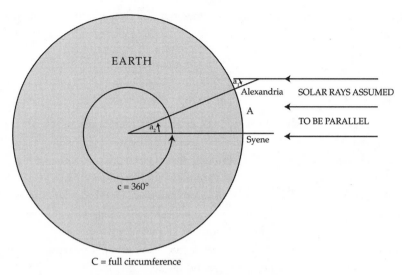

EARTH

Alexandria   SOLAR RAYS ASSUMED

A

TO BE PARALLEL

Syene

c = 360°

C = full circumference

Fig. 5.9. Eratosthenes' calculation of the earth's circumference.

circumference a century earlier; his value of 252,000 stades (very close to the modern figure) became widely known and was never lost.[20] Eratosthenes' procedure was simple. He identified two Egyptian cities, one due north of the other: Syene in the south and Alexandria in the north. He knew from experience that a vertical rod (a gnomon) at Syene cast no shadow at the summer solstice, while on the same day at Alexandria a gnomon cast a shadow of 7.5°. He recognized that if he knew the distance between Syene and Alexandria, he could create an equation that would yield the circumference of the earth. In fig. 5.9, draw a line representing a continuation of the gnomon to the center of the earth. The angle thus formed (a_2) is equal by inspection the angle (a_1) cast by the gnomon at Alexandria. Eratosthenes "measured" (according to ancient sources) segment A of the earth's circumference, the distance between Syene and Alexandria, at 5250 stades (a unit of measure based on the length of a studium). This allowed him to set up the equation: C (the circumference of the earth) is to segment A of the earth's surface as 360° is to 7.5°—an equation that yields a circumference of 252,000 stades.

$$\frac{C \text{ (earth's circumference)}}{A \text{ (Syene to Alexandria)}} = \frac{360°}{7.5°}$$

$$\frac{C \text{ (earth's circumference)}}{5250 \text{ stades}} = \frac{360°}{7.5°}$$

$$C = 252{,}000 \text{ stades}$$

If, as has been estimated, a stade equaled about 600 Roman feet and a Roman foot equaled about 11.5 U.S. inches, then 252,000 stades equals about 24,000 U.S. miles.

HELLENISTIC PLANETARY ASTRONOMY

The acquisition of Babylonian numerical astrology and its sibling, astronomy, in the third and second centuries B.C. provoked a radical transformation in Greek astronomical theory. For the first time Greek astronomers were confronted with the *idea* of models that would enable more or less exact numerical predictions. How Babylonian astronomy reached Greek lands is shrouded in historical mist, but it is clear from the appearance of Babylonian numerical schemes in subsequent Greek astronomy that it was Babylonian influence that made the difference. From that moment to the present, planetary astronomy has been simultaneously geometrical and numerical.

The most prominent Greek recipient of Babylonian influence was the astronomer Hipparchus (fl. 140 B.C.), born in Nicea in Asia Minor and active as an astronomer primarily on the island of Rhodes, just off the coast of Asia Minor. We know from later reports that Hipparchus wrote a number of astronomical works, but the only one to survive is his commentary on a nontechnical astronomical poem, the *Phaenomena*, by the early third-century poet, Aratus. What we know of his astronomical achievements comes primarily from Claudius Ptolemy, whose astronomical efforts were heavily dependent on his.

Hipparchus ranged over all aspects of contemporary astronomy—mathematical, observational, instrumental, and theoretical.[21] His mathematical achievements included creation of methods that made possible a general solution to problems in plane trigonometry. He invented a "diopter" for measuring the apparent diameters of the sun and moon. And he was the probable inventor of stereographic projection, the crucial element of the astrolabe. He may also have invented this instrument, employed ubiquitously throughout the Middle Ages and into the Renaissance for astronomical observations and calculations. On the observational front, he discovered the precession of the equinoxes, established the times of rising and setting of the major constellations for a given location, determined the times of the equinoxes and solstices, and made stellar observations that culminated in a star catalogue. Employing observations going back five hundred years or more, he calculated the average length of the lunar month to within one second of the modern value. He worked on the numerical parameters of solar and lunar theories and developed methods for predicting solar and lunar eclipses for a given location.

Although he was critical of existing planetary theories for their empirical failures, we know of no attempt on his part to create theoretical alternatives. His greatest achievement was to take Babylonian numerical astronomy seriously; to unite its numerical methods with the exclusively geometrical astronomy that had, to this point, dominated Greek astronomical thought; and thereby to provoke a radical reorienting of the astronomical endeavor. No longer would geometrical models of the cosmos survive on geometrical ingenuity, philosophical connections, general plausibility, or aesthetic charm; to be taken seriously, theoretical results would hereafter need to pass the test of quantitative, empirical confirmation.

To see the fruit of this transformation, we turn to the work of Claudius Ptolemy (fl. A.D. 150), the late-Hellenistic astronomer, affiliated with the Museum in Alexandria, Egypt, who has come to symbolize Hellenistic astronomy.[22] It may help to guard against the foreshortening of time, seen from a distance, if we remind ourselves that Ptolemy lived some three hundred years after Hipparchus and five hundred years after Eudoxus. This means not only that he was the beneficiary of the theoretical advances made during those intervening centuries (especially those of Hipparchus), but also that he had access to centuries of astronomical observation, both Greek and Babylonian; and even relatively crude observational data, spread over a sufficiently long span of time, are capable of yielding remarkably precise theoretical conclusions (such as Hipparchus's calculation of the average length of the lunar month).

It would be surprising if the mathematical sophistication of Hellenistic mathematics were not reflected in Hellenistic mathematical astronomy. Ptolemy, coming near the end of that Hellenistic world, brought to planetary astronomy a level of mathematical power that Eudoxus, five hundred years earlier, could not have imagined. Ptolemy's models shared the same geometrical aim as those of Eudoxus—to discover some combination of uniform circular motions that held promise of accounting for the observed positions (including the apparent variations in speed and direction) of the planets. But Ptolemy's models, bearing the imprint of Hipparchus and his Babylonian sources, were also required to make accurate, quantitative predictions of planetary positions, past or future. And they were built with very different geometrical components.

Ptolemy retained the requirement of uniform circular motion, but instead of applying this requirement to spheres, he followed Apollonius of Perga (third century B.C.) and Hipparchus in proposing an astronomy of circles. Let us see how uniform motion in circles can be employed to predict with a reasonable degree of quantitative precision the complicated, clearly nonuniform,

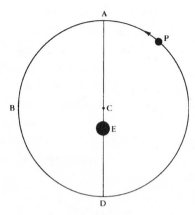

Fig. 5.10. Ptolemy's eccentric model.

motion of the planets. Let circle ABD (fig. 5.10) be the orbit of the planet, and allow the planet, P, to move uniformly around it. If the motion of P is uniform, the planet will sweep out equal angles about the center, C, in equal times. Now if the center of uniform rotation, C, corresponds with the point of observation—that is, if the earth is located at C—then the motion of P will not only *be* uniform, but also *appear* uniform. However, if the center of uniform rotation and the point of view do not coincide—if the earth, for example, is located at E—then the motion of the planet will appear nonuniform, seeming to slow as it approaches A and seeming to speed up as it approaches D. This is what we will call the "eccentric model."

The eccentric model is sufficient for dealing with simple cases of nonuniform motion, such as that of the sun around the ecliptic and the resulting inequality of the seasons. For more complicated cases, Ptolemy found it necessary to introduce the "epicycle-on-deferent" model (fig. 5.11). Let ABD be a "deferent" (or carrying) circle, and draw a small circle (an epicycle) with its center situated on the circumference of the deferent at A. The planet P moves uniformly, counterclockwise, around the epicycle; meanwhile the center of the epicycle moves uniformly, counterclockwise, around the deferent circle. The observer on the earth at E sees the composite of two uniformly circular motions. The exact characteristics of this composite motion will depend on the particular values chosen—the relative sizes of the two circles and the speeds and directions of motion—but it is clear that the model has considerable potential. When P is on the outside of the epicycle, as in figure 5.11, the apparent motion of the planet (as viewed from the earth) will be the sum of its motion around the epicycle and the motion of the epicycle around the deferent, and the planet will at this point have its maximum apparent speed.

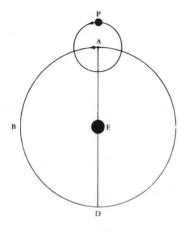

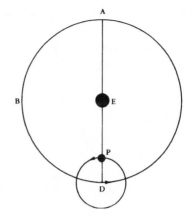

Fig. 5.11 *(above left)*. Ptolemy's epicy-cle-on-deferent model with the planet on the outward side of the epicycle.

Fig. 5.12 *(above right)*. Ptolemy's epicy-cle-on-deferent model with the planet on the inward side of the epicycle.

Fig. 5.13 *(right)*. Retrograde motion of a planet explained by the epicycle-on-deferent model. As the epicycle moves counterclockwise on the deferent the planet moves counterclockwise on the epicycle. The actual path traversed by the planet is represented by the heavy line.

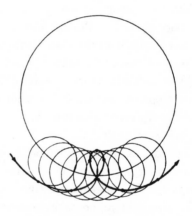

When P is on the inside of the epicycle, as in figure 5.12, its motion on the epicycle and the motion of the epicycle on the deferent are opposed (as viewed from the earth), and the apparent motion of the planet is determined by their difference; if the motion of P about the epicycle should be the greater of the two, the planet will appear to reverse itself and undergo a period of retrograde motion. This retrograde motion is illustrated in figure 5.13.

Both of these models are based firmly on the requirement that the real planetary motions—that is, the component motions—must be uniform and circular. Greek astronomers have been criticized for their "dogmatic" commit-ment to uniform circular motion, on the grounds that a priori assumptions (of this or any other sort) are unjustified, or at least unbecoming, in a scientist. Is such criticism justified? The truth is that scientists, ancient or modern, begin *every* investigation with strong commitments regarding the nature of the universe and very clear ideas about which models may legitimately be employed to represent it. In Ptolemy's case, the requirement of uniform circular motion was justified above all by the nature of the inquiry; his goal was not simply to set forth the relevant observational data, in order to describe the planetary motions in all of their complexity, but to reduce the complex planetary motions to their simplest components—to discover the real order underlying apparent disorder. And the simplest motion, representing the ultimate order, is, of course, uniform circular motion.

But many other considerations would have reinforced the restriction to models based on uniform motion in a circle—for example, the force of common sense and the sanction of tradition, for the cyclic, repetitive nature of celestial phenomena had always suggested that celestial motion must be fundamentally uniform and circular. Furthermore, without uniform circular motion quantitative prediction would have been impossible, for the "trigonometric" methods available to Ptolemy were not readily applicable to any other kind of motion. In addition, there were aesthetic, philosophical, and religious considerations: the special character of the heavens demanded the most perfect of shapes and motions for heavenly bodies. Finally, it is instructive to note that when Copernicus broke with Ptolemy 1,400 years later, it was not because he resented the commitment to uniform circularity, but (in large part) because he believed that Ptolemy had not lived up to this commitment.

In any case, the eccentric and epicycle-on-deferent models, based on uniform circular motion, were extremely powerful tools. But they had their limits, and their inability to account for certain planetary motions demanded the addition of yet another geometrical construction, the "equant," which could be added to the eccentric and epicycle-on-deferent models. Let AFB (fig. 5.14) be a circle, with center at C, while the earth is situated off-center, at E. So far, this is simply a replication of the eccentric model. But here's the difference: the planet, instead of moving uniformly about the deferent circle (that is, sweeping out equal angles in equal times) as viewed from the center, moves uniformly (sweeping out equal angles in equal times) as viewed from a noncentral point, Q. For example, as the planet moves through arc AF (fig. 5.14), it sweeps out the right angle AQF in (let us say) three years. In the next three years, it will again sweep out a right angle, FQB. So if measured by the angle swept out as seen from the equant point Q, the planet is moving uniformly. But as measured by distance traveled around the deferent (linear speed), it is evident by inspection that the planet travels further in the second

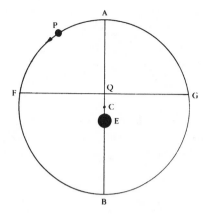

Fig. 5.14. Ptolemy's equant model.

three-year interval than in the first. Moreover, as viewed from the earth E, on the other side of the center from the equant point, the apparent variations in speed are exaggerated. It is clear, then, that the planet gradually increases its speed in passing from A to B, then gradually slows down in passing from B back to A. Ptolemy has added a new tool to his astronomical toolbox that will assist in predicting the variations in speed of planet A, while retaining at least the fiction of uniform circular motion—uniform, that is, as viewed from equant point Q. Whether this diluted version of uniformity was legitimate is a question that Copernicus raised in the sixteenth century, and one that motivated his search for an alternative planetary system. In the meantime, it enabled Ptolemy to complete the development of successful planetary models. In Ptolemy's mind, predictive success outweighed any considerations that might have dictated the stronger version of uniformity.

The three geometrical constructions—eccentric circle, epicycle-on-deferent, and equant—were all effective ways of employing uniform circular motion (whether that uniformity was strict or not so strict) to account for apparent irregularity in the heavens. But they achieved their full power through combination. The eccentric circle and epicycle-on-deferent models could be easily combined by defining a deferent that was eccentric to the earth. An equant point could be added, so that the center of the epicycle would sweep out equal angles in equal times as measured from that noncentral point. It was even possible to take the center of the eccentric circle and move it in a small circle around the earth—something Ptolemy had to do to make his lunar model work.[23] The most typical model for a planet (applicable to Venus, Mars, Jupiter, and Saturn) is illustrated in figure 5.15, where ABD is an eccentric deferent, with center at C; the earth is at E and the equant point at Q. The planet moves

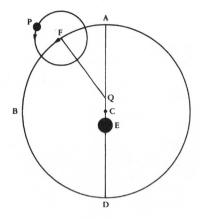

Fig. 5.15. Ptolemy's model for the superior planets. Line QF sweeps out equal angles in equal times as viewed from the equant point Q.

uniformly around the epicycle; the center of the epicycle moves uniformly (as measured by the angle swept out) about the equant point, Q; and the resulting motion is viewed from the earth, E. Models such as this, developed with appropriate variations for all of the remaining planets, proved extraordinarily successful in predicting observed planetary positions. Indeed, it was precisely their degree of success that gave them such longevity. If it works, don't fix it.

It may appear that Ptolemy was engaged in a purely mathematical exercise. He titled the treatise in which these mathematical models were presented *Mathematical Syntaxis* (or *Mathematical System*). Moreover, in the preface to this treatise he announced that speculation about divine causation of celestial motion or about the material nature of things leads only to "guesswork"; if the goal is to achieve certainty, the mathematical way is the only way. And at several points in the work he argued that astronomical models should be chosen on the basis of mathematical simplicity—apparently without concern for physical plausibility.

Nonetheless, if we look closely we see that nonmathematical considerations did, in fact, enter into the analysis. Ptolemy presented physical arguments for the centrality and fixity of the earth—which for him was not merely a mathematical hypothesis, but an important physical belief. He made claims about the nature of the heavens—arguing that, unlike terrestrial substance, they offer no hindrance to motion. And in another treatise, the *Planetary Hypotheses*, he tried to work out a materialized version of his mathematical models.[24] Thus, although Ptolemy was committed to a mathematical approach, his mathematical analysis did not exclude physical concerns, but functioned within the framework of traditional natural philosophy.

For all of this interest in the physics of the cosmos, the *balance* of Ptolemy's astronomical work was weighted toward the numerical/geometrical approach that he had learned from Hipparchus. And it was as a mathematician of the heavens, committed to "saving the phenomena" by mathematical means, that he influenced the Middle Ages and the Renaissance. Indeed, Aristotle and he came to symbolize the two poles of the astronomical enterprise—the former occupied especially with causal questions and the physics of the cosmos, the latter an accomplished builder of mathematical models.

THE SCIENCE OF OPTICS

Light and vision were objects of investigation and speculation in ancient Greece since at least the fifth century B.C. Vision has been universally regarded as the sense through which we gain most of our knowledge of the world in which we live, while light is both the instrument of vision and the conveyor, in the form of sunlight, of heat and life. Any well-developed philosophy of nature is therefore obliged to deal with these optical topics. Early efforts dealt exclusively with the physical aspects of light and sight. Euclid introduced a mathematical approach, which remained influential, eventually joining forces with physical approaches. A productive division of labor was thus achieved, which survives to the present.

The atomists attributed sight to the reception in the eye of a thin film of atoms (a *simulacrum*) issuing from the surface of visible objects. This film, bearing the shape and color of its source, interacts with soul atoms in the eye or mind to produce visual sensation. According to Plato (in his *Timaeus*), fire issues from the observer's eye and coalesces with sunlight to form a medium, stretching from the visible object to the eye, through which "motions" originating in the visible object are transmitted to the eye and ultimately to the soul. And Aristotle argued that a potentially transparent medium is brought to a state of actual transparency when illuminated by a luminous body, such as the sun; light is simply this actualized state of the medium. Thereafter, colored bodies in contact with the actualized medium produce further changes, which are transmitted to the observer's eye and result in visual perception of those bodies.[25]

We first see an attempt at a geometrical theory of vision in the generation after Aristotle. Euclid (fl. 300) wrote a book entitled *Optica*, in which he defined the act of vision and developed a theory of visual perspective. He argued that rectilinear rays emerge from the observer's eye in the form of a cone, with its vertex in the eye and its base on the visible object. Things on which rays fall are seen. Having thus defined the visual cone, Euclid employed it to develop a geometrical theory of perspective. One of the postulates of the *Optica* asserts

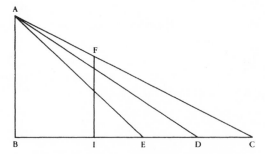

Fig. 5.16. The geometry of vision according to Euclid. A is the observer's eye, and AEC the visual
cone emanating from the eye. C, the most distant
of the observed points, is viewed by ray AC, which
occupies a higher position within the visual cone
(note where it passes through intersecting plane FI)
than does ray AD, by which point D is observed.

that the apparent size of an observed object is a function of the angle under
which it is perceived; another maintains that the perceived location of an
observed object depends on the location within the visual cone of the ray by
which it is observed; things observed by rays higher within the cone appear
higher to the observer. The propositions of the book go on to analyze the appearance of an object as a function of its spatial relationship to the observer.
For example, figure 5.16 shows how a more distant object intercepts a higher
ray in the visual cone and thus appears higher. This is an interesting and impressive piece of mathematical analysis, and it would prove very influential.
But we should not merely be impressed by the mathematics; we should also
note that the theory skips over many aspects of the process of seeing that
people like Aristotle considered of fundamental importance—namely, the
medium, the physical connection between the object and the eye, and the act
of perception. In short, if you were interested only in optical questions that
could be addressed geometrically, Euclid's theory was a brilliant achievement;
if you were interested in any of the nongeometrical features of vision, Euclid's
theory was of little use.[26]

The greatest Hellenistic text on geometrical optics was produced 450 years
after Euclid, by Ptolemy—best known as an astronomer, of course, but also
author of one of the most important works on optics written before Newton.
Ptolemy's *Optica* survives only in an incomplete version, but this version
amply reveals the nature of Ptolemy's achievement.[27] Ptolemy did not follow Euclid's narrow, geometrical approach to optics. Rather, he attempted to

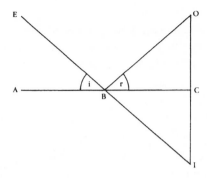

Fig. 5.17. Vision by reflected rays
according to Ptolemy.

create a comprehensive theory that combined Euclid's geometrical theory of vision with a thorough analysis of the physical and psychological aspects of the visual process and, in one celebrated case, an experiment that yielded quantitative results. Ptolemy adopted Euclid's theory of the visual cone (a geometrical theory) and applied it to both monocular and binocular vision. But Ptolemy's visual rays were physically real—a visual flux issuing from the observer's eye, which takes on the color of objects with which it makes contact and transmits that color, along with visual information regarding shape, size, and location, back to the observer's eye and, ultimately, to the Governing Faculty in the brain. These physical aspects of Ptolemy's theory did not detract from his geometrical achievement; and the geometrical portions of his text proved of major importance in teaching people how to think geometrically about vision and light.[28]

All of Ptolemy's *Optica* is devoted to vision, but vision, as he understands it, includes vision by rays reflected at mirror surfaces or refracted at transparent interfaces, and problems of reflection and refraction occupy the majority of his book. Others, including Euclid and Hero, had already written about mirrors; and Ptolemy built on their achievement. He presented a comprehensive account of reflection, which we can best explain by reference to figure 5.17. Let ABC be a plane reflecting surface, O an observed point, and E the eye. Ptolemy argued, first, that incident ray EB (remember that rays travel from the observer's eye to the observed point) and reflected ray BO define a plane that is perpendicular to the plane of the mirror; second, that the angle of incidence, i, is equal to the angle of reflection, r; and, third, that the image of O is located at I, where the extension of the ray emanating from the eye intersects the perpendicular dropped from the observed point to the reflecting surface. (In effect, the observer does not "know" that his or her visual ray has been bent by reflection at the mirror and therefore judges the object to be on the rectilinear extension of that ray.) Ptolemy applied similar rules to reflection from spherical and cylindrical mirrors, both concave and convex. He developed an impressive set of theorems dealing with the location, size, and form of images produced by reflection. It is interesting and important that he devised experiments by which to test his theory.

If Ptolemy's theory of reflection was based on the work of Euclid and Hero, his theory of refraction broke fresh ground. The basic phenomena of refraction—the illusion of the "bent" stick, half submerged in water, for example—had long been known. But Ptolemy gave refraction a thorough mathematical analysis, coupled with experimental investigation. If a ray passes obliquely from one transparent medium to another—the two media having different optical densities—it is bent at the interface in such a way as to lie closer to the perpendicular in the denser medium. Thus, in figure 5.18, if ABC is a plane transparent interface between air above and water below, DBF a perpendicular to that interface, E the observer's eye, and O the observed point, the angle of incidence, EBD, is always greater than the angle of refraction, OBF. And the image of O will be located at I, where the rectilinear extension of incident ray EB intersects perpendicular OG, drawn from the observed point to the refracting surface.

Is there some fixed mathematical ratio between the angles of incidence and refraction? Ptolemy thought there must be and undertook an ingenious experimental investigation to find it. He used a bronze disk, with its circumference marked off in degrees, to measure angles of incidence and the corresponding angles of refraction in three different pairs of media (air and water, air and glass, water and glass). (See fig. 5.19.) He did not discover the desired ratio, and

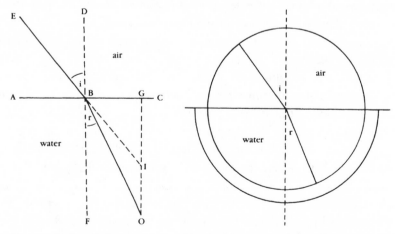

Fig. 5.18. Ptolemy's theory of refraction.

Fig. 5.19. Ptolemy's apparatus for measuring angles of incidence and refraction.

certainly did not discover the modern sine law, but he did find a mathematical pattern in the data—or perhaps chose or adjusted the data to make them conform to a reasonable mathematical pattern.[29] He also passed on to future generations a thorough understanding of the basic principles of refraction and a clear and persuasive example of quantitative experimental investigation.

THE SCIENCE OF WEIGHTS

The science of weights, or of the balance, was a third subject that yielded to mathematical analysis during the Hellenistic period. Indeed, it yielded more completely than the other two. In both astronomy and optics, the level of mathematization was impressive; but in both subjects there remained important physical questions for which no mathematical answer could be found. In the science of the balance beam, by contrast, the physical seemed almost completely reducible to the mathematical.[30]

The central problem was to explain the behavior of the balance beam or lever—the fact that the beam is in balance when the weights suspended from its ends are inversely proportional to their distances (only horizontal distance counts) from the point of support or rotation. Thus a weight of 10 (fig. 5.20) on one end of a beam will balance a weight of 20 on the other end if the former is twice as far as the latter from the fulcrum. One of the earliest surviving explanations is found in a book of *Mechanical Problems*, attributed to Aristotle but actually a later product of the peripatetic school. There we find a "dynamic" account of this static phenomenon: the author explains that if a balanced beam should be set in motion, the velocities of the moving weights would be inversely proportional to the magnitude of those weights. In figure 5.21, in the time required for a weight of 20 to move distance b, the weight of 10 will move a distance of

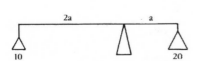

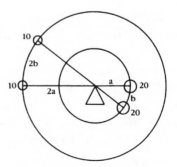

Fig. 5.20. The balance beam in a state of balance.

Fig. 5.21. The dynamic explanation of the balance beam.

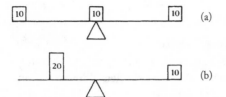

Fig. 5.22. Archimedes' static proof of the law of the lever.

2b. The explanatory notion operating here is that the greater velocity of the one moving body exactly compensates for the greater weight of the other.

A "static" proof of the law of the lever was produced in a treatise attributed to Euclid, and far more elegantly by Archimedes in his *On the Equilibrium of Planes*. Archimedes successfully reduced the problem to one of geometry. Except for the claim that the weights do have weight, physical considerations make no appearance. The balance beam becomes a weightless line; friction is ignored; the weights are applied to a single point of the beam and act in a direction perpendicular to it. Moreover, the demonstration based on these assumptions is Euclidean in form. Two premises provide the basis for the proof: that equal weights at equal distances from the fulcrum (and on opposite sides of it) are in equilibrium; and that equal weights situated anywhere on a lever arm may be replaced by a double weight at a point midway between them (that is, at their center of gravity). Both premises are established by appeal to geometrical symmetry and intuition. In its simplest form, the proof asserts that the beam in figure 5.22a, supporting three identical weights of magnitude 10, is in equilibrium by the principle of symmetry. However, we have agreed that two of the weights can be replaced by a weight of 20 located midway between them, as in figure 5.22b. It follows that a weight of 20 is in equilibrium with a weight of 10 when the former is half as far as the latter from the fulcrum; this result can be easily generalized to yield the law of the lever.

Archimedes' *On the Equilibrium of Planes* contains much more than this; and other works of his, such as *On Floating Bodies*, are also devoted to the solution of mechanical problems. But this examination of his proof of the law of the lever reveals the thoroughness and the extraordinary skill with which he geometrized nature. Many scientific problems continued to resist solution by mathematical methods, but Archimedes remained as a symbol of the power of mathematical analysis and a source of inspiration for those who believed that mathematics was capable of ever greater triumphs. His works had limited influence during the Middle Ages, but in the Renaissance they became the basis of a powerful tradition of mathematical science.[31]

6 ❃ *Greek and Roman Medicine*

EARLY GREEK MEDICINE

The evidence for Greek medicine is spotty, and uncertainty will always remain regarding many of the particulars of Greek healing practices. We have only literary sources for the period prior to the fifth century B.C. Then we have several bodies of writings, explicitly medical but with severe temporal restrictions, that inform us about medical theory and practice in the classical and Hellenistic periods. Obviously these medical treatises convey the views and opinions of learned physicians, many of whom were interested in theoretical issues, such as the connection between medicine and philosophy; but at various points they also offer revealing glimpses of the vast substratum of popular medical belief and practice that must have served the majority of the population.

It is clear that Greeks of the Bronze Age (3000–1000 B.C.) had contact with their Near Eastern neighbors, and we have concrete evidence of the influence of Egyptian medical belief and practice—healing practices, ranging from basic surgery and the use of internal medicines to religious incantations and dream healing. And healing practitioners of varied qualifications must have worked at many levels, for diverse clienteles, utilizing the full range of available medical remedies and techniques.[1]

From the ancient Greek poets Homer and Hesiod, we can glean incidental information about the nature of medical practice toward the end of this period. In Homer's *Iliad* and *Odyssey*, the gods are implicated as causes of plague, and they may be prayed to for healing. Hesiod, too, considers disease to be of divine origin.[2] Homer mentions healing incantations and pharmaceutical remedies, including some of explicit Egyptian origin. He describes a variety of wounds and, in some cases, their treatment. And he makes clear that healers were regarded as members of a distinct craft or profession—professionals in the sense that they had special skills, the exercise of which was a full-time occupation.

Fig. 6.1. Relief of Asclepius, the god of healing. National Archeological Museum, Athens. Alinari/Art Resource N.Y.

The religious side of healing is most clearly manifest in the cult of Asclepius, the god of healing (fig. 6.1). Asclepius, already referred to by Homer as a great physician, was subsequently deified and in the fourth and third centuries became the focus of a popular healing cult. Temples to Asclepius were built in many places—hundreds have been identified—and to these the sick flocked for cures. Central to the therapeutic process was the healing vision or dream, which was supposed to occur as the supplicant slept in a special dormitory.

Healing could occur during the dream, or advice received in the dream might lead to subsequent healing. In addition, the visitor to a temple of Asclepius could expect to bathe, to offer prayers and sacrifices, and to be the recipient of purgatives, dietary restrictions, exercise, and entertainment. And, of course, it was necessary to thank the gods with a suitable offering. A number of tablets have been found at Epidaurus, the center of the cult (fig. 6.2), attesting to cures alleged to have occurred there. According to one of them, a certain Anticrates of Cnidus, who had been blinded by a spear, came to Epidaurus seeking a cure. "While sleeping he saw a vision. It seemed to him that the god pulled out the missile and then fitted into his eyelids again the so-called pupils. When day came he walked out sound."[3] Religious practices such as these remained a very significant part of ancient medicine to the end of the Roman period.

HIPPOCRATIC MEDICINE

In the fifth and fourth centuries, a new, more secular, and more learned medical tradition grew up alongside traditional healing practices—a medical tradition influenced by contemporary developments in philosophy and associated with the name of Hippocrates of Cos (ca. 460–ca. 370 B.C.; fig. 6.3). It is uncertain whether any of the writings (numbering between sixty and seventy) now called "Hippocratic writings" or the "Hippocratic corpus" were, in fact, written by somebody named "Hippocrates." We can only affirm that they are a loosely affiliated body of medical writings, composed for the most part between about 430 and 330 B.C., later collected and attributed to Hippocrates because they shared what were by then regarded as "Hippocratic" traits. What were some of these traits?[4]

Most prominently, the Hippocratic writings represented learned medicine. The very fact that they were "writings" already makes the point that the authors were literate. Their works were the end products of a quest for understanding. Many of the Hippocratic authors defended points of view on the nature of medicine as an art or science, on the nature and causes of disease, on the relationship of the human frame to the universe more generally, and on the principles of treatment and cure. They were engaged in what we must broadly define as natural philosophy—either as original thinkers, philosophers applying themselves to fundamental causal questions about health and disease, or as practicing physicians borrowing from the philosophical tradition. They stood at the intersection of the healing craft and the enterprise of natural philosophy. Hippocratic physicians may not have achieved unanimity on any of

Fig. 6.2 *(above)*. The theater at
Epidaurus (4th c. B.C.), center of
the healing cult of Asclepius. The
theater seats about 14,000 people.
Foto Marburg/Art Resource N.Y.

Fig. 6.3 *(left)*. Hippocrates (Roman
copy of a Greek original). Museo
di Ostia Antica.

these fundamental questions, but they shared the determination to proceed in a learned manner. Even those Hippocratic authors who expressed resentment over the intrusion of philosophy into medicine did not escape its influence.[5]

What view of the medical profession do the Hippocratic writings present? We need to remember that medical practice was completely unregulated in antiquity: many kinds of healers competed for acceptance and prestige—and, of course, for business. Medicine was not studied in formal medical schools but generally learned through apprenticeship to a practicing physician. One of the concerns of the Hippocratic writings was to establish standards, drive out charlatans, and create a climate of opinion favorable to learned medicine. The stress on successful prognosis in the Hippocratic writings was designed not merely to enhance the physician's success as a healer, but also to improve his image, and thus to advance his career. Finally, the Hippocratic Oath was an attempt at self-regulation among medical practitioners.

Prominent in a number of the Hippocratic writings are theories of health and disease. What is most striking, at a general level, is the sharply reduced presence (but not total disappearance, as has sometimes been claimed) of theoretical elements of a magical and supernatural sort. Nobody doubted the existence of the gods, and nature itself could be regarded as divine, but intervention by the gods was generally ruled out as a direct cause of disease or health. We see this in various Hippocratic works, including the treatise *On the Sacred Disease* (which does not correspond exactly to any modern ailment, but includes the symptoms of epilepsy and perhaps stroke and cerebral palsy), where the author expressed the opinion

> that those who first called this disease "sacred" were the sort of people we now call witch-doctors, faith-healers, quacks and charlatans. These are exactly the people who pretend to be very pious and to be particularly wise. By invoking a divine element they were able to hide their own failure to give suitable treatment and so called this a "sacred" malady to conceal their ignorance of its nature.[6]

The author proceeds to offer his own naturalistic account, based on the blockage of "veins" by phlegm from the brain. What is important here is the author's assumption that nature acts uniformly; whatever the causes may be, they are not capricious, but uniform and universal.

The Hippocratic treatises often associate disease with some imbalance in the body or interference with its natural state. In several of the treatises, disease is associated with the bodily humors or fluids. One version of this theory is worked out in *On the Nature of Man,* where it is argued that four humors—blood, phlegm, yellow bile, and black bile—are basic constituents of the body and that imbalances among these humors are responsible for disease:

> The human body contains blood, phlegm, yellow bile and black bile. These are the things that make up its constitution and cause its pains and health. Health is primarily that state in which these constituent substances are in the correct proportion to each other, both in strength and quantity, and are well mixed. Pain occurs when one of the substances presents either a deficiency or an excess, or is separated in the body and not mixed with the others.[7]

Each of these humors was associated with a pair of the basic qualities: hot, cold, moist, and dry. This scheme linked disease to excess or deficiency of warmth and moisture and led to the conclusion that different humors tend to predominate during different seasons. Phlegm, for example, which is cold, was believed to increase in quantity during the winter; and therefore during the winter phlegmatic ailments were particularly common. Blood was held to predominate in the spring, yellow bile in the summer, and black bile in the autumn. Seasonal factors, of course, were not thought to be the only causes of disease: food, water, air, and exercise also contribute to one's state of health.

If disease is associated with imbalance, then therapy must be directed toward the restoration of balance. Diet and exercise (which together makeup what were called "regimen") were among the most common therapies. Purging the body—through bloodletting, emetics, laxatives, diuretics, and enemas—was another way of redressing an imbalance of bodily fluids. Careful attention to seasonal and climatic factors, and to the natural disposition of the patient, was also part of successful therapy. And through it all, the physician was to keep in mind that nature has its own healing power and that the physician's most basic task is to assist the natural healing process. A considerable part of the physician's responsibility was preventive—the giving of advice on how to regulate diet, exercise, bathing, sexual activity, and other factors that would contribute to the patient's health.

But the learned physician did not merely give advice. He also engaged in what we might regard as the "clinical" side of medical practice. Various Hippocratic treatises offered instruction on examination procedures, diagnosis, and prognosis (prediction of the probable future course of the disease). They identified symptoms to be looked for and interpretations to be employed; the physician was to examine the patient's face, eyes, hands, posture, breathing, sleep, stool, urine, vomit, and sputum; he was to be alert to coughing, sneezing, hiccuping, flatulence, fever, convulsions, pustules, tumors, and lesions. Case histories, which reveal the typical course of a given disease, were supplied, many of them remarkable for their precision and clarity. Consider, for example, the following description of what must have been an epidemic of mumps:

> Many people suffered from swellings near the ears, in some cases on one side only; in others both sides were involved. Usually there was no fever, and the patient was not confined to bed. In a few cases there was slight fever. In all cases the swellings subsided without harm, and none suppurated [i.e., discharged pus], as do swellings caused by other disorders. The swellings were soft, large and spread widely; they were unaccompanied by inflammation or pain, and they disappeared leaving no trace. Boys, young men, and male adults in the prime of life were chiefly affected, and . . . those given to wrestling and gymnastics were specially liable.[8]

On the basis of the symptoms observed, the physician rendered a diagnosis and a prognosis. And finally, if the case was treatable, he prescribed treatment. Treatment, as we have noted, was often dietary or directed toward the regulation of exercise and sleep; it might also include bathing and massage. But there were many specific ailments thought to respond to certain internal or external medicines; hundreds of the latter, mostly herbal, are mentioned in the Hippocratic writings: laxatives, purges, emetics (to induce vomiting), narcotics, expectorants (to promote coughing), salves, plasters, and powders. Finally, the treatment of wounds, fractures, and dislocations was also covered in the Hippocratic writings, with a level of skill that has elicited admiration from modern physicians.

Finally, we must say a word about the principles of inquiry embodied in this medical literature. Once we get beyond the commitment to a critical approach to the healing enterprise and the resolve to employ naturalistic principles of explanation and therapy, unanimity disappears. Some treatises display a strong inclination toward philosophical speculation. The author of *The Nature of Man*, for example, offered a speculative theory of human nature and of health and disease, and from this theory derived several therapeutic principles. However, other treatises within the Hippocratic corpus attacked this speculative approach. The author of *On Ancient Medicine* expressed skepticism about the use of hypotheses in medicine, especially the hypothesis that disease is the result of imbalance among the four qualities; he argued that this theory does not lead to remedies that differ in any significant way from remedies prescribed by other physicians, but simply surrounds them with a fog of "technical gibberish."[9] He, and other Hippocratic authors of skeptical bent, preferred to have physicians proceed cautiously, on the basis of accumulated experience, accepting causal theories only when they were supported by overwhelming evidence. As we have seen, the admonition to proceed experientially bore fruit in the careful diagnostic procedures and the impressive case histories of the Hippocratic corpus. Occasionally we even find a specific observation made to confirm a theoretical conclusion—as in *On the Sacred Disease*, where the author proposed dissection of a goat that had the disease, in order to demonstrate that the ailment results from accumulation of phlegm in the brain.[10]

We must conclude this discussion of Hippocratic medicine with two cautionary reminders. First, when learned medicine appeared, it did not drive out its rivals. Learned medicine was never the only kind of medicine, nor even the most popular, but functioned alongside traditional forms of healing belief and practice. Throughout Greek antiquity (from the fifth century B.C. onward) the sick could turn for help to learned physicians, priestly healers in the temples of Asclepius, midwives, herb gatherers, and bonesetters. Moreover, there can be no doubt that the lines demarcating these various types of healers were vague—so that, for example, temple healing might be closely associated with learned medicine. Furthermore, there is no question that the sick sometimes experimented with alternative types of healing simultaneously or sequentially.

Second, if traditional medical practices continued to exist alongside learned medicine, they were also, to some extent, incorporated within it. That is, we must not inflate learned Greek medicine into an early version of modern medicine. Greek medicine was . . . well, Greek. It had to be fitted into the worldview and philosophical outlook of the ancient Greeks; and that means that it did not exclude all medical beliefs and practices that the modern Western physician would find bizarre or repugnant. Thus dream healings remained a part of medicine, including Hippocratic medicine, throughout antiquity.[11] And although divine interference was ruled out, religious elements did not altogether disappear. To offer the simplest example, in the opening lines of the Hippocratic Oath, the physician swears by Apollo and Asclepius and calls on the gods and goddesses to witness his oath. If we are tempted to dismiss this example on the grounds that it may represent empty ritual (comparable to an atheist or agnostic swearing on the Bible in a courtroom), a more persuasive case may be found in the Hippocratic author who recommends prayers along with regimen.[12] Or to consider a subtler case of religious presence, when the author of *On the Sacred Disease* denies that disease is the result of divine intervention, he is only arguing that every disease has a natural cause; he is not opposed to the view that this natural cause is itself an aspect or a manifestation of divine agency. Most Hippocratic physicians undoubtedly continued to believe that natural things partake of divinity, and that disease is simultaneously natural and divine.

HELLENISTIC ANATOMY AND PHYSIOLOGY

Our sources for Greek medicine are bifurcated. We have the Hippocratic writings, which teach us much about early Greek medicine; and we have a variety of sources from the early Christian era, which give us a good picture of medicine under the Roman Empire. But from the intervening period of four or five hundred years we have only fragments of medical literature. The explanation is not that medicine ceased to be practiced or that medical treatises ceased to be written (although the production of medical treatises doubtless had its ups and downs); it is rather that, for reasons unknown to us, the medical writings of this intervening period did not survive. Developments in medicine during this half millennium must, therefore, be reconstructed from fragmentary descriptions in the work of later authors.[13]

Knowledge of human anatomy and physiology among Hippocratic physicians seems to have been quite limited. There is little evidence of the systematic dissection of human bodies during, or prior to, the period in which the Hippocratic treatises were written—owing, no doubt, to traditional taboos regarding proper burial of the dead and perhaps also to the absence of any good reason for supposing that human dissection could provide beneficial knowledge. Such anatomical knowledge as existed was undoubtedly acquired in the course of surgery or the treatment of wounds or by analogy with animal anatomy (well understood, thanks to Aristotle).

It was an event of great significance, therefore, when the practice of human dissection began in Alexandria in the third century B.C.[14] How this extraordinary innovation came about, we do not exactly know. It was presumably assisted by royal patronage from the Ptolemaic dynasty, which was powerful enough to violate traditional burial taboos if it so desired; it may have had something to do with medical developments that elevated the significance of anatomical knowledge, or with the transplantation of Greek medicine into a new social and religious setting. It does appear to have occurred within a philosophical context in which new questions coming to the fore may have called for new methods of investigation. Whatever the reasons, ancient testimony is virtually unanimous in maintaining that Herophilus of Chalcedon and Erasistratus of Ceos were the first to engage in systematic dissection of the human body; if we are to believe the Roman encyclopedist Celsus and the church father Tertullian, they even engaged in the vivisection of prisoners.

What did these pioneers of human dissection learn? Herophilus (d. ca. 255 B.C.), a native of Asia Minor, studied medicine under Praxagoras of Cos before migrating to Alexandria, where he worked under the patronage of the first two Ptolemaic rulers. His pathological theory and therapeutic practice, insofar as we can tell, seem to have been Hippocratic in character; it was as an anatomist that he broke fresh ground.[15] Herophilus investigated the anatomy of the brain and nervous system, identifying two of the brain's membranes (the dura mater and pia mater) and tracing the connections between the nerves, the spinal cord, and the brain. His distinction between sensory and motor nerves reveals his understanding of the functions of the nervous system. He examined the eye with great care, identifying its principal humors and tunics and creating a technical nomenclature that has survived to the present; he traced the optic nerve from the eye to the brain and argued that it was filled with a subtle pneuma.

Herophilus also explored the organs of the abdominal cavity. He presented careful descriptions of the liver, pancreas, intestines, reproductive organs, and heart. He distinguished veins from arteries by the thickness of their walls. He examined the valves in the heart. He studied the arterial pulse—though he did not understand it as a simple mechanical response to the pumping action of the heart—and employed variations in pulse rhythms as a diagnostic and prognostic tool. He described the ovaries and the Fallopian tubes and wrote a treatise on obstetrics. Even this brief sketch reveals Herophilus's remarkable achievements as a student of human anatomy and physiology.

His work was continued by an approximate contemporary, Erasistratus (b. ca. 304 B.C.), from the island of Ceos, who had studied medicine in Athens within the peripatetic school and at Cos.[16] Erasistratus followed up and improved on Herophilus's investigations of the structure of the brain and the heart. He supplied an excellent description (which Galen quotes for us) of the bicuspid and tricuspid valves and their function in determining one-way flow through the heart. The heart, in Erasistratus's view, functioned as a bellows, expanding to draw blood or pneuma into itself, contracting to expel blood into the veins and pneuma into the arteries. The expansion and contraction of the heart, Erasistratus held, were the result of an innate faculty residing in the heart; he argued, correctly, that expansion of the artery during the arterial pulse was simply a passive response to the expansion and contraction of the heart.

Although we have important pieces of physiological theory from Herophilus (his theory of the pulse, for example), he seems to have been more interested in structure than in function. We find a great deal more physiology in the work of Erasistratus. Apparently influenced by the peripatetic school, Strato in particular, Erasistratus argued that matter consists of tiny particles separated by minute void spaces; he combined this corpuscularianism with the theory of pneuma to explain a variety of physiological processes. His explanations of digestion, respiration, and the vascular systems (of particular importance because of their subsequent influence on Galen) will serve for illustrative purposes.

Erasistratus believed that all tissue in the body contains veins, arteries, and nerves, and that these vessels serve as channels by which various substances fundamental to the functioning of the body are conducted to its various organs. Food enters the stomach, where it is mechanically reduced to juice, which passes through tiny pores in the walls of the stomach and intestines to the liver, there to be converted into blood. Blood is then sent through the veins to all parts of the body, where it is responsible for nourishment and growth. The arteries, Erasistratus maintained, contain only pneuma, inhaled from the atmosphere in respiration and drawn down to the left side of the heart through the vein-like artery (our pulmonary vein); from the heart, pneuma is distributed through the arteries to all parts of the body, endowing these parts with their vital capacities. Finally, the nerves contain a finer form of pneuma, "psychic" (i.e., "mental" or "intellectual") pneuma, produced from arterial pneuma by refinement in the brain and responsible for sensation and motor functions. To explain the movement of these substances throughout the body, Erasistratus appealed to nature's abhorrence of a vacuum: the pumping action of the heart or the consumption or wastage of matter in an organ requires that blood or pneuma be immediately drawn in to occupy the newly created or newly vacated space.

This is a very impressive theory, portions of which were to survive within Western physiological thought for nearly two thousand years. But even in Erasistratus's day, a serious (and apparently fatal) objection was posed—namely, that when an artery (the channel by which pneuma is conducted to all parts of the body) is cut, blood flows out of it. However, Erasistratus was equal to the challenge, arguing that the veins and arteries do not communicate under normal circumstances; however, when an artery is opened, the escaping pneuma creates, or threatens to create, a vacuum; this potential vacuum, in turn, opens up tiny channels (anastomoses) between the veins and arteries, allowing blood to pass temporarily from the veins to the arteries and follow the escaping pneuma out through the wound.

Erasistratus's theory of nutrition and vascular flow led easily to a theory of disease. Erasistratus held that disease is caused principally by the flooding of veins with surplus blood, owing to excessive eating. If, for example, the veins are sufficiently charged with blood, the normally closed channels between the venous and arterial systems may be forced open; blood may then pass into the arteries and be sent through the arterial system to the extremities, where it causes inflammation and fever. It follows, from such a theory of disease, that therapy must be directed toward reducing the quantity of blood. This may be accomplished by limiting the intake of food or (less commonly) through bloodletting.

HELLENISTIC MEDICAL SECTS

Herophilus and Erasistratus attracted substantial attention within the medical world and drew leading physicians and medical theorists into their orbits. These students and observers, though no doubt inspired by the example and the teachings of the two men, seem not to have considered themselves bound to any sort of doctrinal orthodoxy. After all, Herophilus and Erasistratus had themselves disagreed on many issues. A student of Herophilus, Philinus of Cos, wrote a book against certain teachings of Herophilus and the Herophileans, which set off a round of attack and counterattack. Over the next several centuries a substantial polemical literature was produced by the Herophileans and their critics (who became known as "empiricists"). Hellenistic medicine was beginning to break up into rival medical sects, each with its own medical theories and its own methodological program.

In the long run, several groups emerged.[17] One family of sects, descended in part from the Herophileans and Erasistrateans, was already lumped together in antiquity under the rubric "rationalists" or "dogmatists" (names that did not signify in Hellenistic Greece precisely what they signify to us). It must be stressed that neither the "rationalists" nor the "dogmatists" constituted a unified or coherent movement but were divided on many issues; if they were united by anything, it was their general commitment to speculative, theoretical medicine—the attempt to apply to the medical realm the methods of natural philosophy that we have observed in the principal philosophical schools. Some "rationalists" continued to defend human dissection as a

Fig. 6.4. Greek physician, grave relief, 480 B.C. Antikenmuseum Basel, Inv. no. BS 236.

valuable methodological tool, capable of contributing to the formulation of hypotheses regarding the hidden causes of disease; what *all* would have agreed on is the value of physiological theory for the practice of medicine.

Their principal rivals and detractors, the "empiricists," adopted the radically opposite opinion that theoretical speculation, including the quest for physiological knowledge and the hidden causes of disease, was a waste of time; and, especially, that human dissection made no useful contribution to medical knowledge and should be forbidden. The "empiricists" maintained, in short, that the anatomical and physiological tradition developed by Herophilus, Erasistratus, and their theoretically oriented followers was a medical dead end, which was to be avoided. The successful physician should concentrate on visible symptoms and visible causes and recommend therapy on the basis of past experience (his own and that of his predecessors) of the efficacy of various remedies.

In the first century A.D., a third group of physicians, known as "methodists," appeared in Rome, claiming that the "rationalists" and "empiricists" had made medicine unnecessarily complicated—that the intricacies of learned medicine, including anatomy and physiology and the quest for the causes of disease (both hidden and visible), could be dispensed with. The core of "methodist" teaching was that disease depends on tenseness and laxness of the body and that the prescribed treatment follows directly and "methodically" from this premise. This teaching proved very popular among Roman aristocrats, whose support made "methodism" a powerful medical force in Rome and throughout the Hellenistic world. A fourth doctrinal school was that of the "pneumatists," who built a medical philosophy on Stoic principles. And finally, we must mention Asclepiades of Bithynia (fl. 90–75 B.C.), an influential Roman physician, who repudiated humoral theories in favor of atomistic doctrines.

GALEN AND THE CULMINATION OF HELLENISTIC MEDICINE

It was this medical world that Galen entered when, at age sixteen, he decided on a medical career. Born in Pergamum (one of the leading intellectual centers of Asia Minor and, indeed, of the entire Hellenistic world) in A.D. 129, Galen studied philosophy and mathematics before turning to medicine.[18] His travels, in pursuit first of education and later of patronage, illustrate the high level of mobility enjoyed by scholars in the ancient world. Galen studied medicine in Pergamum and Smyrna (both in Asia Minor), then in Corinth on the Greek mainland, and finally in Alexandria. From Alexandria, he returned to Pergamum as physician to the gladiators, then moved to Rome in search of patronage, returned to Pergamum, went back to Italy, and eventually settled in Rome, where he enjoyed the friendship and served the medical needs of the rich and powerful, including the emperors Marcus Aurelius, Commodus, and Septimius Severus. He died after 210. Galen produced an enormous body of writings, the surviving portions of which occupy twenty-two volumes in the standard nineteenth-century edition. These writings, which summed up the knowledge and adjudicated the principal disputes of the ancient tradition of learned medicine, established Galen as the leading medical authority of antiquity, rivaled only by Hippocrates, and gave him unparalleled influence well into the modern period.

Galen was a broadly educated philosopher, informed on all of the major philosophical controversies of antiquity and committed to the integration of medicine and philosophy. He was powerfully influenced by the Hippocratic corpus, by Plato, Aristotle, and the Stoics, and by the anatomical and physiological works of Herophilus and Erasistratus and the medical controversies of the Hellenistic period. He has been described as an eclectic rationalist,[19] more interested in the disease than in the patient, viewing the latter as a vehicle by which to gain understanding of the former. Central to his medical aims was the need to classify diseases—to discover the universals that lay behind the particulars—and to search for their hidden causes. And he was convinced that anatomical and physiological knowledge was essential to the success of this venture.

Hippocratic influence was of critical importance in shaping Galen's medical philosophy, though he felt free to borrow selectively and to interpret the borrowed elements loosely. It furnished his view of the human frame and the task of the physician, his stress on the importance of clinical observation and case histories, his concern with diagnosis and prognosis, and his general therapeutic notions. From the Hippocratic treatise *On the Nature of Man*, Galen took the doctrine of the four humors—the view that the fundamental constituents of the human body are blood, phlegm, yellow bile, and black bile, reducible in turn to the basic qualities hot, cold, wet, and dry. The four humors, he argued, come together to form tissues; tissues combine to form organs; and organs unite to make up the body.

Disease may be connected either with disequilibrium among the humors and their constituent qualities or with the specific state of particular organs; one of Galen's principal innovations in the art of diagnosis was to localize disease by identifying specific afflicted organs. Galen's discussion of fevers illustrates both aspects of his theory of disease. Generalized fevers, he argues, are produced throughout the body by the heat of putrefying humors; localized fevers result from noxious or toxic humors within a specific organ, leading to changes, such as hardening or swelling, and to pain. For purposes of diagnosis, Galen depended especially on pulse and examination of the urine; but he also perceived the need to examine all of the other signs stressed in the Hippocratic corpus. In his *On the Art of Healing* he wrote:

> When you meet the patient, you study the most important symptoms without forgetting the most trivial. What the most important tell us is corroborated by the others. One generally obtains the major indications in fevers from the pulse and the urine. It is essential to add to these the other signs, as Hippocratics taught, such as those that appear in the face, the posture the patient adopts in bed, the breathing, the nature of the upper and lower excretions, . . . presence or absence of headache, . . . prostration or good spirits in the patient, . . . [and] the appearance of the body.[20]

Essential to the successful practice of medicine, Galen believed, was knowledge of the structure and functioning of the individual organs. He preached the importance of anatomical knowledge, but acknowledged that in his day dissection of humans was no longer possible. He urged his reader to be alert to the possibility of fortuitous anatomical observation, through the disintegration of a tomb or the discovery of a skeleton along a roadside; for those who could manage it, he recommended a visit to Alexandria, where the skeleton could still be examined firsthand; but he acknowledged that for the most part human anatomy would have to be inferred by analogy from the dissection of animals whose anatomy resembles that of humans. Galen himself dissected a variety of animals, including a small monkey known as the barbary ape (the macaque). His skill as an anatomist is obvious from several anatomical works, including a guide to dissection, *On Anatomical Procedures*. He supplied excellent descriptions of the bones, the muscles, the brain and nervous system, the eyes, the veins and arteries, and the heart. He borrowed, of course, from the work of Herophilus and Erasistratus; but he did not hesitate to correct these predecessors when he believed that they were in error. Unfortunately, Galen's dissection of animals led to the mistaken attribution to humans of certain anatomical features found only in his animal sources; the most notorious case is that of the rete mirabile, to which we will return. Nonetheless, it was Galen's anatomical works, rather than those of Herophilus and Erasistratus, that survived; and thus it was Galen who supplied Europe with its only systematic account of human anatomy until the Renaissance.

Galen's physiological system had even more complex roots than did his anatomical knowledge. Plato had argued for a tripartite soul, consisting of a superior ("rational") part and two inferior parts (associated with the passions and appetites), lodged, respectively, in the brain, the chest, and the abdominal cavity. Galen adopted this scheme and proceeded to correlate the three faculties of the soul identified by Plato with the three basic physiological functions defined by Erasistratus; the result was a tripartite organizational framework for physiology. In this scheme, the brain (seat of the soul's rational faculties) was identified as the source of the nerves. Following Erasistratus, Galen argued that the nerves contain psychic pneuma, which accounts for sensation and motor functions. The heart (seat of the passions) became for Galen the source of the arteries, which convey life-giving arterial blood (and vital pneuma) to all parts of the body. Finally, the liver (seat of desire or appetite) was made the source of the veins, which nourish the body with venous blood.[21]

These three physiological systems, as Galen developed them, were not totally independent but had interconnections. It may be helpful, therefore, if we work through them from beginning to end—from the initial intake of food to the final distribution of psychic pneuma through the nerves. Food arrives in the stomach, where it is reduced to chyle (the Greek word for juice)—not merely through mechanical action, as Erasistratus believed, but through cooking by the body's vital heat. Chyle passes through the walls of the stomach and intestines into the surrounding mesentery veins, which convey it to the liver. In the liver, chyle undergoes further refinement and cooking to yield venous blood. This venous blood, which is nourishment for the body, moves slowly outward through the veins to the various tissues and organs, where it is consumed. This is the venous system, originating in the liver, conveying venous blood, responsible for nourishment, to all parts of the body.[22]

Venous blood reaches the right side of the heart through the vena cava. From there, some of this venous blood is carried by a major blood vessel (our pulmonary artery, Galen's artery-like vein) to the lungs, which, like all other organs, require nourishment. The remainder of the venous blood seeps slowly through tiny pores in the heavy muscle (the interventricular septum) that divides the right ventricle of the heart from the left. Galen acknowledged that these pores are too small to be visible, but he argued that since the incoming vena cava is larger than the outgoing artery-like vein, some venous blood must go elsewhere; moreover, the disparity in size is too great to be accounted for by the fact that the heart (like every other organ) consumes a certain amount of venous blood as nourishment; and finally, the principle that nature does nothing in vain guarantees that the small pits in the surface of the interventricular septum must lead somewhere. It follows that

> the thinnest portion of the blood is drawn from the right ventricle into the left, thanks to perforations in the septum between them: these [perforations] can be seen for a great part [of their length]; they are like pits with wide mouths, and they get constantly narrower; it is not possible, however, actually to observe their extreme terminations, owing both to the smallness of these and to the fact that when the animal is dead all the parts are chilled and shrunken.[23]

What happens when venous blood reaches the left side of the heart? Here we must introduce Galen's theories of vitality and respiration.[24] Galen joined Plato, Aristotle, and the anonymous author of *On the Heart* (a treatise once considered Hippocratic, but probably Hellenistic) in identifying life with innate heat; moreover, he shared their view that the principal seat of this life-giving heat is the heart. Maintaining the right degree of vital heat is critical, of course, and it is the lungs and respiration that perform this function. On the one hand, the lungs surround the heart and diminish or moderate its heat. On the other, they nourish the "fire" within the heart by sending air to it through the vein-like artery (our pulmonary vein); and by the same mechanism they provide means for the heart to rid itself of the waste products of burning. As the heart expands, air is drawn from the lungs into the left ventricle of the heart; as the heart contracts, soot and smoky vapors are sent in the other direction and exhaled into the atmosphere. Air reaching the left ventricle of the heart during the expansive phase mixes with venous blood that has passed through the interventricular septum—blood that has also been heated and thus vivified by the innate heat in the heart. The product is a finer, purer, and warmer arterial blood, now charged with vital spirit or pneuma, which is conveyed throughout the body by the arteries. In the course of defending this theory, Galen devoted considerable attention to proving, contrary to Erasistratus, that the arteries do indeed contain blood. Thus we have Galen's second major physiological system—the arterial system, rooted in the heart, conveying arterial blood through the arteries, imparting life to the tissues and organs of the body.

The brain, like every other organ, is the recipient of arterial blood. Some of this arterial blood passes into the rete mirabile—a network of fine arteries found in certain ungulates (where it serves cooling functions) and mistakenly attributed by Galen to humans. In passing through the arteries of the rete mirabile the arterial blood is refined and emerges as the finest grade of spirit or pneuma—psychic pneuma. This pneuma is sent to all parts of the body through the nerves and accounts for sensation and motor functions; and thus we have the third of Galen's major physiological systems.

Before leaving Galen's physiology, it is necessary to make one more point. Galen found Erasistratus's attempt to mechanize physiology unpersuasive. In particular, he did not believe that the movement of fluids through the body could be sufficiently explained in terms of pumping action or nature's abhorrence of a vacuum. He accepted that the heart functions as a bellows, drawing air from the lungs during expansion and propelling arterial blood into the arteries during contraction, and that the arteries themselves have active motions that move fluids along. But in addition, he was persuaded that all organs possess nonmechanical faculties by which they attract, retain, and repel fluids on the basis of need. Thus the liver has the capacity to draw to itself the chyle that it requires. Similarly, venous blood moves through the body not because it is pumped, but because the organs of the body attract, retain, and repel it according to their need for nourishment.

Galen's medical system proved exceedingly persuasive, dominating medical thought and teaching throughout the Middle Ages and into the early modern period. Part of its persuasive appeal lay in its comprehensiveness. Galen addressed all of the major medical issues of the day. He could be both practical, as in his pharmacology, and theoretical, as in his physiology. He was philosophically informed and methodologically sophisticated.[25] His work embodied the best of Greek pathological and therapeutic theory. It contained an impressive account of human anatomy and a brilliant synthesis of Greek physiological thought. In short, Galen offered a complete medical philosophy, which made excellent sense of the phenomena of health, disease, and healing.

But there was another reason for Galen's popularity. Galen introduced into his anatomy and physiology a massive dose of teleology, which would endear him to Islamic and Christian readers. Galen was not himself a Christian, and his teleological approach had no Christian roots but was inspired by Plato's *Timaeus* and Aristotle's *The Parts of Animals*, as well as by Stoic thought. Like Aristotle—indeed, more than Aristotle—Galen found evidence of intelligent design in the animal and human frame, and his *On the Usefulness of the Parts of the Body* is a paean of praise to the wisdom and providence of the

Demiurge (a term and a conception obviously borrowed from Plato). In this book Galen wrote:

> And I consider that I am really showing him [the Demiurge] reverence not when I offer him [sacrifices] . . . of bulls and burn incense . . . , but when I myself first learn to know his wisdom, power, and goodness, and then make them known to others. I regard it as proof of perfect goodness that one should will to order everything in the best possible way, not grudging benefits to any creature, and therefore we must praise him as good. But to have discovered how everything should best be ordered is the height of wisdom, and to have accomplished his will in all things is proof of his invincible power.[26]

Galen argued that Nature (or the Demiurge) does nothing in vain; that the structure of the human body is perfectly adapted to its functions, unable to be improved even in imagination. Galen even presented the beginnings of a natural theology—that is, a theory of god or the gods based on evidence found in nature. At the conclusion of *The Usefulness of the Parts*, he called attention to lessons that can be learned about the world soul from the investigation of animal anatomy:

> For when in mud and slime, in marshes, and in rotting plants and fruits, animals are engendered which yet bear a marvelous indication of the intelligence constructing them, what must we think of the bodies above [i.e., the celestial bodies]? . . . Thus, when anyone looking at the facts with an open mind sees that in such slime of fleshes and juices there is yet an indwelling intelligence and sees too the structure of any animal whatsoever—for they all give evidence of a wise Creator—he will understand the excellence of the intelligence in the heavens.[27]

As the reader can easily imagine, Galen's teleology, as well as his desire to fit humans and their diseases into a complete and satisfying (and, of course, ancient) worldview, have not always gone down well with modern scholars. Indeed, Galen has been the target of a certain amount of abuse from medical historians, angry at him for not being modern.[28] Of course, Galen was merely being a second-century Greco-Roman; and if we concentrate on his deficiencies from a modern standpoint, we lose the opportunity to learn from his life and thought what it meant to be a physician in the declining years of Greco-Roman civilization. Galen pulled together several strands of ancient thought: he summed up more than six hundred years of Greek and Roman medicine; at the same time, he fitted that medicine into an ancient philosophical and theological framework. The pervasive teleology in Galen's work is a useful reminder that the question of order and organization in the universe remained a problem of central importance, which every major thinker still felt obliged to address and on which the last word had yet to be said—a problem, indeed, on which the last word has yet to be said today. That the gods figure into Galen's worldview, and even into his medical system, is not a feature to be *regretted*, but one to be *understood* as typical of ancient medicine and philosophy. Galen did not differ substantially from the Hippocratic writers, or from his principal philosophical guides, in his view of the gods. Although he admitted divinity into the medical realm, as in his acknowledgment of the healing power of Asclepius,[29] he did not allow this belief to interfere with the formulation of a medical philosophy restricted to natural causation. Galen certainly believed that behind the admirable design found in living things could be discerned a designer; but this belief had no major influence on his analysis of disease or on his diagnostic and therapeutic procedures.

7 ✸ *Roman and Early Medieval Science*

GREEKS AND ROMANS

Galen's career (examined in the previous chapter) nicely illustrates the interpenetration of Greek and Roman intellectual life. Born and raised in Pergamum in Asia Minor—well within the boundaries of the Roman Empire, but still a stronghold of Greek culture—Galen continued his education in Corinth (on the Greek mainland) and in Alexandria (in Egypt). It was a Greek education that he received—delivered in Greek and based on the Greek classics—and it was the Greek intellectual tradition in which he thus gained membership. But Galen finished his career in Rome, serving Roman emperors and lecturing to Roman audiences. His biography thus raises the question to which the opening sections of this chapter will be devoted: what was the political, cultural, intellectual, and especially scientific relationship between Greece and Rome?

The autonomy and the dynamic political life of the Greek city-states ended with the conquests of Alexander the Great (334–323 B.C.) and the establishment of a Greek empire. However, intellectual life in the successor states, after the division of Alexander's empire among his generals, was the object of sporadic, sometimes generous, patronage and remained vigorous, at least for a time. Meanwhile, Rome grew from an insignificant Etruscan town in the seventh century B.C. to a thriving republic in the fifth and fourth centuries. By 265 B.C. it controlled the Italian peninsula, and by 200 it had enough of a foreign policy and sufficient military might to intervene in Greek affairs during the Second Macedonian War (200–197). Rome gradually extended its influence over Greek lands in the next 150 years; by the death of Julius Caesar in 44 B.C., Rome was in control of virtually the entire Mediterranean basin, including Greece, Asia Minor, and North Africa (see map 7.1).

In the Greek provinces, Roman control did not lead to the collapse of culture and learning. On the contrary, as the Roman writer Horace (d. 8 B.C.)

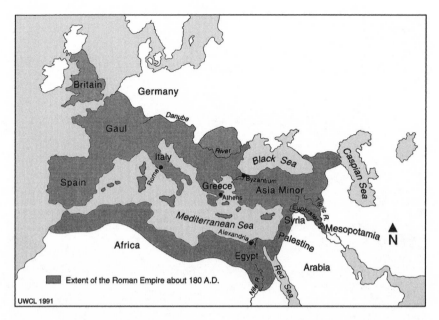

Map 7.1. The Roman Empire.

pointed out so famously, while Rome captured Greece militarily and politically, the artistic and intellectual conquest belonged to the Greeks.[1] As Rome's
power and prosperity increased, its leisured class began to appreciate Greek
achievements in literature, philosophy, politics, and the arts. Any Roman wishing to acquire sophistication in such matters could do no better than to imitate
the Greek achievement—to borrow from a culture of superior accomplishment
in these areas.

The linguistic and geographical barriers that one might expect to have
hindered such borrowing did not, in fact, prove to be a serious problem in
the early years of cultural contact. The ability to read and speak Greek was
common in Italy, which had Greek settlements going back many centuries:
recall, for example, Parmenides and Zeno from the Italian city of Elea and
the Pythagoreans of southern Italy visited by Plato. Rome itself had a community of Greeks by the second century B.C., and bilingualism (at some level)
flourished among the Roman upper classes. With increasing frequency Greek
scholars settled in Rome, voluntarily or as slaves; and it was not difficult to
find Greek teachers willing to expound the content of Greek literature and
philosophy. Study abroad in the Greek provinces was another alternative,
which, for a young Roman with serious scholarly aspirations, became almost

Fig. 7.1. The ancient forum in Rome. Alinari/Art Resource N.Y.

obligatory. By such mechanisms, Rome and its environs acquired a substantial circle of Greek and Roman scholars—all of them in touch with the Greek learned tradition. Eventually, Roman scholars began to convey aspects of the Greek intellectual achievement to a Latin readership. In a few cases, texts were even translated from Greek to Latin.[2]

A number of these circumstances are illustrated by the career of Cicero (106–43 B.C.), a highly educated Roman statesman and man of letters. Cicero studied with Greek teachers, first in Rome, subsequently in Athens and on the island of Rhodes; he learned Greek, of course, and also mastered significant portions of Greek philosophy. He was strongly influenced by Stoicism and the epistemological theories that developed within the Platonic school in the third century. Cicero wrote Latin treatises on a variety of topics and produced a Latin translation of Plato's *Timaeus* (which has not survived).[3]

Support for scholarship was, at the beginning, entirely private. A member of the upper class might devote some of his leisure time to reading and learned discussion; he might have a library, possibly even a substantial one. But anybody who lacked means of his own would need to find a patron. The possible arrangements actually covered a wide spectrum, from distinguished scholars attached to the households of the wealthy to educated Greek-speaking slaves. The obligations of the scholar who made it to the top of this ladder might be to advise or provide intellectual companionship for his patron, or to care for his patron's library; if less fortunate or less capable, he would likely be charged with the education of his patron's children and might also be assigned menial tasks.

The level of discourse in these settings varied. The Roman scholar who wished to proceed at the highest level would do it in Greek. It follows that such scholarly discourse as occurred in Latin (whether spoken or written) fell somewhat below the highest levels of Greek scholarship, which have thus far occupied our attention. Latin was employed when the linguistic limitations of the audience demanded; and it was a lighter, more popular version of Greek learning that appealed to that audience. Certain prominent historians of science, sneering at popularization, as though nothing but "cutting-edge" research merits attention, have been very critical of the Greeks for having developed a popular level of learning, and of the Romans for drawing on it.[4] But this reflects a very narrow outlook. In fact, there must be multiple levels of knowledge and expertise within any scholarly tradition. For every Aristotle, capable of confronting perplexing philosophical or scientific problems in original ways, there were thousands of educated Greeks and Romans whose aspirations did not and could not go beyond grasping what Aristotle had achieved or reconciling Aristotle's views with those of other acknowledged authorities. Inevitably, any program of creative research is accompanied by other programs directed toward preservation, commentary, education, popularization, and transmission. We see this in our own twenty-first-century schools, universities, and mass media.

Given these circumstances, it was natural for scholars setting out to sample and interpret the Greek intellectual achievement for a Roman audience to concentrate on what interested their Roman patrons—not the subtleties of Greek metaphysics and epistemology, nor the advanced details of Greek mathematics, astronomy, and anatomy, but subjects of practical value and intrinsic appeal. A certain amount of mathematics would be included for utilitarian reasons, or as training for the intellect. Medicine hardly needed justification, though Romans were initially suspicious of certain aspects of Greek medicine.

Logic and rhetoric were important in the law courts and the political arena. And Epicurean and Stoic philosophy addressed pressing ethical and religious concerns. But science or natural philosophy, beyond the basics, was rarely valued except as a leisure-time amusement. This state of affairs is vividly illustrated by the fact that for Romans the most celebrated astronomical authority was Aratus of Soli (d. 240 B.C.), whose poem on the constellations and weather prognostication (*The Phaenomena*) was translated into Latin at least four times, while the advanced astronomical works of Eudoxus and Hipparchus remained unavailable or unknown.[5]

Such science or natural philosophy as Romans knew, then, tended to be a limited, popularized version of the Greek achievement. Generations of historians have sought to explain the Roman failure to master the more abstruse or technical aspects of Greek learning in terms of intellectual inferiority, moral weakness, or temperamental defect. It is often said that Romans simply did not have theoretical minds—though it is then quickly added (because everybody has to be good at something) that they made up for this deficiency with administrative and engineering talent.[6] In fact, there is no mystery about the level or the degree of Roman intellectual effort and no reason to be surprised or critical. We need always to remember that the Roman aristocracy regarded learning, except for clearly utilitarian matters, as a leisure-time pursuit. Romans, then, did the obvious thing: they borrowed what seemed interesting or useful. If certain Greeks had devoted their lives to subjects that were abstract, technical, impractical, and (as some no doubt judged) boring, that was no reason for large numbers of Romans to make the same mistake. Members of the Roman upper class had about the same level of interest in the fine points of Greek natural philosophy as the average American politician has in metaphysics and epistemology. At best, their desire was, as the Roman playwright Ennius put it, "to study philosophy, but in moderation."[7] The only surprise is that historians would have expected it to be otherwise.

POPULARIZERS AND ENCYCLOPEDISTS

Having examined the intellectual tastes and motivations of upper-class Romans, we must now investigate in more detail the outcome in the world of Roman learning. We need to find what we can about the major popularizers and survey a few of the most influential books. One of the best known and perhaps most influential of the early popularizers was the Stoic Posidonius (ca. 135–51 B.C.). Born in Syria of Greek parents, Posidonius studied in Athens and subsequently became head of the Stoic school on the island of Rhodes. He exercised a strong indirect influence on Roman intellectual life through his many pupils, one of whom was Cicero; but he also traveled to Rome and impressed the Romans in person. Posidonius was the closest thing to a universal scholar that we can find in the first century B.C. He was interested in history, geography, moral philosophy, and natural philosophy and wrote voluminously on all of these subjects. Among his works (all written in Greek) were commentaries on Plato's *Timaeus* and Aristotle's *Meteorology*; Lucretius borrowed heavily from the latter in writing his *On the Nature of Things*.

Posidonius's works have not survived, and our knowledge of them is therefore secondhand. One of his most influential investigations seems to have been the determination of the earth's circumference—which he calculated first at 240,000 stades (a little smaller than Eratosthenes' estimate), later at 180,000 stades. The importance of this smaller figure is that it was picked up by Ptolemy, passed on to the readers of his *Geography*, and used in the fifteenth century by Christopher Columbus as the basis for his calculation of the distance between Spain and the Indies.

Posidonius exercised substantial influence on Latin writers, such as Varro (116–27 B.C.), thereby helping to determine the form and content of education and scholarship in Latin. Varro, regarded by his Roman admirers as a phenomenal scholar, studied in Rome and Athens; he proved to be a prolific writer, turning out Latin works on a variety of topics (about seventy-five titles, almost all of them now lost). The most important of these was an encyclopedia, *Nine Books of Disciplines*, which was to become a model and source for subsequent Roman encyclopedists. One notable feature of the *Disciplines* was its use of the liberal arts (the subjects deemed suitable for the education of a Roman gentleman) as organizing principles. Varro identified and gave a basic account of nine such arts: grammar, rhetoric, logic, arithmetic, geometry, astronomy, musical theory, medicine, and architecture. Varro's list, narrowed by subsequent writers who omitted the last two arts, came to define the classical seven liberal arts of the medieval schools—the first three arts becoming known as the trivium, the remaining four as the quadrivium.[8]

Varro's contemporary and friend, Cicero, had quite a sophisticated knowledge of Greek philosophy—having studied with the Stoic Posidonius, the Epicurean Phaedrus, and the Platonists Philo of Larissa and Antiochus of Ascalon (fig. 7.2).[9] Cicero was heavily influenced in his intellectual method

Fig. 7.2. Cicero. Museo Vaticano, Vatican City. Alinari/Art Resource N.Y.

by the skeptical tendencies that had developed within the Platonic school; in particular, he became convinced that probability was the most anybody could achieve in philosophical matters and, consequently, that the best way of discovering truth was through the critical sifting of past opinion. The product of this belief was a series of dialogues in which Cicero reported the opinions of his teachers, friends, and earlier writers on a variety of philosophical topics. For the opinions of his predecessors, especially those from the more distant past, Cicero borrowed from existing handbook literature, including the "doxographic" (or "opinion") tradition initiated by Theophrastus.

Cicero thus simultaneously drew on, and contributed to, the popularization movement. He provided his readers with an account of recent and contemporary controversy on the major philosophical issues, including some of the questions that have occupied us in previous chapters—the nature of the underlying reality, the source of order in the universe, the role of the gods, the nature of the soul, and the process of knowing. His own worldview was constructed from a combination of Platonic and Stoic elements, and Cicero became one of the major sources of Stoic philosophy for the Middle Ages and the early modern period. He identified God with nature, nature with fire, and all three (God, nature, and fire) with the active force responsible for the existence, activity, and rationality of the universe. He described the Stoic cosmological cycle of successive conflagration and regeneration (see chap. 4, above). And he advocated the idea of a close parallelism between the macrocosm (god and the universe) and the microcosm (the individual human), arguing that god bears the same relationship to matter in the universe as the human soul bears to the human body. The macrocosm-microcosm analogy was to become a staple of medieval and Renaissance thought and a central theme of astrological writing. Cicero did not devote much attention to the mathematical sciences, which he considered of value principally for their ability to sharpen the wits of young men; however, his discussion of planetary motion in the heavens and his translation of Aratus's astronomical poem, *The Phaenomena*, reveal that he was not totally disinterested or uninformed on such matters.

One of Varro's and Cicero's contemporaries, Lucretius (d. 55 b.c.), wrote a long philosophical poem, *On the Nature of Things*. At one level, this work is a defense of Epicurean natural philosophy, which aims to overcome the fear of death by touting the explanatory power of atoms and the void. However, within this basic Epicurean framework, *On the Nature of Things* is encyclopedic in its scope and popular in its level of presentation and choice of detail. Lucretius discussed the infinity of worlds, their creation and destruction, and such basic astronomical data as the path of the sun around the zodiac, the resulting inequality of days, and the phases of the moon; the mortality of the soul; sense perception, including delusions of the senses; sleep, dreams, and love; mirrors and the reflection of light; the origins of plant and animal life, including a denunciation of teleological explanation in biology; the origins and history of the human race; and extraordinary meteorological and geological phenomena, such as lightning, thunder, earthquakes, rainbows, volcanoes, and magnetic attraction. Lucretius concludes with an account of the great Athenian plague.[10]

Varro, Cicero, and Lucretius represent the flowering of Roman intellectual life in the latter days of the Roman Republic. Others who contributed to this intellectual enterprise were Vitruvius (d. 25 b.c.), a contemporary who wrote on architecture, and several writers of the early imperial period: Celsus (fl. a.d. 25), author of an influential medical encyclopedia, and the Stoic Seneca (d. a.d. 65), who wrote on natural philosophy, including meteorology.[11]

However, the man universally regarded as the pinnacle of the popularization movement is Pliny the Elder (a.d. 23/24–79). Pliny is the central figure in most accounts of Roman science, and we too must briefly examine his work. He was born in northern Italy, into the provincial nobility, and educated in Rome. After a successful military career (the route to advancement for a man of Pliny's social standing), he turned to literary endeavors and ended in the service of the emperors Vespasian and Titus. He wrote several books on the history of Rome and its wars, a book on grammar, and the work on which his fame now rests, the *Natural History*, dedicated to Titus.

The *Natural History* is a remarkable work, which resists easy characterization and really must be read to be appreciated.[12] Pliny had a voracious appetite for information. In the preface to the *Natural History*, he reports that he and his assistants perused two thousand volumes by some one hundred authors and extracted from them twenty thousand facts. It appears that Pliny worked out a system of note cards, so that he could manually sort his twenty thousand pieces of information; the cards were organized by subject and assembled to produce the *Natural History*.[13] The energy with which Pliny proceeded is astounding. His nephew tells us that he rose as early as midnight and worked nearly around the clock, reading or being read to, making or dictating notes. If we are to understand Pliny's achievement, it is important for us to grasp his fascination with factual data. Although in the *Natural History* Pliny sometimes offers explanations of natural phenomena, his goal was not to produce a comprehensive, carefully reasoned natural philosophy, but to create a vast storehouse of interesting and entertaining information—a book, his nephew tells us, "no less varied than nature herself."[14]

Pliny's purpose, then, was to survey the universe and the natural objects that populate it. He devoted seventy-two pages (in a modern English translation) simply to a list of the contents of the *Natural History* and the authorities consulted. Among the subjects treated were cosmology, astronomy, geography, anthropology, zoology, botany, and mineralogy. Pliny had a flair for picking out matters of unusual interest, and he has often been described primarily as a purveyor of marvels. To be sure, natural marvels are not scarce in the pages of the *Natural History*. Pliny reported a series of celestial portents (including multiple suns and moons), thunderbolts called forth by prayers and rituals, the greatest earthquake in human memory (which demolished twelve cities in Asia), human sacrifice among transalpine tribes, a boy said to have been regularly transported to and from school on the back of a dolphin. He also described exotic races of monsters, including the Arimaspi, who have one eye in the center of their foreheads, the Illyrians, who kill with a glance of the evil eye, and the Monocoli, who have only a single leg but manage nonetheless to hop with remarkable swiftness (fig. 7.3).[15]

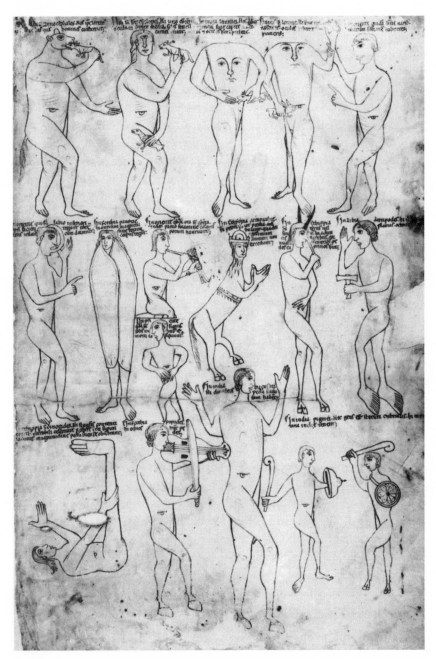

Fig. 7.3. Pliny's monstrous races. British Library, MS Harley 2799, fol. 243r (12th c.). By permission of the British Library.

Just as it would be a mistake to overlook the marvelous element in Pliny's *Natural History*, so it would be a mistake to disregard the more prosaic and commonplace. Pliny's account of astronomy and cosmology is a good example of the latter.[16] He described the celestial and terrestrial spheres and the circles used to map them. He knew that the planets move through the band of the zodiac in a west-to-east direction, and he knew the approximate periods with which they do so; he described planetary retrogressions and reported that Mercury and Venus remain within 22° and 46°, respectively, of the sun. He discussed the motion, phases, and eclipses of the moon; and he understood solar and lunar eclipses as a function of the relative dimensions of the bodies involved and the shadows thus cast. With regard to the dimensions of the earth, Pliny reported Eratosthenes' value of 252,000 stades for its circumference. Pliny thus communicated substantial pieces of cosmological and astronomical knowledge, though not always reliably and certainly not up to the standards of the mathematical astronomer. He was neither borrowing from the tradition of mathematical astronomy (for example, the astronomical sections of the *Natural History* do not reveal the influence of Hipparchus) nor writing for an audience of astronomical specialists. He was merely endeavoring to communicate the essentials to a public not interested in, or equipped to deal with, observational or mathematical complexity.

Pliny was not a typical Roman scholar. Most obviously, nobody matched him for energy and dedication to the task of collecting information. Moreover, his coverage was considerably wider than that of any Roman predecessor (including Varro, who had confined himself to nine arts); in the preface to his *Natural History*, Pliny points out, correctly, that he is the first to attempt to deal with the whole of the natural world in a single work. Although it is sometimes superficial, his *Natural History* serves as a useful measure of what the educated Roman might be expected to know—after Pliny, if not before. And the fact that the *Natural History* survived, while many other popularizing works did not, helped to determine the level and content of early medieval learning.

The Roman encyclopedias that we have looked at were attempts to collect in a single work large quantities of information drawn from many different sources. But Rome also saw the development of a commentary tradition, in which the narrative took its structure and a good part of its content from a single authoritative text. This tradition illustrates the ancient tendency to identify certain venerable, privileged texts as the repositories of knowledge and to measure learning by the ability to read and interpret

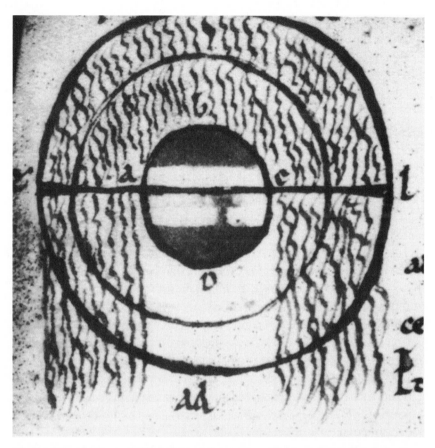

Fig. 7.4. Macrobius on rainfall. A thirteenth-century scribe's attempt to illustrate Macrobius's argument that if we do not assume that all rain falls toward the center of the earth along a radius, we must accept the ridiculous consequence that the portion of it that misses the earth would find itself ascending toward the other hemisphere of the heavens. British Library, MS Egerton 2976, fol. 49v (13th c.). By permission of the British Library. On this illustration and the accompanying argument, see John E. Murdoch, *Album of Science: Antiquity and the Middle Ages*, pp. 282–83.

those texts. An important example of the Roman commentary tradition is the *Commentary on the Dream of Scipio*, by Macrobius (who flourished in the first half of the fifth century, some 350 years after Pliny). This work, which employs Cicero's *Dream of Scipio* as the occasion for an exposition of Neoplatonic philosophy, enjoyed extremely wide circulation during the early Middle Ages. I'll not explore its contents except to note that in it Macrobius set out a comprehensive philosophy of nature, largely Platonic in inspiration, which included substantial sections on arithmetic, astronomy, and cosmology.[17]

One last Roman compiler must be considered because of the window he offers us on the mathematical arts at their best in the schools of the later Roman Empire—and also because his book became one of the most popular school texts of the Middle Ages. Martianus Capella (ca. A.D. 410–439) was probably a North African from the city of Carthage; he thus serves, in addition, to remind us of the strength of the scholarly tradition in the Roman provinces, especially those of North Africa, during the later years of the Empire. The book of his that proved so influential was an allegory, entitled *The Marriage of Philology and Mercury*, in which seven bridesmaids offer surveys of their respective liberal arts to the wedding guests at a celestial marriage ceremony.[18]

The first of the mathematical arts to be presented is geometry. Through the mouth of the bridesmaid Geometry, Martianus briefly surveyed the highlights of Euclid's *Elements*, including most of the definitions, all of the postulates, and three of the five axioms with which that work begins (see chap. 5, above). He discussed and classified plane and solid figures, including Plato's five regular solids; he defined right, acute, and obtuse angles; and he touched on proportionality, commensurability, and incommensurability. But the bulk of this chapter is devoted to a discourse on geography, derived from Pliny and others. Martianus began with proofs of the earth's sphericity; a report of Eratosthenes' figure for its circumference, accompanied by a faulty account of Eratosthenes' method of calculation; and arguments for the centrality of the earth within the universe. He discussed the five climatic zones and the division of the habitable world into three continents (Europe, Asia, and Africa), and proceeded to offer an extremely swift tour of the known world (basically a boiled-down version of Pliny's similar discussion).

Arithmetic came next. Martianus began with an account, heavily Pythagorean in tone, of the first ten numbers, explaining the virtues and associations of each, the deities with which they are connected, and their interrelationships. For example, 3

> is the first odd number [one doesn't count as odd for Martianus], and must be regarded as perfect. It is the first to admit of a beginning, a middle, and an end, and it associates a central mean with the initial and final extremes, with equal intervals of separation. The number three

represents the Fates and the sisterly Graces; and a certain Virgin who, as they say, "is the ruler of heaven and hell," is identified with this number. Further indication of its perfection is that the number begets the perfect numbers six and nine. Another token of its respect is that prayers and libations are offered three times. Concepts of time have three aspects; consequently, divinations are expressed in threes. The number three also represents the perfection of the universe. . . .[19]

Martianus proceeded to the classification of numbers and a discussion of what we would regard as their purely mathematical properties. He defined numbers as prime (integrally divisible by no number except the number 1) or composite; even or odd; plane and solid; perfect, deficient, or superabundant. Perfect numbers, for example, are those for which the sum of the factors equals the number ($1 + 2 + 3 = 1 \times 2 \times 3 = 6$); deficient numbers are those for which the sum of the factors is less than the number ($1 + 2 + 7 < 14$). Martianus also defined and classified various ratios or proportions. For example, the ratio of 8 to 6 is supertertius, because the first number is a third again larger than the second; and the ratio of 6 to 8 is subtertius by similar reasoning.

Martianus began his account of astronomy with a reference to Eratosthenes, Hipparchus, and Ptolemy—astronomers whose reputation he knew, but whose works he had undoubtedly never seen. His chapter on astronomy includes basic cosmological and astronomical information, probably drawn from Varro, Pliny, and other sources.[20] He defined the celestial sphere and its major circles. He described the zodiac, which he breaks into twelve signs of 30° each. He named and catalogued the major constellations. He identified the traditional seven planets and described their principal motions with more sophistication than was typical of handbook literature. For example, he revealed accurate knowledge of the approximate periods of their west-to-east motion around the ecliptic and a good grasp of the retrograde motions of the superior planets. One of the most interesting and influential features of this chapter is Martianus's discussion of the inferior planets, Mercury and Venus, to which he unambiguously assigned orbits that circle the sun, rather than the earth. "Venus and Mercury," he wrote, "do not travel about the earth at all; rather they encircle the sun." The text in question is ambiguous, however, on whether they circle the sun on concentric or intersecting orbits (see fig 7.5). Eleven hundred years later, Copernicus would cite Martianus in support of this feature of his own system.[21]

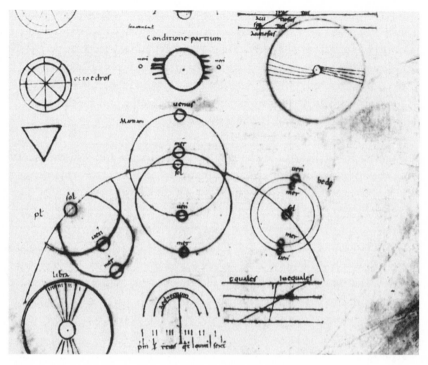

Fig. 7.5. Several attempts to capture Martianus Capella's theory of the motions of Venus and Mercury in relation to the sun. The drawing at center right places Venus and Mercury on nonintersecting, concentric, sun-centered orbits. The central drawing has both planets circling the sun, but on nonconcentric, intersecting orbits. From a ninth-century copy of Martianus's *Marriage of Philology and Mercury,* Paris, Bibliothèque Nationale, MS Lat. 8671, fol. 84r.

TRANSLATIONS

In the early years of cultural contact between Rome and its Greek neighbors (soon to become subjects), there was no problem of scholarly access. Widespread bilingualism, ample opportunity for travel or study abroad, and the easy availability of Greek teachers provided educated Romans with the means by which to participate in the Greek intellectual tradition. For those of lesser linguistic accomplishment or more modest aspirations, there were popularizing works in Latin and a few translations. Among the latter we have already noted Cicero's translations of Plato's *Timaeus* and *The Phaenomena* of Aratus.

Toward the end of the second century A.D., the conditions that had favored scholarship and learning began to weaken. Two centuries of peace and stability gave way, after the death of Emperor Marcus Aurelius (A.D. 180), to political turmoil, civil war, urban decline, and eventual economic disaster. Beginning about 250, attack and invasion from barbarians on the frontiers of the Empire became an additional threat. The results of this mix of events included the loss of economic and political vitality and the general deterioration of living conditions, particularly among the upper classes. Economic problems, exacerbated by what the Romans considered an inadequate supply of slave labor and general depopulation (as a result of plague, warfare, and a declining birth rate), contributed to the loss of leisure—the absolute necessity for serious scholarly endeavor. A further problem that affected scholarship in the West was diminishing communication with the Greek East. Near the end of the third century, and again in the fourth century, the Roman Empire was divided administratively into eastern and western halves. Increasingly, those two halves went their separate ways, and the Latin West gradually lost its vital contact with the Greek East, which was gradually transformed into what is now commonly known as the "Byzantine Empire." Under such circumstances, intellectual continuity between East and West was significantly weakened. In the interest of organizational clarity, we will follow developments in the Roman (or Latin) West for the present and return at the end of this chapter to the Greek East.

As East and West went their separate ways, bilingualism in the western regions of the Roman Empire declined, as did basic literacy, and the problem of access to Greek learning began to be felt. This is not to suggest that the break was complete, but only that the connection became thinner and more precarious. Several men in the later years of the Roman Empire, aware of this growing threat, attempted to lessen its impact by translating some of the more fundamental Greek philosophical literature into Latin. Two of these men are of particular importance for the history of science.[22]

About Calcidius, the first of the two, we know virtually nothing. Even his dates are uncertain, although several lines of argument suggest that he may have lived late in the fourth century.[23] At any rate, he translated Plato's *Timaeus* from Greek to Latin; and it was his version of this treatise, rather than Cicero's, that survived into the Middle Ages and became identified with medieval Platonism. Accompanying the translation was a long commentary, in which Calcidius drew on the doxographic tradition and a variety of late antique philosophers to explain and elaborate Plato's cosmological ideas.

The other translator, Boethius (480–524), lived well over a century later, after Rome had fallen under barbarian rule. Born into the Roman aristocracy, Boethius was active in affairs of state and was appointed to high political office in the regime of Theodoric the Ostrogoth; he was subsequently accused of treason and put to death. We do not know anything about Boethius's education, but his career bears testimony to the continuing existence of substantial fragments of the Greek intellectual tradition within the Roman senatorial class. Boethius set out, as he tells us, to make available to the Latins as many writings by Plato and Aristotle as he could lay his hands on, and also to reconcile their philosophies. He succeeded in translating a number of Aristotle's logical works (which became known collectively as the "old logic"), Euclid's *Elements*, and Porphyry's *Introduction to Aristotle's Logic*. In addition, Boethius wrote handbooks, based on Greek sources, on several of the liberal arts, including arithmetic and music.[24]

By the time Boethius was put to death in 524, the Latin West had been largely cut off from original Greek science and natural philosophy. It possessed Plato's *Timaeus*, some of Aristotle's logical works, and a few other bits and pieces—none of which, in all probability, had a very wide circulation. Beyond that, its knowledge of the Greek achievement existed mainly in the form of commentaries, handbooks, compendia, and encyclopedias. Rome had managed to preserve and transmit the Greek classical tradition only in a thin and limited version.

THE ROLE OF CHRISTIANITY

One piece of the picture has thus far been omitted from consideration. Christianity grew from a small Jewish sect in a remote corner of the Roman Empire into a major religious force in the third century and the state religion by the end of the fourth. This book is not the appropriate place to investigate the details of that extraordinary development.[25] What is important for our purposes is the fact that Christianity came to play a powerful religious role in the late Roman Empire. From this fact follows the question that we must now take up—namely, how did the presence and influence of the Christian church affect knowledge of, and attitudes toward, nature? The standard answer, developed in the eighteenth and nineteenth centuries and widely propagated in the twentieth, maintains that Christianity presented serious obstacles to the advancement of science and, indeed, sent the scientific enterprise into a tailspin from which it did not recover for more than a thousand years. The truth, as we shall see, is dramatically different, far more complicated, and a great deal more interesting.[26]

One of the charges frequently leveled against the church is that it was broadly anti-intellectual—that the leaders of the church preferred faith to reason and ignorance to education. In fact, this is a major distortion. Although Christianity seems at first to have appealed to the poor and disenfranchised, it soon reached out to the upper classes, including the educated. Christians quickly recognized that if the Bible were to be read, literacy would have to be encouraged; and in the long run Christianity became the major patron of education in the Latin West and a major borrower from the classical intellectual tradition. Naturally enough, the kind and level of education and intellectual effort favored by the church fathers was that which supported the mission of the church as they perceived it. But this mission, interestingly, did not include the suppression of scientific investigations and ideas.

When the church developed a serious intellectual tradition of its own, as it did in the second and third centuries, the driving forces were defense of the Christian faith against learned opponents (an enterprise known as "apologetics") and the development of Christian doctrine. For such purposes, the logical tools developed within Greek philosophy proved indispensable. Furthermore, aspects of Platonic philosophy seemed to correlate nicely with, and therefore to support, Christian teaching. For example, Plato had staunchly defended divine providence and the immortality of the soul; better yet, Plato's Demiurge looked very much like a monotheistic answer to the multiple gods of pagan polytheism; and this Demiurge could, with only a little stretching, be viewed as the Christian creator-God. Thus in the second and third centuries we find a series of Christian apologists putting Greek philosophy, especially Platonism, to Christian use.[27]

But this development did not please everybody. Some Christians regarded the Greek philosophical tradition as a source of error rather than truth. For every Plato, author of a philosophy compatible with Christian theology, there were an Aristotle and an Epicurus, whose worldviews were diametrically opposed to Christian doctrine on points of critical importance. Tertullian (ca. 155–ca. 230), a native of Carthage in Roman Africa, denounced philosophy as a source of heresy and warned against those who try to construct Christian doctrine out of Stoic and Platonic materials. However, a more typical attitude was that of Augustine (354–430), another North African, who expressed ambivalent attitudes toward Greek philosophy and science—fearing them as fomenters of heresy but also defending them as the best, if not the only, way of learning about the natural world. Philosophy, in Augustine's influential view, was to be the handmaiden of religion—not to be stamped out, but to be cultivated, disciplined, and put to use.

Now natural philosophy could not be separated from the rest of philosophy and therefore shared the fate of the larger whole of which it was a part. Like philosophy generally, it received mixed reviews from the intellectual leaders of the early church, ranging from mistrust and dislike to appreciation and enthusiasm—the same spectrum of opinion as we find in pagan circles. Augustine, who did so much to determine medieval attitudes, admonished his readers to set their hearts on the celestial and eternal, rather than the earthly and temporal. Nonetheless, he acknowledged that the temporal could serve the eternal by supplying knowledge about nature that would contribute to the proper interpretation of Scripture and the development of Christian doctrine. And in his own works, including his theological works, Augustine displayed a sophisticated knowledge of Greek natural philosophy. Natural philosophy, like philosophy more generally, was to serve handmaiden functions.[28]

Whether this represents a blow against the scientific enterprise or modest, but welcome, support depends largely on the attitudes and expectations that one brings to the question. If we compare the early church with a modern research university or the National Science Foundation, the church will prove to have failed abysmally as a supporter of science and natural philosophy. But such a comparison is obviously unfair. If, instead, we compare the support given to the study of nature by the early church with the support available from any other contemporary social institution, it will become apparent that the church was the major patron of scientific learning. Its patronage may have been limited and selective, but limited and selective patronage is a far cry from opposition.

However, a critic determined to view the early church as an obstacle to scientific progress might argue that the handmaiden status accorded to natural philosophy is inconsistent with the existence of genuine science. True science, this critic might maintain, cannot be the handmaiden of anything, but must possess total autonomy; consequently, the "disciplined" science that Augustine sought is no science at all. In fact, this complaint misses the mark: totally autonomous science is an attractive ideal, but we do not live in an ideal world. And many of the most important developments in the history of science have been produced by people committed not to autonomous science, but to science in the service of some ideology, social program, or practical end; for most of its history, the question has not been whether science will function as handmaiden, but which mistress it will serve.

ROMAN AND EARLY MEDIEVAL EDUCATION

One of the ways in which the church became a patron of learning was through the creation and support of schools. We have already touched upon education in Rome; let us look more closely at the schools in the Roman West, then at the early medieval schools that replaced them.[29]

Rome offered multiple levels of education—the number of students falling off rapidly as they moved upward toward the higher levels. Elementary education was generally a function carried on in the home, presided over by a parent or a tutor, who taught the child (beginning at about age seven) to read, write, and calculate. Organized primary schools also existed for those who needed or preferred them. The education of girls ceased at this point; if a boy were destined for additional education, he would be sent at about age twelve to study Latin grammar and literature (especially poetry) with a grammarian. The study of literature imparted not only writing skills and a knowledge of literary forms, but also, through the content of the works studied, a broad cultural education. Subsequent study, at about age fifteen, called for the skills of the rhetorician in a school of rhetoric. Here the student prepared for a career in politics or law by mastering the theory and techniques of public speaking. To go beyond this level of education—a rare achievement—was to engage in advanced study with a philosopher; this was possible for those of exceptional means or ambition, but it was done exclusively in Greek. Natural philosophy and the mathematical sciences would receive only limited attention in these educational settings: they would probably make an appearance in the teaching of the grammarian or rhetorician; they might figure a little more prominently in the teaching of the philosopher. Seldom would instruction surpass the level achieved in Martianus Capella's *Marriage of Philology and Mercury*.

Roman education began as private enterprise, dependent on the initiative of parents and teachers. Schools were held in a variety of physical settings, including homes, rented shops, public buildings, and the open air. Eventually, a system of municipal and imperial support developed, which established paid positions for teachers in most of the major cities, not only in Italy but also in provinces such as Spain, Gaul, and North Africa. Paid positions were provided for grammarians and rhetoricians, and on occasion for philosophers as well. At its zenith, Rome boasted an educational system that provided an impressive measure of educational opportunity for members of the upper class throughout the Empire.

As the West Roman Empire declined, so did its educational program. Invasion, civil disorder, and economic collapse brought deterioration of the conditions that had favored schools and education. Particularly critical were the loss of urban vitality and decline in the size, affluence, and influence of the upper classes from which the schools had always drawn their support. Disinterest and neglect by the Germanic tribes that overran the Empire in the fourth and fifth centuries were significant contributing factors. However, deterioration was gradual rather than precipitous, particularly in regions bordering the Mediterranean. Roman Britain and northern Gaul quickly lost contact with the classical tradition, but schools and intellectual life continued to exist (if not to thrive) in Rome, northern Italy, southern Gaul, Spain, and North Africa.

The relationship of Christianity to the demise of classical education is an extremely difficult and complex problem. As we have seen, there were leaders of the church who were profoundly worried about the pagan content of the classical tradition and who denounced the schools as a threat. The literature studied in the schools was frequently polytheistic and, by Christian standards, immoral; it certainly did not have the edifying qualities of, say, the Psalms or Jesus' Sermon on the Mount. We might, therefore, expect the church to have moved quickly to establish an alternative Christian educational system; or if it did not, then when Christianity became the state religion we would expect the pagan schools to have been radically transformed into Christian institutions. However, neither event occurred. The fact is that the majority of the early church fathers valued their own classical educations and, while recognizing its deficiencies and dangers, could conceive of no viable alternative to it; consequently, instead of repudiating the classical culture of the schools, they endeavored to appropriate it and build upon it. Substantial numbers of Christians continued to send their children to Roman schools; and educated Christians participated in those schools as teachers of grammar, rhetoric, and philosophy (much as religious people participate in modern secular education), no doubt allowing their Christian beliefs and sentiments to influence the curriculum to a degree, but not departing fundamentally from the classical tradition. As for the clergy, they were drawn from people who had already completed grammatical, and perhaps rhetorical, studies; their theological and doctrinal education would then take place informally, through a process of apprenticeship, or possibly in an episcopal school run by a bishop for the training of converts and prospective clergy.

But participation in the schools was not the same as unqualified enthusiasm and wholehearted support. The church remained ambivalent and divided on the value and appropriateness of classical education and, though willing to use the schools, was not likely to go out of its way to save them from the various forces pushing them towards extinction, especially if an acceptable alternative should present itself. Such an alternative emerged in the fifth century as a by-product of monasticism.

Christian monasticism appeared in the West during the fourth century. Monasteries spread rapidly, providing retreats for Christians who wished to withdraw from the world in the pursuit of holiness. In the sixth century St. Benedict (d. ca. 550) established a monastery at Monte Cassino, to the south

Fig. 7.6. A monk in the monastery library. Florence, Biblioteca Medicea Laurenziana, Codex Amiatinus (7th–8th c.).

of Rome, and drew up rules governing the lives of the monks who settled there—rules that came to be widely adopted within Western monasticism. The Benedictine Rule dictated all aspects of the life of the monk or nun, obliging them to devote the major share of their waking hours to worship, contemplation, and manual labor. Worship included reading the Bible and devotional literature, and that required literacy. The Benedictine Rule also mandated books, tablets, and writing implements for all monks and nuns. Since monasteries accepted young children (committed by their parents to the monastic life), the monastery was obliged to teach them to read—though in the early centuries of monasticism this rarely, if ever, occurred in a formal monastery school. Monasteries also developed libraries and scriptoria (rooms where books needed by the monastic community were produced by copyists).[30]

Initially the education that occurred within the monastery was intended solely to meet the internal needs of the monastic community. It was directed by the abbot or abbess or an educated monk or nun, and it was designed to provide the literacy required for the religious life and thus, ultimately, to promote spirituality. It has frequently been claimed that as the classical schools disappeared, monasteries felt growing pressure from the local gentry and nobility to provide education for their young—that is, for children who were not destined to become monks or nuns—and that monasteries established "external schools" for this purpose. In fact, there is no evidence before the ninth century for the existence of external schools in the monasteries; and thereafter the practice seems to have been exceedingly rare. If we find men with a monastic education performing administrative functions in church and state, this is not because monasteries set out to educate the lay public through external schools; it is because lay students were sometimes admitted to the internal monastic schools, but more especially because the monasteries contained a pool of talent (educated for monastic purposes) that could be tapped for service outside the monastery.[31]

Another question disputed among historians is the degree to which classical learning entered the monasteries—a dispute deriving, perhaps, from differences between monasteries or between medieval writers who addressed the subject of monastic learning. What seems clear is that the monasteries stressed spiritual development and whatever was thought to contribute to that. The Bible was the core of the educational program; biblical commentaries and devotional writings supplemented the biblical text. Classical pagan literature, widely judged to be irrelevant or dangerous, was not prominent. But there were many exceptions; indeed, we often find the use of pagan sources by the very people who were denouncing them. Augustine's admonition for Christians to borrow what is true and useful from pagan literature seems often to have been heeded, and an examination of writings emanating from the monasteries reveals a surprisingly extensive, if selective, knowledge of ancient sources. The mathematical arts of the quadrivium were rarely pursued beyond the most elementary level, but exceptions to this generalization are not hard to find.

A good illustration of the penetration of classical learning into the monasteries is to be found in Ireland from the sixth century onward (a circumstance for which we have no adequate historical explanation). Here we find significant attention given to the classical pagan authors. Some Greek was known, and the mathematical arts of the quadrivium (particularly as applied to the calendar) were well developed.[32]

Another impressive exception to monastic apathy toward classical education was the monastery of Vivarium, founded by a member of the Roman senatorial class, Cassiodorus (ca. 480–ca. 575), upon his retirement from public life. Cassiodorus established a scriptorium in his monastery, arranged for the translation of Greek works into Latin, and made study an essential part of the routine of his monks. He also wrote a handbook of monastic studies, in which he recommended a surprisingly large collection of pagan authors and briefly discussed each of the seven liberal arts. That this was more than lip service is revealed by a treatise on the calendar (still in existence) that appears to have been written at Vivarium during Cassiodorus's lifetime. It is clear that Cassiodorus shared the universal monastic view that secular studies were to be pursued only insofar as they served sacred purposes; where he differed from other leaders of the monastic movement was in his opinion of the range of secular studies capable of making such a contribution.[33]

These exceptions are important; but they do nothing to overturn the generalization that monasteries were dedicated to spiritual pursuits. Learning was cultivated, but only insofar as it contributed to religious ends. Science and natural philosophy were extremely marginal to this enterprise—although not entirely absent. What then is the significance of monasticism for the history of science, and why am I devoting space to it in this book? Was this not the "dark age" of the history of science—an age during which nothing of significance transpired?

There is no question that knowledge of Greek natural philosophy and mathematical science had fallen off precipitously, and few original contributions to it appeared in Western Europe during the early centuries of the medieval period (roughly 400–1000). If we are looking for new observational data

Fig. 7.7. A medieval scribe at work. Oxford, Bodleian Library, MS Bodley 602, fol. 36r (13th c.).

or telling criticism of existing theory, we will find little of it here. Creativity was not lacking, but it was directed toward other tasks—survival, the pursuit of religious values in a barbaric and inhospitable world, and even (on occasion) exploration of the extent to which knowledge about nature was applicable to biblical studies and the religious life. The contribution of the religious culture of the early Middle Ages to the scientific movement was thus primarily one of preservation and transmission. The monasteries served as the transmitters of literacy and a thin version of the classical tradition (including science or natural philosophy) through a period when literacy and scholarship were severely threatened. Without them, Western Europe would not have had more science, but less.

TWO EARLY MEDIEVAL NATURAL PHILOSOPHERS

It may be worthwhile to conclude this examination of science within the Latin West with a pair of illustrations of the early medieval contribution to science or natural philosophy—more specifically, to call attention to two men whose names have become synonymous with early medieval natural philosophy and the medieval worldview.

Isidore of Seville (ca. 560–636) was raised in Spain, then under Visigothic rule, and educated by his older brother (perhaps in a monastic or an episcopal school) before succeeding that brother as archbishop of Seville in the year 600. He was the outstanding scholar of the late sixth and early seventh centuries and illustrates the relatively high level of learning and culture available (but certainly not common) in Visigothic Spain during his lifetime. Isidore's works range widely over biblical studies, theology, liturgy, and history. He wrote two books of special interest to the historian of science: *On the Nature of Things* and *Etymologies*. These works, based on both pagan and Christian sources (including Lucretius, Martianus Capella, and Cassiodorus), communicate a brief, superficial version of Greek natural philosophy. The *Etymologies*, which exists in more than a thousand manuscripts (one of the most popular books of the entire Middle Ages), offers an encyclopedic account of things by way of an etymological analysis of their names. It covers the seven liberal arts, medicine, law, timekeeping and the calendar, theology, anthropology (including monstrous races), geography, cosmology, mineralogy, and agriculture. Isidore's cosmos was geocentric, composed of the four elements. He believed in a spherical earth and revealed an elementary understanding of the planetary motions. He gave an account of the zones of the celestial sphere, the seasons, the nature and size of the sun and moon, and the cause of eclipses. One of the notable features of his natural philosophy is his vigorous attack on astrology.[34]

If there is a certain vagueness regarding Isidore's intellectual formation, that of the Venerable Bede (d. 735) is known in considerably greater detail. At the age of seven, Bede entered the monastery of Wearmouth in Northumbria (northeastern England, near modern Newcastle) and there spent the remainder of his life studying and teaching, first as a student in the monastic school and eventually as the monastic schoolmaster. The monasteries of Northumbria were the direct offspring of Irish monasticism and thus inherited the Irish concern for quadrivial studies and the classics, but they were also in touch with the best of contemporary continental scholarship. Bede, doubtless the most accomplished scholar of the eighth century, wrote on the whole range of monastic concerns, including a series of textbooks for monks. He is best known for his *Ecclesiastical History of the English People*. He also wrote a book, *On the Nature of Things* (based especially on Pliny and Isidore), and two textbooks on timekeeping and the calendar. In the latter, designed to regulate the daily routine of the monks and teach them how to determine the religious calendar, Bede made the most of the limited astronomical knowledge at his disposal and existing treatises on calendrics, to lay a solid foundation for what came to be called the science of "computus," establishing principles of timekeeping and calendar control that were eventually adopted throughout Christendom.[35]

Isidore and Bede are fitting representatives of the tradition of popularization and preservation that we have been tracing in this chapter—men who struggled to preserve the remnants of classical learning and pass them on, in usable form, to the Christian world of the Middle Ages. But is this tradition worthy of the attention we have given it? Does it merit a chapter in a book on the history of early science? If the history of science were simply the chronicle of great scientific discoveries or monumental scientific thoughts, Isidore and Bede would have no place in it; no scientific principles circulate today under their names. However, if the history of science is the investigation of the historical currents that converged to bring us to the present scientific moment—the strands that must be grasped if we are to understand where we came from and how we got here—then the enterprise in which Isidore and Bede were engaged is an important part of the story. Neither Isidore nor Bede was a creator of new scientific knowledge, but both restated and preserved existing scientific knowledge in an age when the study of nature was a marginal activity. They provided continuity through a dangerous and difficult period; in so doing, they powerfully influenced for centuries what Europeans knew about nature and how Europeans thought about nature. Such an achievement may lack the drama of, say, discovering the law of gravitation or devising the theory of natural selection, but to affect the subsequent course of European history is no mean contribution.

LEARNING AND SCIENCE IN THE GREEK EAST

While the classical tradition was slowly declining in the Latin West and natural philosophy was settling into her position as handmaiden of theology and religion, what was happening in the Greek-speaking East?[36] Although the East experienced many of the same misfortunes as the West—invasion, economic decline, and social upheaval—the consequences were less severe. A higher level of political stability was achieved, as the eastern half of the old Roman Empire gradually separated itself from the West, giving rise (eventually) to what we now call Byzantium or the Byzantine Empire, with its capital in Constantinople (present-day Istanbul). That the city of Constantinople did not fall to invaders (Latin crusaders) before 1204, whereas Rome was sacked (by Alaric and the Visigoths) as early as 410, tells us something about the relative levels of stability. Greater social and political stability meant greater continuity in the schools. The tradition of classical studies in the schools was fed by the steady copying of ancient works, as well as translation of new works; and, of course, the East never found itself separated from the original sources of Greek scholarship by a linguistic barrier.

The history of Byzantine science belongs to the history of Byzantine education, with its variable fortunes between Greek antiquity and the Ottoman conquests of the fifteenth century. Schools operated at various levels. At one end of the spectrum, elementary education was organized in theory (but with considerable variety) around the traditional seven liberal arts—the trivium (grammar, rhetoric, and logic) and quadrivium (arithmetic, geometry, astronomy, and music). At the other end of the spectrum, advanced philosophical or theological studies with a master were available in major cities, including Athens, Alexandria, and Constantinople. Despite innumerable variations, education at all levels was oriented toward preservation of existing knowledge in the form of ancient literary classics. This approach to education was reinforced by the civil authorities, who, when they sought literate civil servants, valued erudition and encyclopedic knowledge of the classics over practical or technical expertise.[37]

The works of Plato, Aristotle, and leading Neoplatonists were available, either directly or through commentaries. The fortunes of all three traditions were closely connected to the growth of Byzantine Christianity (Christianity was the official religion of the Roman Empire by the end of the fourth century), which compelled a variety of compromises, but without wholesale subordination of philosophy and the sciences to theology. The most common genre of scholarly writing was commentary on an ancient text, typically employing the archaic structure and vocabulary of the classical period, thereby creating a linguistic barrier between scholarly writing and the language spoken in the marketplace.

Themistius (d. ca. 385), who taught philosophy in Constantinople and served as tutor to the imperial offspring, wrote influential paraphrases and summaries of a variety of Aristotelian works, including the *Physics, On the Heavens*, and *On the Soul*. Simplicius (d. after 533), an Athenian Neoplatonist determined to reconcile Platonism and Aristotelianism, wrote intelligent commentaries on these same three works. And John Philoponus (d. ca. 570), a Christian Neoplatonist who taught in Alexandria, wrote commentaries on Aristotle's *Physics, Meteorology, On Generation and Corruption*, and *On the Soul*. In these commentaries he attempted, in conscious opposition to his contemporary Simplicius, to demonstrate the profound errors propagated by Aristotle, including the celestial-terrestrial dichotomy and the notion of an eternal universe. He also offered a systematic and original refutation of Aristotle's theory of motion, denying Aristotle's explanation of projectile motion and the claim that heavy bodies fall through a medium with speeds proportional to their weights. Through the eventual translation of their works into Arabic and Latin, all three philosophers—Themistius, Simplicius, and Philoponus—helped to determine the subsequent course of Aristotelian natural philosophy.[38]

Instruction in the quadrivial disciplines was elementary, centered on a small number of ancient texts and their commentaries. Calculation was by finger-reckoning; a decimal system was employed, and numbers continued to be represented by letters of the Greek alphabet until Hindu-Arabic numerals made their first appearance in the thirteenth century; even then, the older system was slow to disappear.[39] For geometry the basic text was Euclid's *Elements*, along with its commentaries. Also contributing to geometrical literature were Hero of Alexandria (fl. A.D. 60), known for treatises on practical geometry and the invention of mechanical devices. In the late fifth and early sixth centuries, the Byzantine Empire experienced a surge of geometrical expertise. Eutocius of Ascalon (b. ca. 480) commented on works of Archimedes (ca. 287–ca. 311 B.C.) and the *Conics* of Apollonius of Perga (fl. 200 B.C.). About the same time, Anthemius of Tralles (d. 534) and Isidore of Miletus (fl. 520) demonstrated their geometrical talent as architects for the magnificent Christian church (later Islamic mosque) in Constantinople, Hagia Sophia (completed in 537). Anthemius also wrote a book on paraboloidal burning mirrors, and Isidore, along with a circle of like-minded scholars, took an interest in Archimedes' works and their preservation. The pay-off of this latter activity was the translation of several Archimedean treatises from classical Greek to the current vernacular, thereby contributing to their survival.[40]

Geographical knowledge, often viewed as an offspring of geometry, was based primarily on the *Geographia* of the Roman geographer Strabo (ca. 64 B.C.–A.D. 21). Byzantium saw few geographical or cartographic achievements until the recovery of Ptolemy's *Geography* about 1295 by Maximus Planudes (ca. 1260–1310)—a recovery that revolutionized Byzantine geography and cartography, as it revolutionized Western European geography and cartography when it was translated into Latin about a century later.[41]

Before we leave the subject of Byzantine geography, we must take note of a final Byzantine author: Cosmas Indicopleustes (fl. 540), a self-educated, widely traveled merchant. Converted to Christianity, he was persuaded by certain biblical passages, literally interpreted, that the earth was flat—a flat rectangle, covered with a vault containing the stars and planets. He developed this theory in a book, *Christian Topography*, in which he employed a variety of arguments in defense of his theory of the flat earth. Cosmas was not particularly influential in Byzantium, but he is important for us because he has been commonly used to buttress the claim that all (or most) medieval people believed they lived on a flat earth. This claim (as readers of this book must know by now) is totally false. Cosmas is, in fact, the *only* medieval European known to have defended a flat earth cosmology, whereas it is safe to assume that all educated Western Europeans (and almost one hundred percent of educated Byzantines), as well as sailors and travelers, believed in the earth's sphericity. The myth of pre-Columbian belief in a flat earth, finally laid to rest by Columbus, was the invention of the American essayist Washington Irving, writing in the 1820s.[42]

The history of astronomy in the Byzantine Empire is the history of a pair of books by Ptolemy—his *Almagest* and *Handy Tables*—and the many commentaries on them. Theon of Alexandria (fl. 375) was the best known of the commentators—though probably less famous than his daughter, Hypatia, a superb mathematician and philosopher, who was attacked and killed for her pagan beliefs by a mob of Christians.[43] Byzantine astronomy was enriched in the eleventh century by the acquisition of Islamic and Persian astronomical writings of greater sophistication and accuracy than previous Greek astronomical fare: new and more accurate astronomical data (including an adaptation of the Toledan tables), new and better methods for calculating astronomical events, and an important astronomical instrument—the astrolabe.[44]

A final contribution of Byzantine astronomers to the astronomical tradition came through the efforts of Cardinal Johannes Bessarion (1403–72). Reunion of the Roman Catholic and Greek Orthodox churches after centuries of schism brought Bessarion to Vienna in the 1460s, where he became the friend and patron of two young professors of astronomy: Georg Peurbach (1423–61) and Peurbach's student Johannes Regiomontanus (1436–76). Motivated by the fall of Constantinople to the Turks in 1453, Bessarion campaigned to save as much of the Greek intellectual legacy as possible. One result was a Latin *Epitome of the Almagest*, produced by Peurbach and Regiomontanus—essentially a commentary that improved on all previous commentaries on Ptolemy's *Almagest* and "provided Copernicus with a crucial conceptual stepping stone to the emergence of the heliocentric theory."[45]

A modest level of Byzantine scientific effort was extended to areas such as botany, medicine, and zoology. Perhaps its greatest contribution was preservation of the *De materia medica* of Dioscorides (first century A.D.), containing careful descriptions of more than six hundred plants, along with a variety of animal products, and close to one hundred minerals. Various zoological compositions existed, including a commentary on Aristotle's treatises on animals. But the most widely circulated zoological manual was an anonymous Christian bestiary, the *Physiologus* (ca. A.D. 200)—a collection of animal lore and mythology (discussed more fully below, chap. 13). Finally, a Byzantine contribution that has benefited humankind ever since: hospitals as institutions offering medical care and the possibility of a cure, rather than merely a place to die, appeared in Byzantium as early as the fourth century.[46]

What do we make of all of this? As the world of ancient Greece and Rome declined in late antiquity, it split into a pair of channels—two parallel intellectual streams. In the beginning, Byzantium certainly had an intellectual advantage; but the West leapt ahead with the massive translation efforts from Greek and Arabic into Latin in the eleventh and twelfth centuries. The intellectual cultures characteristic of the two streams also varied, the Byzantines emphasizing preservation of the ancient philosophical and scientific legacy while Latin Christendom (beginning in the twelfth century) was attempting to assimilate an avalanche of new materials. Occasional leakage between the two channels brought intellectual goods from west to east and vice versa. Byzantium delivered several treasures to the West: important criticisms of Aristotelian physics by John Philoponus, the magnificent herbal of Dioscorides, and Ptolemaic geography and astronomy. The latter, in the form of Peurbach and Regiomontanus's *Epitome of the Almagest*, exercised a strong influence on Nicolaus Copernicus and, through him, on the restructuring of the European cosmos. So both traditions contributed, each in its own way, to preservation of the classical tradition—thereby delivering to succeeding generations the legacy that would serve as a foundation and furnish many of the resources (factual and theoretical) that would be deployed to produce the flourishing scientific movement of the sixteenth and seventeenth centuries.[47]

8 ✿ *Islamic Science*

EASTWARD DIFFUSION OF GREEK SCIENCE

Although Greek influence had long extended beyond the Greek homeland, cultural diffusion on a grand scale began with the military campaigns of Alexander the Great.[1] When Alexander conquered Asia and North Africa (334–323 B.C.), he not only acquired territory but also established beach-heads of Greek civilization. His campaigns took him as far south as Egypt, as far east as Bactria in Central Asia (now the far northeastern reaches of Afghanistan) and beyond the Indus River into the northwestern corner of India (see map 4.1). Behind him he left garrisons and dozens of cities named "Alexandria"; successful efforts at colonization enlarged the Greek presence, and in the long run these cities became centers from which Greek culture could emanate into the surrounding regions. Among the notable centers of Greek culture thus established were Alexandria in Egypt and the Kingdom of Bactria in Central Asia.

But conquest and colonization were not the only mechanisms of diffusion. Religion also played an important role. In the millennium after Alexander's conquests, his Asian territories (especially present-day Syria, Iraq, and Iran) proved to be fertile ground for a variety of religious movements. At one time or another Zoroastrianism, Christianity, and Manicheism contended with each other for converts; all three were based on sacred books and thus, of necessity, cultivated literacy and learning. Christianity and Manicheism, in particular, had acquired Greek philosophical underpinnings and thus contributed to the Hellenization of the region.

In the first few centuries of the Christian era, missionary activity led to the establishment of Christian churches through a wide region of western Asia. In the fifth and sixth centuries, reinforcements arrived in the form of dissident Christian sects seeking refuge from persecution. The Christianization of the Roman Empire in the fourth century had led to a series of bitter, sometimes violent, theological disputes and rifts within the Byzantine church. The mid-fifth-century disputes of relevance for our purposes concerned the relationship between Christ's humanity and divinity. Struggle between two factions led to conflict, schism, condemnation, two church councils, and a migration of Nestorian Christians (followers of Nestorius, patriarch of Constantinople) to theologically friendlier territory. In Nisibis, already an important Christian center, just over the Persian border to the east, they established a center of theological higher education, with a curriculum that included Aristotelian logic, owing to its relevance for theological argument. Nisibis later became a center for translation of theological and philosophical texts from Greek to Syriac.[2]

From this foothold in Persia, the Nestorians managed, in the next century, not only to shape Persian Christianity, but also to exercise a broad influence on Persian intellectual life. They managed to acquire positions of power and influence and to impart a taste for Greek culture to the Persian ruling class. We see the results of these developments in the invitation issued by the Persian king Khusraw I about 531, to the philosophers from the Academy in Athens (which had been shut down by a decree of the Byzantine emperor Justinian) to settle in Persia. This same Khusraw is reputed to have been knowledgeable in Platonic and Aristotelian philosophy and to have had Greek philosophical works translated for his use; Nestorian connections are revealed in his treatment by a Nestorian physician. Khusraw II (590–628) had two Christian wives—one of them, at least, a Nestorian before her conversion to Monophysitism—and an influential physician-advisor who also vacillated between the Nestorian and Monophysite sects.[3]

An influential mythology has developed around Nestorian activity in the city of Gondeshapur (often written "Jundishapur") in southwestern Persia. According to the often-repeated legend, the Nestorians turned Gondeshapur into a major intellectual center by the sixth century, establishing what some enthusiasts have chosen to call a university, where instruction in all of the Greek disciplines could be obtained. It is alleged that Gondeshapur had a medical school, with a curriculum based on Alexandrian textbooks, and a hospital modeled on Byzantine hospitals, which kept the realm supplied with physicians trained in Greek medicine. Of greatest importance, Gondeshapur is held to have played a critical role in the translation of Greek scholarship into Near Eastern languages and, indeed, to have been the single most important channel by which Greek science passed to the Arabs.[4]

Recent research has revealed a considerably less dramatic reality. We have no persuasive evidence for the existence of a medical school or a hospital at

Gondeshapur, although there seems to have been a theological school and perhaps an attached infirmary. No doubt Gondeshapur was the scene of serious intellectual endeavor and a certain amount of medical practice—it supplied a string of physicians for the ʿAbbāsid court at Baghdad beginning in the eighth century—but it is doubtful that it ever became a major center of medical education or of translating activity. If the story of Gondeshapur is unreliable in its details, the lesson it was meant to teach is nonetheless valid. Nestorian influence, though not focused on Gondeshapur, did play a vital role in the transmission of Greek learning to Persia and ultimately to the Muslim empire. There is no question that Nestorians were foremost among the early translators; and as late as the ninth century, long after Persia had fallen to Islamic armies, the practice of medicine in Baghdad seems to have been dominated by Christian (probably Nestorian) physicians.[5]

But there is a linguistic shift here, which we must also take into account. Although the content of the education available in Nisibis, Gondeshapur, and other Nestorian centers was predominantly Greek, the language of instruction was not. Teaching was in Syriac (a dialect of Aramaic, the lingua franca of the Near East), adopted by the Nestorians as their literary and liturgical language. Consequently, the teaching program required the translation of Greek texts into Syriac. Such translations were made at Nisibis and other centers of learning, beginning as early as 450. Aristotle's basic logical works and Porphyry's commentary on them were among the earliest translated and the most studied. Some medical literature and elementary mathematical and astronomical treatises were eventually rendered as well.

Several points merit emphasis. First, let us be clear that this is a story about the *transmission* of learning. Our subject (in this section) is not original contributions to natural philosophy, but the preservation and eastward diffusion of the Greek heritage into Asia, where it would subsequently be absorbed into Islamic culture. Second, this process of cultural diffusion was quite slow, but also of very long duration, occupying a period of nearly a thousand years, from the Asian conquests of Alexander the Great (335 B.C.) to the founding of Islam in the seventh century A.D. Third, the story must not be oversimplified to the point where the diffusion of Greek learning is viewed as hanging on the slender thread of Nestorian activity in the city of Gondeshapur or any other religious or ethnic group dwelling in any other specific place. Rather, we must see this as a widespread movement of cultural diffusion, whereby the aristocracies of western and central Asia assimilated broadly and deeply, and by a variety of mechanisms, the fruits of Greek culture. The next step was the further diffusion of those fruits to Islam.

THE BIRTH, EXPANSION, AND HELLENIZATION OF ISLAM

The Arabian peninsula, wedged between Persia to the north and east and Egypt and the Red Sea to the west, had not been touched by Alexander's military campaigns, nor much affected by Byzantine territorial ambitions. Jewish and Christian communities had flourished for a time in the south, but by the seventh century their influence had diminished to a modest level. Except on the southern and northern edges, the population was largely nomadic, although cities had been established around pilgrimage sites and along the major trading routes. It was in one of these cities, Mecca, that Muḥammad was born about 570 (by the Christian calendar) and from which he preached the new religion of Islam. Muḥammad had a series of revelations delivered by the angel Gabriel. The contents of these revelations were dictated to secretaries and presented orally by Muḥammad to his band of followers. Collected by followers, they formed the Koran (or Qur'ān), the holy book of Islam. Its central theme was the existence of a single omnipotent, omniscient god, Allāh, creator of the universe, to whom the faithful (called "Muslims") must submit. Along with religious, legal, and historical literature that grew up around it, the Koran came to define all aspects of Islamic faith and practice; it was the source of later Islamic theology, morality, law, and cosmology, and thus the centerpiece of Islamic education.[6]

Within a few years of Muḥammad's death in 632, his successors had, by military means, extended their rule over the tribes occupying large portions of the Arabian peninsula and begun to extend their reach to the north. They rapidly conquered Byzantium and Persia, which had been weakened by generations of warfare against each other, thus gaining control of major portions of the Near East. In twenty-five years of stunning military success, Islam subjugated almost the whole of Alexander's Asian and North African conquests, including Syria, Palestine, Egypt, and Persia. Within a century, the remainder of North Africa and almost the whole of Spain fell to Muslim armies (map 8.1).

Muḥammad left no male heir or designated successor; consequently leadership of the developing Islamic empire became a matter of bloody dispute. The first caliphs ("successors" of Muḥammad) were chosen from Muḥammad's early followers. In 644 'Uthmān of the Umayyad family became caliph, succeeded in 661 by his cousin Mu'āwiyah, a former governor of Syria. In the interests of security, Mu'āwiyah and his successors ruled from Damascus in Syria, where Umayyad strength was concentrated. Here the Umayyad dynasty,

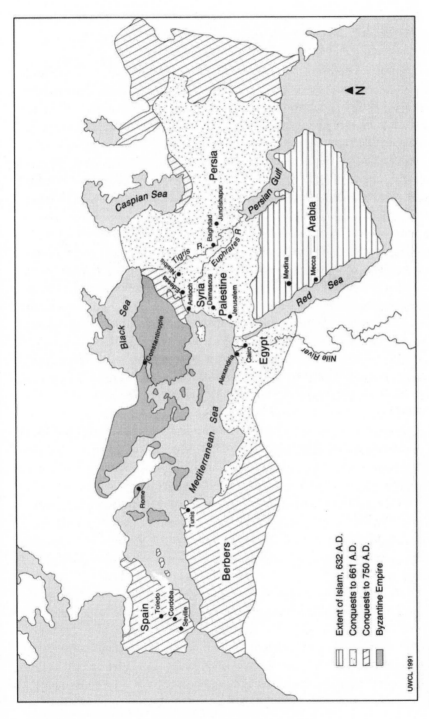

Legend:

☐ Extent of Islam, 632 A.D.

☐ Conquests to 661 A.D.

☐ Conquests to 750 A.D.

☐ Byzantine Empire

UWCL 1991

Map 8.1. Islamic expansion.

which held power for about a century, came into contact with educated Syrians and Persians, whom it used as secretaries and bureaucrats; and thus, on a small scale, began the Hellenization of Islam.

The process of Hellenization accelerated after 750. In that year a new dynasty, the 'Abbāsids (descended from Muḥammad's uncle, al-'Abbās), came to power, with the assistance of several revolutionary groups. The 'Abbāsid caliphs had no intention of remaining in Damascus, where Umayyad sentiment was still strong. In 762, al-Manṣūr (754–75) built a new capital, the city of Baghdad, on the Tigris River at a site chosen, presumably, for geographical, strategic, and astrological reasons. Al-Manṣūr's court in Baghdad was not famous for piety but cultivated a religious and political climate that was relatively intellectual, cosmopolitan, and tolerant.[7] More importantly, the Islamic empire was being transformed from tribal rule into a centralized state, which called for a much more substantial administrative bureaucracy than anything Muḥammad, his immediate successors, or the early Umayyads could have imagined. The staffing of this bureaucracy could hardly be accomplished from among the revolutionaries who had brought the 'Abbāsids to power, and the caliphs had no reasonable alternative to making use of educated Persians and Christians.

The Persian influence is especially apparent in the powerful royal advisors from the Barmak family—formerly from the province of Bactria and recent converts to Islam. Khālid ibn Barmak served al-Manṣūr; and his son Yaḥyā became vizier (chief advisor and tutor of the caliph's heirs) under al-Manṣūr's grandson, Hārūn al-Rashīd (786–809). The Christian influence is most clearly evident in the practice of medicine at court. In 765, al-Manṣūr was treated by a Nestorian physician from Gondeshapur, Jurjīs ibn Bukhtīshūʿ. Jurjīs was apparently successful, for he remained in Baghdad as the caliph's personal physician, becoming a powerful court figure; his son succeeded him, and for eight generations the Bukhtīshūʿ family held the post of court physician. Finally, it is important to note that there were also influences emanating from India in the east; some of these were the long-term result of the earlier Hellenization of northwestern India.

Before we proceed, a word about terminology. Our subject concerns a multireligious, multilingual, multicultural population. It follows that we need to choose our terms carefully when referring to the various peoples, their cultures, and their languages. To refer to the culture as a whole, I will employ the name of its dominant element—thus, "Islam" or "Islamic" (the religion or the culture; context will make clear which is intended). When I wish to refer to the adherent of any of the religions found within greater Islam, I will use the terms "Muslim," "Christian," "Jew," etc. I reserve the word "Arab" for people (soon a minority within Islam) who were descended from the indigenous tribes that had occupied the Arabian peninsula. Finally, I employ "Arabic" to denote the aboriginal language of the Arabs and of the Koran, which also became the primary language of scholarly communication within the Islamic world.

TRANSLATION OF GREEK SCIENCE INTO ARABIC

A crucial element in the Hellenization of Islamic culture was translation of the classical tradition from Greek and Syriac sources into Arabic. How did this extraordinary event occur? How could a tribal culture of limited literacy convert itself, within a century, into a stable political empire that valued learning to the point of investing in this enormous, unprecedented act of intellectual appropriation? An important factor is that the Arab conquerors and their descendants were quickly relegated to minority status as the wars of conquest brought educated members of the conquered peoples (especially Persians and Berbers) into the empire. But the wholesale acquisition and translation of Greek books does not automatically follow. A crucial result of the Islamic conquest was the breakdown of political barriers over a vast region, previously divided into separate political units, including Spain, all of north Africa, Egypt, the Arabian peninsula, Syria, Palestine, and the former Persian Empire as far east as Samarqand and Kabul. One result was a commercial revolution, which brought new ideas and new technologies, as well as new products, to Baghdad. The most important of the new technologies, for present purposes, was paper making, imported from China, which revolutionized the copying of manuscripts by providing an inexpensive alternative to the traditional parchment.

But most important, a unified Islamic state was able to remove the political boundaries that had prohibited free intellectual intercourse. Multilingual scholars were now free to travel. They could acquire books hitherto unavailable and rub shoulders, trade ideas, and debate with intellectual equals interested in similar subjects. This, Dimitri Gutas argues, "led to the transmission of knowledge without translation,"[8] and it explains the remarkable availability, when the need arose, of scholars with both the linguistic skills and the disciplinary knowledge to translate abstruse scholarly literature from Greek into Syriac and from Greek or Syriac into Arabic.

Translators worked in many different places and with a variety of motivations, but the center and initiator of this activity was clearly Baghdad. But why Baghdad? It was not, as we might have supposed, a simple matter of Baghdad's ruling class discovering Greek literature on a variety of practical topics and, on that account, commissioning translations. One very powerful factor was an antecedent tradition of serious intellectual activity. Baghdad was situated in the hinterland of the former Persian Empire, which had significant indigenous intellectual traditions, both religious (Zoroastrian) and secular. For example, an early history of Zoroastrian literature reports that the emperor Sāpūr I (A.D. 241–72) had "collected the non-religious writings on medicine, astronomy, movement, time, ... and other crafts and skills which were dispersed throughout India, the Byzantine Empire and other lands."[9] The immediate predecessors of the 'Abbāsids, the Umayyads, had established a precedent by cultivating the religious sciences, linguistics, and literary studies. Other sources reveal Persian interest in the Greek classics, including Aristotelian logic and Ptolemaic astronomy. In short, the 'Abbāsid court was rich with intellectual opportunity.

There were also political precedents. It was understood at the court of the first 'Abbāsid caliph, al-Manṣūr, that consolidation of the imperial status of his regime would be strengthened by portrayal of the 'Abbāsids as natural successors of the Persian regime (the Sasanids) that they replaced. And one way of accomplishing this was to appropriate Sasanian imperial ideology, in which the ruler was portrayed as a patron of learning—specifically as a patron of the translation of lost or neglected books.

This argument may need further refinement or correction. But what it reveals is the ability of subtle features in a culture to motivate a civilization-changing train of events. Besides the hard historical evidence, what gives this account plausibility is how well it explains the magnitude of energy and funding that were mobilized by the 'Abbāsid court in support of the translation of Greek books into Arabic, and how quickly it acquired the translators to carry out the project—translating in the next 250 years almost all of the scientific works of the classical tradition as we know them today.

Other motivations promoted the translation of specific genres. Astrology was employed not only for the casting of horoscopes but also for political purposes—because, as astrological history, it could be made to buttress 'Abbāsid claims to be the rightful successors of their Persian predecessors. Astrology, of course, required astronomy; and astronomy, in turn, required mathematics. Astronomy was also required for the regulation of Islamic ritual: determination of prayer times, the direction of Mecca, and the beginning of Ramadan (the month for fasting). Medicine was of obvious practical value—a lesson taught, no doubt, by the Nestorian (Christian) physicians who served the ʿAbbāsid court. One of the early commissions—by al-Mahdī, the second ʿAbbāsid caliph and son of al-Manṣūr—was for translation of Aristotle's *Topics* (on the art of argumentation), as an aid to disputation with infidels and successful proselytizing for converts to Islam.[10] Still others commissioned translations to demonstrate intellectual status or support a favorite intellectual pursuit.

The majority of translations were from Greek to Arabic, sometimes with Syriac (the liturgical language of Nestorian and Syriac Orthodox Christians) as intermediary. The translation of Greek and Syriac works into Arabic began under al-Manṣūr and became serious business under Hārūn al-Rashīd, who sent agents to Byzantium in search of manuscripts of important treatises. The required financial support came not merely from the caliphs, but also from courtiers, secretaries and other state functionaries, wealthy families, governors, generals, physicians, and other practicing philosophers and scientists. The Banū Mūsā (three sons of Mūsā) are an illuminating case. Sons of a father who had befriended the caliph al-Maʾmūn, they were raised at court, where they received a superior education that included mathematical subjects. The brothers also acquired a considerable fortune, substantial portions of which they devoted to scholarship and patronizing translation activity.[11]

Faithfully repeated at this point in nearly every historical account of the Greco-Arabic translations (including the first edition of this book) is the story of the *bayt al-ḥikma* (house of wisdom) in Baghdad, described as a research institute, founded by Hārūn's son, al-Maʾmūn (813–33), where massive translation was carried out. Dimitri Gutas has debunked this story as myth, noting that the Arabic expression *bayt al-ḥikma* can be applied to any library or book repository and that no translations can be traced specifically to the *bayt al-ḥikma* in Baghdad.[12] But if there was no translator's institute in Baghdad, the scholar alleged to have been its director, the great translator Ḥunayn ibn Isḥāq (808–73), was very real and does indeed represent Greco-Arabic translation at its best. Ḥunayn was a Nestorian Christian and an Arab, descended from an Arab tribe that had converted to Christianity long before the religion of Islam emerged. Ḥunayn, who studied medicine with the distinguished physician Ibn Māsawayh, was bilingual from childhood in Arabic and Syriac; as a young man he went to the "land of the Greeks," where he acquired a thorough mastery of Greek. Returning to Baghdad, he attracted the notice of a member of the Bukhtīshūʿ family and the Banū Mūsā, who introduced him to al-Maʾmūn. At some point Ḥunayn accompanied an expedition to

Byzantium, in search of manuscripts. He served as translator under several caliphs and finished his career as chief royal physician, replacing one of the Bukhtīshū‘.[13]

Ḥunayn's translating activity is of such critical importance as to deserve our careful attention. Ḥunayn was assisted by his son Isḥāq ibn Ḥunayn, his nephew Ḥubaysh, and others. Many of their translations were collaborative efforts. For example, Ḥunayn might translate a work from Greek to Syriac, after which his nephew would render the Syriac text into Arabic. Ḥunayn's son Isḥāq translated from both Greek and Syriac into Arabic and also revised the translations of his colleagues. And Ḥunayn, besides producing his own translations from Greek to Syriac or Arabic, seems to have insisted on checking the translations of his charges. Ḥunayn and his coworkers were extremely sophisticated in their methods. They understood the need to compare manuscripts whenever possible, in order to weed out errors. Ḥunayn spurned the common translating practice of mechanical, word-for-word substitution, which suffers from severe disadvantages: overlooking the syntactical differences between the languages and ignoring the fact that not every Greek word has a counterpart in Arabic or Syriac. His method was to grasp the meaning of a sentence in the original Greek and construct a replacement of equivalent *meaning* in Arabic or Syriac.

Most of Ḥunayn's translations were medical, with special emphasis on Galen and Hippocrates. He rendered at least ninety-five of Galen's works from Greek to Syriac and thirty-four from Greek to Arabic. He translated some fifteen Hippocratic works. Ḥunayn also translated (or corrected) three of Plato's dialogues, including the *Timaeus*; translated various Aristotelian works (in most cases from Greek to Syriac), including the *Metaphysics*, *On the Soul*, *On Generation and Corruption*, and part of the *Physics*; rendered a variety of other works on logic, mathematics, and astrology; and produced a Syriac version of the Old Testament. Ḥunayn's son Isḥāq translated more Aristotle, as well as Euclid's *Elements* and Ptolemy's *Almagest*. Their coworkers in Baghdad and their contemporaries elsewhere added to these translations. For example, Thābit ibn Qurra (836–901), a trilingual pagan (that is, neither a Christian nor a Muslim nor a Jew) from Harran in northwestern Mesopotamia, who spent most of his career in Baghdad, translated more than a hundred mathematical and astronomical treatises, including works of Archimedes. Translation activity continued at a rapid pace for more than a century after Ḥunayn and Thābit, and in a variety of locations. By the year 1000, nearly the entire corpus of Greek medicine, natural philosophy, and mathematical science had been rendered into usable Arabic versions.

ISLAMIC RECEPTION AND APPROPRIATION OF GREEK SCIENCE

How did the classical tradition fare in Islam—a culture shaped so heavily by religious revelation? Did Islam offer its various scientific communities freedom of thought and action sufficient to permit the flowering of the classical sciences in Islamic soil? Or did the community of intellectuals involved with the natural sciences run into serious theological opposition? These are legitimate questions, to which there is no simple answer. In the first place, it is a serious mistake to suppose that the vast Islamic empire was ever a monolithic, theologically homogeneous state. The distances between Cordoba in Spain, Cairo in Egypt, and Maragha in Central Asia were not merely geographical, but also linguistic, ethnic, cultural, and religious. As the centralized ʿAbbāsid empire fragmented into rival regimes, beginning in the 860s, the diversity became so great as to make the formulation of meaningful generalizations applicable to all of Islam all but impossible. The answer to almost any question we can ask depends on time and place.[14]

But we can take some cautious, preliminary steps. The fact of Islam's foundation in revealed religion surely influenced the reception of the classical tradition. With the passing centuries, Islam saw increasing concern among religious scholars about the legitimacy of what were referred to as the "foreign" or "rational" sciences (as opposed to the "traditional" or "Islamic" disciplines); and there is no question that scholars who preached Platonic or Aristotelian metaphysics were frequently called on to defend themselves against theological backlash from the champions of Islamic orthodoxy. But such cases represent the exceptions rather than the rule. Much of the important philosophical and scientific work was carried out in relatively tolerant urban centers—enclaves where scholars enjoyed considerable intellectual freedom. Many Islamic scientists were protected by powerful patrons. Others were teachers or civil servants. And, of course, most scientific work—especially in the medical and mathematical sciences (the latter including mathematics, optics, astronomy, and astrology)—was theologically benign and of demonstrated practical value. In short, while it might have been imprudent for scholars working on theologically or religiously sensitive topics to shout their opinions from the rooftops, most of them were free to share their ideas and conclusions among friends and colleagues. Their scholarship may have incurred the hearty disapproval of practitioners of the traditional religious disciplines, but the typical outcome was debate rather than condemnation and reprisal.[15]

In an important paper published in 1987, A. I. Sabra proposed three stages in the fortunes of the classical tradition in Islam. In the first stage, he argued, "Greek science entered the world of Islam, not as an invading force ... but as an invited guest"—a guest, moreover, whose Hellenistic worldview experienced "an almost immediate and almost unreserved adoption ... by Muslim members of the household." In the second stage, the guest, now a comfortable member of the community, was the source and inspiration for remarkable scientific achievements by outstanding scholars, who accepted the fundamental assumptions of the classical tradition, took up its unresolved problems, and corrected, refined, and extended its conclusions. This was not the beginning of a new scientific tradition, as some have argued, but a continuation of the Greek classical tradition on Islamic soil and with an Islamic voice. Finally, by the time the third stage made its appearance, the pioneers of Hellenistic science had passed away, to be replaced by a generation of scholars, almost every one of whom "had undergone a thorough Muslim education" and "were imbued with Muslim learning and tradition." The result was the integration of Greek disciplines with traditional learning and Islamic culture more generally. Thus logic became incorporated into theology and law; astronomy became an indispensable tool for the *muwaqqit*, who was responsible for determining the times of daily prayer in his locale; and mathematics became essential for a wide variety of commercial, legal, and scientific purposes. In this stage, which Sabra calls "naturalization," the classical tradition had become fully assimilated and put to use. The guest had become a member of the household in the role of handmaiden.[16]

Did the handmaiden go to school? Elementary schooling in medieval Islam was unregulated and thus dependent on local customs and the abilities and inclinations of the teacher. Instruction occurred typically in a mosque or the teacher's home—students seated in a semicircle around the teacher. Education began with a writing school, where the subjects were reading, writing, and penmanship. With these skills mastered, a child could proceed to a school whose curriculum was based entirely on the Koran, other indigenous religious writings, poetry, and history. Heavy emphasis was placed on memorization. For higher education, there were two alternatives: a madrasa (or "college") primarily for the study of Islamic law and the religious sciences; for everything else, private tutoring or self-study. The madrasas, typically funded by private endowment, were meant to preserve and defend religious knowledge. However, in eastern regions of late medieval Islam we find madrasas with a strong mathematical or mathematical science element in their curricula—most notably the Samarqand madrasa (closely connected to the Samarqand

Fig. 8.1. The ibn Ṭūlūn Mosque (9th c.), Cairo. Foto Marburg/Art Resource N.Y.

observatory), where Euclid's *Elements*, Ptolemy's *Almagest*, and astronomical works by Naṣīr al-Dīn al-Ṭūsī, Qutb al-Dīn al-Shīrāzī, and others were studied. Beginning in the ninth century hospitals were also founded as endowed institutions, followed in the eleventh century by astronomical observatories. These were admirable efforts, known to have provided institutional homes for medicine and astronomy; but the nature and extent of medical instruction in the hospitals is unknown, and many observatories (Maragha and Samarqand are exceptions) disappeared with the death of their patron, their lifetimes measured in decades. This absence of a permanent institutional home for broad higher education in the natural sciences is a factor that must be considered in any comparison between the fate of the classical tradition in medieval Islam and its fate in Western Europe after the twelfth-century revival of learning.[17]

How then do we explain the flowering of Islamic science from the tenth century onward? How was scientific education at the highest level gained by a promising young scholar who had exhausted the resources of the various levels of organized schools? Transmission to the next generation, we must never forget, is one of the crucial obligations of any scientific tradition. The options were to find a teacher or tutor with the expert knowledge that one sought, or to engage in self-education. We are fortunate to have an informative autobiographical account of his own education by one of the leading philosopher-scientists of Islam, the Persian polymath Ibn Sīnā (Avicenna in Latin). Not one to hide his light under a basket, Ibn Sīnā informs his reader that as a child prodigy he first studied the Koran and religious literature under a pair of teachers, completing these studies by age ten, after which he taught himself jurisprudence in sufficient depth to amaze the private tutor hired by the young Ibn Sīnā's father to guide his son's education. Under the tutor he began to read Porphyry's introduction to Aristotle's logical works. The tutor, Ibn Sīnā assures us, "was extremely amazed at me; whatever problem he posed I conceptualized better than he." Ibn Sīnā continued with logic, then Euclid's *Elements* and Ptolemy's *Almagest*—a treatise of such difficulty that Ibn Sīnā had to tutor his tutor. He claims to have been entirely self-educated from that point onward, mastering Aristotle's *Metaphysics*, the "logical, natural, and mathematical sciences"—and also writing prodigiously on medicine (see below, this chapter).[18]

THE ISLAMIC SCIENTIFIC ACHIEVEMENT

This chapter, so far, has been about context and precedent. It is time to investigate some of the scientific achievements that emerged from the Islamic world. The classical tradition arrived not as a finished product but piecemeal, as a work-in-progress; and the literate population of the recipient culture, rather than attempting to pull down the Greek edifice and build from the ground up, applied itself to mastery and advancement of the best and most convincing body of philosophical and scientific knowledge the world had ever seen. This does not mean that originality and innovation were absent; it means that Islamic originality took the form, primarily, of correction, extension, articulation, and application of the Greek heritage to new problems (with some spectacular successes), rather than creation of a new, uniquely Islamic science. This is no pejorative judgment: the great bulk of modern science consists in the correction, extension, and application of inherited scientific principles; a fundamental break with the past is approximately as exceptional today as it was in medieval Islam.

Greek science entered Islam in the form of texts—not as oral traditions conveyed by practicing scientists, not primarily in the form of scientific instruments (the sundial and astrolabe are exceptions), and certainly not as scientific laboratories or other scientific institutions, but *as books*. Although it is clear that oral transmission occurred between teacher and student or in enclaves of scientists (the ʿAbbāsid court in Baghdad in the eighth and ninth centuries, the Maragha observatory in the thirteenth and fourteenth), new scientific achievements were primarily disseminated in the form of books—commentaries or original scientific texts. With this as justification, let us investigate Islamic scientific achievements by looking briefly at some of the important scientific texts and their authors.[19]

Arabic mathematics was a direct descendant of Greek and Indian mathematics. The two earliest surviving Arabic mathematical texts were both written by al-Khwārizmī (ca. 780–ca. 850), a mathematician at the court of the caliph al-Maʾmūn in Baghdad. The second in order of writing, *Concerning Hindu Numbers*, was a thorough account of Indian reckoning, including the decimal place system and what have come to be called "Arabic" or "Hindu-Arabic" numerals. This book gradually replaced older systems of counting and calculating, including finger reckoning and sexagesimal (base 60) arithmetic (see table 1.1, above). Al-Khwārizmī's other book, the *Algebra*, has been celebrated as a landmark in the development of algebra—and so it was if we are careful about how we state the case. Al-Khwārizmī's *Algebra* contains no equations or algebraic symbols, but only geometrical figures and Arabic prose, and it would not be recognized as algebra by a mathematics student of the twenty-first century. Its achievement was to deploy Euclidean geometry for the purpose of solving problems that we would *now* state in algebraic terms (including quadratic equations).[20] This book circulated widely in Western Europe, and contributed greatly, and in the long run, to the development of a true symbolic algebra.

Geometry, the other major branch of mathematics, was dominated by Euclid's *Elements*, which circulated in multiple translations and various commentaries and epitomes and played a central role in several of the mathematical sciences, as we shall see. Islamic trigonometry took its starting point from Ptolemy's *Almagest* (which employed a system of chords) and the Indian *Siddhanta* (which introduced the sine function), adding trigonometric functions, drawing up tables, and applying trigonometry to both plane and spherical surfaces. Finally, applied mathematics or mechanics gave rise to a number of books by engineers, builders, instrument makers, and surveyors.[21]

The mathematical science to which Islam made its greatest contribution was astronomy. We must also acknowledge the presence of its close cousin, astrology, which could not function without astronomical assistance—a fact that undoubtedly contributed to the motivation to pursue astronomical studies.[22]

But our concern here is with mathematical astronomy, which nicely illustrates the relationship between Islamic and Greek science. Islamic astronomy was not founded on primitive astronomical or cosmological notions indigenous to the Muslim world. Rather, it was a direct continuation of the sophisticated cosmologies and mathematical astronomy of the classical tradition.[23] Ptolemaic astronomy and cosmological assumptions prevailed from the beginning. And it was largely within this framework that Muslim astronomers produced a great deal of highly sophisticated astronomical work.[24]

We can grasp the basic contours of Islamic astronomy by looking at achievements in three signature areas: (1) serious efforts to master Ptolemy's *Almagest* and observational programs undertaken to check and correct the parameters of Ptolemaic planetary models (including sun and moon);[25] (2) attempts to create planetary models that yield accurate mathematical predictions, while also remaining consistent with universally held physical principles—something that Ptolemy never achieved; and (3) establishment of astronomical observatories as institutional homes for astronomers and astronomical activity. Let us take these up in that order.

An early observational program, supported by the caliph al-Ma'mūn, was located at observatories (or observation posts?) in Baghdad and Damascus to create a *zīj* (an astronomical table) from which it would be possible to check and rectify the parameters of the major planetary motions. This work, overseen by Yaḥyā ibn Abī Manṣūr (d. 832), produced the first Arabic astronomical tables independent of Greek originals.[26]

Al-Battānī (d. 929, known to Europeans as Albategni), one of the greatest Muslim astronomers, was born in Harran in northwestern Mesopotamia. His practice of astronomy occurred in the city of al-Raqqa along the Euphrates River. He studied the motion of the sun and moon, calculated new values for solar and lunar motions and the inclination of the ecliptic, discovered the movement of the line of apsides of the sun (the shifting of the sun's perigee, or closest approach to the earth, in the heavens), drew up a corrected star catalogue, and gave directions for the construction of astronomical instruments, including a sundial and a mural quadrant. Ibn al-Nadīm, the great Islamic bibliographer (d. ca. 996), reported that

> he composed an important *zīj* containing his own observations of the two luminaries [sun and moon] and an emendation of their motions as given in Ptolemy's *Almagest*. In it . . . he gives the motions of the five planets in accordance with the emendations that he succeeded in making, as well as other necessary astronomical computations. . . . Nobody

is known in Islam who reached similar perfection in observing the stars and in scrutinizing their motions.

The fact that al-Battānī was still being cited (in Latin translation) in the sixteenth and seventeenth centuries by Copernicus and Kepler, among others, testifies to the quality and importance of his astronomical work.[27]

Early in Islamic history, the center of gravity of astronomical activity was in Near Eastern Islam (including Egypt)—witness the two astronomers discussed just above. But some of the most important astronomical achievements emerged from a late flowering of astronomy in eastern Islam. This activity was associated with what has come to be known as the "Maragha school." The prime mover of this astronomical activity was Naṣīr al-Dīn al-Ṭūsī (1201–74), who persuaded his patron, the grandson of Genghis Khan, to build an observatory in Maragha (in present-day northeastern Iran), obtain books for a substantial scientific library, and staff the operation with a librarian and at least ten astronomers. Al-Ṭūsī himself invented what is now called the "Tusi-couple"—a geometrical construction that converts two uniform circular motions into a back-and-forth straight-line motion—put to use by Maragha astronomers and subsequently in the astronomical models of Nicolaus Copernicus (1473–1543).[28] A major goal of the astronomical program of the Maragha astronomers was to find physically plausible substitutes for Ptolemy's equants; and for this purpose the Tusi-couple performed admirably.[29] This was a response to a quest that had its origins in the astronomical efforts of the brilliant mathematician and astronomer Ibn al-Haytham (ca. 965–ca. 1039, known in the West as Alhacen), two-and-a-half centuries earlier, to resolve discrepancies between the mathematical models of Ptolemy's *Almagest* and the physical model presented in Ptolemy's *Planetary Hypotheses* (see fig. 11.9, below).

These efforts reached their culmination in the astronomical models of Ibn al-Shāṭir (ca. 1305–ca. 1375)—regularly treated by historians of astronomy as a member of the Maragha circle, despite his living a hundred years after al-Ṭūsī and in Damascus (the former Umayyad capital), 1,200 miles west of Maragha. But the fact is that scientific theories and agendas are as portable as the books that contain them, and (perhaps owing to an immigrant astronomer) news of the work in Maragha found its way to al-Shāṭir in the Syrian capital. His achievement—one of the crowning achievements of Islamic astronomy—was to produce lunar and planetary models, employing double epicycles, which satisfied the quest articulated by Ibn al-Haytham in the eleventh century for a physically plausible and mathematically accurate substitute for the Ptolemaic equant (fig. 8.2).

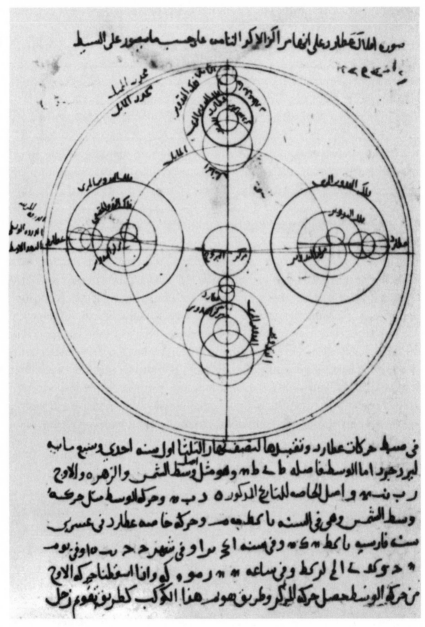

Fig. 8.2. The motion of Mercury according to Ibn al-Shāṭir (14th c.). Oxford, Bodleian Library, MS Marsh 139, fol. 29r.

In retrospect, the stunning aspect of al-Shāṭir's achievement is that mathematically identical counterparts of his models turned up some two hundred years
later in Copernicus's famous astronomical text, written in northern Poland.[30]

The last of what I have called the signature achievements of Islamic astronomy is associated with the institutional homes that made these achievements
possible. Any sustained, systematic program of astronomical observation required at least an observation post: a fixed place equipped with the necessary observational instruments. Better yet, a true observatory: a building,
equipped with observational instruments (including astrolabes, quadrants, a
sundial, perhaps an armillary sphere, and sighting tubes), a library, and a staff
of professional astronomers. Such observatories—institutional homes for
cooperative astronomical observation, research, and instruction—were the
invention of medieval Islam. Questions remain as to whether the Damascus
and Baghdad observatories mentioned above were anything more than
observation posts. There is no such question regarding the later observatories
of Maragha and Samarqand in the eastern reaches of the Islamic world. The
former, built in the latter half of the thirteenth century, included a dome with
an observation aperture, a library reputed to have contained 40,000 volumes
(a suspiciously high figure), astronomical instruments of the highest precision,
and a sizable staff of astronomers, engaged in astronomical observation and
model-building. The most famous products of the Maragha observatory are
Ibn al-Shāṭir's models and various astronomical tables. The Samarqand observatory (built about 1420) is best known for its great underground sextant
of forty-meter radius (see fig. 8.3) used for making meridian observations.
These two observatories developed research traditions resembling those in
a twenty-first-century research institute. Astronomers and would-be astronomers spent variable periods of time at the observatory in Samarqand
(studying, teaching, and doing research), emigrating eventually to various
destinations throughout greater Islam, spreading astronomical knowledge
that was discernible as late as the middle of the sixteenth century and as far
away as Western Europe. Unfortunately, few other observatories are known
to have survived, for long, the death of their original patron; but the example
lived on, to be replicated in sixteenth-century Istanbul and various European
locations.[31]

It was not unusual in Islam for astronomers to take an interest in geometrical optics. This may reflect the opinion of Aristotle and his commentators that vision is the principal sense organ for the acquisition of knowledge
and thus the noblest of the senses. But more important, I suspect, was the
prominence within the classical tradition of texts (by Euclid, Aristotle, and

Fig. 8.3. Underground sextant in Samarqand, built by Ulugh Beg for determining altitudes of celestial bodies as they crossed the meridian.

Ptolemy) dealing with light and sight, which cried out for correction and clarification by scholars equipped with the geometrical skills of the astronomer. The earliest surviving Arabic treatises on optics are by Abū Isḥāq al-Kindī, Qusṭā ibn Lūqā, and (in fragmentary form) Aḥmad ibn ʿĪsā—all written in the second half of the ninth century. The latter two studied reflection in concave mirrors (spherical and conical, respectively). Al-Kindī (d. ca. 866), a prominent neo-Platonist philosopher, astronomer, and astrologer, wrote a thorough and determined critique of Euclid's theory of vision. His most important achievement in this treatise was the revolutionary claim that light does not radiate from a luminous object as a unit, but in all directions from each point on the surface of the object, independently of light radiating from other points. On this foundation, the eleventh-century mathematician and astronomer Ibn al-Haytham built a new theory of vision that synthesized pieces of theory from various predecessors. This new theory, accompanied by a complete system of geometrical optics, was laid out in extravagant detail in his *Book of Optics* (fig. 8.4). Thanks to its translation into Latin around the beginning of the thirteenth century, this book and its author (known in the

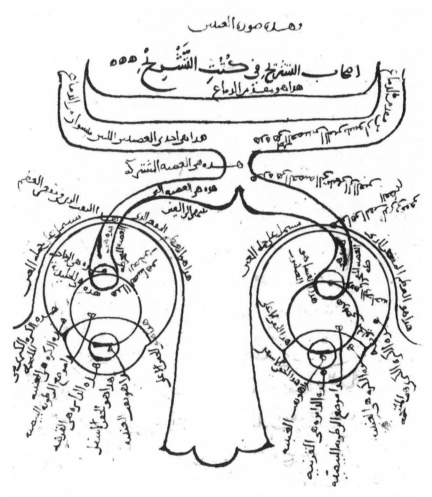

Fig. 8.4. The eyes and visual system according to Ibn al-Haytham. From a copy of al-Haytham's *Book of Optics* made in 1083. Istanbul, Süleimaniye Library, MS Fatih 3212, vol. 1, fol. 81v.

West as Alhacen) became Europe's principal authorities on the science of optics until the seventeenth century.[32] (See chap. 12, below.)

Two other optical achievements are worth mentioning for their demonstration of serious experimentation in the mathematical sciences. Toward the end of the tenth century, Abū Saʿd al-ʿAlāʾ ibn Sahl (fl. 985), a Baghdad mathematician, undertook a brilliant experimental analysis of the refraction of light as it passed from one transparent medium to another through plane or curved interfaces. Measurements of angles of incidence and refraction led Ibn Sahl to the geometrical equivalent of the modern law of refraction (known as Snell's law since the seventeenth century), which relates angles of incidence to corresponding angles of refraction for a given pair of transparent media. This answered a question that had defeated Ptolemy eight hundred years earlier and that would not be seen in the West for another three hundred years.[33]

The second experimental optical achievement—as much meteorological as optical—came about three hundred years after Ibn Sahl's. Kamāl al-Dīn al-Fārisī (1267–1319), a student of al-Shīrāzī and member of the Maragha circle, undertook an analysis of the rainbow. Inspired by Ibn Sīnā's theory that the rainbow results from the reflection of solar light by individual droplets of moisture that make up a cloud, and by Ibn al-Haytham's experiments on burning spheres, Kamāl al-Dīn used a water-filled glass sphere to simulate a droplet of moisture on which solar rays were allowed to fall. Driven by his observations to abandon the notion that reflection alone was responsible for the rainbow (the traditional view, going back to Aristotle), Kamāl al-Dīn concluded that the primary rainbow was formed by a combination of reflection and refraction. The rays that produced the colors of the rainbow, he observed, were refracted upon entering his glass sphere, underwent a total internal reflection at the back surface of the sphere (which sent them back toward the observer), and experienced a second refraction as they exited the sphere. This occurred in each droplet within a mist to produce a rainbow. Two internal reflections, he concluded, produced the secondary rainbow. Location and differentiation of the colored bands of the rainbow were determined by the angular relations between sun, observer, and droplets of mist. Kamāl's theory was substantially identical to that of his contemporary in Western Europe, Theodoric of Freiberg (chap. 11, below). It became a permanent part of meteorological knowledge after publication by René Descartes in the first half of the seventeenth century.[34]

While the areas of greatest Islamic achievement were mathematics and the mathematical sciences, the medical sciences were also prominently cultivated.[35] Islam saw a wide chasm between medical theory and medical practice, which we must bear in mind if we are to avoid misunderstandings. What we know about pre-Islamic medicine is found in a popular genre of literature written by clerics, known as "prophetic medicine." These treatises (which are still in use) represent traditional folk medicine of the sort that we saw in ancient Greece and medieval Europe—including dietary concerns, treatment of fevers and wounds, bloodletting, cautery, and magical remedies, all practiced by local healers. Common diseases seeking treatment were malaria, tuberculosis, conjunctivitis, dysentery, smallpox, leprosy, parasitic infections, and more.[36] After the acquisition of Greek medical theory through translation of the works of Greek medical writers, especially Galen (A.D. 129–ca. 215), the Islamic world gained access to ample medical theory. However, in practice this medical theory worked to the benefit only of the rich and powerful; and it has been convincingly demonstrated that even for the upper classes, the practices and procedures preached in medical writings, and sometimes claimed to have been performed, bore only a distant relationship to medical reality.[37] Once again, what we know about the Islamic achievement is found in a collection of texts.

When the ʿAbbāsid caliphs came to power in the middle of the eighth century, news of the classical medical tradition was already available—emanating, it appears, from Nestorian Christian practitioners. Court patronage was quickly directed toward this new, promising body of knowledge, in the form of support for translators and practitioners. The Nestorian Christian physician Jurjīs ibn Jibraʾil ibn Bukhtīshūʿ was summoned from Gondeshapur to Baghdad by al-Manṣūr to treat the caliph's stomach ailment. Jurjīs remained as court physician, subsequently replaced by his son, then grandson, and onward for eight generations of Bukhtīshūʿ physicians. The medicine they practiced was the Galenic medicine of the classical tradition.

The most important translators of medical writings were Ḥunayn ibn Isḥāq and his team of translators (see above). Ḥunayn also wrote original medical treatises, including *Questions on Medicine*, a popular medical textbook for students, in the form of questions and answers. Two of his books became standard introductory texts in medieval Latin translation: his book on the eye and vision and his *Introduction to Medicine* (fig. 8.5). He also corrected the recent translation of Dioscorides' great pharmaceutical guide, *De materia Medica*—a treatise on herbs and herbal remedies that was influential not only in Islam, but also in Byzantium and Western Europe. Original treatises in Arabic were written by Ḥunayn's teacher, the Nestorian Christian Yūḥannā ibn Māsawayh (d. 857), including many short treatises on medical subjects such as eye diseases, fever, and leprosy.[38]

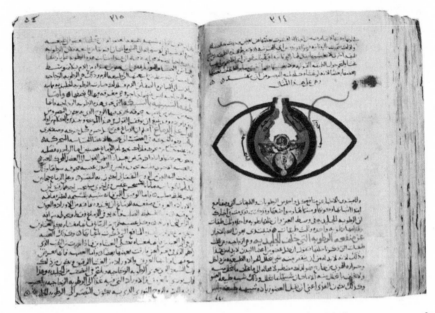

Fig. 8.5. Ḥunayn ibn Isḥāq on the anatomy of the eye, from a thirteenth-century copy of Ḥunayn's *Book of the Ten Treatises of the Eye*, Cairo, National Library.

Textbooks of broader scope began to appear in the tenth century. Undoubtedly the most controversial author of that century was Abū Bakr Muḥammad ibn Zakariyyā al-Rāzī (ca. 854–ca. 925, known in the West as Rhazes)—philosopher, social critic, religious controversialist, critic of Galen and Aristotle, defender of atomism (theologically unacceptable in Islam as a natural philosophy), and physician. Al-Rāzī wrote a general textbook on medicine, *The Book of Medicine for al-Manṣūr* (the latter an Iranian prince), which became very popular and circulated widely in Islam and also in Europe in Latin translation. He also wrote a treatise, *Doubts about Galen*, and at his death left behind a collection of reading notes, personal observations, and medical case histories that present a firsthand portrait of medicine as it was actually practiced—a rather different picture from the lofty prescriptions of Islamic and Greek medical literature in translation. Two other authors wrote influential encyclopedic treatises: ʿAlī ibn al-ʿAbbās al-Majūsī (fl. 983, known to Europeans as Haly Abbas), authored a *Complete Book of the Medical Art*, dedicated to his patron, a Persian ruler. Practicing medicine in Moorish Spain, at the opposite end of the Islamic empire, was Abū al-Qāsim al-Zahrāwī (ca. 936–ca. 1013, Abulcasis to Europeans), a practicing physician and surgeon who wrote a thirty-part medical encyclopedia, the final chapter of which was devoted to surgery, accompanied by illustrations of surgical instruments. In this surgical chapter, al-Zahrāwī dealt with a wide range of surgical procedures, some invasive, some not. These include treatment of wounds, bonesetting, amputation, tracheotomy, cauterization, tonsillectomy, removal of hemorrhoids and cysts, pharmaceutical preparations, an account of the use of wire to connect loose teeth to sound teeth, and more. The surgical chapter was printed in Latin translation in Venice in 1497.[39]

But the greatest of the Islamic medical encyclopedists was the Persian Abū ʿAlī al-Ḥusayn ibn ʿAbdallāh ibn Sīnā, known in the west as Avicenna (980–1037); we have already seen his own account of his education. Ibn Sīnā has been credited with more than one hundred treatises: on arithmetic, geometry, music, astronomy, logic, physics, and philosophy (especially metaphysics). He was one of the great commentators on Plato and Aristotle, creating a Platonized Aristotelianism that was profoundly influential in medieval Europe. His *Canon of Medicine*, which borrowed freely from al-Rāzī, organized medical knowledge under five headings: generalities; herbal and other medical remedies; diseases, head to toe; diseases not specific to a single organ; and compound drugs. So influential was this medical encyclopedia that it was translated into Latin by Gerard of Cremona in the twelfth century, printed five times in the first fifty years of the printing press, and served as a textbook in European medical schools as late as the seventeenth century.[40]

I have claimed that the classical scientific tradition entered the Islamic world as a work-in-progress. Was this true of medicine? Did progress in medical theory or practice continue under Islamic auspices? Yes, although the picture is mixed. The medicine of the classical tradition was certainly a vast improvement over bedouin medicine. But our question is whether Islam saw progress in this learned medicine (solidly Galenic) of the classical tradition. One of the major achievements of Islamic physicians was an outpouring of medical texts aimed at organizing, elaborating, and disseminating classical medical knowledge. One thinks of al-Rāzī (Rhazes), al-Majūsī (Haly Abbas), and Ibn Sīnā (Avicenna), whose medical writings were important enough to earn translation and wide circulation in medieval Europe. Another was the development of new specialties—perhaps most importantly the treatment of eye diseases, including cataracts. The book on this subject by Ḥunayn ibn Isḥāq received wide circulation in both Islam and, after translation into Latin, in medieval Europe.[41]

Institutionalization of medicine was also an important Islamic achievement. The hospital as an institution for treating the sick was created in

Byzantium, but the model of institutionalized medicine, open to all social classes, was elaborated and multiplied in Islam. The earliest recorded hospital (like all Islamic hospitals, a secular charitable foundation) appeared in Baghdad around the year 800. The next century saw the foundation of five more Baghdad hospitals; and hospitals were eventually scattered throughout the Islamic empire. One of them, founded in 982, is reported to have had "twenty-five doctors, including oculists, surgeons, and bonesetters."[42]

The most startling single achievement of Islamic anatomy and physiology emerged from a hospital setting. This was discovery of the pulmonary transit of the blood in the human body by Ibn al-Nafis (d. 1288). Ibn al-Nafis studied medicine in Damascus and subsequently practiced and taught in a hospital in Cairo. He described the pulmonary transit (a crucial element in William Harvey's seventeenth-century theory of circulation of the blood) in two of his books. In the second of these, a commentary on Ibn Sīnā's *Canon of Medicine*, he wrote:

> When the blood in [the right] cavity [of the heart] becomes thin, it must be transferred into the left cavity, where the pneuma is generated. But there is no passage between these two cavities, for the substance of the heart separating them is impermeable. It neither contains a visible passage, as some people have thought, nor does it contain an invisible passage which would permit the passage of blood, as Galen thought.... It must, therefore, be that when the blood has become thin, it is passed into the arterial vein [our pulmonary artery] to the lung, in order to be dispersed inside the substance of the lung, and to mix with the air. The finest parts of the blood are then strained, passing into the venous artery [our pulmonary vein] reaching the left of the two cavities of the heart, after mixing with the air and becoming fit for the generation of pneuma.

Historians have explored the possibility that news of this discovery of Ibn al-Nafis was communicated to the supposed European discoverer of the pulmonary circulation, Realdo Colombo (1510–59) in Padua, Italy, and have deemed it a reasonable possibility but by no means a certainty.[43]

I conclude this discussion of medieval Islamic medicine on an institutional note. One of the primary features of a successful scientific tradition is institutional support. Today the loci of support for scientific research are in research universities, governmental laboratories, endowed research institutions, and business and industrial laboratories. In medieval Islam we see the forerunners of such institutions. The hospitals, undoubtedly in imitation of Byzantine hospitals (the first of which were established in the fourth century), were not merely venues for surgery and treatment of disease and injury, but also institutions where practicing physicians could make a living treating patients, possibly teaching an apprentice physician, or writing and doing research (as the case of Ibn al-Nafis reveals).[44]

THE FATE OF ISLAMIC SCIENCE

The scientific movement in Islam was distinguished and of long duration. Translation of Greek works into Arabic began in the second half of the eighth century; by end of the ninth century translation activity had crested, and serious scholarship was under way. From the middle of the ninth century until well into the fourteenth, we find impressive scientific work in major branches of Greek science, carried forward in widely scattered cities of the Islamic world. If we concentrate our attention on mathematics and astronomy, where many of Islam's greatest achievements lay, we find serious research by capable mathematicians and astronomers as late as the first half of the sixteenth century. This period of Islamic preeminence in the mathematical sciences lasted well over half a millennium—a longer period than the interval between Copernicus and ourselves.

This scientific movement had its origin, for practical purposes, in Baghdad under the ʿAbbāsids, though many other centers of scientific patronage also emerged. Early in the eleventh century, Cairo under the Fatimids came to rival Baghdad. In the meantime, the foreign sciences had made their way to Spain, where the Umayyads, displaced in the Near East by the ʿAbbāsids, built a magnificent court at Cordoba. Under Umayyad patronage, the sciences flourished in eleventh- and twelfth-century Spain. Instrumental in this development was al-Ḥakam (d. 976), who built and stocked an impressive library in Cordoba.[45] Finally, toward the end of the Middle Ages, the center of gravity of mathematical science shifted eastward to Maragha and Samarqand, where the mathematical sciences found institutional support.

A sizable collection of historians has come to hold the view (more often assumed than defended) that Islamic science peaked somewhere between the twelfth and fifteenth centuries and slid downhill toward insignificance just as medieval European science was showing signs of coming to life.[46] This view has its attractions: as early as the ninth century, the ʿAbbāsid empire became the scene of continuous uprisings, as it fragmented into petty, warring principalities. In the West the Christian reconquest of Spain began

Fig. 8.6. Interior of the Great Mosque of Cordoba, middle of the 8th c. Foto Marburg/ Art Resource.

to make serious, if sporadic, headway after about 1065 and continued until the entire peninsula was in Christian hands two centuries later. Toledo fell to Christian armies in 1085, Cordoba in 1236, and Seville in 1248. At the same time, the Islamic empire faced attack in the east by Mongols and Turks. In 1258 the Mongols sacked Baghdad, bringing the ʿAbbāsid empire to an end.

Looked at in the abstract, one might suppose that such political turmoil would have been debilitating and, along with growing religious opposition to Aristotelian natural philosophy and metaphysics, would have resulted in loss of patronage and a decline of scientific activity. But the surprising truth is that many of the conquerors became patrons of the sciences. For example Hulagu Khan, who sacked Baghdad, was also (along with his brother Mangu) a patron of the Maragha observatory; and Ulugh Beg, grandson of

Tamerlane (the great Mongol conqueror of central Asia), saw to the construction, support, and management of the Samarqand madrasa and observatory. In Spain, Christians, Muslims, and Jews had a long history of living and working in harmony until the events leading up to the expulsion of Jews in 1492.[47] Finally, religious opposition was confined to issues with theological import and had little or no effect on the natural sciences. The truth is that the image of decline in the twelfth to fifteenth centuries is not the product of research in manuscript archives, but an assumption made in the absence of research and encouraged for its usefulness as a tool in religious polemics over the relative merits of Islam and Christianity: which religious culture wins the natural science sweepstakes?[48]

Finally, current archival research in the history of Islamic astronomy reveals decisively that at least this specific discipline flourished well into the sixteenth century—producing a continuous flow of knowledgeable, sometimes brilliant, astronomers, scattered throughout greater Islam.[49] As for other sciences, thousands of Arabic, Persian, and Turkish manuscripts remain in libraries from Europe to the Middle East, unexamined. What they may contain we have no way of knowing until we look.

Perhaps the question that we ought to be asking is not "Why or when did Islamic science decline?" or "Which culture outperformed the other in the race for preeminence in the natural sciences?," but "How is it that an intellectual tradition that began in such unpromising circumstances developed an astonishing scientific tradition that endured as long as it did?" The short answer is that by an incredibly complex concatenation of contingent circumstances, the Islamic rulers and their educated subjects took possession of the classical scientific tradition, which contained the resources that allowed the sciences, especially the mathematical sciences and medicine, to flourish. As we have seen, this was a highly diverse, multireligious, multilingual, cosmopolitan culture. And it had an elite of sufficient education and linguistic diversity to keep the natural sciences alive. It may be that this very diversity ensured that there would remain enclaves of educated, theologically tolerant people, where a scientific tradition, foreign in both origin and content, could take root and flourish.

Also not to be overlooked are utilitarian motives. If we place metaphysical questions (as found in Plato's and Aristotle's works, for example) to one side, the remaining disciplines could be cultivated for their practical utility. Logic for law; mathematics for commerce, governmental record-keeping, surveying, and engineering; astrology for casting horoscopes and interpreting omens; astronomy for fixing the calendar and determining the direction of Mecca and the times of daily prayer; and medicine for a wide array of ailments. Fortunately, in the twelfth and thirteenth centuries many of the texts containing these scientific achievements (of both Islamic and Greek origin) came into the possession of Latin Europe, where they were hungrily received, and the process of cultural transmission began again.

9 ✤ *The Revival of*
Learning in the West

THE MIDDLE AGES

To this point, I have employed the expression "Middle Ages" without definition and without specifying exact chronological limits. This may be a case where inexactness is a virtue, since historians themselves do not agree on what the expression means; but the time has come to be a little more definite. The idea of the Middle Ages (or medieval period) first arose in the fourteenth and fifteenth centuries among Italian humanist scholars, who detected a dark middle period between the bright achievements of antiquity and the enlightenment of their own age. This derogatory opinion (captured in the familiar epithet "dark ages") has now been almost totally abandoned by professional historians in favor of the neutral view that takes "Middle Ages" simply as the name of a period in Western history, during which distinctive and important contributions to Western culture were made—contributions that deserve fair and unbiased investigation and appraisal without prior depreciation.

The chronological limits of the Middle Ages are necessarily blurred, because medieval culture (whatever exactly we take it to be) appeared and disappeared gradually, and at different times in different regions. If we must have dates, then the Middle Ages may be taken to cover the period from the end of Roman civilization in the Latin West (500 is a good round number) to 1450, when the artistic and literary revival commonly known as the Renaissance was unmistakably under way. It will be convenient for our purposes to subdivide this period into the early Middle Ages (approximately 500 to 1000), a transition period (1000 to 1200), and the high or late Middle Ages (1200 to about 1450). These are not exactly the standard subdivisions (the "high" and "late" Middle ages are frequently differentiated), but they will serve our purposes.

CAROLINGIAN REFORMS

We have observed (chap. 7) the declining fortunes of the Roman Empire in the Latin West and the appearance of socioreligious structures, such as monasticism, that we think of as characteristically medieval. Western Europe went through a process of deurbanization; the classical schools deteriorated, and leadership in the promotion of literacy and learning passed to monasteries, where a thin version of the classical tradition survived as the handmaiden of religion and theology. Certainly scholarship on scientific subjects declined, as the focus shifted to religious or ecclesiastical matters: biblical interpretation, religious history, church governance, and the development of Christian doctrine. But it does not follow that scientific subjects were abandoned or scientific books burnt. Chap. 7 gave us a look at the scientific interests of Macrobius (late fourth century) in his *Commentary on the Dream of Scipio*; Martianus Capella's treatise on the seven liberal arts; and the Aristotelian translations of Boethius.

But even religion and theological works drew on the classical tradition, especially Greek logic and metaphysics. Boethius established the pattern in his determined effort to think through such problems as divine foreknowledge and the nature of the divine trinity with the help of Aristotelian logic and Platonic and Aristotelian metaphysics. Isidore of Seville attempted to explain the origin of various Christian heresies by their parallels within the philosophical tradition. Even Gregory the Great, an outspoken critic of pagan learning, revealed at many points the implicit or explicit philosophical underpinnings of his theology.[1]

The end of the eighth century saw an important renewal of scholarly activity, associated with Charlemagne (Charles the Great) and his court. In 768 Charlemagne inherited a Frankish kingdom that encompassed portions of modern Germany and most of France, Belgium, and Holland. By the time of his death in 814, Charlemagne had enlarged the kingdom (which historians know as the Carolingian Empire) to include all of Western continental Europe apart from Spain, Scandinavia, and the southern half of Italy—the first serious attempt at centralized government in Western Europe since the disappearance of the Roman Empire (see map 9.1).

Charlemagne was himself literate and dedicated to the spread of learning, not only at court but throughout the realm. His biographer, Einhard (ca. 770–840), described Charlemagne's thirst for knowledge (doubtless inflated) in the following terms:

He avidly pursued the liberal arts and greatly honoured those teachers whom he deeply respected. To learn grammar, he followed [the teaching of] Peter of Pisa, an aged deacon. For the other disciplines he took as his

Map 9.1. Carolingian Empire about 814.

teacher Alcuin of Britain, . . . the most learned man in the entire world.
[Charlemagne] invested a great deal of time and effort studying rhetoric,
dialectic and particularly astronomy with him. He learned the art of calculation [arithmetic] and with deep purpose and great curiosity investigated
the movement of the stars.[2]

As part of a campaign for the improvement of clerical education, Charlemagne
ordered the creation of cathedral and monastic schools, and he invited some
of Europe's finest scholars to join his court. Many of these he subsequently

appointed to important posts as bishops and abbots—positions from which they could advance educational reform. The most important of Charlemagne's appointments brought Alcuin, distinguished scholar and headmaster of the cathedral school at York (in England), as advisor and director of this extensive educational effort and, in his later years, abbot of the monastery of St. Martin of Tours. A beneficiary of the tradition of Irish scholarship (see chap. 7) who could trace his intellectual lineage directly back to Bede, Alcuin (ca. 730–804) became the central figure in a circle of scholars interested in contemporary theological, ecclesiastical, and (where relevant) scientific and philosophical issues.

But the most important contribution of the Carolingian period was the collection and copying of books in the classical tradition. Under Alcuin's leadership books were collected, corrected, and copied—including the Bible and the works of the church fathers and various classical authors. The importance of the copying of classical texts is demonstrated by the fact that our earliest known copies of most Roman scientific and literary texts (also Latin translations of Greek texts) date from the Carolingian period. The recovery and copying of books, combined with Charlemagne's imperial edict mandating the establishment of cathedral and monastery schools, contributed to a wider dissemination of education than the Latin West had seen for several centuries and laid a foundation for future scholarship.[3]

Although this Carolingian revival was motivated primarily by concerns about the low level of clerical literacy, it also welcomed the natural sciences. We see this especially in astronomical studies (broadly defined), which were pursued for intertwined religious and secular purposes. Timekeeping, an aspect of applied astronomy, was required for determining the times that regulated the mandatory, daily monastic ritual of prayers and other communal activities. This was ordinarily accomplished by reference to positions of the sun during the day and certain stars at night. More demanding of astronomical knowledge were calendric problems, represented by the science known as "computus." Computus developed as a set of rules for determining the proper date of Easter and other religious feast days, mandated by church councils to be celebrated on the same day throughout Christendom. Easter, the most important of these holy days, was generally celebrated by the churches of Western Europe on the first Sunday after the first full moon after the spring equinox. This date had to be determined well in advance, so that suitable preparations could be made. Making that determination was a serious exercise in calculation, construction, and use of calendrical (or computus) tables.[4]

Interest in the planetary and cosmological aspects of astronomy was fueled in Carolingian court circles by the books of four Roman and/or medieval authors. Two have already been mentioned: Macrobius's *Commentary on the Dream of Scipio* (first half of the fifth century); and Martianus Capella's *The Marriage of Philology and Mercury* (late fifth century), which became an enormously popular textbook on the liberal arts. The others were Calcidius's *Commentary on Plato's "Timaeus"* (fourth century); and the *Natural History* of Pliny the Elder, the earliest and most sophisticated of the four, but not widely available in Carolingian circles until the ninth century.[5] These four books became the authoritative sources, frequently copied or glossed, offering basic, garden-variety cosmology and planetary astronomy. Topics covered included the zodiac, the ecliptic, planetary stations and retrogradations, the sequence of the planets in space, apsides, variations in planetary latitudes, risings and settings, eclipses, the Milky Way, the size of the earth, climatic zones, temperate zones capable of human habitation, and more. It was universally agreed that the earth was spherical (Martianus reports Eratosthenes' calculation of its circumference) and that southern temperate zones (the antipodes) were inhabited by people (antipodeans) whose "up" and "down" were the opposite of ours.[6] One of the more interesting pieces of content is Martianus Capella's attempt to deal with the sequence in space of the innermost celestial bodies (Moon, Sun, Mercury, and Venus) by giving Mercury and Venus paths that "do not in any way enclose the earth, but instead circle around the sun in a wider path"—in short, circumsolar paths[7] (see fig. 7.5).

The drawing of circumsolar orbits for Mercury and Venus was only one of many Carolingian attempts to illustrate geometrical aspects of the cosmos, making them intelligible to scholars who lacked higher mathematical skills. Pliny's *Natural History*, the most sophisticated of the four authoritative astronomical sources, was the only one to include a full discussion of epicycles and eccentrics. But it included no illustrations, and two of the major challenges confronting Carolingian authors were to make sense of epicycles and eccentrics by creating geometrical illustrations that would shed light on Pliny's prose account.[8] One such illustration (fig. 9.1) deals with planetary apsides (the closest approach to and greatest retreat from the earth, assumed to be the central body) by simply placing them on eccentric circles.[9]

Nothing in the foregoing account of Carolingian achievements in astronomy and other scientific topics represents a revolutionary leap forward. The importance of Charlemagne, his reforms, and the scientific achievements of the Carolingian period is to be found not in novelty, but in the recovery and preservation of important portions of the classical tradition, the establishment

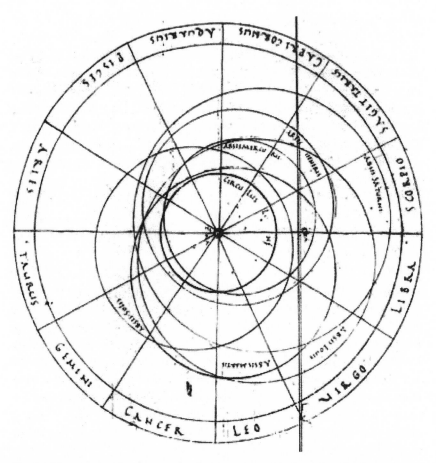

Fig. 9.1. Planetary apsides. Accounting for planetary apsides by giving the planets eccentric orbits about the earth. Oxford, Bodleian Library, MS Canon. Class. lat. 279, fol. 33v. (Bruce Eastwood, *Planetary Diagrams*, p. 30.)

of schools, the spread of literacy, and (in the case of astronomy) attempts at geometrical representations of various planetary phenomena. All of this laid the foundation for a true and nearly full recovery of the classical tradition in the eleventh, twelfth, and thirteenth centuries.

The benefits of the Carolingian educational reforms can be seen in the career of John Scotus Eriugena (fl. 850–75). An Irish scholar attached to the court of Charlemagne's grandson Charles the Bald, Eriugena was undoubtedly the ablest scholar of the ninth century in the Latin West. He had many gifts, including a keen, original mind and rare linguistic talent. He had an excellent command of Greek, probably first acquired in an Irish monastery school but improved after his arrival on the Continent, which he put to use in the translation of several Greek theological treatises into Latin: first, at the request of Charles the Bald, the works of pseudo-Dionysius (an anonymous Christian Neoplatonist of about 500 A.D.); later, works of several Greek church fathers. Eriugena also produced original and sophisticated theological treatises, in which he developed the Neoplatonism of pseudo-Dionysius and attempted a synthesis of Christian theology (with a Greek slant) and Neoplatonic philosophy. His *On Nature*, an attempt at a comprehensive account of created things, contains a well-articulated (and, of course, thoroughly Christian) natural philosophy. Finally, he wrote a commentary on Martianus Capella's textbook of the liberal arts, *The Marriage of Philology and Mercury*, presumably in connection with his teaching responsibilities. Eriugena had an immediate impact on an entourage of disciples, and through them a continuing influence on Western thought.[10]

An important transitional figure between the Carolingian revival of the eighth and ninth centuries and the "Renaissance" of the twelfth century was a tenth-century monk from the monastery of Saint-Géraud at Aurillac in south-central France. Gerbert of Aurillac (ca. 945–1003) enjoyed a meteoric career, explained by a combination of intellectual talent and political opportunism. Born in humble circumstances, he received an impressive education at Aurillac and later in northern Spain, where he studied for a period of about three years. He returned from Spain to the important cathedral school at Reims in northern France, first as a student of logic, subsequently as headmaster. From Reims he moved to northern Italy as abbot of the monastery at Bobbio, back to Reims as archbishop, and back to Italy as archbishop of Ravenna. Finally, in 999 his patron Otto III (Saxon emperor) saw to his election as Pope Sylvester II.

It has been customary to concentrate on Gerbert's role as the first, or one of the earliest, and certainly the most important, initiators of fruitful intellectual contact between Islam and Latin Christendom. But before examining that vitally important aspect of his career, we must note that he also contributed to an older tradition of scholarship—the recovery and dissemination of the classical liberal arts, especially Aristotelian logic as transmitted through Boethius and other Latin sources. At Reims Gerbert lectured on various logical works of Aristotle, Cicero, Porphyry, and Boethius; he also composed at least one logical treatise of his own. Nonetheless, Gerbert's fame rests on his pursuit of the mathematical quadrivium. A fortuitous set of circumstances brought him into contact with Arabic mathematical treasure—the first scientifically important contact by a northern European and the beginning of the recovery of the thick version of the classical tradition, including Islamic additions, that

Fig. 9.2. Personification of the quadrivium. From left to right: music, arithmetic, geometry, and astronomy. From a ninth-century copy of Boethius's *Arithmetic*, Bamberg, MS Class. 5 (HJ.IV.12), fol. 9v.

would revolutionize study and practice of the natural sciences in Christian Europe. To understand the circumstances that brought this about, we need to look briefly at the history and culture of Islamic Spain.

The final step in Islamic military and political expansion took place with the conquest of Spain. A band of North African Muslims crossed into Spain in 711, marched northward against insignificant resistance, and with the help of major reinforcements gained control, before the end of the decade, of all of Spain except a narrow strip along the northern border. From 756 to the early eleventh century, Muslim Spain (al-Andalus) was ruled by an Umayyad dynasty, which built a magnificent capital in the south-central Andalusian city of Cordoba, patronized education and the copying and translating of books, built a library reputed to contain 400,000 volumes (probably an exaggeration, but vastly larger than any library in Christian Europe), and encouraged scholarship on the classical heritage. This has become known as the "golden age" of al-Andalus. Crucial to our story is the further fact that the substantial indigenous Jewish and Christian populations of al-Andalus were not only tolerated, but also accepted as major contributors with a seat at the table of learning, and even government. Al-Andalus had become a multiethnic, multireligious, and multilinguistic state.[11]

Gerbert left his monastery in Aurillac in the year 967 to travel southward in the company of the count of Barcelona, over or around the Pyrenees (350 miles as the crow flies) to the town of Vich in Catalonia (the far northeastern corner of the Spanish peninsula). His purpose was to study mathematics with Atto, bishop of Vich. Catalonia (in Christian hands by this time) is reputed to have been an intellectual backwater, far from centers of Islamic learning in Cordoba and Toledo, with no reputation for quadrivial studies; and on this account some scholars have belittled the significance of Gerbert's Catalonian pilgrimage. However, recent research has revealed that in the second half of the tenth century Catalonia was no longer the island of cultural and scientific isolation that some scholars have supposed. By the middle of the tenth century Catalonia was flourishing politically and culturally—and about to experience an influx of scientific ideas from the south. Among the agents of change were harmonious relations and regular cultural and scientific exchanges with al-Andalus.

An important figure in this Catalonian renaissance was Abū Yusuf Ḥasdāy ben Isḥāq ben Shaprūt, physician, scholar, leader of the Andalusian Jewish community, and also an important figure (embassador and advisor to two caliphs) at the Andalusian court in Cordoba. Ḥasdāy is known to have carried (or arranged for the carrying of) various Arabic mathematical and astronomical texts to Catalonia, several of which are known to have fallen into Gerbert's hands. One of these was a book on the armillary sphere by a Jewish philosopher affiliated with the Fatimid court in Tunisia, Dunāsh ibn Tamīm al-Qarawī. In it Dunāsh laid out a program for teaching astronomy that made use of a spherical globe and an armillary sphere—arguing that the

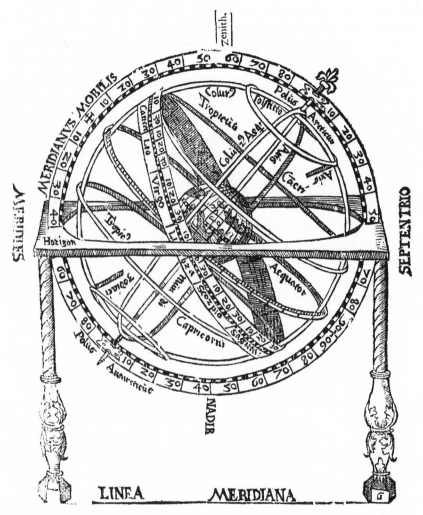

Fig. 9.3. A sixteenth-century armillary sphere. From Petrus Apianus and Gemma Frisius, *Cosmographia* (Antwerp, 1584).

three-dimensionality of these teaching aids enhances understanding of the celestial motions (fig. 9.3). The importance of this is that we find Gerbert employing an identical instructional program in his teaching career, after return to France—the first northern European, as far as we know, to do so.[12]

Gerbert's subsequent career provides eloquent testimony to his mastery of the mathematical sciences, surpassing that of any European predecessor, and to his familiarity with Islamic achievements in mathematics and astronomy. His correspondence, despite the turbulent political and religious context within which much of it was written, is laced with references to mathematics, astronomy, manuscripts to be copied or corrected (including Pliny's *Natural History*), translated books, and works to be obtained (including those of Boethius and Cicero). In one letter, Gerbert requests a book on multiplication and division by Joseph the Spaniard (probably identical with Ḥasdāy); in another, he solicits a book on astronomy translated from the Arabic by Lupitus (archdeacon of the cathedral of Barcelona); in yet another, he announces the discovery of an astronomical work that he believes to have been written by Boethius. He praises his patron, Otto III, for his interest in numbers. He instructs friends and associates on how to solve various arithmetical and geometrical problems. He follows Dunāsh's program for astronomical education, promoting the use of spherical models. He presents instructions for the construction of astronomical models and on use of the abacus for multiplication and division (employing Arabic numerals). In short, Gerbert used his influential posts as teacher and church dignitary to advance the cause of mathematical science in the West and to stimulate serious and productive scientific interchange between Islam and Christian Europe.

THE SCHOOLS OF THE ELEVENTH AND TWELFTH CENTURIES

When Gerbert died in 1003, Western Europe was on the eve of political, social, and economic renewal. The causes of this renewal were numerous and complex. One of them was the emergence of stronger monarchies, capable of administering justice and reducing the level of internal disorder and violence. At the same time, secure borders were restored after the Viking and Magyar invasions of the ninth and tenth centuries. Indeed, having been on the receiving end of aggression for so many centuries, Europeans were about to turn around and become the aggressors, driving the Muslims out of Spain and dispatching armies of crusaders to rescue the Holy Land.

Political stability led to the growth of commerce and increased affluence. The extension of a money economy to the countryside enhanced trade in agricultural products. Technological developments played a critical role in supplying necessities and producing sources of wealth. The refinement and spread of the water wheel, for example, gave rise to a minor industrial revolution; and agricultural innovations, such as crop rotation and the invention of the horse collar and the wheeled plow (combined, possibly, with improved climatic conditions), led to a major increase in the food supply.[13] One of the most dramatic results of these changes was a population explosion; exact figures are not available, but between 1000 and 1200 the population of Europe may have doubled, tripled, or even quadrupled, while the city-dwelling portion of this population increased even more rapidly.[14] Urbanization, in turn, provided economic opportunity, allowed for the concentration of wealth, and encouraged the growth of schools and intellectual culture.

It is widely agreed that a close relationship exists between education and urbanization. The disappearance of the ancient schools was associated with the decline of the ancient city; and educational invigoration followed quickly upon the reurbanization of Europe in the eleventh and twelfth centuries. The prototypical school of the early Middle Ages was a monastic school (fig. 9.4)—rural, isolated from the secular world, and dedicated to narrow educational objectives (even if those objectives were sometimes stretched by external pressures). With the shift of population to the cities in the eleventh and twelfth centuries, urban schools of various sorts, which to this point had been minor contributors to the educational enterprise, emerged from the shadow of the monastery schools and became major educational forces. This

Fig. 9.4. A grammar school scene. The master threatens his pupils with a club. Paris, Bibliothèque Nationale, MS Fr. 574, fol. 27r (14th c.).

development was assisted by reform movements within monasticism, aimed at reducing monastic involvement in the world and reemphasizing the spiritual nature of the monk's calling. Among the urban schools that came into prominence at this time were cathedral schools, schools run by parish clergy, and a wide variety of public schools, both primary and secondary, not directly linked to ecclesiastical needs but open to anybody who could afford them.[15]

The educational aims of the new urban schools were far broader than those of the monastery schools. The emphasis of the teaching program varied from one school to another, according to the vision and specialty of the master who directed it, but in general, urban schools expanded and reoriented the curriculum to meet the practical needs of a diverse and ambitious clientele, which would go forth to occupy positions of leadership in church and state. Even the cathedral schools, which resembled the monastery schools in having exclusively religious aims, based their curricula on a broader conception of the range of studies that would contribute to religious ends. And if the educational ambitions of a master or his students went beyond what could be sustained within the framework of the cathedral school, they might detach themselves from the cathedral and operate independently of its authority. Indeed, it was quite possible for "schools" to be migratory, rather than geographically fixed, and to follow the itinerary of a charismatic master whose teaching, wherever it might take place, bound the students together.[16] The product of these new arrangements was a rapid broadening of the curriculum: logic, the quadrivial arts, theology, law, and medicine came to be cultivated in the urban schools to a degree unheard of within the monastic tradition. The new schools multiplied in number and size; at their best, they emitted an aura of intellectual excitement that attracted the ablest masters and students into their orbits.

In France, some of the most robust schools were attached to (or operated in the shadows of) cathedrals in regions influenced by the Carolingian reforms of the ninth century. Laon was an early leader, with a significant cathedral school by 850 and a strong reputation in theology as late as the eleventh and twelfth centuries. In the tenth century Gerbert was attracted to the cathedral school at Reims as student and master. In the twelfth century, schools at Chartres, Orleans, and Paris emerged as leading centers for the liberal arts. The most famous of the twelfth-century schools is the cathedral school of Chartres—though the degree and duration of its preeminence have been called into question.[17] Schools in nearby Paris certainly flourished about the same time, offering instruction in a wide range of subjects, including the liberal arts. Outside of France, the leading schools were less apt to have any connection with cathedrals: Bologna became renowned for advanced legal instruction (by private teachers) early in the twelfth century, and by the end of the century Oxford (which had no cathedral) gained a reputation for studies in law, theology, and the liberal arts.

Several characteristics of these schools are important for our purposes. First, they witnessed a determined effort to recover and master the Latin classics (or Greek classics available in ancient Latin translations), surpassing anything seen in the early Middle Ages. Bernard of Chartres spoke for his age when he pictured his generation as dwarfs standing on the shoulders of giants, able to see farther not by virtue of individual brilliance but through mastery of the classics. Among the favorite Roman authors were the poets Virgil, Ovid, Lucan, and Horace. Cicero and Seneca were valued as moralists, and Cicero and Quintilian as models of eloquence. The logical works of Aristotle and his commentators (especially Boethius) were carefully studied and applied to all manner of subjects. Legal studies were critically dependent on the recovery of the *Digest*, a summary of Roman legal thought. And Martianus Capella, Macrobius, and Plato (through Calcidius's translation of the *Timaeus* and accompanying commentary) served as the principal sources for cosmology and natural philosophy. None of this is to suggest that the pagan classics displaced the Christian sources that had formed the core of religious education; rather, the newly recovered sources took their place alongside the Bible and the writings of the church fathers; it was assumed that these bodies of literature were mutually compatible and that recovery of the ancient classics was simply a matter of expanding the sources from which one might legitimately learn.[18]

Second, the urban schools, like European society more generally, saw a marked "rationalistic" turn—that is, an attempt to apply intellect and reason to many areas of human enterprise: attempts, for example, to bring order and rationality to commercial practices and the administration of church and state through record keeping and the development of accounting and auditing procedures. One historian has described this as a "managerial revolution."[19] The same confidence in human intellectual capacity pervaded the schools, where philosophical method was applied with increasing zeal to the whole of the curriculum, including biblical studies and theology.

The application of reason to theology was not new. As we have seen, the earliest Christian apologists undertook a reasoned defense of the faith; and scholars of the early Middle Ages (inspired by the example of Boethius) made a persistent effort to apply Aristotelian logic to knotty theological problems. The difference in the eleventh and twelfth centuries was in the lengths to which theologians were willing to go in the application of philosophical method. Anselm of Bec and Canterbury (1033–1109) is an excellent

Fig. 9.5. The chained library of Hereford Cathedral (England).

example.[20] Though perfectly orthodox in his theological beliefs, Anselm was prepared to stretch the limits of theological methodology: to explore what unaided reason could achieve in the theological realm, to ask whether certain fundamental theological doctrines were true as judged by rational or philosophical criteria. His best-known piece of theological argumentation is a proof of God's existence (known as the "ontological proof") in which he places no reliance on biblical authority. Anselm's purpose was entirely constructive; clearly he applied philosophical method to doctrines about the existence and attributes of God not because he doubted the doctrines but in order to buttress them and make them evident to nonbelievers. At first glance this may not seem particularly daring, but in fact the risks were serious: if reason can prove theological claims, presumably it can also disprove them. This is not a problem as long as reason arrives at the "right" answer; but what shall we do if, having committed ourselves to reason as the arbiter of truth, we find reason and theology in opposition?[21]

A generation after Anselm, Peter Abelard (ca. 1079–ca. 1142), a brilliant, restless, and abrasive student and teacher in the schools of northern France (including Paris and Laon), extended the rationalist program begun

Fig. 9.6. Hugh of St. Victor teaching in Paris. Oxford, Bodleian Library, MS Laud. Misc. 409, fol. 3v (late 12th c.).

by Anselm. In various works, he defended theological positions considered dangerous by his contemporaries and was twice condemned by the religious authorities. Abelard's best known book was entitled *Sic et non* (roughly translatable as *Yes and No* or *Pro and Con*); in this source book for students, he assembled conflicting opinions of the church fathers on a series of theological questions. He was using conflicting opinions to pose problems, which must then become the objects of philosophical investigation; in his view, the road to belief passes through doubt. There can be no question that Abelard intended to reason about, and in support of, the faith: he wrote on one occasion that he did not "wish to be a philosopher if it meant rebelling against [the Apostle] Paul, nor an Aristotle if it meant cutting [himself] off from Christ."[22] There is also no question that he was perceived by those of more conservative outlook, such as the monastic reformer Bernard of Clairvaux, who thundered against him, as a dangerous champion of philosophical method. The fact that Abelard attracted a following of enthusiastic students must have confirmed Bernard's worst fears.

In the work of Anselm and Abelard and like-minded contemporaries, we can see the makings of a confrontation between faith and reason. Anselm and Abelard raised in a compelling way such questions as: How does one "know" in the theological realm? Are the rational methods employed in other school subjects (logic, natural philosophy, and law) applicable also to theology, or does theology submit to some other master? How are conflicts between reason (Greek philosophy) and revelation (the truths revealed in the Bible) to be resolved? Worries about questions such as these jeopardized the intellectual revival and established an agenda for philosophers and theologians of the thirteenth and fourteenth centuries. The wholesale translation of Greek and Islamic philosophical and scientific literature, which was about to begin, would only intensify the problem. We will return to this subject below (chap. 10).

NATURAL PHILOSOPHY IN THE TWELFTH-CENTURY SCHOOLS

Natural philosophy did not hold center stage in the twelfth-century schools, but it did benefit from the general intellectual ferment. The determination among scholars to master the Latin classics extended to the classics of natural philosophy—Plato's *Timaeus* with Calcidius's commentary, Martianus Capella's *Marriage of Philology and Mercury*, Macrobius's *Commentary on the Dream of Scipio*, Seneca's *Natural Questions*, Cicero's *On the Nature of the Gods*, and the works of Augustine, Boethius, and John Scotus Eriugena. Most of these texts have a Platonic tilt (few of Aristotle's works were available), and the scholars who read and analyzed them were inevitably drawn to the Platonic conception of the cosmos. Plato's *Timaeus*, source of the most coherent discussion of cosmological and physical problems then available, and also the repository of Plato's own words, became the central text. That position of centrality, in turn, gave the *Timaeus* the power to shape the agenda and content of twelfth-century natural philosophy. This does not mean that the Platonism of the twelfth century was pure or entirely without rival: certain Stoic ideas managed to elbow their way into the Platonic milieu; toward the end of the century Aristotle's physical and metaphysical works began to make their presence felt; and in the thirteenth century Platonic philosophy would retreat before an Aristotelian onslaught. But for the time being, Plato held the position of leadership.[23]

However, Plato was a versatile guide, and Platonic leadership could mean many different things. The *Timaeus* is first and foremost an account of the formation of the cosmos by the divine craftsman. The obvious and most pressing task, therefore, was to apply all of the cosmology and physics that could be gleaned from Plato and the other ancients to elucidation of the Genesis account of creation. Science was expected to function as handmaiden.

A number of excellent scholars applied themselves to this project in the twelfth century. One of them was Thierry of Chartres (d. after 1156), a teacher of international renown at Chartres and (perhaps) Paris. Thierry wrote a commentary on the six days of creation, in which he managed to read the content of Platonic cosmology (along with pieces of Aristotelian and Stoic natural philosophy) into the biblical text. One of the major needs was to explain the specific sequence of God's creative activity as described in Genesis. According to Thierry, the four elements were created by God in the first instant of time; everything that followed was a natural unfolding of the order inherent in that initial creative act. Once created, fire immediately began to rotate (owing to its lightness, which forbids rest), while also illuminating the air, thus accounting for day and night (the first day of creation). During the second rotation of the fiery heaven, the fire heated the waters below, causing them to ascend as vapors until suspended above the air, forming what the biblical text refers to as the "waters above the firmament" (second day). Reduction in the quantity of water below, owing to vaporization, caused dry land to emerge from the seas (third day). Further heating of the waters above the firmament led to the formation of the celestial bodies, composed of water (fourth day). Finally, heating of the land and the lower waters brought forth plant, animal, and human life (fifth and sixth days).[24]

This has been a very brief and incomplete survey of Thierry's commentary, but it is enough to reveal the nature of the philosophical program on which he and various contemporaries had, under Platonic inspiration, embarked. Thierry's cosmology may not be sophisticated by modern standards. But the important thing is that, following Plato's lead, it restricts direct divine intervention to the initial moment of creation; what happens thereafter is the result of natural causation, as the elements move and interact in the manner proper to them, and as seeds (the "seminal causes" of Stoic philosophy, borrowed through Augustine) implanted in created things undergo a natural process of development. Even the appearance of Adam and Eve, and of subsequent humans, did not call for miraculous intervention.

This naturalism is one of the most salient features of twelfth-century natural philosophy. It is found in commentaries on the days of creation (perhaps the best place for a natural philosopher to display his naturalistic inclinations), but also in more general treatises on natural philosophy by scholars such as William of Conches, Adelard of Bath, Honorius of Autun, Bernard

Fig. 9.7. God as architect of the universe. Vienna, Österreichische Nationalbibliothek, MS2554, fol. 1v (13th c.).

Sylvester, and Clarembald of Arras (most of whom were connected with the schools of northern France). These men differed on matters of cosmological and physical detail, of course, but they shared a new conception of nature as an autonomous, rational entity, which proceeds without interference according to its own principles. Among scholars we see a growing awareness of natural order or natural law and a determination to see how far natural principles of causation would go in providing a satisfactory explanation of the world.[25]

An outspoken advocate of the new naturalism was William of Conches (d. after 1154), who studied and taught at Chartres or Paris or both, before joining the household of Geoffrey Plantagenet, where he tutored the future King Henry II of England. William developed an elaborate cosmology and physics based on Platonic principles (with important additions from some newly translated sources). In his *Philosophy of the World*, William lashed out at those who appeal too readily to direct divine causation:

> Because they are themselves ignorant of nature's forces and wish to have all men as companions in their ignorance, they are unwilling for anybody to investigate them, but prefer that we believe like peasants and not inquire into the [natural] causes [of things]. However, we say that the cause of everything is to be sought.... But these people, ... if they know of anybody so investigating, proclaim him a heretic.

William's purpose, as he made clear elsewhere, was not to deny divine agency, but to declare that God customarily works through natural powers and that the philosopher's task is to push these powers to their explanatory limit. Adelard of Bath (fl. 1116–42) made the same point about the same time, urging that only when "human knowledge completely fails should the matter be referred to God." And a little later, Andrew of St. Victor, discussing interpretations of biblical events, advised that "in expounding Scripture, when the event described admits of no natural explanation, then and then only should we have recourse to miracles."[26]

This may seem a sensible position, but it was also a dangerous one. How could a strong commitment to the search for natural causes (the position of these twelfth-century natural philosophers) avoid slipping into outright denial of the miraculous—a totally unacceptable outcome for a Christian scholar? Would scholars be capable of maintaining the delicate balance between belief and unbelief called for by this position? William of Conches addressed the problem directly, pointing to the difference between acknowledging that it was within God's power to perform some act and maintaining that God actually had performed it; surely God did not do everything of which he was capable. William added that his philosophical position (and that of his fellow "naturalists") did not detract from divine power and majesty, since whatever comes to pass is ultimately of divine origin: "I take nothing away from God; all things that are in the world were made by God, except evil; but he made other things through the operation of nature, which is the instrument of divine operation." Indeed, study of the physical world enables us to appreciate "divine power, wisdom, and goodness."[27] Searching for secondary causes is not a denial, but an affirmation, of the existence and majesty of the first cause.

Several other philosophical maneuvers could also help to relieve the tension. It was possible to reconcile the reality of miracles with the fixity of nature by acknowledging that miracles represent genuine suspensions of the usual laws of nature, while maintaining that these suspensions were planned by God from the original creation and built into the cosmic machinery, so that they remain perfectly natural in the larger sense. Moreover, one could speak about a fixed natural order without infringing on divine omnipotence and freedom by arguing (a) that God had unlimited freedom to create any kind of world he wished, but (b) that in fact he chose to make this world and, having completed his creative activity, was not going to meddle with the product. This latter distinction was to become crucial to developing thought on the subject in the thirteenth and fourteenth centuries.[28]

Some modern readers will be tempted to view all of this as unacceptable intrusion of theology into the scientific realm. However, if we wish to understand the twelfth century, it is important for us to realize that twelfth-century bystanders would have been apt to see these developments in precisely the opposite light—as a possibly dangerous intrusion of philosophy into the theological realm. What was new and threatening was not theological presence in philosophical precincts, where it had always made itself quite at home, but the flexing of philosophical muscle in territory where theology had hitherto ruled without challenge. To critics of the twelfth-century naturalists, it appeared that philosophy might be about to throw off her handmaiden status.

Let us briefly examine several other aspects of twelfth-century natural philosophy. The *Timaeus* and supporting sources not only promoted the idea of a fixed natural order; they also made humans a part of that order, governed by the same laws and principles, so that an exploration of human nature was understood to be continuous with exploration of the universe more generally. Frequently the point was made even more forcefully, through the macrocosm-microcosm analogy: humans not only belong to the cosmos but are actually miniatures of it. It follows that the cosmos and the individual person are linked by structural and functional similarities, which bind them into a tight unity. For example, just as the cosmos consists of the four elements, animated by the world soul (the exact nature of which gave rise to considerable debate in the twelfth century), so the human is a composite of body (the four elements) and soul.

Having made humankind part of the natural order, twelfth-century scholars increasingly took an interest in the "natural man" and his capacities—that is, humans as they are, independently of divine grace. (Thus historians sometimes write of twelfth-century "humanism.") In this connection, there was a strong tendency to affirm the value of human reason; as part of the natural order, and therefore sympathetic to its rhythms and harmonies, reason was regarded as a particularly suitable instrument for the exploration of the cosmos.[29]

Closely associated with the macrocosm-microcosm analogy was the science of astrology. Astrology had fallen into disrepute during the early Middle Ages, owing to the opposition of the church fathers. Augustine attacked it as a form of idolatry (since it had traditionally been associated with worship of planetary deities) and for its tendency to lead to fatalism and the denial of free will. But under the influence of twelfth-century Platonism, as well as an influx of Arabic astronomical and astrological literature in translation, astrology was restored to a position of at least quasi respectability. In the *Timaeus*, the Demiurge is credited with making the planets or celestial gods, but then delegating to them the responsibility of bringing forth subsequent forms of life in the lower regions. This suggestive account, coupled with the idea of cosmic unity, the macrocosm-microcosm analogy, and certain long-known correlations between celestial and terrestrial phenomena (the seasons and the tides), and augmented by newly translated Arabic astrological works, led to a resurgence of interest and belief in astrology. This is not the place to enter into a detailed analysis of astrological theory or practice (treated in chap. 11, below). What is important for our purposes is to notice that twelfth-century astrology had nothing to do with the supernatural; quite the contrary, it flourished among the naturalists of the twelfth century precisely because it entailed the exploration of the natural forces that link heaven and earth.[30]

Finally, did the mathematical tendencies of Platonic philosophy influence twelfth-century thought, as we might expect? Yes, but in a form that may surprise modern readers. Mathematics was employed in the first half of the twelfth century, not to quantify natural laws or to provide a geometrical representation of natural phenomena, but to answer questions that we would regard as metaphysical or theological. This is an exceedingly abstruse subject, which we cannot go into deeply, but one example may help to point the way. Following Boethius, twelfth-century scholars saw the theory of numbers (specifically, the relationship of the number 1 to the remaining numbers) as a vehicle for understanding the relationship between divine unity and the multiplicity of created things. It is the latter point that Thierry of Chartres was addressing when he wrote that "the creation of number is the creation of things." Mathematics also served in the twelfth century as a model of the axiomatic method of demonstration. A broader conception of the scientific uses of mathematics would have to await the translation and assimilation of Greek and Arabic mathematical science later in the century.[31]

THE TRANSLATION MOVEMENT

The revival of learning began as an attempt to master and exploit traditional Latin sources. However, before the end of the twelfth century it was transformed by the infusion of new books, containing new ideas, freshly translated from Greek and Arabic originals. This new material, first a trickle and eventually a flood, radically altered the intellectual life of the West. Up to this point, Western Europe had been struggling to reduce its intellectual losses; hereafter, it would face the altogether different problem of assimilating a torrent of new ideas.[32]

The separation between East and West had never been total, of course. There were always travelers and traders, and near the borders bilingual (or multilingual) people would be numerous. There were also diplomatic contacts between Byzantine, Muslim, and Latin courts: an early and significant case was the exchange of ambassadors (both of them scholars) between the courts of Otto the Great in Frankfurt and ʿAbd al-Raḥmān in Cordoba, which occurred about 950. Another kind of contact is illustrated by Gerbert's pilgrimage to northern Spain in the 960s to study Arabic mathematical science. Considered individually, such events may seem of small significance; but added together, they planted in Western minds an image of Islam and (to a lesser extent) Byzantium as repositories of intellectual riches. It became clear to Western scholars wishing to enlarge the body of knowledge in Latin Christendom that they could do no better than to make contact with these intellectually superior cultures.

The earliest translations from the Arabic—several treatises on mathematics, the astrolabe, and the armillary sphere—were made in the tenth century in Spain. A century later a North African named Constantine (fl. 1065–85) made his way to the monastery of Monte Cassino in southern Italy. There he began to translate medical treatises from Arabic to Latin, including the works of Galen and Hippocrates, which would supply the foundation of medical literature on which the West would build for several centuries.[33] These early translations whetted the European appetite for more. Beginning in the first half of the twelfth century, translation became a major scholarly activity, with Spain as the geographical focus. (Contact with the Middle East as a result of the Crusades had a minimal impact on translations.) Spain had the advantage of a brilliant Arabic culture, an ample supply of Arabic books, and communities of Christians and Jews who had been allowed to practice their religion freely under Muslim rule and who would now help to mediate between the two (or three) cultures. As the Christian reconquest of Spain gathered steam, centers of Arabic culture and libraries of Arabic books fell into Christian hands; Toledo, the most important center, fell in 1085, and in the course of the twelfth century the riches of its library began to be seriously exploited, thanks in part to generous patronage from the local bishops.

Some of the translators were native Spaniards, fluent in Arabic from childhood: such a man was John of Seville (fl. 1133–42), a Christian, who translated a large number of astrological works; another was Hugh of Santalla (fl. 1145), from one of the Christian states in the north of Spain, who translated texts on astrology and divination; yet another, and one of the ablest, was Mark of Toledo (fl. 1191–1216), who translated several Galenic texts. But others came from abroad: Robert of Chester (fl. 1141–50) came from Wales; Hermann the Dalmatian (fl. 1138–43) was a Slav; and Plato of Tivoli (fl. 1132–46) was Italian. These men came to Spain, presumably without prior knowledge of Arabic. Once there, they found a teacher, learned Arabic, and began to translate. Occasionally they joined forces with a bilingual native (perhaps a Christian or a Jew who knew Arabic and Latin, Hebrew, or the Spanish vernacular) and proceeded to translate cooperatively.

The greatest of the translators from Arabic to Latin was undoubtedly Gerard of Cremona (ca. 1114–87).[34] Gerard, from northern Italy, came to Spain in the late 1130s or early 1140s in search of Ptolemy's *Almagest*, which he had been unable to locate elsewhere. He found a copy in Toledo, remained to learn Arabic, and eventually rendered it into Latin. But he also discovered texts on all manner of other subjects, and over the next thirty-five or forty years (perhaps with the help of a crew of assistants)[35] produced translations of many of these. His output is absolutely astonishing: at least a dozen astronomical texts, including the *Almagest*; seventeen works on mathematics and optics, including Euclid's *Elements* and al-Khwārizmī's *Algebra*; fourteen works on logic and natural philosophy, including Aristotle's *Physics*, *On the Heavens*, *Meteorology*, and *On Generation and Corruption*; and twenty-four medical works, including Avicenna's great *Canon of Medicine* and nine Galenic treatises. The total comes to seventy or eighty books, all translated carefully and literally by a man who possessed a good command of both the subject matter and the languages.

Translation of Greek originals had never entirely ceased: recall Boethius in the sixth century and Eriugena in the ninth. But translation from the Greek resumed and dramatically accelerated in the twelfth century. Italy was the principal location, especially the south (including Sicily), which had always had Greek-speaking communities and libraries containing Greek books. Italy also benefited from ongoing contact with the Byzantine Empire. One of the important early translators was James of Venice (fl. 1136–48), a legal scholar in touch with Byzantine philosophers, who translated a collection of Aristotle's works. A series of important works on mathematics and mathematical science appeared in Greco-Latin translations about the middle of the century: Ptolemy's *Almagest* (whether before or after Gerard's translation from the Arabic cannot be determined) and Euclid's *Elements*, *Optics*, and *Catoptrics* (the former a prerequisite for any attempt to master the mathematical sciences).

Greco-Latin translating activity continued in the thirteenth century, most notably in the work of William of Moerbeke (fl. 1260–86). Moerbeke set out to provide Latin Christendom with a complete and reliable version of the Aristotelian corpus, revising existing translations where he could, producing new translations from the Greek where that was required. Moerbeke also translated some of the major Aristotelian commentators, a variety of Neoplatonic authors, and some mathematical works by Archimedes.[36]

Finally, a word about the motivation for the translations and the selection of materials to be translated. The aim was clearly utility, broadly defined. Medicine and astronomy led the way in the tenth and eleventh centuries; early in the twelfth century the emphasis seems to have been on astrological works, along with the mathematical treatises required for the successful practice of astronomy and astrology. Both medicine and astrology rested on philosophical foundations; and it was at least partly to recover and assess those foundations that attention was directed, beginning in the second half of the twelfth century and continuing through the thirteenth, toward the physical and metaphysical works of Aristotle and his commentators, including the Muslims Avicenna (Ibn Sīnā) and Averroes (Ibn Rushd). Once the full scope of Aristotle's works became known, of course, it became clear that his philosophical system was applicable to an enormous range of scholarly issues treated in the schools.[37]

By the end of the twelfth century, Latin Christendom had recovered major portions of the Greek and Arabic philosophical and scientific achievement;

in the course of the thirteenth century many of the remaining gaps would be filled. These books were copied and spread quickly to the great educational centers, where they contributed to an educational revolution. In the next chapter we will examine some of the struggles provoked by the newly translated materials.

THE RISE OF UNIVERSITIES

The typical urban school in the year 1100 was small, consisting of a single master or teacher and perhaps ten or twenty pupils. By the year 1200 these schools had grown dramatically in number and size. We have almost no quantitative data, but in leading educational centers like Paris, Bologna, and Oxford, students undoubtedly numbered in the hundreds (perhaps the high hundreds). Some idea of the explosion in the school population can be gained from the fact that more than seventy masters taught in Oxford between 1190 and 1209.[38] An educational revolution was in progress, driven by European affluence, ample career opportunities for the educated, and the intellectual excitement generated by teachers like Peter Abelard. Out of the revolution emerged a new institution, the European university, which would play a vital role in promoting the natural sciences. Let us briefly examine the process.

The absence of documentary evidence makes it impossible for us to trace in detail the steps by which universities came into existence. But what is clear is that the great expansion of educational opportunity at the elementary level (where instruction was offered in Latin grammar, the art of chanting, and basic arithmetic) led to demand among the intellectually ambitious for higher studies. Certain cities, such as Bologna, Paris, and Oxford, acquired a reputation for advanced studies in the liberal arts, medicine, theology, or law, and to these cities teachers and students gravitated in large numbers. Once there, a teacher would set up shop under the auspices of an existing school or as an independent, freelance teacher—advertising for students and teaching them individually or in groups for a fee (rather like a modern teacher of music or dance). Instruction would generally take place in quarters provided by the teacher.

With numerical expansion came the need for organization—to secure rights, privileges, and legal protection (since many of the teachers and students were foreigners, without the rights of the local citizenry), to gain control of the educational enterprise, and generally to promote their mutual well-being. Fortunately, there was an organizational model ready at hand in the guild structure that was developing at the same time within various trades and crafts; it was therefore natural for teachers and students to organize them-selves similarly into voluntary associations or guilds. Such a guild was called a "university [*universitas*]"—a term that originally had no scholarly or edu-cational connotations but simply denoted an association of people pursuing common ends. It is important to note, therefore, that a university was not a piece of land or a collection of buildings or even a charter, but an association or corporation of teachers (called "masters") or students. The fact that a uni-versity owned no real estate (to begin with) made it extremely mobile, and the early universities were thus able to use the threat of packing up and moving to another city as leverage to secure concessions from the local municipal authorities.

It is impossible to assign a precise date to the founding of any of the early universities for the simple reason that they were not founded, but emerged gradually out of preexisting schools—their charters coming after the fact. It is customary, however, to see the masters of Bologna as having achieved uni-versity status by 1150, those of Paris by about 1200, and those of Oxford by 1220 (map 9.2). Later universities were generally modeled on one or another of these three.[39]

Among the aims of these corporations were self-government and mo-nopoly—which amount to control of the teaching enterprise. Gradually the universities secured varying degrees of freedom from outside interference and thus the right to establish standards and procedures, to fix the curriculum, to set fees and award degrees, and to determine who would be permitted to teach or study. They managed this by virtue of high-level patronage from popes, emperors, and kings, who offered protection, guaranteed privileges, granted immunity from local jurisdiction and taxation, and generally took the side of the universities in a variety of power struggles. The universities were consid-ered vital assets, which needed to be carefully nurtured and (if circumstances dictated) judiciously disciplined. The extraordinary thing is how effective the nurturing proved to be, and how rarely and benevolently the disciplinary function was exercised. There were certain episodes, as we shall see, in which the church intervened decisively; but for the most part the universities man-aged the rare and remarkable feat of securing patronage and protection with only minimal interference.[40]

As the universities grew in size, internal organization was required. There were variations, of course, but Paris (the preeminent university in northern Europe) will serve to illustrate. Four faculties or guilds emerged at Paris: an undergraduate faculty of liberal arts (by far the largest of the four) and three graduate facilities—law, medicine, and theology. The liberal arts were

Map 9.2. Medieval universities and dates of founding.

considered preparatory for work in the graduate faculties, admission to which ordinarily depended on completion of the course of study in the arts. Because the masters in the arts faculty were more numerous than the teachers in the other faculties, they came to control the university.

A boy came to the university at about age fourteen, having previously learned Latin in a grammar school. In northern Europe, matriculation in the university generally conferred clerical status. This does not mean that students were priests or monks (or intending to become priests or monks), but simply that they were under the authority and protection of the church and had certain ecclesiastical privileges. The student enrolled under a particular master (the apprenticeship model should be kept in mind), whose lectures he followed for three or four years before being examined for the bachelor's (young man's) degree. If he passed the examination, he became a bachelor of arts, with the status of a journeyman apprentice, and was permitted to give certain types of lectures under the direction of a master (rather like the modern teaching assistant), while continuing his studies. At about age twenty-one, having heard lectures on all of the required subjects, he could take the examination for the MA (master of arts) degree. Passage of this examination brought the student full membership in the arts faculty, with the right to teach anything in the arts curriculum.

The universities were enormously large by comparison with Greek, Roman, or early medieval schools, but they fell far short of the mammoth public universities of the present. There were wide variations, of course, but a typical medieval university was comparable in size to a small American liberal arts college—with a student population falling somewhere between about 200 and 800. The major universities were considerably larger: Oxford probably had between 1,000 and 1,500 students in the fourteenth century; Bologna was of similar size; and Paris may have peaked at 2,500 to 2,700 students.[41] It is evident from these figures that university-educated people were but a minuscule fraction of the European population, but their cumulative influence over time should not be underestimated; that German culture, for example, was profoundly shaped by the more than 200,000 students who passed through the German universities between 1377 and 1520 seems indisputable.[42]

It would be a mistake to suppose that most of these students emerged from their university experience with a degree; the great majority dropped out after a year or two, having acquired sufficient education to meet their needs, run out of money, or discovered themselves unsuited to the academic life. Substantial numbers died before completing their studies—a reminder of the high mortality rates of the Middle Ages.[43] The student who did earn the M.A. was often required to teach for two years (because of a chronic shortage of teachers in the arts faculty); he might simultaneously embark on a degree program in one of the graduate faculties, which promised to lead to far more lucrative employment. Few masters of arts actually made an entire career of teaching in the faculty of arts. The program of studies in medicine (leading to the master's degree or doctorate—there was no difference) required five or six years beyond the M.A. degree; in law, about seven or eight additional years; and in theology, somewhere between eight and sixteen years of further study. This was a long and demanding program, and those who completed the master's degree in any of the graduate faculties belonged to a small scholarly elite.

We come at last to the curriculum. This evolved as the Middle Ages progressed, of course, but certain generalizations are possible.[44] First, it came to be understood that the seven liberal arts no longer provided an adequate

SCHOLA ASTRONOMIAE ET RHETORICAE

Fig. 9.8 *(above)*. Mob Quad, Merton College, Oxford. Dating from the fourteenth century, this is the oldest complete quadrangle in Oxford.

Fig. 9.9 *(left)*. Doorway to one of the late medieval schools, presently part of the Bodleian Library, University of Oxford.

framework within which to conceive the mission of the schools. Grammar declined in significance, yielding its place in the curriculum to a greatly expanded emphasis on logic. The mathematical subjects of the quadrivium, never prominent in the medieval schools, retained their low profile (with some exceptions, to be dealt with below). The arts curriculum was rounded out by the three philosophies: moral philosophy, natural philosophy, and metaphysics. And, of course, medicine, law, and theology came to be viewed as advanced subjects, covered in graduate faculties and requiring study in the arts as a prerequisite.

Second, where does this leave the subjects that we think of as scientific? We will look at the content of the various sciences in subsequent chapters; here the question concerns their place in the curriculum. The quadrivial arts were generally taught, but rarely stressed. Arithmetic and geometry, between them, occupied perhaps eight to ten weeks in the curriculum of the typical medieval undergraduate; but those who wanted more could frequently obtain it, at least in the larger universities. Astronomy was more highly cultivated, either as the art of timekeeping and establishment of the religious calendar or as the theoretical substructure for the practice of astrology (frequently in connection with medicine). The teaching texts were Greek and Arabic books in translation (including, on occasion, Ptolemy's *Almagest*) or new books written expressly for the purpose. The average level of astronomical knowledge must have been quite low, but there were times and places where the subject was taught with skill and sophistication; and there can be no doubt that the universities produced some highly proficient astronomers (see chap. 11).

If the mathematical sciences remained generally inconspicuous, Aristotelian natural philosophy became central to the curriculum. From modest beginnings late in the twelfth century, Aristotle's influence grew until, by the second half of the thirteenth century, his works on metaphysics, cosmology, physics, meteorology, psychology, and natural history became compulsory objects of study. No student emerged from a university education without a thorough grounding in Aristotelian natural philosophy. And finally, we must note that medicine had the good fortune of being cultivated within its own faculty.[45]

Third, one of the most remarkable features of this curriculum was the high degree of uniformity from one university to another. Before the universities emerged, different schools generally represented different schools of thought. In ancient Athens, for example, the Academy, the Lyceum, the Stoa, and the Garden of Epicurus were committed to the propagation of rival and (to some extent) incompatible philosophies. But the medieval universities, while differing somewhat in emphasis and specialty, developed a common curriculum consisting of the same subjects taught from the same texts.[46] This was largely a response to the sudden influx of Greek and Arabic learning through the translations of the twelfth century, which supplied European scholars with a standard collection of sources and a common set of problems. It was also connected, as both cause and effect, with the high level of mobility of medieval students and professors. Professorial mobility was facilitated by the *ius ubique docendi* (right of teaching anywhere) conferred on the master by virtue of completing his course of study. Thus a scholar who earned his degree at Paris could teach at Oxford without interference and, perhaps more importantly, without acquiring a case of intellectual indigestion; this was possible only because subjects taught at the one did not differ markedly in form or content from those same subjects as taught at the other. For the first time in history, there was an educational effort of international scope, undertaken by scholars conscious of their intellectual and professional unity, offering standardized higher education to generations of students.

Fourth, this standardized education communicated a methodology and a worldview based substantially on the intellectual traditions traced in the early chapters of this book. Methodologically, the universities were committed to the critical examination of knowledge claims through the use of Aristotelian logic. And the system of belief that emerged from the application of this method integrated the content of Greek and Arabic learning with the claims of Christian theology. We will deal below (especially in chap. 10) with struggles over the reception of the new learning and the form and content of the resulting synthesis. At present it is sufficient to note that these struggles were won by the liberal party, which wished to enlarge the store of European learning by assimilating the fruits of Greek and Arabic scholarship. Thus in the medieval universities, Greek and Arabic science (almost in their entirety) at last found a secure institutional home.

Finally, it must be emphatically stated that within this educational system the medieval master had a great deal of freedom. The stereotype of the Middle Ages pictures the professor as spineless and subservient, a slavish follower of Aristotle and the church fathers (exactly how one could be a slavish follower of both, the stereotype does not explain), fearful of departing an iota from the demands of authority. Broad theological limits did exist, of course, but within those limits the medieval master had remarkable freedom of thought and expression; there was almost no doctrine, philosophical or theological, that was not submitted to minute scrutiny and criticism by scholars in the medieval university. Certainly the master who specialized in the natural sciences would not have considered himself restricted or oppressed by either ancient or religious authority.

10 ❋ *The Recovery and Assimilation of Greek and Islamic Science*

The educational revival of the eleventh and twelfth centuries was broadened and transformed in the course of the twelfth century by the acquisition of new texts. In 1100, the revival could still be construed as an attempt to recover and master the Latin classics: Roman and early medieval authors, including the Latin church fathers, and a few Greek sources (Plato's *Timaeus* and parts of Aristotle's logic, for example) that existed in early Latin translations. A trickle of new translations from both the Greek and the Arabic had begun to flow, but their impact was still modest. A hundred years later, this trickle had become a torrent, and scholars found themselves struggling to assimilate and organize a body of new learning overwhelming in scope and magnitude.

The existence of these new resources was the central feature of intellectual life in the thirteenth century, setting an agenda that would preoccupy the best scholars of the century. The task was to come to terms with the contents of the newly translated texts—to master the new knowledge, organize it, assess its significance, discover its ramifications, work out its internal contradictions, and apply it (wherever possible) to existing intellectual concerns. The new texts were enormously attractive because of their breadth, their intellectual power, and their utility. But they were also of pagan origin; and, as scholars gradually discovered, they contained material that was theologically dubious. It was thus a sobering intellectual challenge that thirteenth-century scholars confronted; their approach to the new material and their skill in dealing with it would contribute permanently to the shape of Western thought.

Most of the translated works were theologically benign. The very fact that a text was translated tells us that somebody thought its usefulness outweighed any potential dangers. Technical treatises on all manner of subjects (mathematics, astronomy, statics, optics, meteorology, and medicine) were, in fact, received with unqualified enthusiasm: they were obviously superior to anything previously available on their respective subjects; in many cases they filled an intellectual void; and they contained no unpleasant philosophical or theological surprises. Thus Euclid's *Elements*, Ptolemy's *Almagest*, al-Khwārizmī's *Algebra*, Ibn al-Haytham's *Optics*, and Avicenna's *Canon of Medicine* were peacefully added to the corpus of Western knowledge. The process by which these and other technical treatises were mastered and assimilated will be treated in subsequent chapters.

Insofar as there was trouble, it appeared in broader subject areas that impinged on worldview or theology—subjects such as cosmology, physics, metaphysics, epistemology, and psychology. Central to these subjects were the works of Aristotle and his commentators, which successfully addressed a multitude of critical philosophical problems while promising untold future benefits from the proper employment of their methodology. The explanatory power of the Aristotelian system was obvious, and it proved exceedingly attractive to Western scholars. But it bestowed its benefits at a certain price, for Aristotelian philosophy inevitably touched on many issues already addressed by the blend of Platonic philosophy and Christian theology that had gradually become entrenched over the previous millennium. Unlike treatises on narrower, more technical subjects, Aristotelian philosophy did not fill an intellectual vacuum, but invaded occupied territory. This led to a variety of skirmishes, which ended (as we shall see) in a negotiated settlement. Let us examine the steps by which this occurred.

ARISTOTLE IN THE UNIVERSITY CURRICULUM

Most of Aristotle's works and some of the commentaries on them—especially those of the eleventh-century Muslim Avicenna (Ibn Sīnā)—were available in translation by 1200. We know very little about their early circulation or their use in the schools, but they seem to have made an appearance at both Oxford and Paris during the first decade of the thirteenth century. At Oxford, no obstacles arose in the next few decades to the slow, but steady, growth of Aristotelian influence.[1] At Paris, however, Aristotle ran into early trouble: allegations were made that pantheism (roughly, the identification of God with the universe) was being taught by masters of arts under Aristotelian inspiration. The outcome of these charges was a decree, issued by a council of bishops meeting in Paris in 1210 and reflecting conservative opinion in the faculty of theology, forbidding instruction on Aristotle's natural philosophy within the faculty of arts at Paris. This decree (still applicable only to Paris) was renewed in 1215 by the papal legate Robert de Courçon.[2]

Pope Gregory IX became directly involved in 1231, in the course of promulgating regulations governing the University of Paris. Gregory acknowledged the legitimacy of the ban of 1210, and renewed it, specifying that Aristotle's books on natural philosophy were not to be read in the faculty of arts until they had been "examined and purged of all suspected error." Gregory explained himself ten days later, in a letter appointing a commission to act on the matter: "Since the other sciences should serve the wisdom of holy Scripture, they are to be appropriated by the faithful insofar as they are known to conform to the good pleasure of the Giver." However, it had come to Gregory's attention that "the books on natural philosophy that were prohibited in a provincial council at Paris . . . contain both useful and useless matter." Therefore, "in order that the useful not be contaminated by the useless," Gregory admonished his newly appointed commission to "eliminate all that is erroneous or that might cause scandal or give offense to readers, so that when the dubious matter has been removed, the remainder may be studied without delay and without offense."[3]

What is noteworthy is that Gregory acknowledged both the utility and the dangers of Aristotelian natural philosophy. Aristotle remained under the ban until purged of error; but once error had been eliminated, scholars were encouraged to put him to use. It is also important to note that the commission appointed by Gregory seems never to have met, perhaps because one of its leading members, the theologian William of Auxerre, died within the year, and no purged version of Aristotle has ever been discovered. The subsequent acceptance of Aristotle was based on a complete, uncensored version of his works.

Various documents address the fate of Aristotle's works during the next twenty-five years. They reveal that the bans of 1210, 1215, and 1231 were partially successful for a time, but that they began to lose their effectiveness around 1240. One reason for this may have been Gregory IX's death in 1241, after which his regulations of a decade earlier may have lost some of their coercive power; another may have been a growing awareness among the Parisian masters of arts that they were losing ground (and reputation) to their counterparts at Oxford and other universities. We should also reckon with the possibility that the teaching of Aristotle's logic (not covered by the bans), ready availability of Aristotle's works on natural philosophy (despite the ban on teaching them), and the recovery of new Aristotelian commentators (especially Averroes) elevated Aristotle's reputation to the point where the juggernaut of Aristotelian philosophy was unstoppable. And, of course, we need to remember that it had always been legitimate for theologians to make such use of Aristotle as they saw fit. Whatever the causes, Aristotle's works on natural philosophy seem to have become the subject of lectures in the faculty of arts in the 1240s or shortly before; one of the earliest to teach them was Roger Bacon.[4] About the same time, more liberal attitudes toward the use of Aristotle were penetrating the theological faculty at Paris, as we see in a growing tendency to allow Aristotelian philosophy to shape theological speculation and thought. By 1255 the tables had turned completely, for in that year the faculty of arts passed new statutes making mandatory what was apparently already the practice—namely, lectures on all known works of Aristotle. Aristotle's natural philosophy had not merely created a place for itself in the arts curriculum; it had become the principal ingredient.

POINTS OF CONFLICT

It is time to pinpoint the features of Aristotelian philosophy that were worth worrying or fighting about. But first we must note that the content of Aristotle's philosophy, as understood by his Western readers, was in a state of flux. Because Aristotle was exceedingly difficult to follow, readers naturally availed themselves of whatever explanatory aids they could acquire; fortunately, commentators in both late antiquity and medieval Islam had paraphrased Aristotle or explained difficult points in the various Aristotelian texts, and the works of these commentators were progressively translated along with Aristotle's own works and used wherever Aristotle was seriously studied. In the closing decades of the twelfth century and the early decades of the thirteenth, the principal commentator was the Muslim Avicenna (Ibn Sīnā, 980–1037), who presented a Platonized version of Aristotelian philosophy.[5] The charge of pantheistic teaching at Paris made in 1210 undoubtedly reflects the inroads of Avicenna's Neoplatonic reading of Aristotle. However, beginning about 1230, the commentaries of Avicenna began to be displaced by those of the Spanish Muslim Averroes (Ibn Rushd, 1126–98).[6] There is no question that Averroes too was capable of extending or distorting Aristotle's meaning, and that he did so on occasion, but in general the switch from Avicenna's guidance to that of Averroes meant a return to a more authentic, less Platonized, version of Aristotelian philosophy. So influential did Averroes become in the West that he came to be known simply as "the Commentator."

What was there in the Averroistic (or more authentic) reading of Aristotle that caused trouble? Some specific claims appeared (with varying degrees of clarity) to violate orthodox Christian doctrine; and beneath these claims there lay a general outlook, rationalistic and naturalistic in tone, which struck some

Fig. 10.1. The beginning of Avicenna's *Physics* (*Sufficientia*, pt. II), Graz, Universitätsbibliothek, MS 11.482, fol. 111r (13th c.)

observers as antithetical to traditional Christian thought. The simplest way of discussing these issues will be to begin with the specific claims.

A prominent feature of the Aristotelian cosmos was its eternity, defended by a variety of arguments in a variety of Aristotle's works. Bearing, as it did, on the doctrine of creation, this was a claim that could hardly be overlooked by Aristotle's Christian readers. Aristotle's position was that the cosmos did not come to be and cannot cease to be. The elements, he argued, have always behaved according to their natures; consequently, there cannot have been a moment when the universe as we know it came into being, and no moment will come when it ceases to be; it follows that the universe is eternal. Aristotle thus repudiated the evolutionary cosmology of the pre-Socratic philosophers.[7]

From a Christian standpoint, however, this is an intolerable conclusion. Not only does the Bible contain an account of creation in the opening chapters of Genesis, but the absolute dependence of the created universe on the Creator was fundamental to Christian conceptions of God and the world. Consequently, among Aristotle's thirteenth-century Christian commentators, we find an unbroken string of attempts to resolve this problem.[8] We will consider some of the arguments below.

Another problem, also bearing on the relationship between Creator and creation, is that of determinism. The question of deterministic tendencies in Aristotle's natural philosophy is a very thorny one. What needs to be stated here is that the universe as he described it contains unchangeable natures, which are the basis of a regular cause-and-effect sequence. This is particularly evident in the heavens, where that which is will always be. Moreover, Aristotle regarded the deity, the Prime Mover, as eternally unchanging and therefore incapable of intervening in the operation of the cosmos; the cosmic machinery thus runs inevitably and unchangeably onward, initiating a chain of causes and effects that descend into and pervade the sublunar realm. The danger here is that within the Aristotelian framework no room can be found for miracles.[9] And finally, attached to Aristotelian philosophy were astrological theories, which threatened the freedom of human choice (essential to Christian teaching on sin and salvation) if celestial influences could be shown to act on the will.

All of these deterministic tendencies or elements were viewed in the thirteenth century as a challenge to Christian doctrine—particularly divine freedom and omnipotence, divine providence, and miracles. Aristotle's Prime Mover, who does not even know of the existence of individual humans and certainly does not intervene on their behalf, is a far cry from the Christian God, who knows when the sparrow falls and numbers the very hairs on our head.[10]

As a final example of troublesome Aristotelian ideas, we turn to the nature of the soul. Aristotle argued that the soul is the form or organizing principle of the body—the full actualization of the potentialities inherent in the matter of the individual. It follows that the soul cannot have independent existence, since form, even if it can be distinguished from matter, cannot exist independently of matter. To suppose that the soul might be separated from the body would be as foolish as to suppose that the sharpness of an axe could be separated from the matter of the axe. At death, therefore, when the individual disintegrates, its form or soul simply ceases to be.[11] Such a conclusion is clearly incompatible with Christian teaching on the immortality of the soul.

The immortality of the individual soul was called into question also by another psychological doctrine developed by Averroes, as he attempted to work out certain difficulties in Aristotle's epistemology. The full Averroistic theory, known as "monopsychism," is extremely intricate. What is important for us is his claim that the immaterial and immortal part of the human soul, the "intellective soul," is not individual or personal but a unitary intellect shared by all humans. It would seem to follow that what survives death is not personal but collective; immortality is preserved, but not personal immortality. The violation of Christian teaching is again clear.[12]

Such claims as these were not isolated pieces of philosophy, but manifestations of basic attitudes toward reason and its proper relationship to faith and theology; they entered Western Europe as specific instances of an outlook and a methodology. The champions of the new Aristotelianism were inclined to enlarge the scope of rational activity, naturalistic explanation, and Aristotelian demonstration; philosophy was their game, and they wished to display its virtues in every intellectual arena. When philosophy entered the theological faculty and began to influence theological method, coming to rival biblical studies as the focus of theological education, traditionalists understandably reacted in anger and frustration. Charges of intellectual arrogance and vain curiosity became commonplace. Were the articles of the faith to be tested by the content and methods of pagan philosophy? Were the teachings of Christ, the Apostle Paul, and the church fathers subordinate to those of Aristotle?

A particularly acute example of this outlook in the realm of natural philosophy was the tendency to restrict analysis to causal principles discoverable through the exercise of human observation and reason, without regard for the teachings of biblical revelation or church tradition. Divine or supernatural causation was never denied, but it was placed (by the more aggressive proponents of the new methodology) outside the province of natural philosophy. This naturalism, the seeds of which are visible in such twelfth-century thinkers as William of Conches (see chap. 9, above), blossomed under the stimulus of Aristotle and his commentators. Perhaps the most threatening manifestation of these naturalistic inclinations was the growing tendency of some philosophers to make a distinction between "speaking philosophically" and "speaking theologically" and, what is much worse, to acknowledge that philosophical and theological methods might lead to incompatible conclusions.

The advocates of the new methods no doubt saw the introduction of philosophical rigor into theological debate as a great step forward. But to the traditionalists, this seemed to be a serious case of insubordination and a violation of traditional distinctions between the enterprises of philosophy and theology. Viewed in the worst light, it appeared that Jerusalem was being asked to yield to the authority of Athens.

Before considering thirteenth-century attempts to resolve these difficulties, we must briefly examine the institutional framework within which the attempts took place. The debates over the new Aristotle were scholarly in nature, and all of the participants were products of the universities. Many were active teachers; others were university alumni who had ascended to positions of leadership and authority in the church. It will help us to understand the persistent medieval tendency to mingle philosophy and theology if we grasp the career patterns of university scholars: virtually all theologians had studied philosophy in the arts faculty before embarking on theological studies;

Fig. 10.2. The Basilica of St. Francis, Assisi. Begun within a few years of Francis's death in 1226, to contain his tomb, this church became the "head and mother" of the Franciscan order and an important pilgrimage site. Courtesy of Christopher Kleinhenz.

moreover, theological students frequently found themselves teaching simultaneously in the faculty of arts, as a means of support. Consequently, some of the most influential philosophical treatises of the Middle Ages were written by scholars who were teaching philosophy while studying theology.[13]

Some of the leading figures, by the middle of the century, were Franciscans or Dominicans—members, that is, of the mendicant orders founded early in the thirteenth century. The mendicants were "regular clergy," because they lived under a *regula* or rule (which included a vow of poverty), unlike "secular clergy" (such as parish priests) who did not. In contrast to the monastic orders, which stressed withdrawal from the world in pursuit of personal holiness, the mendicants were committed to an active ministry within an urban setting; this eventually propelled them into the educational arena, including the universities, where they became actively involved in all of the great philosophical and theological controversies.

These institutional details contributed in subtle ways to the intellectual developments with which we are concerned. The struggles surrounding the new learning were not purely ideological, but were complicated by disciplinary and institutional affiliations and rivalries. Philosophers and theologians were united by the educational experience of the arts faculty; but this did not prevent them from skirmishing periodically over disciplinary boundaries. Within theology, the mendicants were locked for a time in a power struggle with secular (nonmendicant) theologians at the University of Paris over the right to hold professorial chairs. And within the mendicant orders, Franciscans and Dominicans developed somewhat different philosophical loyalties and characteristic approaches to the problem of faith and reason. If we wish to gain a nuanced understanding of the course of events, we need to be sensitive to these disciplinary and institutional undercurrents.

RESOLUTION: SCIENCE AS HANDMAIDEN

Despite the dangers that we have recounted, Aristotelian philosophy proved too attractive to ignore or suppress permanently. Since the translations of Boethius early in the sixth century, Aristotle's name had been synonymous with logic, and that logic had insinuated itself deeply into scholarship on almost every subject; now an enlarged corpus of Aristotelian logic was available and ready to go to work. Aspects of Aristotelian metaphysics had also filtered through the literature of the early Middle Ages, and now, with access to the full Aristotelian text, Western scholars had in their hands a powerful engine for understanding and analyzing their universe. Form, matter, and substance, actuality and potentiality, the four causes, the four elements, contraries, nature, change, purpose, quantity, quality, time, and space—Aristotle's discussion of all of these topics, and more, furnished a persuasive conceptual framework through which to experience and talk about the world. In his various psychological works Aristotle treated the soul and its faculties, including sense perception, memory, imagination, and cognition. He also offered a cosmology, in which the universe was convincingly mapped and its operation explained, from the outermost heaven to the earth at the center. Aristotle gave an account of motion, of what we would call matter theory, and of meteorological phenomena that went far beyond anything previously available. And finally, he offered a biological corpus unmatched for size and for descriptive and explanatory detail. It was inconceivable that these intellectual treasures would be simply repudiated; and there was never a serious movement with that as its aim. The problem was not how to eradicate Aristotelian influence, but how to domesticate it—how to deal with points of conflict and to negotiate boundaries, so that Aristotelian philosophy could be put to work on behalf of Christendom.

The process of reconciliation began as soon as the works of Aristotle and his commentators became available. An early attempt was made by Robert Grosseteste (ca. 1168–1253), a formidable Oxford scholar and first chancellor of the university (fig. 10.3). Though not himself a Franciscan, Grosseteste was the first lecturer in the Franciscan school at Oxford, thereby exercising a formative influence on the intellectual life of the order. Grosseteste's commentary on Aristotle's *Posterior Analytics*, written probably in the 1220s, was one of the earliest efforts to deal seriously with Aristotle's scientific method.[14] Grosseteste was also acquainted with Aristotle's *Physics*, *Metaphysics*, *Meteorology*, and biological works; he reveals their influence in his commentary on the *Physics* and in a series of short treatises on various physical subjects. However, Grosseteste's intellectual formation was strongly shaped by Platonic and Neoplatonic influences, and also by some of the newly translated works on mathematical science; and in his physical works we find a rather uneasy juxtaposition of Aristotelian and non-Aristotelian elements. Grosseteste's cosmogony (his account of the origin of the cosmos), for example, while set within a broadly Aristotelian framework, should be seen primarily as an attempt to reconcile Neoplatonic emanationism—the idea that the created universe is an emanation from God, as light is an emanation from the sun—with the biblical account of creation ex nihilo.[15]

Important aspects of Grosseteste's program were continued by a younger Englishman, Roger Bacon (ca. 1220–ca. 1292). An admirer of Grosseteste (but probably never his student), Bacon was inspired by Grosseteste's scholarly example, especially his mastery of the mathematical sciences. The details of Bacon's education are obscure, but it is clear that he studied at both Oxford and Paris. He began to teach in the faculty of arts at Paris in the 1240s, where he was one of the first to lecture on Aristotle's books on natural philosophy—specifically, the *Metaphysics*, *Physics*, *On Sense and the Sensible*; probably *On Generation and Corruption* (which deals with theory of matter), *On the Soul*, and *On Animals*; and perhaps *On the Heavens*.[16] Later he joined the Franciscan order and spent the remainder of his life in study and writing.

We will deal with various aspects of Bacon's scientific thought in subsequent chapters; what is important here is his campaign to save the new learning from its critics. Bacon's major scientific writings were not "pure" pieces of philosophy or science, but passionate attempts to persuade the church hierarchy (these works were addressed to the pope) of the utility of the new learning—not just Aristotelian philosophy, but the totality of new literature on natural philosophy, mathematical science, and medicine.

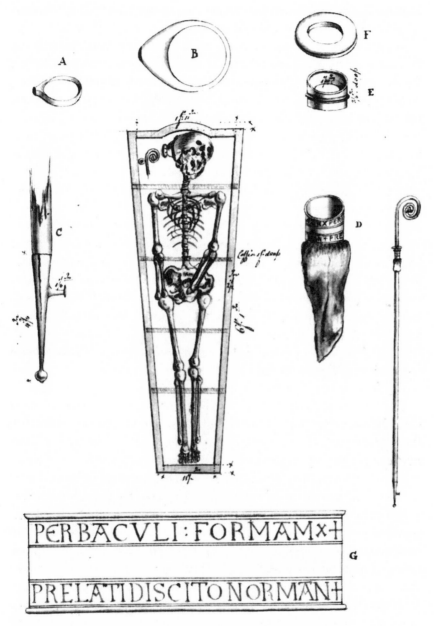

Fig. 10.3. The skeleton of Robert Grosseteste. Sketched when Grosseteste's tomb in Lincoln Cathedral was opened in 1782, this is one of a very few drawings of a medieval scholar made, so to speak, "from life." Shown with the skeleton are the other items found in the coffin, including the episcopal ring and remains of the pastoral staff. For a fuller description, see D. A. Callus, ed., *Robert Grosseteste, Scholar and Bishop*, pp. 246–50. By permission of The Natural History Museum, London.

Bacon argued that the new philosophy is a divine gift, capable of proving the articles of the faith and persuading the unconverted, that scientific knowledge contributes vitally to the interpretation of Scripture, that astronomy is essential for establishing the religious calendar, that astrology enables us to predict the future, that "experimental science" teaches us how to prolong life, and that optics enables us to create devices that will terrorize unbelievers and lead to their conversion. The object of Bacon's campaign was to take the handmaiden formula of Augustine and apply it to new circumstances, in which the quantity of purported knowledge waiting to be enlisted as handmaiden was vastly larger and more complicated.[17] The natural sciences were thus justified by their religious utility. There is "one perfect wisdom," Bacon argued in his *Opus maius*,

> and this is contained in holy Scripture, in which all truth is rooted. I say, therefore, that one discipline is mistress of the others-namely, theology, for which the others are integral necessities and which cannot achieve its ends without them. And it lays claim to their virtues and subordinates them to its nod and command.[18]

Theology does not oppress the sciences, in Bacon's view, but puts them to work, directing them to their proper end.

As for the points of alleged conflict with Christian belief, Bacon dismissed them as problems arising from faulty translation or ignorant interpretation; if philosophy is truly God-given, there can be no genuine conflict between it and the articles of the faith. To reinforce this point, Bacon marshaled the authority of Augustine and other patristic writers who urged Christians to reclaim philosophy from its pagan possessors. And just in case these arguments failed, he shouted down his critics with a blast of rhetoric about the wonders of science.

Despite Bacon's enthusiasm, a cautious attitude toward the new philosophy, especially the new Aristotle, became typical of the Franciscan order about midcentury. One of the people most instrumental in shaping this attitude was the Italian Franciscan Bonaventure (ca. 1217–74). Bonaventure studied both the liberal arts and theology at the University of Paris, then remained to teach theology from 1254 to 1257, resigning to become minister general of the Franciscan order. There can be no doubt that Bonaventure respected Aristotelian philosophy, drawing his logic and much of his metaphysics from it; but, like Grosseteste and Bacon, he was heavily influenced by Augustine and the Neoplatonic tradition, and in his thought we find a rich synthesis of Aristotelian and non-Aristotelian elements.

Bonaventure certainly agreed with Bacon on the validity and applicability of Augustine's handmaiden formula: pagan philosophy was an instrument, to be used for the benefit of theology and religion. But he was much more cautious than Bacon about the utility of philosophy and more sharply aware of the risks of promoting it. He was pessimistic about the capacity of reason alone, without the assistance of divine illumination, to discover truth; and as a result, he was apt to keep philosophy on a short leash and quick to abandon Aristotle or his commentators on any issue where they departed from the teachings of revelation. Thus he flatly rejected any possibility of an eternal world; he defended the immortality of the individual soul, dismissing monopsychism and arguing that each soul is itself a substance (a composite of spiritual form and spiritual matter) that survives the dissolution of the body; and he vigorously fought any suggestion of astrological determinism. Finally, in opposition to Aristotelian naturalism, Bonaventure stressed God's providential participation in every case of cause and effect.[19]

In the careers of Grosseteste, Bacon, and Bonaventure we can see several important tendencies of the early and middle years of the thirteenth century: growing knowledge of the Aristotelian corpus, a mixture of admiration and suspicion about its contents, and a tendency to read various Augustinian or Platonic ideas into the Aristotelian text. Two Dominicans active in the middle and later years of the century, Albert the Great and Thomas Aquinas, contributed to a much fuller mastery of Aristotelian philosophy and a more open attitude toward its claims.

German by birth, Albert the Great (ca. 1200–1280) was educated at Padua and the Dominican school in Cologne (fig. 10.4). In the early 1240s he was sent to Paris to study theology, becoming master of theology in 1245. For the next three years he occupied one of the two Dominican professorships at Paris. Thomas Aquinas studied under him during this period, and when Albert was called back to Cologne in 1248 to reorganize the Dominican school there, Thomas accompanied him. Most of Albert's Aristotelian commentaries were composed after his departure from Paris; they are not (except for his commentary on Aristotle's *Ethics*) the products of Albert's teaching, but extracurricular writings intended for the benefit of Dominican friars.[20]

Albert was the first to offer a comprehensive interpretation of Aristotle's philosophy in Western Christendom; on these grounds, he is often regarded as the effective founder of Christian Aristotelianism. This should not be taken to mean that Albert achieved philosophical purity; some of his early commentaries were devoted to Neoplatonic authors, and to the end of his life he retained allegiance to portions of Platonic philosophy; moreover, he was always

Fig. 10.4. Albert the Great. Fresco by Tommaso da Modena (1352), located in the Monastery of San Niccolò, Treviso. Alinari/Art Resource N.Y.

ready to correct or discard Aristotelian doctrines that he considered false and to introduce pieces of truth found elsewhere. Nonetheless, Albert perceived the profound significance of Aristotelian philosophy and set out to interpret the whole of it for his fellow Dominicans. In the prologue to his commentary on Aristotle's *Physics*, he explained:

> Our purpose . . . is to satisfy as far as we can those brethren of our order who for many years now have begged us to compose for them a book on physics in which they might find a complete exposition of natural

science and from which also they might be able to understand correctly the books of Aristotle.[21]

Albert responded not only with a *Physics* commentary but with commentaries on, or paraphrases of, all available Aristotelian books—an output that occupies twelve fat volumes (more than eight thousand pages) in the nineteenth-century edition of Albert's works. Included in these commentaries are long digressions, in which Albert lays out the results of his own investigations and reflections. Nobody before Albert had given the Aristotelian corpus this kind of painstaking attention, and few have done it since.

His purpose in doing so was to exhibit and make available the explanatory power of Aristotelian philosophy, which he regarded as the necessary preparation for theological studies. He had no intention of releasing Aristotelian philosophy from handmaiden status, but he did mean to give it substantially larger responsibilities. Among Albert's contemporaries, only Roger Bacon had as grand a vision of the importance of the new learning for the practice of theology; but, apart from his youthful lectures on Aristotle at Paris, Bacon devoted his best efforts to the mathematical sciences (especially optics) and the writing of propaganda on behalf of the new learning in general, while Albert committed himself to the work of mastering and interpreting the Aristotelian corpus. Historians have tended to honor those who broke with the Aristotelian tradition; Albert deserves our attention and respect as the man who put Western Christendom fully in touch with the Aristotelian tradition.

At the same time, Albert recognized his obligation to supplement the Aristotelian text on subjects that Aristotle had overlooked or explored superficially, and to correct Aristotle wherever he was wrong; though mightily impressed by Aristotle's achievement, Albert was never tempted to become his slave. To this end, Albert read everything he could lay his hands on: he was heavily dependent on Avicenna; he knew the works of Plato, Euclid, Galen (to a limited degree), al-Kindī, Averroes, Constantine the African, and a host of other Greek, Arabic, and Latin authors. And he brought these other sources to bear, whenever they were relevant, on the problems that he confronted as he interpreted the Aristotelian text.[22]

Albert was also a remarkably acute firsthand observer of plant and animal life: for example, he corrected Avicenna regarding the mating of partridges on the basis of personal observation and reported that he had visited a certain eagle's nest six years in a row. He was surely the best field biologist of the entire Middle Ages.[23] His intellectual energy was boundless, and his nontheological writings (less than half the total) include works on physics, astronomy, astrology, alchemy, mineralogy, physiology, psychology, medicine, natural history, logic, and mathematics. The authority with which he could address any subject explains why Albert was referred to as "the great" already during his lifetime; it also helps to explain why Roger Bacon (who was intolerant of intellectual rivals) viewed him with such hostility.

What did Albert have to say about the sensitive Aristotelian doctrines that had led to the banning of Aristotle early in the century and still threatened the acceptance of his works? On the critical problem of the eternity of the world, Albert never wavered in his commitment to the Christian doctrine of creation. His early view was that philosophy is incapable of addressing this issue definitively, so that one's obligation is simply to accept the teaching of revelation. Later he became convinced that the idea of an eternal universe is philosophically absurd, so that philosophy could settle the matter without theological assistance. In neither case did philosophy (properly practiced) and theology find themselves at odds.

Albert devoted substantially more attention to the nature of the human soul and its faculties. The trick was to account for the soul as a separate immortal substance, independent of the body and able to survive its death, while also accounting for the unification of the soul with the body, as the agent of perception and vitality. Albert could see no way of defending the immortality of the soul without denying Aristotle's claim that the soul is the form of the body. In its place, he substituted the opinion of Plato and Avicenna, that the soul is a spiritual and immortal substance, separable from the body. It was not necessary, however, to repudiate Aristotle totally: Albert argued that although soul is not actually the form of the body, It performs the functions of form.[24]

Finally, how did Albert respond to the "rationalism" of Aristotelian philosophy—the commitment, that is, to the application of philosophical method in all areas of human enterprise? In setting himself the task of showing his colleagues how to look at the world through Aristotelian eyes, Albert was committing himself to a fairly strong form of the rationalist program. He proposed to distinguish between philosophy and theology on methodological grounds and to find out what philosophy alone, without any help from theology, could demonstrate about reality. Moreover, Albert did nothing to diminish or conceal the "naturalistic" tendencies of the Aristotelian tradition. He acknowledged (with every other medieval thinker) that God is ultimately the cause of everything, but he argued that God customarily works through natural causes and that the natural philosopher's obligation was to take the latter to their limit. What is remarkable is Albert's willingness to adhere to this methodological prescription even in his discussion of a biblical miracle—Noah's flood. Noting that some people wish to confine the discussion of floods (including Noah's) to a statement of divine will, Albert pointed out that God employs natural causes to accomplish his purposes; and the philosopher's task is not to investigate the causes of God's will, but to inquire into the natural causes by which God's will produces its effect. To introduce divine causality into a philosophical discussion of Noah's flood would be a violation of the proper boundaries between philosophy and theology.[25]

Albert's program of understanding and disseminating Aristotelian philosophy, while respecting its utility for theology and religion, was continued by his pupil Thomas Aquinas (ca. 1224–74). Thomas, who was born into the minor nobility in south-central Italy, received his first education at the ancient Benedictine Abbey of Monte Cassino (founded by Benedict of Nursia in the sixth century); he continued his studies in the faculty of arts at the University of Naples, where he was introduced to Aristotelian philosophy. After joining the Dominican order, Thomas was sent to Paris, earning the theological doctorate there in 1256. He devoted the remainder of his life to teaching and writing, including two stints teaching theology at Paris, 1257–59 and 1269–72.

Like Albert, Thomas hoped to solve the problem of faith and reason by defining the proper relationship between pagan learning and Christian theology.[26] Against those who would dismiss philosophy as contrary to the faith, he argued that

> even though the natural light of the human mind [i.e., philosophy] is inadequate to make known what is revealed by faith, nevertheless what is divinely taught to us by faith cannot be contrary to what we are endowed with by nature. One or the other would have to be false, and since we have both of them from God, he would be the cause of our error, which is impossible.[27]

Aristotelian philosophy and Christian theology, though methodologically distinct, are compatible roads to truth. Philosophy employs the natural human faculties of sense and reason to arrive at such truths as it can. Theology offers access to truths given by revelation that go beyond our natural capacities to discover and understand. The two roads may sometimes lead to different truths, but they never lead to contradictory truths.

Does this mean that philosophy and theology are equals? Certainly not. Thomas points out that theology is to philosophy as the complete to the incomplete, the perfect to the imperfect. If this is so, why go to the trouble of doing philosophy? Because it renders vital services to the faith. In the first place, it can demonstrate what Thomas calls "the preambles of the faith"— certain propositions, such as God's existence or his unity, which faith takes for granted as its starting points. Second, philosophy can elucidate the truths of the faith by the use of analogies drawn from the natural world; Thomas refers to the doctrine of the trinity as a case in point. And third, philosophy can disprove objections to the faith.[28]

This may seem like a simple reassertion of the Augustinian handmaiden formula, but in fact Thomas has subtly but significantly altered its content. The handmaiden named "philosophy" is still subordinate to the theological enterprise, and therefore still a handmaiden; but in Thomas's view, she has amply demonstrated her usefulness and her reliability, and he therefore offers her enlarged responsibilities and elevated status. Thomas also thinks she will do her job better if relieved of overly close theological supervision. Philosophy and theology both have their spheres of competence, he argues, and each can be trusted within its proper sphere: if we wish to know the details or causes of planetary motion, for example, we must look to the philosophers; on the other hand, if we wish to understand the divine attributes or the plan of salvation, we must be prepared to enter the precincts of theology. In Thomas there is a respect for the philosophical enterprise, and a determination to employ it whenever possible, that takes him beyond the Augustinian position and places him in the forefront of the liberal or progressive wing of theologians in the second half of the thirteenth century.

Despite the methodological differences between philosophy and theology, there are regions where they overlap. For example, the existence of the Creator is known both by reason and by revelation; it can be proved by the philosopher, but it is also given to us by Scripture, as expounded by the theologian. What rules govern the relations of theology and philosophy in such cases? The fundamental principle is that there can be no true conflict between theology and philosophy, since both revelation and our rational capacities are God-given. Any conflict must, therefore, be apparent rather than real—the result of bad philosophy or bad theology. The remedy in such cases is to reconsider both the philosophical and the theological argument.

How did this prescription work itself out in Thomas's practice? In particular, how successfully was it applied to the troublesome Aristotelian doctrines enumerated in the preceding section of this chapter? The short answer is that Thomas confronted all of the problems raised by Aristotelian philosophy, and did so with impressive rigor. He directly addressed the

Aristotelian controversies in two books: *On the Eternity of the World* and *On the Unicity of the Intellect, against the Averroists* (concerned with monopsychism and the nature of the soul). His position on the eternity of the world was that we know, thanks to revelation, that the world was created at a point in time, but that philosophy cannot settle the matter one way or the other. Those (like Bonaventure) who argued that the eternity of the world was philosophically absurd were wrong, for it is not contradictory to maintain that the universe is created (that is, dependent on divine creative power for its existence) and yet that it has existed eternally. On the nature of the soul, Thomas agreed with Aristotle that the soul is the substantial form of the body (that which combines with the matter of the body to produce the individual human being); but he argued that this form is a special kind of form, capable of existing independently of the body and therefore imperishable. He also claimed that this solution was compatible with Aristotle's own thought.[29]

This, then, is Thomas's solution to the problem of faith and reason. He has made room for both, subtly merging Christian theology and Aristotelian philosophy into what we may call "Christian Aristotelianism." In the process it was necessary for Thomas to Christianize Aristotle by confronting and wrestling with the Aristotelian doctrines that appeared to conflict with the teachings of revelation, and correcting Aristotle where he had fallen into error. At the same time he "Aristotelianized" Christianity, importing major portions of Aristotelian metaphysics and natural philosophy into Christian theology. In the long run (by the nineteenth century), Thomism came to represent the official position of the Catholic church; in the short run, as we shall see, Thomas was viewed by theologians of more conservative persuasion as a dangerous radical.

RADICAL ARISTOTELIANISM AND THE CONDEMNATIONS OF 1270 AND 1277

Albert the Great and Thomas Aquinas were the leaders of a liberal movement that favored a robust philosophy. But however robust philosophy might become, it would, in their view, forever remain a handmaiden; reason would never be allowed to prevail over revelation. Albert and Thomas pushed philosophy as far as it would go, but they never gave up on a philosophical problem until reason and faith had been harmonized.

But how robust can a handmaiden become before she begins to think about insubordination or insurrection?[30] When biblical miracles are reduced to their natural causes, as in Albert's discussion of Noah's flood, aren't things already out of hand? These were the concerns of conservative theologians, observing developments at Paris. And, as it turns out, their fears were not entirely groundless. Our evidence is fragmentary, but it is apparent that while Albert and Thomas were harmonizing philosophy and theology, certain masters of arts began to teach dangerous philosophical doctrines, without regard for their theological consequences. These were committed philosophers, aggressively practicing their trade and recognizing no need to yield, or even pay attention, to any outside authority. The harmonization of philosophy and theology was not their problem.

The best known of this radical faction, and its leader, was Siger of Brabant (ca. 1240–84). Siger, a brash young master of arts, began his teaching career defending the eternity of the world and Averroistic monopsychism, with its dangerous implications for personal immortality. His aim was to do philosophy without so much as a glance at theological teaching on any of the subjects touched, and he maintained that the conclusions he reached were the necessary and inevitable conclusions of philosophy, properly practiced. After the appearance of Thomas's treatise *On the Unicity of the Intellect*, which was directed specifically at his teaching, Siger modified his position on the nature of the soul, bringing it into conformity with orthodox Christian teaching.[31] Older and wiser after this run-in with the theologians, Siger was thereafter careful to make clear that although his philosophical conclusions were not wrong, but in fact necessary philosophical conclusions, nonetheless they need not be true; when it came to truth, he affirmed the articles of the faith. Historians have been divided over whether to take this profession of faith at face value or to conclude that Siger was merely attempting to placate the ecclesiastical power structure. Either way, the dangerous implications of Siger's public position are clear: philosophical inquiry properly conducted can lead to conclusions that contradict those of theology.

The position of the radicals is nicely illustrated by a little treatise, *On the Eternity of the World*, written by Boethius of Dacia (fl. 1270), a member of Siger's circle. One of the most striking characteristics of this work is its rigorous separation of philosophical and theological argument. Boethius systematically assembled and then refuted the philosophical arguments that had been used to defend the Christian doctrine of creation against the Aristotelians. He proceeded to demonstrate that the philosopher, speaking as a philosopher, had no alternative but to defend the eternity of the world. He made clear, however, that according to theology and faith he himself accepted the doctrine of creation, as any Christian must.

Fig. 10.5. Cathedral of Notre Dame, Paris, built in the twelfth and thirteenth centuries.

Boethius thus yielded in the end to the articles of the faith, but in the meantime he displayed an intensely rationalistic orientation. He argued that there is no question capable of rational investigation that the philosopher is not entitled to investigate and resolve. "It belongs to the philosopher," he wrote,

> to determine every question that can be disputed by reason; for every question that can be disputed by rational arguments falls within some part of being. But the philosopher investigates all being—natural, mathematical, and divine. Therefore it belongs to the philosopher to determine every question that can be disputed by rational arguments.

Boethius went on to argue that the natural philosopher cannot even consider the possibility of creation, because to do so would introduce supernatural principles that are out of place in the philosophical realm. Likewise the philosopher denies the resurrection of the dead, because according to natural causes, to which the natural philosopher limits himself, such a thing is impossible.[32]

This is an attempt, impressive for its rigor, to follow philosophical argument to its logical conclusion, without regard for the faith, while still acknowledging the ultimate authority of theology. However, nobody should be surprised to learn that the theological faculty and the religious authorities were neither convinced nor pleased, but regarded Siger, Boethius, and their group as a growing menace. If philosophy was consistently going to reach conclusions at odds with the faith, it could no longer be regarded as a faithful handmaiden; rather, it began to appear as a hostile force and a threat requiring decisive action.

That decisive action came in two condemnations issued by the bishop of Paris, Etienne Tempier, in 1270 and 1277. The former condemned thirteen philosophical propositions allegedly taught by Siger and his fellow radicals in the faculty of arts. Coming, as it apparently did, with the encouragement of both Bonaventure and Thomas Aquinas, this condemnation represents a reaction by the theological establishment to the activities of the radical fringe in the faculty of arts. By 1277, the menace seemed broader and more serious: it was clear by this time that the earlier condemnation had not snuffed out radical Aristotelianism, and conservatives within the faculty of theology moved with increased vigor to meet what they perceived as a growing threat. Indeed, there was a tendency among the conservatives to view everybody significantly more liberal than themselves as dangerous. The result was the issuance (on the third anniversary of Aquinas's death) of a greatly enlarged list of forbidden propositions, 219 in all, the teaching of which was declared grounds for excommunication. Included in this list were fifteen or twenty propositions drawn from the teaching of Aquinas. Let us pause to examine the content of some of the condemned propositions and the import of Tempier's action.[33]

The obviously dangerous elements of Aristotelian philosophy are all represented on Tempier's two lists of forbidden propositions: the eternity of the world, monopsychism, denial of personal immortality, determinism, denial of divine providence, and denial of free will. The rationalistic tendencies of Siger and the radicals are also explicitly targeted: after 1277 it is forbidden, for example, to proclaim the right of philosophers to resolve all disputes on subjects to which rational methods are applicable; or to maintain that reliance on authority never yields certainty. The naturalism of the Aristotelian tradition also figured prominently in the condemnation of 1277: Tempier condemned the opinion that secondary causes are autonomous, so that they would continue to act even if the first cause (God) were to cease to participate; also the claim that God could not have created a man (an obvious reference to Adam) except through the agency of another man; and the methodological principle that natural philosophers, because they restrict their attention to natural causes, are entitled to deny the creation of the world.

This is the sort of list of condemned propositions we might have expected. But in addition the condemnation of 1277 included a heterogeneous assortment of other propositions that impinge in various ways on natural philosophy. Several astrological propositions were condemned: that the heavens influence the soul as well as the body and that events will repeat themselves in 36,000 years, when the celestial bodies return to their present configuration. Also forbidden was the claim that the celestial spheres are moved by souls. A particularly important set of condemned propositions—important because of their implications for fourteenth-century debates—dealt with things that God allegedly could not do, because Aristotelian philosophy had demonstrated their impossibility. It was apparently being argued by philosophers that God could not have created additional universes (Aristotle had argued that multiple universes are impossible); that God could not move the outermost heaven of this universe in a straight line (because a vacuum, which Aristotelian philosophy ruled out, would be left behind in the vacated space);[34] and that God could not create an accidental quality without a subject (for example, redness without something to be red). All of these propositions were condemned in 1277 on the grounds that they fly in the face of divine freedom and omnipotence. The position of Tempier, or of those who formulated the list of propositions on his behalf, was that Aristotle and the philosophers must not be allowed to place a lid on God's freedom or power to act; God can do anything that involves no logical contradiction, including create multiple universes or qualities without subjects.

What can we learn from these events? The condemnation of 1277 has been much discussed and its importance often inflated or misunderstood. In the first place, condemnations of theologically unacceptable theses circulating within the universities was not uncommon. The University of Paris saw at least sixteen such cases in the thirteenth and fourteenth centuries. The condemnation of 1277 differed in its auspices, the range of condemned teachings, and the anonymity of the targeted scholars; whereas the typical censure was internal to the university and directed at specific named masters, the condemnation of 1277 emanated from the local bishop and was aimed at specific ideas or propositions and anybody who held them.[35]

Pierre Duhem, writing early in the twentieth century, saw the condemnation of 1277 as an attack on entrenched Aristotelianism, especially Aristotelian physics, and therefore as the birth certificate of modern science. This is a clever interpretation, and it is not entirely wrong: there can be no doubt (as we shall see below) that the condemnations encouraged scholars to explore non-Aristotelian physical and cosmological alternatives.[36] But to place the emphasis here is to miss the primary significance of the condemnations. Duhem viewed the condemnations as the key event in the shattering of Aristotelian orthodoxy, but in 1277 no such orthodoxy existed; the boundaries and the power relationship between Aristotelian philosophy and Christian theology were still being negotiated, and the degree to which Aristotelianism would acquire the status of orthodoxy was not yet clear.

Or to express the same point a little differently, the condemnations of 1270 and 1277 are significant not so much for the effect they had on the future course of natural philosophy as for what they tell us about what had already occurred. Coming toward the end of nearly a century of struggle over the new learning, the condemnations represent a conservative backlash against liberal and radical efforts to extend the reach and secure the autonomy of philosophy, especially Aristotle's philosophy. They reveal the extent of that reach and the power of the opposition—the fact that a sizable and influential group of traditionalists was not yet ready to accept the brave new world proposed by the liberal, and especially the radical, Aristotelians. Thus, to put the event in its proper light, the condemnations represent a victory not for modern science but for conservative thirteenth-century theology. The condemnations were a ringing declaration of the subordination of philosophy to theology.

They were also an attack on Aristotelian determinism and a declaration of divine freedom and omnipotence. We have noted that a number of the propositions condemned in 1277 dealt with things that God could not do—for example, endow the heaven with rectilinear motion (on the grounds that a vacuum, which Aristotelian philosophy forbids, would thereby be created in the vacated space). In condemning this proposition, Tempier certainly did not mean to debate a point of natural philosophy with Aristotle but to announce that whatever the natural state of things might be (and we may presume that he accepted Aristotle's account of this), God has the power to intervene, should he wish; a vacuum may not exist naturally, but it can surely exist supernaturally; it may not exist in this universe, but a free and omnipotent God could have created a different universe.[37] Aristotle had attempted to describe the world not simply as it is, but as it must be. In 1277 Tempier declared, in opposition to Aristotle, that the world is whatever its omnipotent Creator chose to make it.[38]

What were the implications of these theological points for the practice of natural philosophy? In the first place, certain articles in the condemnations raised new and pressing questions, which required further analysis. For example, the claim that God could supernaturally create qualities without subjects (important because it impinged on the doctrine of transubstantiation)[39] provoked serious debate about a fundamental point in Aristotelian metaphysics—the nature and relationship of accidental qualities and their subjects. The anti-astrological article condemning the idea that history will repeat itself every 36,000 years, when the planetary bodies return to their original configuration, provoked Nicole Oresme (ca. 1320–82) to write an entire mathematical treatise in which he explored questions of commensurability and incommensurability and demonstrated the unlikelihood of all the planetary bodies returning to their original configuration within a finite period of time. Articles of the condemnation about the celestial movers gave rise to animated debates about this important feature of cosmic operation. And the articles that stressed God's unlimited creative power gave license to all manner of speculations about possible worlds and imaginary states of affairs that it was evidently within God's power to create. This led to an avalanche of speculative or hypothetical natural philosophy in the fourteenth century, in the course of which various principles of Aristotelian natural philosophy were clarified, criticized, or rejected.[40]

Second, many of the articles in the condemnations were motivated by concern over the element of necessity that Aristotle had attached to his natural philosophy—the claim that things cannot be otherwise than as they are. When Aristotelian necessity was forced to yield before claims of divine omnipotence, other Aristotelian principles immediately became vulnerable. For example, the mere possibility that God could create other universes outside our own requires a conception of space outside our universe compatible with that possibility. Consequently, in the aftermath of the condemnations, many scholars came to agree that there must be a void space, perhaps even an infinite void space, outside the cosmos suited to receive these possible universes. Likewise, if it is supernaturally possible for the outermost heaven or perhaps the entire cosmos to be moved in a straight line, it follows that motion must be the kind of thing that can be meaningfully applied to the outermost heaven or the cosmos as a whole. But Aristotle had defined motion in terms of surrounding bodies, and there is nothing outside the outermost heaven to surround it. It is evident, therefore, that Aristotle's definition of motion requires revision or correction.[41]

THE RELATIONS OF PHILOSOPHY AND THEOLOGY AFTER 1277

The condemnations are important as benchmarks in the gradual assimilation of Aristotelian philosophy by medieval Christendom. They reveal the strength of conservative sentiment in the 1270s and signal a provisional conservative victory. But it may be well to pause and consider precisely what had been won.

In the first place, even the most conservative of those involved in promulgating the condemnations were not aiming for the elimination of Aristotelian philosophy. Their purpose was merely to administer a healthy dose of discipline, which would unforgettably remind philosophy of her handmaiden status, while at the same time settling certain points of dispute. Second, although this was, strictly speaking, a local victory (since Tempier's decree was formally binding only in Paris), its influence was, in fact, substantially wider. For one thing, Paris was the premier European university for theological studies (the only one on the Continent at the time), and a decree such as this would inevitably have reverberations throughout Christendom. For another, the pope was known to be in touch with Parisian developments, concerned about the dangers of radical Aristotelianism, and possibly willing to intervene on behalf of the conservatives. Moreover, eleven days after Tempier promulgated his 1277 decree, the archbishop of Canterbury, Robert Kilwardby, issued a smaller, but in many respects similar, condemnation applicable to all of England. And in 1284 Kilwardby's decree was renewed by his successor as archbishop of Canterbury, the Franciscan John Pecham, an old adversary of Aquinas and one of the leaders among the traditionalists.

We have no exact knowledge about the force of the condemnations late in the thirteenth century or early in the fourteenth; we can assume that their power to compel obedience and to shape philosophical thought varied widely. By 1323 Thomas Aquinas's reputation had recovered to the point where Pope John XXII could elevate him to sainthood; and in 1325 the bishop of Paris revoked all articles of the condemnation of 1277 applicable to Thomas's teaching. Nonetheless, we can still detect the shadow of the condemnations a century after their promulgation. John Buridan, a Parisian master of arts and twice rector of the university, who flourished around the middle of the fourteenth century, was one of many who continued to wrestle with the difficulties posed by the condemnations. Indeed, on several occasions Buridan revealed a sharp awareness of the threat of theological censure (particularly acute for masters of arts) when his scholarly labors carried him into theological territory. In his *Questions on Aristotle's Physics*, where he found it necessary to comment on the movers of the celestial spheres, he concluded by declaring his willingness to yield to theological authority: "this I do not say assertively but [tentatively], so that I might seek from the theological masters what they might teach me in these matters." And in 1377, a full century after the condemnation, the distinguished Parisian theologian Nicole Oresme defended his opinion that the cosmos is surrounded by an infinite void space by advising potential critics that "to say the contrary is to maintain an article condemned in Paris."[42]

Meanwhile, Aristotelian philosophy had come to stay. It became firmly entrenched in the arts curriculum and came more and more to dominate undergraduate education; in 1341 new masters of arts at Paris were required to swear that they would teach "the system of Aristotle, his commentator Averroes, and the other ancient commentators and expositors on Aristotle, except in those cases that are contrary to the faith." At the same time Aristotelian philosophy was becoming an indispensable tool for practitioners of the advanced disciplines of medicine, law, and theology, and increasingly served as the foundation of serious intellectual effort on any subject.[43]

It does not follow, however, that an enduring solution to the problem of faith and reason had been found. There has been no adequate historical analysis of fourteenth-century developments, and it is not possible at this time even to draw an adequate sketch. However, a few modest generalizations are possible.

First, there was a rapid growth of epistemological sophistication and a broad retreat from the ambitious claims made on behalf of philosophy (by liberal and radical Aristotelians) in the thirteenth century. The ability of philosophy to measure up to traditional Aristotelian standards of certainty or successfully address certain subject matters was increasingly called into question, as skeptical tendencies asserted themselves. In particular, the ability of philosophy to address theological doctrine was drastically curtailed. For example, John Duns Scotus (ca. 1266–1308) and William of Ockham (ca. 1285–1347), while not seeking a total separation between philosophy and theology, diminished their area of overlap by questioning the ability of philosophy to address articles of the faith with demonstrative certainty. Deprived of its ability to achieve certainty, philosophy no longer threatened theology, at least to the same degree; the articles of the faith were not open to philosophical demonstration, but were to be accepted by faith alone. In short, a workable peace was produced by compelling philosophy and theology to disengage—to acknowledge their methodological differences and, on that basis, to accept different spheres of influence. In the case of natural philosophy, this was clearly a smaller sphere.[44]

Second, theologians and natural philosophers in the fourteenth century became heavily preoccupied with the theme of divine omnipotence—a traditional theme within Christian theology, but one whose importance had been reemphasized by the condemnations. If God is absolutely free and omnipotent, it follows that the physical world is contingent rather than necessary; that is, there is no necessity that it should be what it is, for it is dependent solely on the will of God for its form, its mode of operation, and its very existence. The observed order of cause and effect is not necessary, but freely imposed by divine will. A fire has the power to warm, for example, not because fire and warmth are necessarily connected, but because God chose to connect them, endowing fire with this power and choosing continually to concur as it performs its warming function. God is free, however, to introduce exceptions: when Shadrach, Meshach, and Abednego were cast into the fiery furnace without harm, as chapter 3 of the Old Testament book of Daniel recounts, this miracle represented a perfectly permissible decision on God's part to suspend the usual order.[45]

This much is widely accepted by historians; but two divergent lines of argument have developed from it. According to one, if nature does not have its own permanently assigned powers but owes its behavior at every moment to the (possibly capricious) divine will, the idea of a fixed natural order is severely compromised, and serious natural philosophy becomes impossible. According to the other, acknowledgment that God could have created any world he wished led fourteenth-century natural philosophers to perceive that the only way to discover which one he did create is to go out and look—that is, to develop an empirical natural philosophy, which helped to usher in modern (i.e., seventeenth-century) science. Both arguments require brief comment.

The former, which sees the doctrine of divine omnipotence as a destructive influence on natural philosophy, exaggerates the level of divine interference envisioned by medieval natural philosophers—none of whom believed that God meddled frequently or arbitrarily with the created universe. A formula that was regularly invoked distinguished between God's absolute and ordained power. When we consider God's power absolutely or in the abstract, we acknowledge that God is omnipotent and can do as he wishes; at the moment of creation there were no factors other than the law of noncontradiction limiting the kind of world he might create. But in fact we recognize that God chose from among the infinity of possibilities open to him and created this world; and, because he is a consistent God, we can be confident that he will (but for a rare exception)[46] abide by the order thus established, and we need not worry about perpetual divine tinkering. In short, the infinite scope of God's activity guaranteed by the doctrine of divine omnipotence (God's absolute power) was, for practical purposes, restricted to the initial act of creation, and natural philosophers could proceed in their pursuits with confidence that physical reality was not going to undergo capricious change in mid-investigation. What was at issue, therefore, was God's activity within the existing order (his ordained power). This formula was attractive precisely because it safeguarded absolute divine omnipotence without sacrificing the kind of regularity required for serious natural philosophy.[47]

The latter argument, which finds the origins of seventeenth-century experimental science in the doctrine of divine omnipotence, is plausible enough on the face of it. We might expect medieval natural philosophers to have recognized that the behavior of a contingent world cannot be inferred with certainty from any known set of first principles and therefore to have launched an effort to develop empirical methodologies. The only trouble with this conclusion is that the historical record does not appear to bear it out. The ringing proclamation of divine omnipotence and nature's contingency, whether in the condemnations or in the writings of philosophers and theologians, was not accompanied or quickly followed by a dramatic increase in the exploitation of experiment in the natural sciences. That had to wait until the seventeenth century.[48] Natural philosophers and theologians of the Middle Ages continued to believe that both the world and the proper method for exploring it were more or less as Aristotle had described them—though they were prepared, just as they had always been, to read Aristotle critically, to question this or that detail of Aristotelian natural philosophy or methodology, and even (when the opportunity presented itself) to perform experiments. The full development of a systematic program of experimentation was still centuries away; when it emerged, it may have owed something to the theological doctrine of divine omnipotence, but there are more likely sources, including belief in the "fall" of humankind in the garden of Eden and the resulting drastic loss of human intellectual capacity—a loss that, in the view of some, could be fully or partially ameliorated through the systematic employment of observation and experiment.[49] Many other factors also contributed, including examples of experimentation in the literature of ancient and medieval science.

11 ❋ *The Medieval Cosmos*

Previous chapters have examined the medieval reception of the new learning and the struggles surrounding its assimilation in the thirteenth and fourteenth centuries. This and the next two chapters will undertake a survey of the cosmology, astronomy, physics, and natural history that emerged from these struggles. To organize the material, I will work my way down from the top—from the outermost reaches of the cosmos to the earth at its center. I will also employ the distinction (familiar to Aristotle and his medieval followers) between the organic and inorganic realms. The present chapter begins with the basic architecture and mathematics of the cosmos, emphasizing the heavens but touching also on the structure of the terrestrial region and a few peripheral matters. The next chapter will deal with the behavior of inanimate things in the sublunar realm. And the following chapter will turn to the domain of living creatures.[1]

THE STRUCTURE OF THE COSMOS

Examination of early medieval and twelfth-century cosmologies in chaps. 7 and 9, above, revealed that encyclopedic writers of the early Middle Ages communicated a modest assortment of basic cosmological information, drawn from a variety of ancient sources, especially Platonic and Stoic. These writers proclaimed the sphericity of the earth, discussed its circumference, and defined its climatic zones and division into continents. They described the celestial sphere and the circles used to map it; many revealed at least an elementary understanding of the solar, lunar, and other planetary motions. They discussed the nature and size of the sun and moon, the cause of eclipses, and a variety of meteorological phenomena.

This picture was enriched in the twelfth century by renewed attention to the content of Plato's *Timaeus* (and Calcidius's commentary on it) and by early contact with Greek and Arabic books in translation. One of the results was increased emphasis (surpassing that of the early church fathers) on reconciling Platonic cosmology and the biblical account of creation. Another novelty was the frequent claim, among twelfth-century authors, that God limited his creative activity to the moment of creation; thereafter, they held, the natural causes that he had created directed the course of things. Twelfth-century cosmologists stressed the unified, organic character of the cosmos, ruled by a world soul and bound together by astrological forces and the macrocosm-microcosm relationship. In an important continuation of early medieval thought, twelfth-century scholars described a cosmos that was fundamentally homogeneous, composed of the same stuff from top to bottom: Aristotle's quintessence or aether and his radical dichotomy between the celestial and terrestrial regions had not yet made their presence felt.[2]

Thierry of Chartres appeared in chap. 9 as illustrative of some of these features of twelfth-century cosmology. Another representative of the same tradition, more useful to us because he wrote more voluminously, is Robert Grosseteste (ca. 1168–1253), one of the celebrated figures of medieval science.[3] Grosseteste is also important as an illustration of the continuation of Platonic currents into the thirteenth century; for although he was educated late in the twelfth century, his major writings date from the first half of the thirteenth.

Central to Grosseteste's cosmology was light: the cosmos came into existence when God created a dimensionless point of matter and its form, a dimensionless point of light.[4] The point of light instantaneously diffused itself into a great sphere, drawing matter with it and giving rise to the corporeal cosmos. Subsequent radiation and differentiation (as the light made a return trip toward the center) gave rise to celestial spheres and the characteristic features of the sublunar region. In his early writings, Grosseteste seems to have accepted the idea of a world soul—an idea from which he later retreated. The theme of microcosm and macrocosm is fundamental to Grosseteste's works: humanity represents the pinnacle of God's creative activity, simultaneously mirroring the divine nature and the structural principles of the created cosmos. Finally, Grosseteste shared the early medieval and twelfth-century belief in a homogeneous cosmos: the heavens in his cosmology are made of finer (specifically, more rarefied) stuff than are terrestrial substances, but the difference is quantitative rather than qualitative.[5]

Cosmology, like so many other subjects, was transformed by the wholesale translation of Greek and Arabic sources in the twelfth and thirteenth centuries. Specifically, the Aristotelian tradition gained center stage in the thirteenth century and gradually substituted its conception of the cosmos for that of Plato and the early Middle Ages. This is not to suggest that Aristotle and Plato disagreed on all of the important issues. On many of the basics they were in full accord. Aristotelians, like Platonists, conceived the cosmos to be a great (but unquestionably finite) sphere, with the heavens above and the earth at the center. Nobody representing either school of thought doubted that the cosmos was unique. Although nearly everybody acknowledged that God could have created multiple worlds, nobody seriously believed that he had done so. And all (clinging to a literal interpretation of Genesis) departed from Aristotle by acknowledging that the cosmos had a beginning in time—although, as we have seen, some Aristotelians of the thirteenth century were prepared to argue that this could not be established by philosophical argument.

But where Aristotle and Plato disagreed, the Aristotelian world picture gradually displaced the Platonic. One of the major differences concerned the issue of homogeneity. Aristotle divided the cosmic sphere into two distinct regions, made of different stuff and operating according to different principles. Above the moon are the celestial spheres, bearing the fixed stars, the sun, and the remaining planets. This celestial region, composed of aether or the quintessence (the fifth element), is characterized by unchanging perfection and uniform circular motion. Below the moon is the terrestrial region, formed out of the four elements. This region is the scene of generation and corruption; of birth, death, and decay; and of transient (typically rectilinear) motions. Other Aristotelian contributions to the cosmological picture were his elaborate system of planetary spheres and the principles of causation by which celestial motions produced generation and corruption in the terrestrial realm.

A variety of Aristotelian features, then, merged with traditional cosmological beliefs to define the essentials of late medieval cosmology—a cosmology that became the shared intellectual property of educated Europeans in the course of the thirteenth century. Universal agreement of such magnitude emerged not because the educated felt compelled to yield to the authority of Aristotle, but because his cosmological picture, as a whole, offered a persuasive and satisfying account of the world as they perceived it. Nonetheless, certain elements of Aristotelian cosmology became the objects of criticism and debate, and it is here, in the attempt to flesh out and fine-tune Aristotelian cosmology and bring it into harmony with biblical teaching and the opinions of other authorities, that medieval scholars made their cosmological contributions. It would be impossible in a single chapter, or even a single book, to deal fully with medieval cosmological thought (Pierre Duhem devoted ten volumes to the subject), and we must limit ourselves to the most important and most hotly debated questions.[6]

Before entering the cosmos, let us pause just outside it: what, if anything, exists there? All agreed that no material substance is found outside the cosmos; if the cosmos is held to contain all the material substance that God created, this conclusion is unavoidable. But what about space devoid of corporeal substance? Aristotle had explicitly denied the possibility of place, space, or vacuum outside the world, and this conclusion was generally accepted until a reevaluation of the issue was provoked by the condemnation of 1277. Two articles in the condemnation bore directly on the problem. In one of them it was declared that God has the power to create multiple worlds, in the other that God is capable of endowing the outermost heaven with rectilinear motion. Now if an additional cosmos could be deposited outside ours, then it must be possible for void space to exist there, capable of receiving it; likewise, a heavenly sphere in rectilinear motion would inevitably vacate one space and move into another. Most writers were satisfied to acknowledge the *possibility* that God could create a void space outside the cosmos; a few, like Thomas Bradwardine (d. 1349) and Nicole Oresme (ca. 1320–82), in the fourteenth century, argued that he had actually done so. Bradwardine identified this void space with God's omnipresence and argued that since God is infinite, extracosmic void space must likewise be infinite.

Christian considerations seem to have been paramount in this modification of Aristotelian cosmology, but Stoic influence is also apparent. The notion of extracosmic void came to the West with a Stoic pedigree. Western scholars even borrowed specific Stoic arguments, such as the often repeated thought experiment about somebody situated at the very periphery of the material universe, at the outermost limit of all material substance, thrusting an arm upward, beyond that periphery. It seemed obvious that the arm must be received by a space that had hitherto been empty. Through a combination of Christian and Stoic influence, then, an important modification was imposed on Aristotelian cosmology—a modification that was to figure prominently in cosmological speculations through the end of the seventeenth century and beyond.[7]

As we enter the cosmos, we immediately encounter celestial spheres. How many of these exist, what is their nature, and what are their functions? There were seven known planets—Moon, Mercury, Venus, Sun, Mars, Jupiter, and Saturn, generally thought to be arranged in that order. In the simplified version of the cosmos preferred by medieval writers on cosmology, which ignored most of the astronomical details, each planet required a single sphere to

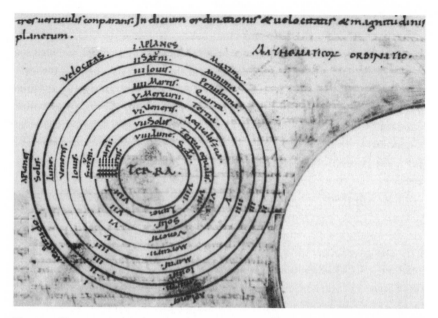

Fig. 11.1. The simplified Aristotelian cosmology popular in the Middle Ages. Paris, Bibliothèque Nationale, MS Lat. 6280, fol. 20r (12th c.).

account for its motion (fig. 11.1). In addition, according to Aristotle, outside the planetary spheres, defining the outer limit of the cosmos, is the sphere of the fixed stars, or the primum mobile. Several problems arose as medieval scholars thought about this outermost sphere.

One of them was to define its place. The place of a thing, according to Aristotle, is determined by the body or bodies that contain it. But if the sphere of fixed stars is itself the outermost body, there is nothing outside it to serve as container. The inevitable conclusion of this line of argument, that the primum mobile is not in a place, was too paradoxical to be accepted by all but a few of the most obstinate minded. Various solutions were therefore proposed, including an attempt to redefine place so as to allow it to be determined by the contained, rather than the containing, body.[8]

Another problem for Aristotle's outermost sphere grew out of the account of creation in the book of Genesis, where a distinction was made between the "heaven (*caelum*)" created on the first day and the "firmament (*firmamentum*)" created on the second—obviously two different things, since created on different days. Moreover, the biblical text states that the firmament separates waters beneath it from waters above it; the waters beneath the firmament could be equated with the sphere of water in the terrestrial region, but the waters above the firmament apparently constituted yet another celestial sphere. Discussion of this problem led some Christian commentators to postulate three spheres beyond the seven planetary spheres: the outermost of these, the invisible and motionless empyreum, served as the abode of the angels; next came the aqueous or crystalline heaven, perfectly transparent, consisting of water (possibly in a hard or crystallized form but more likely fluid, and possibly water only figuratively); and then the firmament, bearing the fixed stars. The total number of heavenly spheres, for those who accepted this line of argument, came to ten. In time, all three outer spheres were assigned cosmological and astronomical functions; some scholars, wishing to account for an additional stellar motion, postulated an eleventh sphere as well. It is important to note, once again, the mutual interaction between cosmology and theology in these discussions: Aristotelian cosmology was adjusted to meet the demands of biblical interpretation; at the same time, the biblical account absorbed the fundamentals of Aristotelian cosmology, with its medieval modifications, and took substantial portions of its meaning from contemporary cosmological theory.[9]

Medieval cosmologists were, of course, interested in the substance or "material cause" of the celestial region. Many writers of the early Middle Ages, drawing on the Stoic tradition, supposed the heavens to consist of a fiery substance. After the recovery of Aristotle's works, some version of Aristotle's opinion that the heavens were formed out of the quintessence or aether (a perfect, transparent substance not subject to change) was generally accepted. There were debates about the nature of this aether—for example, whether it was a composite of form and matter. Among those who admitted the existence of form and matter in the heavens, some argued that the matter of the heavens was similar in kind to terrestrial matter, while others maintained that the two matters must be totally different. Whatever the nature of the aether might be, everybody agreed that it was divided into distinct spheres, in perfect contact (for otherwise there would be void spaces), each rotating frictionlessly in its proper direction, bearing a planet with its characteristic speed. No interstices or gaps existed within individual spheres. Seldom did a writer inquire whether they were fluid or hard; each alternative found support among the few who addressed this issue. The planets were judged to be small spherical regions of greater density or lucidity in the transparent, lucid aether.[10]

A much more hotly debated question was the nature of the celestial movers. Aristotle had identified a set of "Unmoved Movers" as the causes of celestial motion—the objects of desire of the planetary spheres, which do their best to imitate the changeless perfection of the Unmoved Movers by rotating with eternal, uniform circular motion. The Unmoved Movers are thus final, rather than efficient, causes. In medieval Christendom the Unmoved Mover of the uppermost movable sphere (the "Prime Mover") was customarily identified with the Christian God; but the identity of the additional Unmoved Movers was a more troublesome problem. It would have been easy to identify them with the planetary deities described in Plato's *Timaeus*; but to acknowledge the existence of any deity besides the Creator would have been a clear case of heresy within the Christian tradition, and it was therefore important for Christian scholars to distance themselves from such notions by assigning the Unmoved Movers a status well short of divinity. A common solution was to conceive of them as angels or some other kind of separated intelligences (minds without bodies). However, alternative solutions that dispensed entirely with angels and intelligences were devised. Robert Kilwardby (ca. 1215–79), archbishop of Canterbury, assigned the celestial spheres an active nature or innate tendency to move spherically. The Parisian professor John Buridan (ca. 1295–ca. 1358) argued that there is no need to postulate the existence of celestial intelligences, since they have no scriptural basis; it is possible, therefore, that the cause of celestial motion is an impetus or motive force, analogous to the impressed force that moves a projectile (see below, chap. 12), which God imposed on each of the celestial spheres at the moment of creation.[11]

The packing principle followed in this system—thick planetary spheres, packed contiguously, without wasted space—made it possible to calculate the size of the various planetary orbits and ultimately the dimensions of the cosmos. To get the calculation started, an estimate of the size of the innermost sphere, that of the moon, was needed. Several Muslim astronomers, including al-Farghānī and Thābit ibn Qurra in the ninth century and al-Battānī in the ninth or tenth, performed the calculation, borrowing the required data from Ptolemy's *Almagest*, with modifications. In the West, Campanus of Novara (d. 1296), gave his version of the calculation, assigning a figure of 107,936 miles to the radius of the inner surface of the moon's sphere (the moon's closest approach to the earth) and 209,198 miles to the radius of the outer surface of the moon's sphere (the moon's farthest retreat from the earth). Similar calculations for Mercury and Venus produced a "theoretical" distance for the sun that accorded roughly with the parallax calculated for the sun by astronomers in antiquity. Continuation of the computation for the superior planets yielded a radius of 73,387,747 miles for the outside of Saturn's sphere and the inside of the stellar sphere. These figures, or figures close to them, prevailed until revised by Copernicus in the sixteenth century.[12]

MATHEMATICAL ASTRONOMY

The analysis thus far has assumed that the heavens consist of a simple set of tightly nested, concentric spheres. This seems to have been Aristotle's view; it was articulated and vigorously defended by Averroes (Ibn Rushd) in Islamic Spain; and it had a number of important Western adherents. However, as circulation of the Latin translation of Ptolemy's *Almagest* (that great testament to the power of mathematical astronomy) spread, beginning late in the twelfth century, the attempt to describe the heavens became a great deal more complicated. Some medieval cosmological writers looked for ways of modifying their cosmology to take into account the eccentric deferents and epicycles of Ptolemaic astronomy—an attempt, obviously, to harmonize cosmology and mathematical planetary astronomy.

Before proceeding, we need to pause and consider an important distinction, proposed in Greek antiquity, restored to prominence in the later Middle Ages and Renaissance, and revived early in the twentieth century by the French physicist and philosopher turned historian of science, Pierre Duhem (1861–1916). The distinction is between two ways of viewing astronomical models. On the "realist" view, astronomical models are expected to represent physical reality and answer to the physical criteria of the physicist or cosmologist. On the "instrumentalist" or "fictionalist" view, astronomical models are nothing more than convenient instruments—mathematical fictions useful for predicting planetary positions but without any necessary *physical* truth value. To allow physical or cosmological considerations to interfere with the mathematical models of the astronomer would violate the norms of astronomical science.[13] The thirteenth and fourteenth centuries were the scene of considerable controversy over this issue, as scholars explored the status of theoretical claims or looked for compromise positions. The astronomers, who inevitably demanded quantitative results, had no choice but to retain Ptolemaic models. For some of the more philosophically inclined, the idea of creating a quantitatively exact astronomy on Aristotelian principles remained an elusive dream.[14]

If we cannot sharply distinguish between astronomy and cosmology on methodological grounds, is there any justification for treating them as distinct enterprises or disciplines? Yes! One of the best ways of distinguishing medieval disciplines is to forget about their formal definitions and to examine them as textual traditions. The cosmological questions that occupied the opening sections of this chapter tended to appear in commentaries on certain texts—Aristotle's physical works (especially *On the Heavens* and *Metaphysics*),

Fig. 11.2. Astrolabe, Italian, ca. 1500.
Diameter: 4.25 inches. London,
Science Museum, Inv. no. 1938–428.
Reproduced by permission of the
Trustees of the Science Museum.

John of Sacrobosco's *Sphere*, Peter Lombard's *Sentences*, and the creation account in Genesis.[15] The mathematical analysis of the heavens belonged to a different textual tradition, springing from Ptolemy's *Almagest* and other works of mathematical astronomy produced during the Hellenistic period. Reinforcing this separation may have been the fact that the serious mathematical astronomer required esoteric mathematical skills that were out of reach for the average natural philosopher engaged in cosmological pursuits.

During the early Middle Ages, the West was without access to the most important Greek sources of mathematical astronomy—the works of Hipparchus, Ptolemy, and others. Astronomy was certainly understood to be a mathematical art, a member of the mathematical quadrivium, but the amount of mathematical astronomy actually known to early medieval scholars was minimal. Authors such as Pliny, Martianus Capella, and Isidore of Seville offered an elementary description of the celestial sphere and its major circles; of the seven planets and their west-to-east motion through the band of the zodiac, including retrograde motion; and of the sun-linked movement of Mercury and Venus. The ability to deal with problems of chronology and the calendar was also a well-developed art. But knowledge of Ptolemaic models or any other scheme for the practice of serious mathematical astronomy was nonexistent.[16]

The state of Western astronomical knowledge was radically altered in the tenth and eleventh centuries by contact with Islam, principally through Spain.

Gerbert of Aurillac (ca. 945–1003) is famous for his contribution (above, chap. 9), but other pilgrims also contributed to knowledge of the scientific treasures available in Islamic and (later, after the reconquest) Christian Spain. One of the most important pieces of astronomical treasure was the astrolabe, accompanied by the mathematical knowledge required to put it to use. Several treatises on the construction and use of the astrolabe, translated from Arabic to Latin, circulated in northern Europe in the eleventh century. This versatile instrument, in turn, was responsible for a reorientation of Western astronomy, away from qualitative and toward quantitative concerns.[17]

The astrolabe was a hand-held instrument consisting of a graduated circle and a sighting rule (the alidade), which pivoted about a pin and allowed observations of the altitude of a star or planet, and a set of circular brass plates that fit into a brass body or "mother" and made the astrolabe into an astronomical computer (see figs. 11.3 and 11.4). The mathematical principle

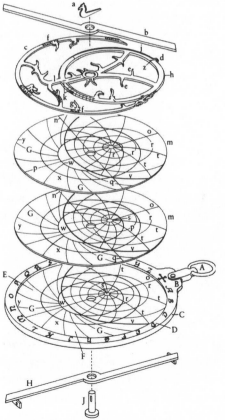

Fig. 11.3. An "exploded" view of the astrolabe. Courtesy of J. D. North. Originally published in J. D. North, *Chaucer's Universe*, p. 41.

a Horse (a wedge ending in a horse's head)
b Rule
c Rete
e Star pointers
h Ecliptic circle
k Lines demarcating the signs of the zodiac
m Climates
r Almucantars (circles of equal altitude)
s Zenith
t Lines of equal azimuth
v Horizon line
w Tropic of Cancer
x Equator
y Tropic of Capricorn
C Mother
G Hour angle lines
H Alidade (with sighting holes)
J Pin

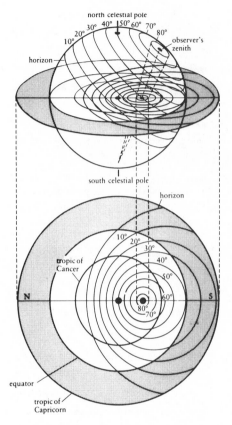

Fig. 11.4. Stereographic projection of
the almucantars. The circles of equal
altitude (top) are projected onto a
horizontal plane passing through the
equator of the celestial sphere, as they
would be seen by an observer situated at the south celestial pole. These
circles of equal altitude or almucantars,
along with lines of equal azimuth were
primary features of the "climate" of
the astrolabe. Courtesy of J. D. North.
Originally published in J. D. North,
Chaucer's Universe, p. 53.

that made it possible for the astrolabe to function as a computer was stereographic projection, by which the spherical heavens could be projected (for
convenience) onto a set of flat plates (see figs. 11.3 and 11.4). The uppermost
plate (the "rete"), designed to represent the rotating heavens, contained a star
map (limited to a few of the most prominent stars) and an eccentric circle representing the ecliptic (figs. 11.2 and 11.3); much of this plate was cut away, so
that the user could see through it to a plate fixed beneath, called the "climate."
The "climate" bore the projection of a fixed coordinate system defined for the
latitude of the user, consisting of a horizon line, circles of equal altitude, and
lines of equal azimuth, as well as the celestial equator, tropic of cancer, and
tropic of capricorn (fig. 11.3). The rete could then be rotated over the climate
to simulate the rotation of the heavens with respect to a terrestrial observer;
the position of the sun on the ecliptic could be marked, and a variety of useful
calculations then became possible.

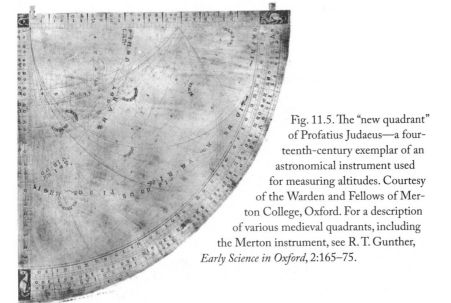

Fig. 11.5. The "new quadrant" of Profatius Judaeus—a fourteenth-century exemplar of an astronomical instrument used for measuring altitudes. Courtesy of the Warden and Fellows of Merton College, Oxford. For a description of various medieval quadrants, including the Merton instrument, see R. T. Gunther, *Early Science in Oxford*, 2:165–75.

Although astronomical instruments and tables of astronomical data were necessary for the practice of mathematical astronomy, they were not sufficient. A third requirement was astronomical theory. The instructions accompanying a set of astronomical tables might offer a glimpse of its theoretical underpinnings, but this was limited in quantity and confusing. Treatises of theoretical astronomy, presenting the mathematical models that lay behind the data and the calculations, were needed; and again they were supplied through translation, in this case from both Arabic and Greek texts. Al-Farghānī's elementary handbook of Ptolemaic astronomy was translated in 1137 by John of Seville as *The Rudiments of Astronomy*. In the second half of the twelfth century the more technical astronomical works of Thābit ibn Qurra, Ptolemy, and others became available: Ptolemy's *Almagest* was rendered into Latin twice, once from a Greek original and subsequently (by Gerard of Cremona) from an Arabic version. Astrological texts that appeared about the same time contributed to the interest in astronomical theory and calculations. Indeed, the astrologer's need for astronomical calculations, along with the growing connection between astrology and medicine, helps to explain the growth of astronomical studies.[18]

By the end of the twelfth century, the most important astronomical texts were available in Latin. The history of Western astronomy from this point onward is a story of growing mastery and increasing dissemination of astronomical knowledge, primarily within the universities. One of the necessities in the universities was for textbooks that would bring the complexities of Ptolemaic astronomy within the reach of students. An introductory treatise such as al-Farghānī's *Rudiments of Astronomy* could, of course, be put to use; but teachers in the universities soon produced books of their own. One of the earliest and most popular was *The Sphere* of Johannes de Sacrobosco (John of Holywood), written at Paris about the middle of the thirteenth century. This work, which continued to be commented upon and used as a university textbook as late as the seventeenth century, contained an elementary account of spherical astronomy and a few brief remarks on planetary motions. For example, Sacrobosco described the west-to-east motion of the sun around the ecliptic at the rate of about 1°/day. He noted that each of the planets except the sun is carried around on an epicycle, which in turn is carried on a deferent circle, and explained how the epicycle-on-deferent model accounts for retrograde motion. He also attributed lunar and solar eclipses, respectively, to the shadows cast by the earth and moon. Beyond this his planetary astronomy did not go.[19]

Sacrobosco's *Sphere* was obviously meant to convey only the most elementary astronomical knowledge. Another treatise, *Theorica planetarum* (*Theory of the Planets*), composed a little later by an anonymous author, possibly also a Parisian teacher, raised the discussion of planetary astronomy to a substantially higher level. The *Theorica* sketched the basic Ptolemaic theory for each of the planets, supplementing the description with geometrical diagrams. For example, the motion of the sun around the ecliptic was explained as the result of uniform west-to-east motion about an eccentric deferent circle at the rate of 59′8″ (just short of 1°) per day. Meanwhile that eccentric circle is carried uniformly east-to-west at the rate of one full rotation per day by the stellar sphere. In the model for the superior planets—Mars, Jupiter, and Saturn—the planet P (fig. 11.6) moves uniformly around the epicycle from west to east,

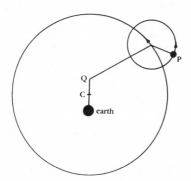

Fig. 11.6. The model for one of the superior planets, according to the *Theorica planetarum*.

while the center of the epicycle moves in the same sense around the deferent. The motion of the epicycle around the deferent is uniform with respect
to equant point Q; the center of the deferent is halfway between the equant
point and the center of the earth.[20] The *Theorica* seems quickly to have become
the standard textbook of astronomical theory, firmly establishing Ptolemaic
mathematical models against any possible rivals and fixing astronomical terminology for several centuries.

One serious necessity that the thirteenth-century *Theorica planetarum* did
not convey was the quantitative content of Ptolemaic astronomy or the means
for making actual astronomical calculations. This function came to be served
by the *Toledan Tables*; then, after about 1275 the *Alfonsine Tables* (prepared
at the court of Alfonso X of Castile), which frequently accompanied the
Theorica. The *Alfonsine Tables* (fig. 11.7) served as the standard guide to the
practice of mathematical astronomy until confronted by new competitors
in the sixteenth century.[21] Translated tables were treasuries of quantitative
astronomical information, but they had been constructed for earlier eras and
locations other than the ones where they were to be used. Consequently
they required adaptation—work carried out by a number of twelfth-century
scholars, including Raymond of Marseilles and Robert of Chester. In their
work we have the beginnings of a genuine Western tradition of mathematical
astronomy.

While a modest amount of elementary astronomical knowledge may have
become quite common among those with a university education, advanced
knowledge of the sort represented by the *Toledan* or *Alfonsine Tables*, or even
the *Theorica*, was considerably less common. Seldom did the universities require astronomical knowledge for a degree in arts, though some instruction
in astronomy was frequently available—usually in the form of lectures on
the *Theorica*, occasionally through lectures on the *Almagest*. The result, varying with time and place, was a small, but active and growing, community of
competent astronomers becoming more sophisticated—sufficiently knowledgeable, for example, to recompute astronomical tables for their own location. And out of this medieval tradition would emerge, in the fifteenth and
sixteenth centuries, astronomers of the stature of Johannes Regiomontanus
and Nicolaus Copernicus.[22]

The problem of bringing Ptolemy's mathematical astronomy into harmony
with cosmological considerations remained—the problem that had occupied
Ibn al-Haytham and members of the Maragha school in Islam (see chap. 8,
above). It appeared to astronomers that the eccentric and epicyclic circles of
Ptolemaic astronomy were incompatible with Aristotelian concentric spheres

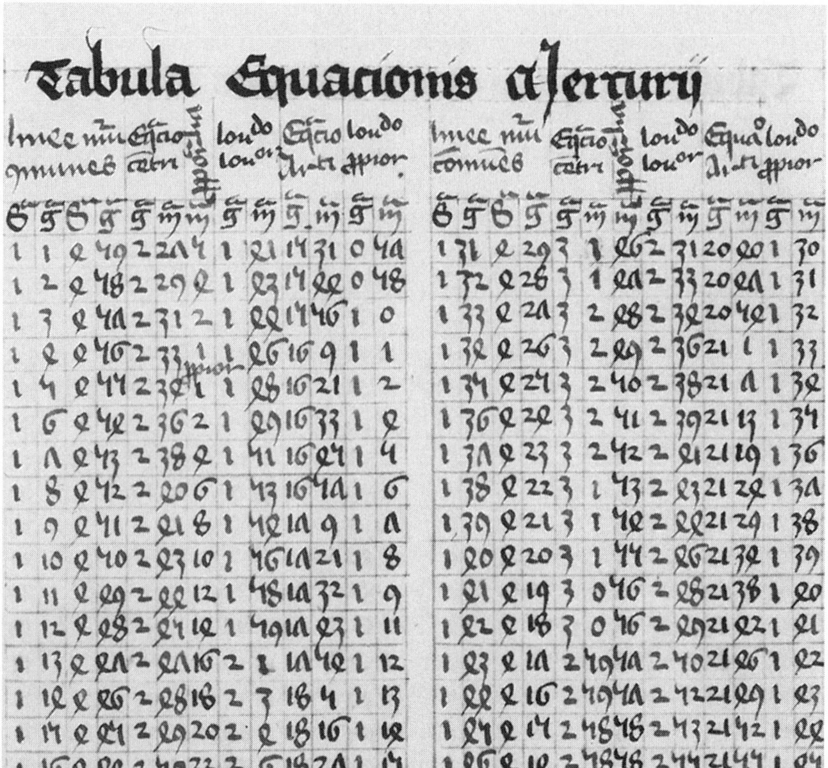

Fig. 11.7. The *Alfonsine Tables*. A page from the table for Mercury. Houghton Library, Harvard University, f MS Typ 43, fol. 46r (ca. 1425). By permission of the Houghton Library.

Fig. 11.8. An astronomer observing with an astrolabe. Paris, Bibliothèque de l'Arsenal, MS 1186, fol. 1v (13th c.).

and the principles of Aristotelian natural philosophy; anybody who needed help perceiving the magnitude of the problem could obtain it from Averroes' attack on the paraphernalia of Ptolemaic astronomy. The only system to achieve success from a quantitative standpoint seemed questionable from a physical or philosophical perspective. There was considerable thrashing about on this issue in the thirteenth and fourteenth centuries, as scholars explored the status of theoretical claims or looked for compromise positions. The mathematical astronomers, who inevitably demanded quantitative results, had no choice but to retain the Ptolemaic models. But how do you harmonize the Ptolemaic mathematical models with Aristotelian cosmological principles?

One widely accepted answer was the physical-sphere version of the Ptolemaic system. Developed first by Ptolemy in his *Planetary Hypotheses*, but lost until revived by Ibn al-Haytham in the eleventh century (fig. 11.9), this model first appeared in Western Europe in the thirteenth century. Roger Bacon, writing in the 1260s, was the first Western scholar known to have given this scheme a lengthy discussion. After him we see a flurry of interest

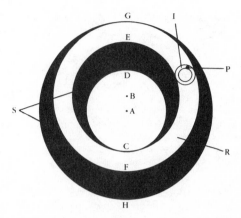

Fig. 11.9. Ibn al-Haytham's physical-sphere model of the Ptolemaic deferent and epicycle. The thickened space, S, is bounded by spherical surfaces CD and GH. A is the center of the universe, where the earth is situated. Cutting through the sphere and eccentric to it is a ring R, centered on B and bounded by surfaces CE and FG. Situated within the ring is the epicycle I, bearing the planet P. The entire sphere rotates about its center A on a daily basis, carrying the ring with it; meanwhile, the epicycle "rolls" through the ring in the sidereal period of the planet (the time required for the planet to complete one circuit of the ecliptic), and through it all the planet is carried around the rotating epicycle. Similar thickened spheres are required for each of the remaining planets.

in it among Franciscans, including Bernard of Verdun and Guido de Marchia. Bernard criticized Bacon's theory, but Guido (fl. 1292–1310), in his *Tractatus super planetorbium*, defended the idea of physical spheres (fluid, in his opinion) and extended it, with important modifications, to the entire ensemble of planetary bodies. The idea continued to attract interest in the fourteenth and fifteenth centuries. By the fifteenth century, the theory of physical spheres had reached Vienna, where it appeared in the influential *Theoricae novae planetarum* (printed in 1454 and frequently thereafter) of Georg Peurbach (1423–61)—a book that contributed to a revival of astronomy that, before the end of the century, would shape the astronomical education of Nicolaus Copernicus (1473–1543).[23]

ASTROLOGY

The history of astrology has suffered from a tendency among historians to judge the practice of astrology harshly, as an example of primitive, irrational, or superstitious ideas, promoted by fools and charlatans. There were charlatans, of course, as medieval critics themselves never tired of pointing out. But medieval astrology also had a serious scholarly side, and we must not allow our attitude toward it to be colored by the low regard in which astrology is held today. Medieval scholars judged astrological theory and practice by medieval criteria of rationality and by the contemporary evidence to which they had access. We must do the same if we hope to understand the importance and the changing fortunes of astrology during the Middle Ages.[24]

It will help if we begin by distinguishing between (1) astrology as a set of beliefs about physical influence within the cosmos and (2) astrology as the art of casting horoscopes, determining propitious moments, and the like. The former was a respectable branch of natural philosophy, the conclusions of which were rarely called into question. The latter, by contrast, was vulnerable to a variety of objections (empirical, philosophical, and theological) and remained a subject of contention throughout the Middle Ages. Although this account will touch upon astrology in the second sense, it is astrology as an aspect of cosmic physics that will be our primary concern.

There were compelling reasons for believing that the heavens and the earth were physically connected. First, there were observational data that made the connection obvious: nobody could doubt that the heavens were the major source of light and heat in the terrestrial region; the seasons were plainly connected with solar motion around the ecliptic; the tides were apparently connected with lunar motion; and it seemed clear enough, once the compass made its appearance (late in the twelfth century), that the poles of the celestial sphere exercised a magnetic influence on certain minerals.

Observational arguments of this sort were reinforced by traditional religious beliefs. The association of the heavens with divinity and the understanding that divinity exercised influence in the terrestrial realm were prominent features of ancient religions. The belief that stellar and planetary events were omens (signs rather than causes) of terrestrial events was widespread in ancient Mesopotamia, where reading the omens became a specialized art, demanding a measure of astronomical knowledge. Such beliefs were gradually enriched and transformed by the addition of new elements, including the notion that the celestial configuration at the time of a person's conception or birth could be used as a means of predicting certain details of that person's life (see above, chap. 1).[25]

Within Greek culture, astrological ideas obtained support from a variety of philosophical systems. In the *Timaeus*, Plato's Demiurge explicitly delegated to the planets or planetary deities the task of bringing into existence things in the sublunar realm; and this suggested the possibility of an ongoing relationship. Plato also stressed the unity of the cosmos, including parallels between the cosmos as a whole (the macrocosm) and individual humans (the microcosm). In Aristotle's cosmos the Unmoved Mover was the source not simply of the motions of the celestial spheres, but also of motion and change in the sublunar realm. Aristotle argued, in his discussion of meteorological phenomena, that the terrestrial region "has a certain continuity with the upper [celestial] motions; consequently all its power is derived from them." Elsewhere he attributed seasonal changes, as well as all generation and corruption in the terrestrial realm, to the motion of the sun around the ecliptic. Finally Stoics, with their vision of an active, organic cosmos characterized by unity and continuity, seem to have embraced and defended the science of astrology. It should be clear, then, that astrology in its physical or cosmological form was the rational and empirical investigation of the causal connections between the heavens and the earth. Almost any ancient philosopher would have considered it extraordinarily foolish to deny the existence of such connections.[26]

Ptolemy is an excellent case in point—excellent not only because he addressed the question fully and clearly, but also because he exercised a powerful influence on both the Islamic and Western astrological traditions. In his astrological handbook, the *Tetrabiblos*, Ptolemy acknowledged that astrological prognostications cannot match the certitude of astronomical demonstrations; nonetheless, he affirmed the existence of celestial forces and the validity of astrological prognostications of a general sort. It is apparent to everybody, he argued,

> that a certain power emanating from the eternal ethereal substance . . . permeates the whole region about the earth. . . . For the sun . . . is always in some way affecting everything on the earth, not only by the changes that accompany the seasons of the year to bring about the generation of animals, the productiveness of plants, the flowing of waters, and the changes of bodies, but also by its daily revolutions furnishing heat, moisture, dryness, and cold in regular order and in correspondence with its positions relative to the zenith. The moon, too, . . . bestows her effluence most abundantly upon mundane things, for most of them, animate or inanimate, are sympathetic to her and change in company with her. Moreover, the passages of the fixed stars and the planets through the sky often signify hot, windy, and snowy conditions of the air, and mundane things are affected accordingly.

The practitioner who understands these influences and who has also mastered the celestial motions and configurations ought to be able to predict a wide variety of natural phenomena:

> If, then, a man knows accurately the movements of all the stars, the sun, and the moon, . . . and if he has distinguished in general their natures as the result of previous continued study . . . ; and if he is capable of determining in view of all these data, both scientifically and by successful conjecture, the distinctive mark of quality resulting from the combination of all the factors, what is to prevent him from being able to tell on each given occasion the characteristics of the air from the relations of the phenomena at the time, for instance, that it will be warmer or wetter? Why can he not, too, with respect to an individual man, perceive the general quality of his temperament from the atmosphere at the time of his birth, as for instance that he is such and such in body and such and such in soul, and predict occasional events, by use of the fact that such and such an atmosphere is attuned to such and such a temperament and is favourable to prosperity, while another is not so attuned and conduces to injury?[27]

A certain amount of anti-astrological sentiment surfaced within Hellenistic philosophy and subsequently within both the Islamic and Christian traditions. The object of attack, however, was not belief in the reality of celestial influence, but the threat of determinism and (among the church fathers) the assignment of divinity to the stars and planets. The most influential voice within Christendom was that of Augustine (354–430), who attacked vulgar astrology as a fraudulent enterprise, practiced by impostors. But his greatest concern was for what he regarded as the tendency toward fatalism or determinism within serious astrological theory. At all costs, the freedom of the will must be protected, for otherwise there would be no human responsibility. Augustine appealed frequently to the "twins problem" (not original with him), pointing out that twins, conceived at the same instant and born almost simultaneously, often experience dramatically different fates. But Augustine opened the door to the possibility of physical influence, as long as it was held to affect only the body, when he wrote:

> It is not entirely absurd to say, with reference only to physical differences, that there are certain sidereal [i.e., stellar] influences. We see that the seasons of the year change with the approach and the receding of

the sun. And with the waxing and waning of the moon we see certain kinds of things grow and shrink, such as sea-urchins and oysters, and the marvelous tides of the ocean. But the choices of the will are not subject to the positions of the stars.[28]

The anti-astrological polemics of Augustine and other church fathers helped to create a climate of opinion hostile to astrology during the early Middle Ages. In early medieval literature we find regular condemnation of the practice of horoscopic astrology—often accompanied, however, by admission of the reality of celestial forces and their influence on a variety of terrestrial phenomena.[29]

The flowering of Platonic philosophy and the recovery of Greek and Arabic astrological writings in the twelfth century led to a resurgence, in Christendom, of interest in astrology and a more favorable attitude toward its doctrines. Any suggestion of astrological determinism remained anathema, of course, but assertions about the reality of stellar and planetary influence and the possibility of successful astrological prognostication now became commonplace. For example, in his influential *Didascalicon* (written in the late 1120s) Hugh of St. Victor (d. 1141) expressed approval of the "natural" part of astrology, which deals with the "temper or 'complexion' of physical things, like health, illness, storm, calm, productivity, and unproductivity, which vary with the mutual alignments of the astral bodies." An anonymous author writing near the end of the twelfth century or early in the thirteenth noted that "we do not believe in the deity of either the stars or the planets, nor do we worship them, but we believe in and worship their Creator, the omnipotent God. However, we do believe that the omnipotent God endowed the planets with the power that the ancients supposed came from the stars themselves." Another author from the twelfth century, addressing the issue of determinism, wrote that "the stars ... can produce an aptitude for having wealth, never the fact of having it."[30]

The translation of astrological treatises from Greek and Arabic was of critical importance in shaping these new attitudes. The major works were Ptolemy's *Tetrabiblos*, translated in the 1130s, and the *Introduction to the Science of Astrology* of Albumasar (Abū Maʿshar), translated twice in the 1130s and 1140s; these were accompanied by various smaller astrological tracts and joined eventually by works of Aristotle that addressed the question of celestial influence. The *Tetrabiblos* offered a defense of astrological belief and introduced its readers to some of the technical principles of the art. For example, it identified the various planets with specific terrestrial effects: the sun heats and dries, the moon chiefly humidifies, Saturn principally cools but also dries, and Jupiter heats and humidifies in moderation; the influence of certain planets is favorable, that of other planets unfavorable; certain planets are masculine, others feminine. The *Tetrabiblos* also explained how the powers of the planets are strengthened or weakened according to their geometrical relationship to the sun (their "aspect"). It assigned specific qualities to the signs of the zodiac. And it explained the general traits of people dwelling in different regions of the terrestrial globe by "familiarity" or sympathy between those regions and the planets and signs of the zodiac that govern them.[31]

The contribution of Albumasar's *Introduction* was to elaborate on the astrological principles found in Ptolemy's *Tetrabiblos* and other astrological literature (including Persian and Indian sources), but more especially to establish astrology on proper philosophical foundations by integrating traditional astrological lore with Aristotelian natural philosophy (fig. 11.10). In practical terms, what this meant was the adoption of Aristotle's metaphysics of matter, form, and substance, as well as Aristotle's claim that the celestial bodies are the source of all motion in the terrestrial region and the agents of generation and corruption. Through planetary influence, forms are imposed on the four elements to produce the physical substances of daily experience; changes in the planetary configuration bring about a perpetual cycle of transmutations, birth and death, coming to be and passing away. Aristotle's account of generation and corruption had concentrated on the motion of the sun around the ecliptic; while attaching priority to the sun, Albumasar (following long astrological tradition) brought the remaining planets, as well as their geometrical relationship to the sun and to the signs of the zodiac, into the causal picture.[32]

The Aristotelianizing of astrology was furthered by the acquisition of Aristotle's own works in the course of the twelfth century. During the thirteenth, astrological belief took root and became a standard part of the medieval worldview. Astrology also became closely associated with the practice of medicine: no reputable physician of the later Middle Ages would have imagined that medicine could be successfully practiced without it.[33] Philosophers and theologians continued to worry about astrological determinism—a subject that surfaced in the condemnation of 1277—and astrological practitioners were regularly denounced as charlatans. But even the most vigorous opponents of astrology were ready to acknowledge the reality of celestial influence. Nicole Oresme (d. 1382), who wrote whole books attacking astrology, admitted that the part of astrology that deals with large-scale events, such as "plagues, mortalities, famine, floods, great wars, the rise and fall of

Fig. 11.10. The Arabic astrologer Albumasar or Abū Maʿshar, presumably holding his *Introduction to the Science of Astrology*. Paris, Bibliothèque Nationale, MS Lat. 7330, fol. 41v (14th c.).

kingdoms, the appearance of prophets, new religions, and similar changes, . . . can be and is sufficiently well known but only in general terms. Especially we cannot know in what country, in what month, through what persons, or under what conditions, such things will happen." As for the influence of the heavens on health and disease, "we can know a certain amount as regards the effects which ensue from the course of the sun and moon but beyond this little or nothing." Astrology as an aspect of natural philosophy would flourish until the seventeenth century and beyond.[34]

THE SURFACE OF THE EARTH

A close examination of natural phenomena in the terrestrial realm will appear below (chap. 12). Here we must touch on various macroscopic features of the sublunar region that bear on the larger cosmological issues to which this chapter has been devoted.

We enter the terrestrial region by descending beneath the lunar sphere. This is the region of the four elements, which (in the idealized model) are arranged in concentric spheres, each in its proper place: first fire, then air, followed by water, and finally earth at the center. Two of the elements—fire and air—are intrinsically light and naturally ascend; the other two—water and earth—are intrinsically heavy and naturally descend. The elements are continuously transmuted one into another owing to the influence of the sun and other celestial bodies. Thus, for example, water is transformed into air in the process that we know as evaporation; conversely, air can be transformed into water to produce rain.

The fiery and aerial spheres were held to be the scene of various other meteorological phenomena, such as comets, shooting stars, rainbows, lightning, and thunder. Comets were considered to be atmospheric phenomena, the burning of a hot and dry exhalation that has ascended from the earth into the sphere of fire. Rainbows, it was generally held, are produced when sunlight is reflected from the droplets of water in a cloud. Various authors introduced refraction of light into the process; and early in the fourteenth century Theodoric of Freiberg (d. ca. 1310), expanding on analyses offered by Robert Grosseteste and Roger Bacon, and experimenting with water-filled glass globes, discovered that the colors of the primary rainbow were the product of two refractions and an internal reflection in each of the innumerable droplets that make up the surface of a cloud (see fig. 11.11). The theory was duplicated about the same time by Theodoric's Muslim contemporary, Kamāl al-Dīn al-Fārisī, a member of the Maragha school (see chap. 8, above).[35]

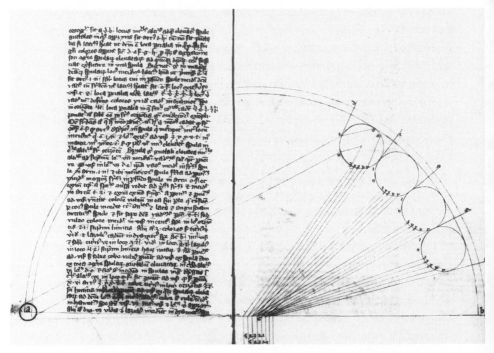

Fig. 11.11. Theodoric of Freiberg's theory of the rainbow. The sun is shown at the lower left, a set of raindrops at the upper right, and the observer is located at the lower center. The drawing aims to demonstrate how two refractions and an internal reflection (see the uppermost drop) within individual drops can produce the observed pattern of colors. Basel, Öffentliche Bibliothek der Universität, MS F.IV.30, fols. 33v–34r (14th c.).

At the center of everything is the sphere of the earth. Every medieval scholar of the period agreed on its sphericity, and ancient estimates of its circumference (about 252,000 stades) were widely known and accepted.[36] The terrestrial landmass was typically divided into three continents—Europe, Asia, and Africa—surrounded by sea. Sometimes a fourth continent was added. Beyond such basics, knowledge of the surface features of the earth and their spatial relationships varied radically with time, place, and individual circumstances.

Geographical knowledge existed in many forms during the Middle Ages, and we must be careful not to indulge the modern tendency to identify it exclusively with maps or maplike mental images.[37] Of course, medieval people had firsthand, experiential knowledge of their native region. Knowledge of more distant places could come from travelers, of which there were many sorts: merchants, craftsmen, laborers, pilgrims, missionaries, warriors, troubadours, itinerant scholars, civil and ecclesiastical officials, even fugitives and the home-less. For the few fortunate enough to have access to libraries, books such as Pliny's *Natural History* or Isidore of Seville's *Etymologies* offered geographical knowledge of a more exotic sort and on a grander scale in the form of writ-ten descriptions. Pliny and Isidore communicated a substantial collection of geographical lore (some of it mythological) through use of the "periplus"—a sequential list of the cities, rivers, mountains, and other topographical fea-tures encountered as one navigated a coastline. This information was usually accompanied by interesting historical, cultural, and anthropological detail. Drawing on earlier compilations, Pliny and Isidore led their readers on swift tours of the periphery of the European and African continents.[38] Toward the end of the Middle Ages, new travel literature began to enrich the store of such knowledge.

Traditional literary sources also dealt with climate, dividing the terrestrial globe into climatic zones or "climes." In a typical scheme, there were five of these: two frigid zones (the arctic and antarctic) around the poles, a temper-ate zone adjacent to each of these, and a torrid zone straddling the equator and (according to some sources) divided into two distinct rings by a great equatorial ocean. The torrid zone was considered uninhabitable on account of its heat—though some scholars disputed this claim. Medieval Europeans, of course, found themselves living in the northern temperate zone. On the opposite side of the earth, in the southern temperate zone, are the antipodes. Whether the antipodes are inhabited by antipodeans (people who walk upside down) was a matter of dispute.

There is a natural tendency for those of us familiar with modern maps to organize our geographical knowledge spatially, by the use of map coor-dinates, thereby reducing geography to geometry. But this was not true of medieval people, most of whom had never seen a map of any kind, let alone a map constructed on geometrical principles. Such maps as medieval people produced were not necessarily intended to portray in exact geometrical terms the spatial relationships of the topographical features indicated on them, and the notion of scale was almost nonexistent. Their function may have been symbolic, metaphorical, historical, decorative, or didactic. For example, the thirteenth-century Ebstorf map employs the world as a symbol of the body of Christ. And a representation of the terrestrial globe in a fifteenth-century manuscript illustrates the division of the world into three continents, each ruled by one of Noah's sons.[39] If, therefore, we wish to avoid misrepresenting medieval aims and achievements, we must be careful not to regard medieval maps as failed attempts at modern mapping.

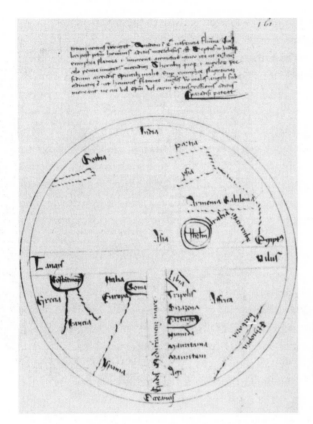

Fig. 11.12. A T-O map. Paris, Bibliothèque Nationale, MS Lat. 7676, fol. 161r (15th c.).

Among the most numerous, most interesting, and most studied medieval maps are the mappaemundi, or world maps. The most common form of mappamundi was the T-O map, associated with Isidore of Seville, which gave a schematic representation of the three continents—Europe, Africa, and Asia. In figure 11.12, the "T" inserted within the "O" represents the waterways (the Don and Nile Rivers and the Mediterranean Sea) believed to divide the known landmass into its major parts: Asia at the top of the map, Europe at the lower left, and Africa at the lower right. Nonschematic versions of the T-O map, which departed from the rigid T-O diagram in order to incorporate a variety of geographical detail, were also produced (see fig. 11.13). Another common type of map was zonal, featuring the climatic zones as its organizing principle.[40]

Medieval mapping took a mathematical turn (and thereby a turn toward modern cartography) in the form of portolan charts, embodying the practical knowledge of sailors and designed to facilitate travel by sea. These maps, invented perhaps in the second half of the thirteenth century, offered a "realistic"

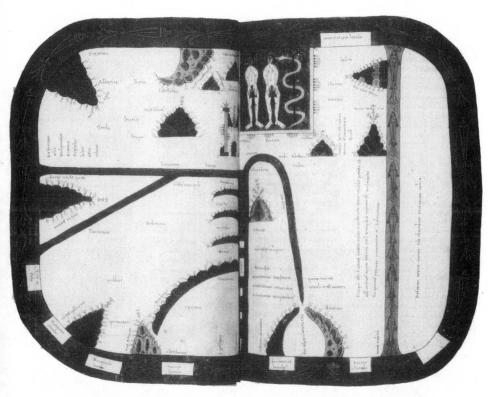

Fig. 11.13. A modified T-O map, the Beatus map (early 12th c.). A fourth continent is shown at the far right. London, British Library, MS Add. 11695, fols. 39v–40r. By permission of the British Library. For further discussion, see J. B. Harley and David Woodward, eds., *The History of Cartography*, vol. 1, plate 13.

representation of the coastline and employed a network of "rhumb lines" arranged around a compass rose to convey the distances and directions between any two points (see fig. 11.14). First applied to the Mediterranean Sea, portolan charts were later produced for the Black Sea and the Atlantic coastline of Europe. The use of portolan charts made possible more adventurous voyages of exploration, which in turn greatly expanded European geographical knowledge. Cartography was decisively transformed, finally, by the acquisition of Ptolemy's *Geography*, translated into Latin early in the fifteenth century, which taught Europeans the mathematical techniques by which to represent a spherical body on a two-dimensional surface.[41]

If mapmaking seems impressive for its practicality, we may do well to restore balance to this section by concluding with an investigation of a question that will appear (at least on the surface) to have no practical application

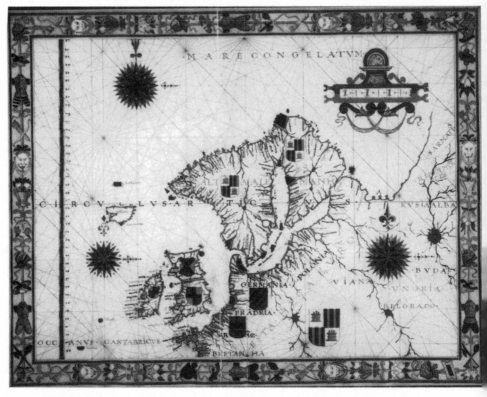

Fig. 11.14. A portolan chart by Fernão Vaz Dourado (ca. 1570). The Huntington Library, HM 41(5).

whatsoever—namely, whether the earth rotates on its axis and what would happen if it were to do so. Aristotle had convincingly presented the grounds for believing that the earth is stationary; and although all medieval scholars agreed with this, several thought the arguments for a rotating earth worth exploring. In looking at this question, they joined good ancient company, for the idea had never entirely disappeared from ancient cosmological and astronomical literature: Aristotle, Ptolemy, and Seneca all discussed it. The most searching explorations of the implications of a rotating earth came in the fourteenth century from two Parisian professors, John Buridan (d. ca. 1358) and Nicole Oresme (d. 1382).

There was no thought here of removing the earth from the center of the cosmos. What Buridan and Oresme had in mind was simply a daily rotation of the earth about its axis. The obvious advantage of postulating such a rotational motion was that doing so would eliminate the necessity of assigning a daily rotation to each of the celestial spheres; it would mean replacing many fast motions with a single slower one (tangential motions in both cases), an economy that nearly everybody could appreciate. Buridan pointed out that astronomers observe relative, rather than absolute, motions and that assigning the earth a daily rotation would have no effect on astronomical calculations. Consequently the question of a rotating earth could not be decided on astronomical grounds but must depend on physical arguments. Buridan devised the perfect test, pointing out that an arrow shot vertically upward (on a windless day) from the surface of a rotating earth would not return to its starting point, because while it was in the air the earth would be moving beneath it; since an arrow shot vertically upward does return to its starting point, he argued, we can be certain that the earth is stationary.[42]

A fuller investigation of the problem was produced a few years later by Oresme (fig. 11.15). One of the most acute natural philosophers of the medieval period, Oresme began by replying to the standard objections to a rotating earth. He argued that all we ever perceive is relative motion and, therefore, that observation cannot settle the issue. He replied to Buridan's arrow argument, maintaining that on a rotating earth, while the arrow is moving vertically upward and then vertically downward, it would also accompany the earth in its horizontal motion; the arrow would therefore remain above the point on the earth from which it was shot and return eventually to its starting point. He reinforced this argument with a shipboard example similar to the one that Galileo would use in the seventeenth century to defend the relativity of motion:

> Such a thing appears to be possible in this way, for, if a man were in a ship moving very rapidly eastward without his being aware of the movement, and if he drew his hand in a straight line downward along the ship's mast, it would seem to him that his hand was moving with a rectilinear [downward] motion; so, according to this opinion it seems to us that the same thing happens with the arrow which is shot straight down or straight up. Inside the boat so moved, there can be all kinds of movements—horizontal, criss-cross, upward, downward, in all directions—and they seem to be exactly the same as those when the ship is at rest. Thus, if a man in this boat walked toward the west less rapidly than the boat was moving toward the east, it would seem to him that he was moving west when actually he was moving east; and similarly as in the preceding case, all the motions here below would seem to be the same as though the earth were at rest.

Oresme proceeded to argue that scriptural passages seeming to teach the fixity of the earth can be interpreted as an accommodation on the part of the biblical text, "which conforms to the customary usage of popular speech."[43]

Fig. 11.15. Nicole Oresme. Paris, Bibliothèque Nationale, MS Fr.565, fol. 1r (15th c.). The large instrument is an armillary sphere, a teaching aid that offers a physical representation of the ecliptic, celestial equator, and other celestial circles.

Having thus refuted the objections to a moving earth, he completed the argument by presenting the positive case—a set of arguments for the economy of moving the earth instead of all the heavens.

This is a powerful and (for Copernicans like ourselves) convincing argument for the rotation of the earth on its axis. Did it convince Oresme's contemporaries? No; and, in fact, it apparently did not even convince Oresme. His argument represented the best philosophical or rational argument for the mobility of the earth that Oresme could construct. But the doctrine of divine omnipotence guaranteed that it was at best a probable argument, which could not be allowed to place limits on God's creative freedom; for all we know, God might prefer an uneconomical world. In the end, therefore, Oresme accepted the traditional opinion that the earth is fixed, supporting it with a quotation from Psalm 92:1: "For God hath established the world, which shall not be moved."[44] Apparently this scriptural passage would not yield (as the others had) to the principle that Scripture accommodates itself to popular speech.

Historians have been unsure how to interpret Oresme's apparent turnabout. Many have been tempted to suppose that he saw that he was heading for theological trouble and decided to save himself with a disclaimer. In fact, Oresme took the trouble to explain what he was doing, and surely we should take his own account seriously. His purpose, he stated, was to offer an object lesson for those who would impugn the faith by rational argument. His success at formulating persuasive philosophical arguments for an idea as "opposed to natural reason" as the rotation of the earth demonstrated the unreliability of rational argument and, therefore, the caution to be used where rational argument touches the faith, as this one did. We will never know for certain what his purpose was, but I do not believe that we can lightly dismiss his own version.[45]

12 ❄ *The Physics of the Sublunar Region*

Medieval physics was not a primitive version of modern physics and cannot be legitimately judged by comparison with its modern namesake. Certainly there is overlap between the two, but medieval physics, deeply engaged with Aristotelian metaphysics and natural philosophy devoted much more attention to fundamental issues that we would classify as "metaphysical" or "philosophical": concerning the fundamental stuff of the universe, the elements and their constituents, the sources of motion and change, and the like.

The medieval natural philosopher (as I will generally refer to him) took his starting point from the text of Aristotle's *Physics* and other works, and devoted his career to clarification of ambiguities, disputation about difficult or contentious portions of the text, and original application or extension of Aristotelian principles. But he was emphatically *not* a slave to the Aristotelian text, as a widespread myth would have it; rather, he was typically a gifted reader and interpreter of the texts of Aristotle and his commentators (including critics), eager to display his logical and creative powers in discussion and debate. There were theological corners of which he needed to be wary, but otherwise the medieval professor had the freedom to go where reason and experience led.

MATTER, FORM, AND SUBSTANCE

But let us abandon generalities and sample medieval physics as it was practiced on the ground and as it existed in the minds of the historical actors who devoted their careers to it. What were the fundamental explanatory principles of medieval physics? After the reception and assimilation of Aristotle's philosophy in the twelfth and thirteenth centuries, the principles in question were broadly Aristotelian—though obscurity, incompleteness, and inconsistency in the various Aristotelian texts where these principles were set forth left plenty of room for further articulation of the theory and for discussion and debate about the fine points. Let us begin with a brief review of some of the basics.[1]

According to Aristotle, all objects in the terrestrial realm ("substances" he called them) are composites of form and matter. Form, the agent, bearer of the properties of the individual thing, combines inseparably with matter, the passive recipient of the form, to produce a concrete corporeal object. If the object in question is a "natural" object (as opposed to one produced artificially, by a craftsman), it also has a nature (determined primarily by its form but secondarily by its matter), which disposes it to certain kinds of behavior. Thus fire naturally communicates warmth, rocks naturally fall (if lifted out of their natural place), babies naturally grow and mature, and acorns naturally develop into oak trees. These natures we discern through long and persistent observation. Because natures are the determining factors in all cases of natural change, they are necessarily of great interest to the natural philosopher.[2]

Aristotle's medieval followers, contemplating this scheme, identified two kinds of form—one of them associated with essential properties, the other with incidental properties. The defining characteristics of a thing, which make it what it is, are conveyed by what came to be called its "substantial form." Substantial form combines with absolutely propertyless first matter to give being or existence to a substance and to endow it with those properties that make it the kind of thing it is. However, besides essential properties, every substance also has properties of an incidental or accidental sort, associated with "accidental form." Thus the family dog may be short-haired or long-haired, lean or fat, friendly or ferocious, housebroken or not—the results of different accidental forms—and yet retain the characteristics (supplied by its substantial form) that enable us to identify it unmistakably as a dog.

Aristotle's theory of form, matter, and substance is nicely exemplified by his theory of the elements. Aristotle accepted the position of his predecessors, Plato and the pre-Socratics: that the familiar materials or substances of everyday experience are complex rather than simple. That is, sensible things in the sublunar world are compounds or mixtures, reducible to a small set of fundamental roots or principles, called "elements." Aristotle adopted Empedocles' and Plato's list of four elements—earth, water, air, and fire—and argued that these combine in various proportions to produce all of the common substances. Aristotle agreed with Plato that the four elements are not fixed and immutable, but undergo transmutations; and the scheme that explained how this was possible was his theory of form and matter.

Each of the elements, he argued, is a composite of form and matter; since the matter in question is capable of assuming a succession of forms, the elements can be transformed into one another. The forms instrumental in producing the elements are those associated with hot, cold, wet, and dry—the four primary or "elemental" qualities. Primary matter informed by coldness and dryness yields the element earth; primary matter informed by coldness and wetness yields water; and so forth. But the situation is not static: if the quality of dryness in a piece of the element earth yields to wetness through the action of a suitable agent, that piece of earth will cease to exist, and an appropriate amount of the element water will take its place. Aristotle argued that such transformations are occurring constantly, and the elements are therefore constantly being transmuted one into another. Changes of this kind proved capable of accounting for many of the familiar phenomena that we associate today with the disciplines of chemistry and meteorology.[3]

The matter-form theory was elaborated in Islam by Avicenna (Ibn Sīnā, 980–1037) and Averroes (Ibn Rushd, 1126–98) in ways that would prove influential in the West. The two Islamic commentators thought it impossible to derive the elements from the imposition of the elemental forms directly on primary matter. An intermediate step was required, which would first invest the primary matter with three-dimensionality. To this end, they developed the notion of "corporeal form," which must first be imposed on primary matter to yield a three-dimensional body. The elements emerge, then, when this three-dimensional body (a kind of secondary matter) receives the forms of the four elements. The idea of corporeal form was transmitted to Christendom, where it proved both influential and controversial. We have seen its adoption by Robert Grosseteste, who identified corporeal form with light.[4] Aristotle's scholastic commentators in medieval Islam and medieval Christendom found much more to wrestle with in Aristotle's theory of matter and form, but these basics are sufficient for our needs.[5]

COMBINATION AND MIXTURE

One very important class of phenomena to which the theory of matter, form, and substance was applicable was what we would today identify as "chemical combination." The centrality of this class of phenomena is apparent when we recall that, according to Aristotle, all substances encountered in the real world, are compounds of the four elements. It comes as no surprise, therefore, that Aristotle should have inquired into the nature of chemical combination and the status of the original ingredients in a compound. He distinguished between a mechanical aggregate, in which the small particles of two substances are situated side by side without loss of individual identity, and a true blending of the ingredients into a homogeneous compound in which the original natures disappear. He called the latter a "mixt" or "mixture" (I will employ the Latin terms *mixtio* for the process and *mixtum* [plural, *mixta*] for the product, in order to preserve the technical meaning that Aristotle had in mind). It is this latter kind of combination that he considered applicable to mixing of the elements.

In a mixtum, according to Aristotle, the individual natures of the ingredients are replaced by a new nature that permeates the compound down to its smallest parts. The properties of the mixtum represent an averaging of the properties of the ingredients. If, for example, we combine a wet and a dry element (say, water and earth), the wetness or dryness of the resulting compound will fall on a scale that runs between the extremes of wetness and dryness, at a point determined by the relative abundance of those two qualities. Although the original elements no longer have actual existence in the mixtum, Aristotle made remarks that suggested that they maintain a virtual or potential presence that permits them to exercise some kind of continuing influence.[6]

Aristotle's discussion left a number of challenges for his commentators. One was to recast the theory of combination or mixtio in the language and conceptual framework of matter and form, for those terms do not appear in Aristotle's account. In the course of that effort it was necessary to inquire how the new substantial form of the mixtum emerges from the forms of the constituent elements. Another problem of critical importance was to determine in what sense the forms of the original elements continue to exist in the mixtum; since it was acknowledged that when the mixtum is destroyed the elements out of which it was formed reappear, it seemed evident that they survive in some way within the mixtum. Debates on these matters became extremely intricate, and I must limit myself to a few remarks.

Everybody agreed that the substantial forms of the constituent elements are replaced by a new substantial form of the mixtum. But how does this come about? This question provoked intricate debates among Aristotelian commentators, but the usual solution was to invoke higher powers—celestial forces or celestial intelligences, possibly even God himself—assigning to them the responsibility for infusing the new substantial form into the primary matter when the preconditions had been met. Another problem associated with mixta was the necessity of finding some way of allowing the elements to lurk in the mixtum potentially or virtually, awaiting a suitable opportunity to reassert themselves. This too became the basis of lively debate among late medieval natural philosophers.[7]

A final question with which we must deal has to do with the physical divisibility of corporeal substances—say, wood or stone or organic tissue. Is there
a limit to the process of division, and what are the properties of the smallest
pieces? Are they anything like atoms? Aristotelian commentators developed a
theory of what came to be called *minima* or *minima naturalia* (smallest natural
parts). The theory acknowledged that in principle divisibility should be endless; however small the piece before you, there is no physical reason why you
cannot divide it again. But it was argued that there is nonetheless a smallest
quantity of each substance, below which it will no longer be that substance
because the form of the substance cannot be preserved in a smaller quantity.

Attempts were made in the Middle Ages to construe the theory of minima
as a variant of atomism. It is true that both theories acknowledged the par-
ticulate structure of matter, but they were otherwise far apart. The particles
of the atomists were unbreakable least parts; the minima of the Middle Ages
were divisible, though if divided they would lose their identity. All atoms were
of identical stuff, differing only in size and shape; minima were as different
as the substances to which they belonged. In the atomist vision, properties in
the macroscopic world did not, in general, have exact counterparts in the mi-
croscopic world: atomists did not explain the redness of a flower, for example,
by the redness of its constituent particles. Rather, the atomist program was
to reduce the qualitative richness of the world of sense experience to austere,
qualitatively bare atoms (characterized only by size, shape, motion, and possibly weight). Minimists, by contrast, continued the Aristotelian program,
assigning to the least parts precisely the properties of the whole to which they
integrally belonged: minima of wood are still wood.[8]

ALCHEMY

In today's public mind, alchemy was mysterious, secretive nonsense, practiced by quacks. It is true that in some of its manifestations it became an all-
embracing mystical philosophy, linked to the spiritual transformation of the
alchemical practitioner. The imposition of religious meanings on alchemy
was especially true in the later Middle Ages and Renaissance, when alchemy
came to serve religious functions. But from antiquity to the Renaissance
and beyond, alchemy was the practice of serious metallurgy and chemical
technology. And we will limit ourselves to this part of the story: alchemy
as both theory of matter and the procedures of laboratory experimentation,
directed primarily toward the transmutation of base materials into precious
metals.

Looming over any discussion of alchemy is the question: how could presumably intelligent people (literate and therefore educated) during the Middle Ages take seriously the nonsense of transmutation of base metals into gold? Doesn't such silliness place them firmly outside the bounds of serious science and, therefore, outside the limits of this book? No! Quite the contrary. It is difficult to imagine how people who lacked our knowledge of plant and animal physiology could have *doubted* the reality of transmutation. Consider the case of a plant or tree, which transforms water and soil nutrients into a delicate blossom or succulent fruit; or the even more extraordinary case of a lamb, which has the ability to convert water and grass into wool and flesh. The transformation of one metal into another seems, by comparison, a considerably less challenging feat. There was also theory to support belief in the possibility of transmutation. Aristotle had declared the fundamental unity of all corporeal substance, portraying the four elements as products of prime matter endowed with pairs of the four elemental qualities: hot, cold, wet, dry. Alter the qualities, and you transmute one element into another. Alter the proportions of the elements in a mixtum (or compound), and you transform the mixtum into a different substance.

It is widely agreed by historians that alchemy had Greek origins, perhaps in Hellenistic Egypt. Greek texts were subsequently translated into Arabic and gave rise to a flourishing and varied Islamic alchemical tradition.[9] Most of the Arabic alchemical writings are by unknown authors, many of them attributed pseudonymously to Jābir ibn Hayyān (fl. 9th–10th c., known in the West as Geber). Important, along with this Geberian (or Jabirian) corpus, was the *Book of the Secret of Secrets* by Muḥammad ibn Zakariyyā al-Rāzī (d. ca. 925). Beginning about the middle of the twelfth century, this mixed body of alchemical writings was translated into Latin, initiating (by the middle of the thirteenth century) a vigorous Latin alchemical tradition. Belief in the ability of alchemists to produce precious metals out of base metals was widespread but not universal; from Avicenna onward, a strong critical tradition had developed, and much ink was devoted to polemics about the possibility of transmutation.

According to a widely held theory of transmutation deriving from the Geberian corpus, all metals are compounds or mixta of sulphur and mercury.[10] The mixtio of sulphur and mercury was conceived as a process of maturation or purification that takes place naturally in the earth, under the influence of heat. The particular metal that emerges depends on all of the factors that go into the maturation process, including the purity and homogeneity of the sulphur and the mercury, their proportions in the mixtum, and the degree of heat.

Now it was the aim of the alchemist to short-cut and accelerate the process of purification—to reproduce in a short time, by artifice, what nature took perhaps a thousand years to accomplish in the womb of the earth. The goal and endpoint of the process, if perfectly carried out, was gold; imperfection or shortfall gave rise to one of the other metals.

According to another theory, prominent in the Geberian corpus, substances consist of both internal (or occult) and external (or manifest) qualities. For example, silver is cold and dry externally but hot and moist internally. The appropriate alchemical processes are capable of drawing out and making manifest the formerly occult, internal qualities of heat and moisture—thereby turning silver (cold and dry externally) into gold (hot and moist externally). According to yet another account, the alchemist's task was to reduce a base metal to prime matter by stripping off its substantial and accidental forms, then adding forms in such a way as to reconstitute the metal as one of the precious metals—all of this by following the appropriate alchemical recipes.[11]

It was a feature of all of the competing theories that transmutation must be performed ultimately by "nature," imitating what occurs within the earth. The alchemist's responsibility is to discern and then to create the circumstances that will enable nature to act. This was no simple matter, and one of the main functions of an alchemical manual was to specify the required "laboratory" techniques, which included calcination, sublimation, fusion, distillation, solution, coagulation, precipitation, and crystallization (fig. 12.1).[12]

The most influential of all medieval alchemical writings in the Latin West was the *Summa perfectionis* authored by an otherwise unknown Franciscan friar, Paul of Taranto, but attributed to Geber—an obvious attempt by its author to clothe himself in the mantle of the Geberian corpus.[13] What is important (indeed, revolutionary) about this book is the author's adoption of a corpuscular interpretation of the elements. This flies in the face of standard Aristotelian theory of the elements, which was totally opposed to any sort of atomism or corpuscularianism, but reveals the influence of the *Meteorologica* attributed to Aristotle (perhaps spuriously).[14] According to Paul, the four elements—earth, water, air, and fire—exist in the form of tiny corpuscles. Corpuscles of earth (for example) bond with one another in tight clusters to form larger corpuscles of earth. Though very tightly bonded, the original tiny corpuscles retain their identity within the larger corpuscle. These larger corpuscles of earth then form tight clusters with larger corpuscles of the other elements in various proportions to form the material of which metals and other substances consist. Because the smaller corpuscles retain their identities within the larger corpuscles, they may be called upon as a second

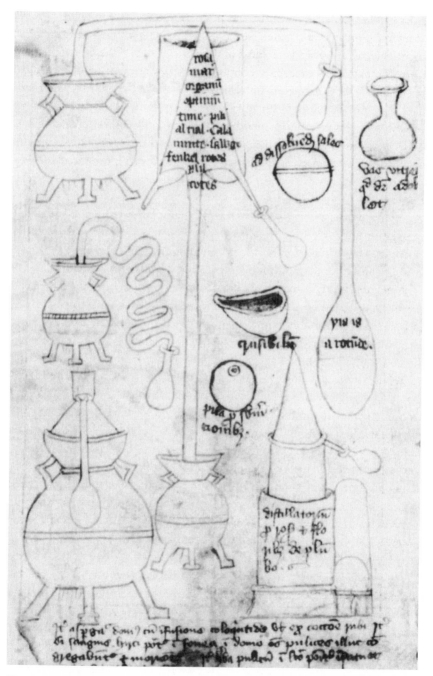

Fig. 12.1. Alchemical apparatus, including furnaces and stills. British Library, MS Sloane 3548, fol. 25r (15th c.). By permission of the British Library.

causal and explanatory level. The properties of a given substance—a metal, for example—can be attributed either to the original tiny corpuscles or to the larger clusters. If we set aside some of the details, we have here a corpuscular theory of matter resembling that of the ancient atomists, differing in that the "atoms" of Paul of Taranto do not represent the end of the process of division, but are themselves composites of still smaller entities.

This is all very interesting, but its full importance becomes apparent when we examine the sort of defense given by Paul of Taranto to his theories. A single, but remarkable, example must suffice. Defending the theory of two layers of causation outlined just above, Paul argues as follows:

> [When metals are resolved into their fundamental constituents by cal-cination], if there were a complete resolution to the simple elements and not to certain [intermediate] mineral or metallic principles which are nearer than the first simple bodies [Aristotle's four elements], the metal or such and such a body would no more return from them upon [its exposure] to fire than anything else made up of the simple elements, and gold would no more return from gold than would stone or wood [return from gold], especially since fire is a common agent, behaving alike towards all and each. But since these [metals and minerals] return just the same as before, it is manifest that they were only resolved to certain components of theirs and not to the simple elements or to the prime matter as those aforesaid [philosophers] mistakenly assert.

This is only one of many cases that could be examined. What is important is Paul's adoption of a form of alchemical practice that demands confirmation or disconfirmation of alchemical theory by contrived chemical experimentation.[15]

There is more to the story. The *Summa perfectionis* of Paul of Taranto continued to be influential well into the seventeenth century. Daniel Sennert (1572–1637), professor of medicine at the University of Wittenberg, became an outspoken proponent of both the corpuscularian chemistry of the *Summa perfectionis* and its experimental methodology. And evidence suggests that his example influenced Robert Boyle's scientific practices. Boyle, in turn, was a leading figure in the shaping of seventeenth-century science, both theoretically and methodologically. I think it best to conclude by allowing Newman (the source of much of the above) to speak for himself: "Daniel Sennert's *De chymicorum cum Aristotelicis et Galenicis consensu ad dissensu* actually provided a dramatically convincing experimental demonstration that matter at the micro-level is corpuscular, thereby paving the way for the flagrantly anti-Aristotelian 'mechanical philosophy' of Robert Boyle and his compatriots." Commenting further on Sennert's experimental practices, Newman writes that "despite their seemingly mundane character, Sennert's operations provided a powerful basis for the increasingly experimental corpuscular theory of the seventeenth century."[16] Alchemy, the paradigmatic pseudo-science of the Middle Ages (along with astrology), may have made a powerful contribution to the "new" science of the seventeenth century.

CHANGE AND MOTION

Historians frequently contrast the static character of the Aristotelian universe with the dynamism of the atomic philosophy. It is easy to see what they have in mind. In Aristotle's sublunar realm, natural motion ceases when the moving object reaches its natural place, and violent motion comes to an end when the external force no longer acts. If we could put everything in its natural place and get rid of external movers, Aristotle's world would come to a total halt. By contrast, the world of the atomists is in a permanent state of motion—atoms moving, colliding, and forming temporary clusters in an eternal maelstrom.

However, the impression that Aristotle's cosmos is static comes from restricting our attention to one kind of change—change of place or "local motion."[17] Look beneath the surface, not at the location of an object, but at the *nature* of the object, and the true dynamism of the Aristotelian cosmos becomes apparent. For Aristotle, natural things are always in a state of flux; it is part of their essential nature to be in transition from potentiality to actuality. This is no doubt most obvious in the biological realm, where growth and development are inescapable, but Aristotle's biological studies powerfully shaped his entire philosophy of nature. His definition of nature, as the inner source of change found in all natural bodies, may well have had biological origins, but it was applicable to both the organic and the inorganic realms. The central object of study in Aristotle's natural philosophy, then, was change in all of its forms and manifestations. Aristotle stated bluntly in his *Physics* (book 3) that if we are ignorant of change, we are ignorant of nature.[18] If the gross objects that fill the Aristotelian cosmos seem to prefer rest over motion, beneath the surface they are seething with change.

Aristotle and his medieval followers identified four kinds of change: (1) generation and corruption, (2) alteration, (3) augmentation and diminution, and (4) local motion. Generation and corruption occur when individual things (that is, substances) come into existence and go out of existence. Alteration is change of quality, as when a cold object becomes warm. Augmentation and diminution refer to quantitative change—that is, change of size, as in rarefaction and condensation. And local motion is change of place—the kind of change that seventeenth-century scientists elevated to a position of centrality that it did not have within Aristotelian physics.

When we examine Aristotle's theory of local motion, therefore, we are looking at one aspect of his theory of change. It was change in general that interested Aristotle and his commentators, and local motion was but one of several varieties and by no means the most fundamental. It will save us from a great deal of confusion if we keep this in mind. Features of Aristotelian and medieval theories of motion that seem strange and idiosyncratic when examined from the standpoint of modern dynamics frequently take on quite a different appearance when we judge them in the light of the questions they were intended to answer.

This brings us face to face with an important and difficult methodological issue. The customary way of approaching medieval theories of motion is to carry the conceptual framework of modern dynamics back to the Middle Ages and use it as a grid through which to view medieval developments. This procedure has the enormous advantage of keeping us on familiar intellectual ground; it has the disadvantage of bringing into focus only those medieval developments that resemble some piece of modern dynamics. The alternative is to adopt a medieval perspective—an approach that has the obvious advantage of fidelity to the system of ideas that we are endeavoring to understand, but one that may be nearly impossible to carry out in practice. The intellectual framework of medieval theories of motion is a conceptual jungle, suitable only for hardened veterans and certainly no place for day trips from the twenty-first century. Faced with a choice between viewing this jungle from the safe distance of the seventeenth or twenty-first century and not viewing it at all, most historians of medieval science have understandably chosen the former alternative. My own view is that we must make some pragmatic compromises in the effort to find a middle way. In the pages that follow, we will take several short excursions into regions of the medieval jungle judged safe for tourists, in order to give some sense of the lay of the land. We will also examine certain medieval developments important for their later influence, endeavoring to describe those developments in ways that will help the reader to grasp the medieval framework out of which they grew.

THE NATURE OF MOTION

When an ancient or medieval natural philosopher turned his attention to any area of inquiry, the first thing he wanted to know was what things (relevant to the inquiry) exist. This is a question about the entities that populate the universe. Once he had settled this question, he could move on to others, such as: What is the nature of the things that exist? What kind of existence do they have? How do they change? How do they interact? And how do we know about them? If the object of study was motion, the first task would be to figure out whether motion exists and, if so, what sort of thing it is.

Aristotle had addressed this question—the nature of motion—with enough ambiguity to give his commentators plenty to chew on. In Islam, the two great Aristotelian commentators, Avicenna (Ibn Sīnā) and Averroes (Ibn Rushd), both joined the fray. And in the West the problem was reopened by Albert the Great. We cannot probe the fine points of this extremely technical debate, but we can reveal the broad outlines by calling attention to two prominent alternatives that emerged by the end of the thirteenth century, and a few of the arguments employed to adjudicate between them. According to one opinion, which came to be designated by the phrase *forma fluens* (flowing form), motion is not a thing separate or distinguishable from the moving body, but simply the moving body and its successive places. When Achilles runs a race, the existing things are Achilles and the objects that define the places successively occupied as he runs; no additional entity is involved, and the word "motion" denotes not an existing *thing*, but merely the *process* by which Achilles comes to occupy successive places. This view was developed by Averroes and Albert the Great. The alternative opinion, known by the name *fluxus formae* (flow of a form), maintained that in addition to the moving body and the places it successively occupies, there is some *thing* inherent in the moving body, which we may call "motion."[19]

We can perhaps begin to perceive the rationale behind this debate by examining a pair of famous arguments, one for each of the alternatives. William of Ockham (ca. 1285–1347) defended the *forma fluens* opinion with characteristic logical rigor. In Ockham's view, "motion" is an abstract, fictional term—a noun that corresponds to no really existing entity. This was not an attempt on Ockham's part to deny that things move, but simply a declaration that *motion* is not a *thing*. The way to get clear on this, Ockham argued, is to consider a sentence such as the following: "Every motion is produced by a mover." A naive reader might suppose that the noun "motion" stands for a real thing (a substance or a quality), for nouns often serve that function. However, we can replace this sentence with another that has identical dynamic content, but different implications for the *nature* of motion: "Each thing that is moved is moved by a mover." Here the noun "motion" has disappeared, and with it the implication that motion might be a real thing. But how are we to choose between the two sentences and the alternative worlds they describe? On the basis of economy! Although the two sentences make the same dynamic claim (things move only if moved by movers), the world in which motion is not an existing thing is a more economical world, because there are fewer things in it; consequently, employing Ockham's time-honored "razor," we should, unless there are convincing arguments to the contrary, regard it as the real world.[20]

An altogether different set of considerations led John Buridan, in the fourteenth century, to defend the *fluxus formae* view. In his commentary on Aristotle's *Physics*, Buridan answered the now familiar question—whether local motion is a thing distinct from the moved object and the places it successively occupies—by reference to theological doctrine. The theological starting point of Buridan's argument was the assumption that God, by his absolute power, could have endowed the cosmos as a whole with a rotational motion had he so wished. Buridan knew this by virtue of the principle that God can do anything that involves no self-contradiction; moreover, one of the articles of the condemnation of 1277 (five decades or more in the past, but still to be taken seriously, at least in Paris) explicitly affirmed God's power to accomplish the analogous feat of moving the entire cosmos in a straight line. But if we adopt the *forma fluens* view that motion is nothing more than the moving object and the places it successively occupies, a serious problem arises. Aristotle had defined place in terms of surrounding bodies. Since the cosmos is not surrounded by anything (for any container would have to be considered part of the cosmos), it seems to have no place. If the cosmos has no place, it obviously cannot change places; and if it does not change places, it cannot be said to move. However, this conclusion is incompatible with the starting point of the argument—the unquestionable assumption that God is capable of giving the cosmos a rotational motion. The solution, Buridan thought, was to adopt the broader *fluxus formae* conception of motion. If motion is not simply the moving body and its successive places, but an additional attribute of the moving body analogous to a quality, then the cosmos could possess this attribute even in the absence of place, and the difficulty would be at least partially overcome. The implication of this theory—that motion is a quality, or something that could be treated as a quality—became quite common among natural philosophers in the second half of the fourteenth century.[21]

MATHEMATICAL DESCRIPTION OF MOTION

Today the application of mathematics to motion needs no defense. Theoretical mechanics, the parent discipline of theories of motion, is mathematical by definition, and to anybody with a grasp of modern physics the mathematical way would seem to be the only way. But perhaps it is only by hindsight and from a modern perspective that this conclusion is obvious; it would not have seemed plausible to many who worked within the Aristotelian tradition. We must remember that Aristotle and his medieval followers regarded motion as one of four kinds of change and that their analysis of change was not meant to focus on local motion, but rather to be applicable to all four classes of change. We also need to recognize that there is nothing obviously mathematical about most kinds of change. When we observe sickness yielding to health, virtue replacing vice, and peace emerging from war, no numbers or geometrical magnitudes leap out at us. The generation or corruption of a substance and the alteration of a quality are not obviously mathematical processes, and it is only by heroic efforts over the centuries that scholars have found ways of placing a mathematical handle on a few kinds of change, including local motion. Let us investigate the early stages of this process in the Middle Ages.

The mathematization of nature, of course, had ancient proponents, including the Pythagoreans, Plato, and Archimedes; and early success was achieved in the sciences of astronomy, optics, and the balance (see chap. 5, above). It was inevitable that the success of these efforts would provide encouragement for those interested in mathematizing other subjects. Indeed, Aristotle himself was responsible for a primitive beginning of the mathematical analysis of motion in his *Physics*, where distance and time, both quantifiable, were employed as measures of motion. Aristotle argued that the quicker of two moving objects covers a greater distance in the same time or the same distance in less time, while two objects moving with equal quickness traverse equal distances in equal times.[22] A generation after Aristotle, the mathematician Autolycus of Pitane (fl. 300 B.C.), took a further step, defining a uniform motion as one in which equal distances are traversed in equal times. It is important to note that in these ancient discussions distance and time were taken as the critical measures of motion, to which a numerical value might be assigned, while "quickness" or speed never achieved that status, remaining a vague, unquantified conception.[23]

The impact of this mathematical analysis in medieval Europe can first be seen in the work of Gerard of Brussels, a mathematician who may have taught at the University of Paris in the first half of the thirteenth century. For our purposes the most important thing about Gerard's brief *Book on Motion* is the restriction of its contents to what we now call "kinematics"—the purely mathematical description of motion—as opposed to "dynamics," which is concerned with causes. This is an important distinction (bearing a resemblance to the distinction between "instrumentalism" and "realism" in astronomy), which will serve as one of the organizing principles for the remainder of our discussion of medieval theories of motion. For the moment, Gerard of Brussels is important as a harbinger of the kinematic tradition that was to develop in the Latin West.[24]

This tradition flowered among a group of distinguished fourteenth-century logicians and mathematicians affiliated with Merton College, Oxford, between about 1325 and 1350. This group included Thomas Bradwardine (d. 1349), subsequently appointed archbishop of Canterbury; William Heytesbury (fl. 1335); John of Dumbleton (d. ca. 1349); and Richard Swineshead (fl. 1340–55). To begin with, members of the Merton group made explicit the distinction between kinematics and dynamics that was implicitly present in Gerard's *Book on Motion*, noting that motion can be examined from the standpoint either of cause (dynamics) or of effect (kinematics). The Merton scholars proceeded to develop a conceptual framework and a technical vocabulary for dealing with motion kinematically. Included in this conceptual framework were the ideas of "velocity" and "*instantaneous* velocity," both treated as mathematical concepts to which magnitude could be assigned.[25] The Mertonians distinguished between uniform motion (motion at constant velocity) and nonuniform (or accelerated) motion. They also devised a precise definition of uniformly accelerated motion identical to our own: a motion is uniformly accelerated if its velocity increases by equal increments in equal units of time. Finally, the Merton scholars developed a variety of kinematic theorems, several of which we will examine below.[26]

Before we do that, we must consider the philosophical underpinnings of this kinematic achievement. The emergence of velocity as a new measure of motion, to go along with the ancient measures (distance and time), is a development that needs to be explained. Velocity, after all, is quite an abstract conception, which did not force itself on the observer of moving bodies but had to be invented by natural philosophers and imposed on the phenomena. How did this come about? The answer is found in the philosophical analysis of qualities and their strength or intensity. The fundamental idea was that qualities or forms can exist in various degrees or intensities: there is not just a single degree of warmth or cold, but a range of intensities or degrees running from very cold to very hot. Moreover, it was acknowledged that forms or qualities can vary within this range; that is, they can be strengthened and weakened, or, to employ the technical medieval terminology, undergo intensification and remission.[27] Now when this general discussion of qualities and their intensification and remission was transferred to the particular case of local motion (motion being conceived as a quality or something closely analogous to a quality), the idea of velocity quickly emerged. The intensity of the quality of motion—that which measured its strength or degree—could be none other than swiftness or (to employ the technical medieval Latin term) "velocitas." Intensification and remission of the quality of motion must then refer to variations in velocity.

Reflection about qualities, their intensity, and their intensification thus led the Mertonians to a new distinction: between the *intensity* of a quality (defined above) and its *quantity* (how much of it there is). An example will help us to understand this distinction: it is obvious enough in the case of heat that one hot object can be hotter than another; this is a reference to the intensity of the quality, what we call "temperature."[28] But we also have a conception of the quantity of heat—how much of it there is. If we have two objects at the same temperature, one of them twice as big, that larger object clearly has twice the "quantity" of this quality of heat. For fourteenth-century mathematicians, it followed that all qualities should submit to a similar analysis, possessing both a quantity (how much of the quality) and an intensity (the degree or strength of the quality). For heat, we have temperature (intensity) and calories (quantity); for weight, heaviness (quantity) and density or specific gravity (intensity); and so on. Could the same analysis be successfully applied to motion? Yes, it could, as we shall see.[29]

News of the Merton College achievements in the analysis of qualities was transmitted quickly to other European intellectual centers. In the process, the analysis was enriched and clarified by the development of geometrical representation. The original analysis of qualities at Merton College was carried out verbally, in much the same way as we have been analyzing it. However, the advantages of geometrical analysis were recognized, and fairly elaborate systems of geometrical representation were eventually worked out. One of the first to develop such a system was Giovanni di Casali, a Franciscan from Bologna (who had also spent time in Cambridge), writing about 1351; a far more elaborate geometrical analysis was formulated by Nicole Oresme (d. 1382) at the University of Paris later in the same decade. An examination of Oresme's scheme may prove as illuminating for us as it no doubt did for his medieval readers.

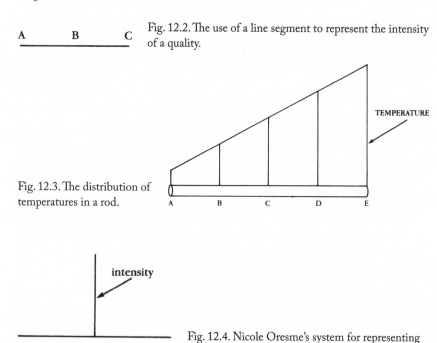

A   B   C   Fig. 12.2. The use of a line segment to represent the intensity
                       of a quality.

Fig. 12.3. The distribution of
temperatures in a rod.

TEMPERATURE

A   B   C   D   E

intensity

subject line or extension

Fig. 12.4. Nicole Oresme's system for representing
the distribution of any quality in a subject.

The first step was to represent the intensity of a quality by means of a line
segment—a relatively easy step for medieval scholars brought up on Aristotle
(who employed lines to represent time) and Euclid (who used lines to represent numerical magnitudes). If line segment AB (fig. 12.2) represents a
given intensity of some quality, then line segment AC represents twice that
intensity. This is fine, but it has not yet gotten us very far. The critical next step
was to employ this line to represent the intensity of the quality at any point
of the subject. Take a rod AE (fig. 12.3), heated differentially, so that the heat
increases uniformly from one end to the other. At point A and at whatever
intervals you choose, erect a vertical line representing the intensity of heat
at that point. If (as we have postulated) the temperature increases uniformly
from A to E, then the figure will reveal a uniform lengthening of the vertical
lines. Now Oresme made the system a good deal more abstract by substituting
a horizontal line for the drawing of the rod (fig. 12.4). This has the effect of
creating a generalized system of representation in which the horizontal line
(called the "subject line" or the "extension") represents the subject, whatever it
might be, while vertical lines represent the intensity of any quality we choose
at the points of the subject where they are erected.

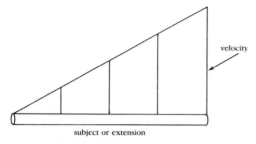

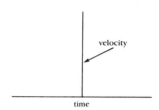

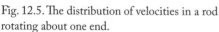

Fig. 12.5. The distribution of velocities in a rod rotating about one end.

Fig. 12.6. Velocity as a function of time.

What Oresme has produced is a form of geometrical representation—an obvious forerunner of modern graphing techniques—in which the shape of the figure (as in fig. 12.3) informs us about variations in the intensity of a quality over its subject. But how do we make the transition from qualities in general to motion in particular? One way is to consider a body, the different parts of which move with different velocities; a rod held by a pin at one end and rotated about that pin would be a good example. In such a case, we can draw the rod horizontally and erect a perpendicular at any point, indicating the angular velocity of that point. The result will be a distribution of velocities in a subject, as in figure 12.5.

But there is another case, more difficult because it requires more abstract treatment. Suppose we have a body that moves as a unit, all of its parts having the same velocity, but a velocity that varies over time. The way to understand this, Oresme explained, is to see that here the subject line is not the extension of a corporeal object, as in the examples above, but the duration of a local motion. Time becomes the subject, represented by the horizontal line. This gives us a primitive coordinate system in which velocity can be plotted as a function of time (see fig. 12.6). Oresme proceeded to discuss various configurations of velocity with respect to time. Uniform velocity will be represented by a figure in which all the vertical lines are of equal length—that is, a rectangle. Nonuniform velocity requires verticals of variable length. Within this category of nonuniform motion, we have uniformly nonuniform velocity (our uniformly accelerated motion), represented by a triangle, and nonuniformly nonuniform motion (nonuniformly accelerated motion), represented by a variety of other figures, the shapes of which are determined by the specific pattern of nonuniformity (see fig. 12.7). Finally, how did Oresme deal with that other feature of qualities noted above—their total quantity? He

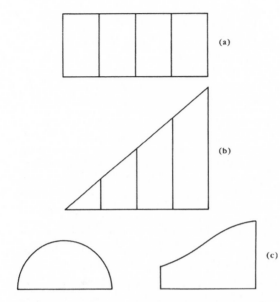

Fig. 12.7. The representation
of various motions.

(a) Uniform motion.
(b) Uniformly nonuniform
    motion (= uniformly ac-
    celerated motion).
(c) Nonuniformly nonuni-
    form motions (= non-
    uniformly accelerated
    motions).

identified the total quantity of motion with the distance traversed; and this, he argued must be represented by the area of the figure.

Oresme has very cleverly put geometry to work on behalf of the representation of motions of all varieties. He and those who followed him were not content with having created the geometrical tools. They proceeded to use them to illustrate and prove kinematic theorems applicable to uniform or uniformly accelerated motion. The most important case was the latter, represented in figure 12.7(b). This case was of special interest in the fourteenth century, not because it was identified with any particular motion in the real world but because it offered a substantial mathematical challenge. Let us examine two important theorems applicable to uniformly accelerated motion that emerged from these efforts.

The first had already been stated, without geometrical proof or illustration, by the Merton scholars; it is now known as the "Merton rule" or the "mean-speed theorem." This theorem seeks to find a measure for uniformly accelerated motion by comparing it with uniform motion. The theorem claims that a body moving with a uniformly accelerated motion covers the same distance in a given time as if it were to move for the same duration with a uniform speed equal to its mean (or average) speed. Expressed in numerical terms, the claim is that a body accelerating uniformly from a velocity of 10 to a velocity of 30 traverses the same distance as a body moving uniformly for that same period of time with a velocity of 20. Now Oresme provided a simple but elegant

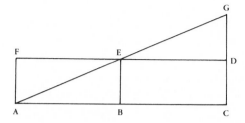

Fig. 12.8. Nicole Oresme's geometrical proof of the Merton rule.

geometrical proof of this theorem (fig. 12.8). The uniformly accelerated motion can be represented by triangle ACG and its mean speed by line BE. The uniform motion that is to be compared with the uniformly accelerated motion must therefore be represented by rectangle ACDF (the altitude of which is BE, the mean speed of the uniformly accelerated motion). The Merton rule claims simply that the distance traversed by the accelerated motion is equal to the distance traversed by the uniform motion. Since, in Oresme's diagrams, distance traversed is measured by the area of the figure, we can prove the theorem by showing that the area of triangle ACG equals the area of rectangle ACDF. A glance at the two figures will reveal that this is so.[30]

The second theorem, like the first, aimed to elucidate the mathematical properties of uniformly accelerated motion by means of a comparison involving distances traversed. In this case, the distance covered in the first half of a uniformly accelerated motion (beginning from rest) is compared to the distance covered in the second half of the same motion; the claim was that the latter is three times the former. To prove this theorem geometrically, we need merely show that the area of quadrangle BCGE (fig. 12.8), which represents the distance covered in the second half of the time, BC, is three times the area of triangle ABE, representing the distance traversed in the first half of the time, AB. Once again, inspection will establish that this is true.[31]

Finally, two general points: First, we must remind ourselves that medieval kinematics was a totally abstract endeavor—much like modern mathematics. It was claimed, for example, that *if* a uniformly accelerated motion were to exist, *then* the Merton rule would apply to it. Never did a medieval scholar identify an instance of such motion in the real world. Is there a satisfactory explanation for such seemingly odd behavior? Yes, there is. Given the technology available in the Middle Ages (particularly for the measurement of time), demonstrating that a particular motion is uniformly accelerated would have been a considerable feat. Even in the twenty-first century, imagine the challenge of proving with precision that a motion is uniformly accelerated, using any or all of the resources available in your local hardware store. But perhaps more importantly, the medieval scholars who developed this kine-matic analysis were mathematicians and logicians; and no more than modern mathematicians and logicians would they have thought of moving their place of labor from the study to the workshop.

Second, out of this purely intellectual labor came a new conceptual frame-work for kinematics and a variety of theorems (the Merton rule, for example) that figured prominently in the kinematics developed in the seventeenth century by Galileo—through whom they entered the mainstream of mod-ern mechanics. Proposition 1, theorem 1 of Galileo's analysis of uniformly accelerated motion in his *Two New Sciences* is the Merton rule (or mean-speed theorem). It is implausible to suppose that Galileo was ignorant of his fourteenth-century forerunners.[32]

THE DYNAMICS OF LOCAL MOTION

Having dealt at length with medieval kinematics—the effort to describe mo-tion mathematically—I conclude this discussion of medieval mechanics with a brief account of contributions to the *causal* analysis of motion. The start-ing point of all dynamical thought in the Middle Ages was the Aristotelian principle that moved things are always moved by a mover. We must first get clear on what this principle was taken to mean in the Middle Ages. We will then look at attempts to identify the mover in several particularly difficult cases of motion. And finally we will examine attempts to quantify the rela-tionship between the force or power of a mover and the resulting velocity of the moved body.

Aristotle, readers will recall, divided motion into two categories: natural and forced. A natural motion, by which an object moves toward its natural place, apparently arises from an internal cause or principle: the nature of the body. A motion in any other direction must be a forced motion, produced by the application of an external force in continuous contact with the moved body. This seems clear enough in broad outline, but problems arose when medieval scholars attempted to identify the mover in natural motion and in one particularly troublesome case of forced motion.

In his *Physics*, where he gave an account of the mover for natural motion, Aristotle vacillated, suggesting first that natural motion may result from an internal cause, the nature of the body, but arguing later that the nature of the body cannot be the whole story and that the participation of an external mover is also required. Aristotle's ambivalence posed an obvious problem for his medieval followers, who felt compelled to inquire whether or not it is sufficient to affirm that the body is moved by its own nature. Avicenna and Averroes considered this explanation unacceptable on the grounds that it did not distinguish sufficiently between that which is moved (the body) and that which moves it (the nature of the body). They discovered what seemed to them an adequate alternative in the form-matter distinction, proposing that the form of the body is the mover, while its matter is the thing moved. In the West, Thomas Aquinas repudiated this solution, reminding his readers that matter and form are inseparable and cannot be treated as distinct things. Aquinas argued instead (reviving one of Aristotle's proposals) that the mover in the case of natural motion is whatever generated the body outside its natural place to begin with; thereafter the body requires no mover but simply does what comes naturally: moving toward its natural place. The debate over this issue continued through the later Middle Ages, with no clear victor.[33]

The particular case of forced motion that proved troublesome was that of projectiles; the problem was to explain their continued motion after they lose contact with the original projector (e.g., the hand that threw the rock). Aristotle had assigned causation to the medium, arguing that the projector simultaneously projects the projectile and endows the surrounding medium with the power to produce motion; this power is transmitted from part to part in such a way that the projectile is always surrounded by a portion of the medium capable of moving it. It was clear, according to this account, that an external force, continuously in contact with the projectile, is required.

The first major opposition to Aristotle's explanation came in the commentary on Aristotle's *Physics* by the sixth-century Alexandrian Neoplatonist philosopher John Philoponus (d. after 575), to whom it seemed that the medium serves as resistance rather than mover and who doubted that it could serve both functions simultaneously. As a Neoplatonist and a dedicated anti-Aristotelian, Philoponus launched a broad attack on Aristotelian natural philosophy, including the notion that forced motions require external movers. He proposed, rather, that all motions, natural and forced alike, are the result of internal movers. Therefore, when a projectile is hurled, the projector impresses on the projectile an "incorporeal motive force," and this internal force is responsible for its motion.[34] If this seems an improbable answer, consider the motion of living things, which are apparently moved by internal, rather than external, forces.

Although Philoponus's impressed motive force had radically anti-Aristotelian origins, it was eventually absorbed into the medieval Aristotelian tradition. Philoponus's commentary on Aristotle's *Physics* had an influential career in Arabic translation and seems to have had an indirect impact on medieval Latin thought, although the details of transmission remain to be fully traced.[35] In the thirteenth century, theories bearing a close resemblance to that of Philoponus were discussed and rejected by Roger Bacon and Thomas Aquinas. In the fourteenth century, the theory of impressed force was defended, first by the Franciscan theologian Franciscus de Marchia (fl. 1320), subsequently by John Buridan (ca. 1295–ca. 1358) and others. Let us examine Buridan's version of the theory, often considered its most advanced form.

Buridan employed a new term, "impetus," to denote this *internal* impressed motive force—terminology that remained standard down to the time of Galileo. Buridan described impetus as an internal quality whose nature it is to move the body in which it is impressed, and took pains to distinguish this quality from the motion it produces: "Impetus is a thing of permanent nature distinct from the local motion with which the projectile is moved. . . . And it is probable that impetus is a quality naturally present and predisposed for moving a body in which it is impressed." In defense of his impetus theory, Buridan pointed to the analogous case of a magnet, which is able to impress in iron a quality capable of moving that iron toward the magnet. Like any quality, impetus is corrupted by the presence of opposition or resistance, but otherwise retains its original strength. Buridan took a first step toward quantifying impetus by declaring its strength to be measured by the velocity and the quantity of matter of the body in which it inheres. He also extended the explanatory range of the impetus theory beyond simple projectile motion, arguing that motion in the heavens might plausibly be explained by God's imposition of an impetus on the celestial spheres at the moment of creation; because the heavens offer no resistance, this impetus would not be corrupted, and the celestial spheres would be moved (as observation reveals they are) with an eternally unchanging motion. Finally, he explained the acceleration of a falling body by the assumption that as the body falls its weight continually generates additional impetus in the body; as the impetus increases, so does the velocity of the falling body.[36]

The theory of impetus became the dominant explanation of projectile motion until the seventeenth century, when a new theory of motion, which denied that force (either internal or external) is required for the continuation of unresisted motion, gradually won acceptance. There have been many attempts to view the theory of impetus as an important step in the direction of modern dynamics; for example, attention has often been called to the quantitative resemblance between Buridan's impetus (velocity × quantity of matter) and the modern concept of momentum (velocity × mass). No doubt there are connections, but we must note that Buridan's impetus was the *cause* of the continuation of projectile motion, whereas momentum is the *measure* of a motion that requires no cause for its continuation so long as no resistance is encountered. In short, Buridan was still working within a conceptual framework that was fundamentally Aristotelian; and this meant that he was a world (or worldview) away from those natural philosophers in the seventeenth century who formulated a new mechanics on the basis of a new conception of motion and inertia.

QUANTIFICATION OF DYNAMICS

One question remains. Is it possible to quantify the dynamic relations between force, resistance, and velocity? Many medieval scholars believed that it was. The problem went as far back as Aristotle, who had made a brief and preliminary stab at quantitative analysis, defending a variety of propositions such as the following: the greater the weight (of a falling body), the swifter its motion; the greater the resistance (encountered by a falling body), the slower its motion; and the smaller a moved object, the more rapidly a given force will move it. Historians have managed, by concerted effort, to extract a mathematical relationship from these claims, attributing to Aristotle the view that velocity is proportional to the force and inversely proportional to the resistance. Expressed in modern terms, this becomes

$$v \propto F/R.$$

This relationship is unquestionably useful as an economical means of conveying a substantial piece of Aristotelian dynamics, which explains why it continues to be repeated. But it is also potentially misleading and must be employed with great caution. For one thing, Aristotle had no clear conception of velocity as a technical, quantifiable term. For another, the relationship clearly does not hold for all values of F and R. Consider the case of a body at rest, acted upon by a force and an equal and opposite resistance. Common sense tells us (as it would have told Aristotle) that no motion will result, but in such a case the velocity given by the above proportionality is not 0, but 1.

Aristotle's dynamic ideas had clear implications for the possibility of motion in a void. If it is true that the swiftness of a falling body is a function of the resistance it encounters, then in a vacuum, where there is no resistance at all, there would be nothing to retard the motion of the body. In that case it would move with infinite speed. But in an infinitely swift motion, no time would be required for an object to move from point A to point B, from which it follows that the object is at both points at the same time—a physical absurdity. The conclusion that Aristotle drew from this absurdity was the impossibility of void space.

Aristotle's use of his theory of motion to prove the impossibility of a void provoked a broad attack from John Philoponus. Responding to the crucial assumption of Aristotle's argument against the existence of void space—namely, that speed of motion is inversely proportional to the density of the medium (so that zero density meant infinite speed)—Philoponus points out that we can't test it experimentally because we have no means of determining the relative densities of media. The only recourse is to think about the *cause* of motion, rather than *resistance* to motion. Take two bodies, one twice the weight of the other, falling through the same medium. It would be plausible to suppose that the heavier body would descend a given distance in half the time required by its lighter companion. Now, this happens to be a proposition that can be put to an experimental test—which, it seems clear, Philoponus did. He writes that

> our view may be corroborated by actual observation more effectively than by any sort of verbal argument. For if you let fall from the same height two weights, one many times as heavy as the other, you will see that the ratio of the times required for the motion does not depend on the ratio of the weights, but that the difference in time is very small. And so, if the difference in the weights is not considerable, that is, if one is, let us say, double the other, there will be no difference, or else an imperceptible difference, in time.[37]

Here we have a genuine, sixth-century anticipation of Galileo's notorious (but unlikely) experiment of dropping objects from the leaning tower of Pisa, and with similar results.[38]

If Aristotle's theory is false, what is the truth? Philoponus urged his reader to think about falling bodies in the following way. The efficient cause of the descent of a falling body is weight. In a void, where there is no resistance, the sole determinant of motion will be the weight of the body; consequently, heavier bodies will traverse a given distance more rapidly (that is, in less time) than will lighter bodies; and, of course, none of them will move with infinite speed, as Aristotle had supposed. (Philoponus did not state that the rapidity of motion in the void would be directly proportional to the weight, but presumably he expected this to be assumed.) Now in a medium, the resistance of the medium slows the motion by a certain amount, and the net effect of this slowing is to close the gap in swiftness between heavier and lighter bodies, leading to the observed results described in the quotation above.

Philoponus's views were developed and defended in Islam by Avempace (Ibn Bājja, d. 1138). Avempace, in turn, was attacked by Averroes; and through Averroes the controversy was transmitted to the West, where it was pursued by the fourteenth-century Mertonian Thomas Bradwardine. But with Bradwardine there was a difference. Whereas all of his predecessors had been interested primarily in the nature and *causes* of motion, Bradwardine was determined to view the problem in *mathematical* terms. This meant that he had to begin by giving a mathematical formulation of each of the alternatives—three of which he was able to identify. Bradwardine expressed these alternatives in words rather than mathematical symbols, but the following formulas adequately capture his intent.

1. First theory (no doubt meant to represent the opinion of Philoponus and Avempace):
   $V \propto F - R$
2. Second theory (suggested by a passage in Averroes):
   $V \propto (F - R)/R$
3. Third theory (representing the traditional interpretation of Aristotle):
   $V \propto F/R$

Bradwardine was able to refute each of these theories by calling attention to one or more absurd or unacceptable consequences. The first theory, for example, fails on the grounds that it contradicts Aristotle's claim that doubling both the force and the resistance will leave the velocity unchanged. The second theory fails on several grounds, including the claim (one needs to be a fourteenth-century Mertonian mathematician to follow the argument) that if it were true, everything would move with the same speed. And the third theory fails on the grounds that it does not predict zero velocity when the resistance is equal to or larger than the force.[39]

In place of these discredited theories Bradwardine proposed an alternative "law of dynamics." There is no easy modern way of stating Bradwardine's "law." To remain close to Bradwardine's own account would lead us more deeply into the medieval theory of the compounding of ratios than we can afford to go. Perhaps the simplest modern way of expressing the mathematical relationship that Bradwardine had in mind is to state that according to his "law" velocity increases arithmetically as the ratio F/R increases geometrically. That is, in order to double the velocity, we must square the ratio F/R; to triple the velocity, we must cube the ratio F/R; and so on. Or consider the following numerical example:

Begin with the third theory, ($V \propto F/R$). Apply a force (F_1) of strength 4, then a force (F_2) of strength 16, to a body that offers a resistance (R) of 2. Calculate the F/R ratios.

$$F_1/R = 4/2 = 2$$

$$F_2/R = 16/2 = 8$$

What will be the ratio of the velocities V_2 to V_1? Not 4 (the *multiplier* that must be applied to 2 in order to yield 8) but 3 (the *power* to which 2 must be raised to yield 8).[40]

Four points need to be made about Bradwardine's achievement. First, it evades all of the negative consequences of the three foregoing proportionalities. Second, we make Bradwardine's "law" more complicated than it really was by expressing it, as we have done above, in modern terms. We need to understand that in the medieval mathematical tradition within which Bradwardine was working, the way to talk about the compounding or increase of ratios was through the language of addition. Therefore, the operation that we refer to as the multiplication of two ratios would have been, in Bradwardine's terminology, the addition of one ratio to the other; and what we refer to as the squaring of the ratio F/R would have been, in his terms, the doubling of F/R. Consequently, instead of relating geometrical increases in the ratio F/R to arithmetical increases in the velocity (as we did above), Bradwardine would merely have stated that to "double" the velocity you must "double" the ratio of F to R. In short, Bradwardine proposed not some esoteric mathematical relationship, but (as one historian has recently put it) "the least complicated expression available to him."[41]

Third, Bradwardine's formulation of a "law of dynamics" proved influential. Its implications were brilliantly worked out by Richard Swineshead and Nicole Oresme in the fourteenth century, and the law continued to be discussed as late as the sixteenth century.[42] Fourth, whatever our precise evaluation of Bradwardine's achievement, we must acknowledge that his enterprise was unmistakably a mathematical one. It is true that his refutation of the alternatives included appeal to everyday experience, but it is clear that his primary aim was to satisfy the criteria of mathematical coherence. In short, Bradwardine neither discovered nor defended his "law" by experimental means; nor is it clear what benefits an experimental approach, if he had been inclined to adopt it, could have rendered. The task undertaken by medieval scholars was the formulation of a conceptual and a mathematical framework suitable for analyzing problems of motion. Surely this was the first order of business, and medieval scholars executed it brilliantly. The further task of interrogating mother nature, in order to find out whether she would accept the conceptual framework thus formulated, was left to future generations.

THE SCIENCE OF OPTICS

The decision to treat optics (or *perspectiva*, as it came to be called in Latin Christendom) in this chapter is arbitrary, for optics was a discipline of exceedingly broad reach, affiliated in one way or another with many subjects, including mathematics, physics, cosmology, theology, psychology, epistemology, biology, and medicine. But it will fit here well enough.[43]

The works of Aristotle, Euclid, and Ptolemy, which had dominated Greek thought about light and vision, were all translated into Arabic and gave rise to a substantial Islamic tradition of optical studies. The various Greek approaches to optical phenomena were taken seriously, defended, and extended. But the major achievement of Islamic optics was the successful integration of these separate and incompatible Greek optical traditions into a single comprehensive theory.

Most Greek optical thought was narrowly focused, guided by one or another relatively narrow set of criteria. Aristotle gave lip service to the geometry of light and vision, but devoted his attention primarily to the physical nature of light and the physical mechanism of contact in visual perception between the observed object and the observing eye; neither mathematical analysis nor anatomical or physiological issues occupied a significant place in his theory. Specifically, he argued that the visible object produces an alteration of the transparent medium; the medium instantaneously transmits this alteration to the observer's eye, with which it is in contact, to produce sensation. This is an "intromission" theory—so called because the agent responsible for vision passes from the observed object to the eye. The Greek atomists, who also demanded a physical account of vision, identified a different causal agent—a thin "skin" or "simulacrum" of atoms "peeled" from the outer surface of the object, rather than an alteration of the transparent medium—but joined Aristotle in the belief that a causal theory must be an intromission theory.

Euclid's concerns, by contrast, were almost exclusively mathematical; the aim of his *Optics* was to develop a geometrical theory of the perception of space, based on the visual cone, with only minimal concern for the nonmathematical aspects of light and vision. According to his theory of vision, radiation emanates from the eye in the form of a cone; perception occurs when the rays within the cone are intercepted by an opaque object. The perceived size, shape, and location of the object are determined by the pattern and location of intercepted rays. Because it holds that radiation issues from the eye, we refer to this as an "extramission" theory.

Finally, physicians such as Herophilus and Galen were preoccupied with the anatomy of the eye and the physiology of sight. Galen revealed a firm grasp of the mathematical and causal issues, but made his principal contribution to visual theory through an analysis of the anatomy of the eye and the participation of the various organs forming the visual pathway in the act of vision.

The Islamic contribution, as I have indicated, was to produce a merger of these disparate Greek theories. The primary architect of the merger was the brilliant mathematician and mathematical astronomer Alhacen[44] (Ibn al-Haytham, ca. 965–ca. 1040)—although Ptolemy, the last great optical writer of antiquity, had pointed the way. Our analysis of Alhacen's achievement will be simplified if, for the moment, we lay aside the anatomical and physiological concerns of the medical tradition and restrict our attention to the mathematical and physical aspects of vision.

To begin, it is important to notice that ancient theories of vision with mathematical aims (those of Euclid and Ptolemy) invariably assumed the extramission of light from the eye, while theories with physical plausibility as their primary concern (if we may judge from the works of Aristotle and the atomists) tended to assume the intromission of light into the eye.[45] If there was any doubt about this correlation, it would have been removed for the attentive reader of Aristotle's works by the discovery that on the one occasion when he attempted a mathematical analysis of optical phenomena (in his theory of the rainbow), Aristotle employed an extramission theory of vision.[46]

What Alhacen achieved, then, was twofold. First of all, he attacked the extramission theory with a compelling set of arguments. For example, he called attention to the ability of bright objects to injure the eye (noting that it is the nature of injury to be inflicted from without) and inquired how it would be possible, when we observe the heavens, for the eye to be the source of a material emanation that fills all of the space up to the fixed stars. Having discredited the extramission theory as a useful account of vision, he salvaged its major useful feature, the visual cone, for incorporation in his own new version of the intromission theory. Along with the visual cone came the geometrical power of the extramission theory, which was thereby coupled for the first time with the satisfying physical explanations provided by the intromission theory. This might seem a simple step, but consider some of the obstacles.[47]

In the first place, ancient writers offered no theory of radiation adequate for Alhacen's purposes. In ancient sources, intromitted radiation was generally presented as a holistic process in which the visible object radiates as a coherent unity. Rays were not thought to proceed independently from individual points (as in modern optical theory); rather, the object as a whole was supposed to send a coherent image or power through the medium to the eye. For the atomists, this was rather like a mask stripped from the face of its wearer, moving through space as a coherent whole, shrinking as it goes, in order to enter the eye of the observer. For Aristotle, as we have noted, it was a qualification of the medium originating in the visual field, wholistically entering the observer's eye.[48] There was no way of imposing a visual cone on such a conception of the process of radiation. However, a new conception of radiation had been formulated centuries earlier by the Arab philosopher al-Kindī (d. ca. 866) and adopted (or independently invented) by Alhacen. Al-Kindī and Alhacen conceived radiation as an *incoherent* process, in which individual points or small parts of the luminous body radiate not as a coherent group but each one independently of the others and in all directions (see fig. 12.9).

This was an important innovation, but it raised new problems for anybody hoping to defend an intromission theory of vision. Can an incoherent process

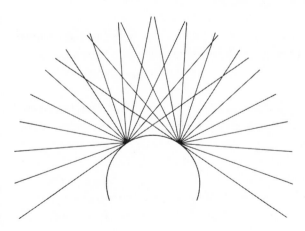

Fig. 12.9. Incoherent radiation from two points of a luminous body.

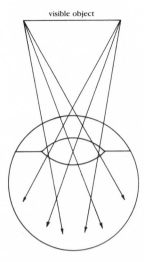

visible object

Fig. 12.10. Rays issuing from the end points of the visible object mixing within the eye. For the sake of simplicity, the bending of rays by refraction at the various interfaces has not been shown.

of radiation from visible objects account for the coherent visual perception that all normal-sighted people experience? If every point of the visible object radiates in all directions, then surely every point in the eye will receive radiation from every point in the visual field (see fig. 12.10). This should lead to total confusion rather than clear perception. What we require to explain our perceptual experience is a one-to-one correspondence, each point of the sensitive humor or organ of the eye (identified by Galen and his followers as the crystalline humor or lens) responding to radiation from one point in the visual field; and if possible, the pattern of recipient points in the eye should exactly replicate the pattern of radiating points in the visual field, thus explaining the correspondence between the world out there and the world as we see it.

Alhacen's solution to this problem was to argue that although every point in the visual field does indeed send radiation to every point in the eye, not all of this radiation is strong enough to be vision-producing. Only one ray from each point in the visual field, he noted, falls on the eye perpendicularly (see fig. 12.11); all others fall obliquely and are refracted. As a result of refraction these other rays are weakened to the point where they play only an incidental role in the process of sight. The primary sensitive organ of the eye, the crystalline humor or lens, pays attention to the perpendicular rays, and these form a visual cone, with the visual field as base and the center of the eye as vertex (see fig. 12.11). Alhacen has thus achieved his aim: by successfully importing the visual cone of the extramissionists into the intromission theory, he has combined the advantages of the extramission and intromission theories; he has united the mathematical and physical approaches to vision in a single theory.

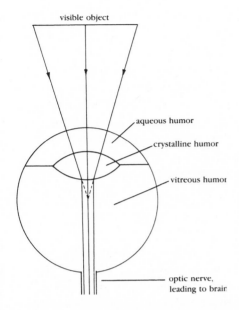

Fig. 12.11. The visual cone and the eye in Alhacen's intromission theory of vision. Rays from the object that fall obliquely on the eye (and undergo refraction) are not shown, since they enter only incidentally into the process of vision.

It is important to add, though we do not have space to delve into it, that he also incorporated the anatomical and physiological ideas of the Galenic tradition (fig. 12.11 conveys his basic conception of the anatomy of the eye), thus producing a unified visual theory, answering to all three kinds of criteria.

Visual theory may have been the centerpiece of Alhacen's optics, but his interests extended to the whole range of optical phenomena. He analyzed the nature of the radiation associated with light and color, distinguishing between naturally luminous objects and those that shine with derived or secondary light. He considered the physics of reflection and refraction. He continued and extended the mathematical analysis of radiating light and color, dealing in sophisticated fashion with problems of image formation by reflection and refraction. And he offered a serious and influential discussion of the psychology of visual perception.

Alhacen's *Optics* exercised a powerful influence on Western optics after its translation into Latin late in the twelfth century or early in the thirteenth. But it was not the only influential source. Plato's *Timaeus* had long been available; not only did it address vision, but it had also given rise to a substantial tradition of Neoplatonic optical thought. The optical works of Euclid, Ptolemy, and al-Kindī, translated in the second half of the twelfth century, revealed the promise of a mathematical approach to optics at a time when Alhacen's *Optics* was not yet available. Writings by Aristotle, Avicenna, and Averroes left the definite impression that the real problems were physical and psychological, rather than mathematical. And a variety of sources, including a small work by Ḥunayn ibn Isḥāq, conveyed the anatomical and physiological content of the Galenic tradition. As in so many other areas, Western scholars found themselves suddenly enriched with a splendid new body of knowledge—but one that contained conflicting ideas and tendencies. The problem confronting Western scholars was to reconcile and harmonize, reworking this perplexing intellectual heritage into a coherent and unified philosophy of nature.[49]

Among the first to undertake such efforts in the field of optics were two distinguished Oxford scholars: Robert Grosseteste in the 1220s and 1230s and Roger Bacon in the 1260s. Working early in the century, Grosseteste (ca. 1168–1253) was handicapped by an imperfect knowledge of the optical sources listed above, and his optical writings were valuable primarily as inspiration. It was Roger Bacon (ca. 1220–ca. 1292), inspired by Grosseteste but with the advantage of a full mastery of the optical literature of Greek antiquity and medieval Islam, who determined the future course of the discipline.

Bacon followed the broad outlines of optical theory as it was developed by Alhacen, adopting the latter's intromission theory of vision in almost all of its details. He was extraordinarily impressed by Alhacen's successful mathematical analysis of light and vision, and in his own works he effectively communicated the promise of the mathematical approach to future generations. But Bacon (like many of his generation) was convinced that all of the Greek and Islamic authorities were in fundamental agreement, and he was therefore committed to showing that all (or almost all) who had written about light and vision were of one mind. This meant that he would have to reconcile the optical teachings of so diverse a group as Aristotle, Euclid, Alhacen, and the Neoplatonists. Two examples will serve to illustrate how he managed this feat.[50]

Regarding the direction of radiation (from or toward the eye—the point of contest between the extramissionists and the intromissionists), Bacon agreed with Alhacen and Aristotle that vision occurs only through intromitted rays. What then of the extramitted rays advocated by Plato, Euclid, and Ptolemy? Clearly they could not be responsible for vision, but they could still exist and play an auxiliary role in the visual process—that of preparing the medium to receive the rays emanating from the visible object and of ennobling the incoming rays to the point where they could act on the eye. Regarding the nature of the radiation, Bacon accepted the Neoplatonic conception of the universe as a vast network of forces, in which every object acts on objects in its vicinity through the radiation of a force or likeness of itself. Moreover, he conceived this universal force to be the instrument of all causation and, on this basis,

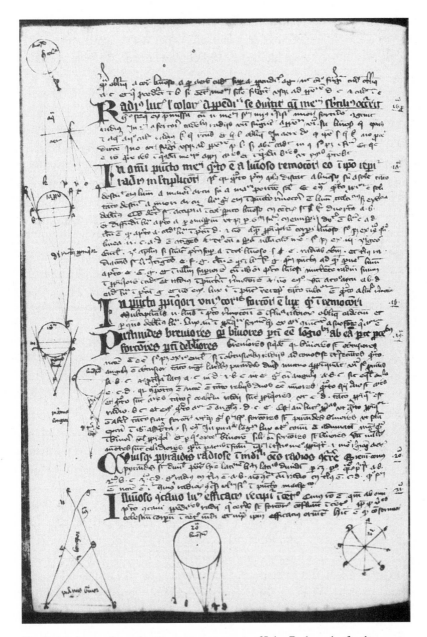

Fig. 12.12. A page from the *Perspectiva communis* of John Pecham, by far the most popular optical text in the medieval universities. Kues, Bibliothek des St. Nikolaus-Hospitals, MS 212, fol. 240v (early 15th c.)—a manuscript that once belonged to Nicholas of Cusa. Refraction of light is represented in the upper left hand corner, various patterns of radiation in the remaining diagrams.

developed (what proved to be) an influential philosophy of nature. As for light and color, Bacon argued that they (and any other visible agents discussed by the optical authors) were likenesses of their sources and thus manifestations of this universal force.[51]

Bacon's was not the only voice addressing optical problems in the second half of the thirteenth century, but it was in large part through his influence and that of two younger contemporaries—another English Franciscan, John Pecham (d. 1292), and a Polish scholar named Witelo (d. after 1281), connected with the papal court—that Alhacen's optical theories, including his combined physical-mathematical-physiological approach, became dominant in Western thought (fig. 12.12). When theories of light and vision appeared in fourteenth-century natural philosophy (as they frequently did, especially in epistemological discussions), they were almost always derived from the tradition of Alhacen and Bacon. When Johannes Kepler began to think about visual theory in the year 1600 (efforts that led eventually to his invention of the theory of the retinal image), he took up the problem where Bacon, Pecham, Witelo, and Alhacen had left it.[52]

13 ❋ *Medieval Medicine and Natural History*

THE MEDICAL TRADITION OF THE EARLY MIDDLE AGES

Medieval medicine was an outgrowth and continuation of the ancient medical tradition (examined above, chap. 6). Medieval medical practitioners were heirs to Greek and Roman theories of health and disease, diagnostic techniques, and therapeutic procedures. But access to this ancient legacy was partial and sometimes precarious, and the portions of it that were available in medieval Islam and Christendom had to be adapted to new cultural circumstances that profoundly shaped their development and use.[1]

It is difficult to get a clear picture of early medieval medicine in the West.[2] The social and economic chaos that accompanied the disintegration of the Roman Empire probably did not seriously affect the craft side of healing—the treatment of wounds and common ailments, midwifery, bonesetting, the preparation and distribution of familiar remedies, and the like. Especially in rural areas and on the domestic front, people skilled in the healing arts continued to practice their craft more or less as local healers had always done. What suffered from Roman collapse was the learned, and especially the theoretical or philosophical, component of medicine. To what extent that meant a loss of health benefits for the European population is unclear. In any case, the decline of schools and the gradual disappearance of facility in Greek increasingly deprived the West of the learned aspects of the Greek medical tradition, so that the number of medical practitioners with a command of the learned traditions of ancient medicine declined precipitously.

This is not to suggest that the West was totally cut off from Greek medical knowledge. Medicine received a certain amount of coverage in early Latin encyclopedias—those of Celsus, Pliny, and Isidore of Seville, for example.[3] Moreover, by the middle of the sixth century, a small collection of Greek medical writings was available in Latin translation. But Greek medical literature covered a broad spectrum of medical interests, from theoretical to practical, and the translated works tended toward the practical. Included were several works by Galen and Hippocrates, a collection of excerpts from Greek medical sources assembled by Oribasius (fl. 4th c. A.D.), a handbook for midwives by Soranus (1st c. A.D.), and the great pharmacopeia (*De materia medica*) of Dioscorides (fl. A.D. 50–70).

The practical, therapeutic orientation of early medieval medicine is nicely illustrated by Dioscorides' *De materia medica* and the pharmaceutical tradition it spawned (fig. 13.1). Containing descriptions of some nine hundred plant, animal, and mineral products alleged to have therapeutic value, Dioscorides' work was one of the monumental achievements of Hellenistic medicine. Translated into Latin in the sixth century, it enjoyed only a limited circulation during the early Middle Ages, perhaps because it was too comprehensive to be useful—containing, as it did, descriptions of many substances unavailable to early medieval Europeans. Far more popular was a shorter, illustrated herbal entitled *Ex herbis femininis*, based on Dioscorides but containing descriptions of only seventy-one medicinal plant substances, all available in Europe. Many additional collections of medical recipes were produced in the course of the early Middle Ages.[4]

Who were the medical practitioners able to make use of these texts? In Italy, the Roman pattern of secular, nonreligious medicine persisted, although it no doubt experienced quantitative decline. Publicly salaried physicians could still be found in early sixth-century Italy under Ostrogothic rule. Alexander of Tralles (a Greek physician) is known to have practiced in Rome in the second half of the sixth century. And a variety of evidence points to the continued existence of lay medical practice at royal courts (for example, that of the Frankish king Clovis at the end of the fifth century) and in major cities (Marseilles and Bordeaux) outside Italy.[5]

But increasingly the most hospitable settings for medical practice seem to have been religious ones, particularly monasteries, where care of sick members of the community was an important obligation. Our earliest evidence comes from Cassiodorus (ca. 480–ca. 575), founder of a monastery at Vivarium, who instructed his monks to read Greek medical works in Latin translation, including the works of Hippocrates, Galen, and Dioscorides (possibly a reference to *Ex herbis femininis*). Other evidence reveals a high level of medical practice, including the use of secular medical literature, in such monastic centers as Monte Cassino, Reichenau, and St. Gall.[6] It is probable that substantial medical expertise could be found in most monasteries, except the very smallest, throughout the Middle Ages. And although

Fig. 13.1. A page from a Greek manuscript of Dioscorides' *Materia medica*. Paris, Bibliothèque Nationale, MS Gr. 2179, fol. 5r (9th c.).

the medicine practiced within the monastery was intended primarily for members of the monastic community, there is no doubt that on occasion it was made available to others—pilgrims, visitors, and the surrounding population.

The presence, in a monastic environment, of secular medical literature and the medical practices linked with it raises an obvious question, which we must now consider: how did the traditions of Greek and Roman secular medicine interact with Christian ideas about healing? There is no simple answer, but we can begin to make sense of the complex reality if we keep in mind (1) that a philosophical tension *did* emerge between the naturalism of the medical tradition (the assumption that only natural causes are at work) and supernaturalist traditions (miraculous healing) within Christianity; (2) that most people (including literate people) were not philosophically inclined, and therefore few ever noticed the tension; and (3) that for those who did, there were various ways of easing or resolving the tension, short of repudiating one kind of healing or the other.

The sources of tension are obvious enough. As medieval Christianity matured, it became common for sermons and religious literature to teach that sickness is a divine visitation, intended as punishment for sin or a stimulus to spiritual growth. The cure, in either case, would seem to be spiritual rather than physical. Moreover, medieval Christianity developed a widespread tradition of miraculous cures, associated especially with the cult of saints and relics. And to complete the picture, we have the concrete evidence of religious leaders denouncing secular medicine for its inability to produce results.[7]

It is fairly easy to inflate such beliefs and attitudes into a general portrayal of the Christian church as an implacable opponent of Greek and Roman medicine, resolutely committed to belief in supernatural causation and to the exclusive use of supernatural remedies. Unfortunately, such attempts seriously misrepresent the historical reality. Although it is true that sickness was widely understood to be of divine origin, this did not rule out natural causes, for most medieval Christians shared the view, common since the Hippocratic writers, that an event or a disease could be simultaneously natural and divine (see above, chap. 6). Within a Christian context, it made perfectly good sense to believe that God customarily employs natural powers to accomplish divine purposes. For example, plague could be explained both as divine retribution for sin and as the result of an unfavorable conjunction of planets or corruption of the air.[8] As for the practice of medicine and the use of natural remedies, all Christian writers would have agreed that cure of the soul is more important than cure of the body, and a few spoke out against any use of secular medicine. Bernard of Clairvaux (1090–1153), writing to a group of monks in the twelfth century, expressed views that had existed for centuries:

> I fully realize that you live in an unhealthy region and that many of you are sick. . . . It is not at all in keeping with your profession to seek

for bodily medicines, and they are not really conducive to health. The use of common herbs, such as are used by the poor, can sometimes be tolerated, and such is our custom. But to buy special kinds of medicines, to seek out doctors and swallow their nostrums, this does not become religious [i.e., monks].[9]

But the vast majority of Christian leaders looked favorably on the Greco-Roman medical tradition, viewing it as a divine gift, an aspect of divine providence, the use of which was legitimate and perhaps even obligatory. Basil of Caesarea (ca. 330–79) spoke for many of the church fathers when he wrote that "we must take great care to employ this medical art, if it should be necessary, not as making it wholly accountable for our state of health or illness, but as redounding to the glory of God." Even a writer as hostile to Greco-Roman learning as Tertullian (ca. 155–ca. 230) revealed his appreciation of the value of Greco-Roman medicine. The denigrating accounts of conventional medicine that appear in saints' lives served an obvious polemical function—namely, to authenticate and magnify the power of the saint in question by demonstrating how he or she had healing abilities that transcended those of the secular healer. That we cannot take such denunciations as representative of the views of the author (let alone the remainder of medieval society) toward secular medicine is evident from the fact that many of these same authors, in other contexts or even in the same context, reveal a large measure of respect for conventional healing practices. What the church fathers were eager to denounce was not the use of secular medicine, but the tendency to overvalue it and the failure to recognize and acknowledge its divine origin.[10]

While defending the church against the charge of having repudiated the medical tradition, we must be careful to avoid the opposite error. There is no question that early medieval Christians believed in healing miracles and that they availed themselves of both religious healing and secular medicine, sometimes simultaneously, sometimes sequentially (fig. 13.2). In the fourth and fifth centuries the cult of saints became a dominant feature of European culture. Shrines were established around the tomb or some relic (perhaps a bone) of a saint; and these became pilgrimage sites of enormous drawing power. One of the features of these sites that contributed most powerfully to their attraction was the report of miraculous cures produced there. A single example will serve to illustrate: Bede (d. 735), in his *Ecclesiastical History of the English People*, recounted many stories of miraculous healings, including that of a monk on the island of Lindisfarne (off the northeastern

Fig. 13.2. The miraculous healing of a leg. Paris, Bibliothèque Nationale, MS Fr. 2829, fol. 87r (late 15th c.). For discussion of this illustration, see Marie-José Imbault-Huart, *La médecine au moyen âge à travers les manuscrits de la Bibliothèque Nationale*, p. 182.

coast of England), suffering from palsy, who was brought to the tomb of Cuthbert:

> falling prostrate on the corpse of the man of God, he prayed with godly earnestness that through his help the Lord would become merciful unto him: and as he was at his prayers, . . . he felt (as he himself was afterwards wont to tell) as though a great wide hand had touched his head in the place that suffered and with that same touch placed pressure on all parts of the body that had been sore vexed with sickness; and little by little the pain receded and health returned all the way down to his feet.[11]

Similar tales from the medieval period could be multiplied without end.

If the church was neither the enemy of the Greco-Roman medical tradition nor its single-minded supporter, how are we to characterize its attitude and influence? A familiar approach would be to weigh the factors on each side of the equation—both the opposition and the support offered by the church—and to argue that on balance the church was a force for good or ill, as the case might be. But such a conclusion would be simplistic. We will come closer to the truth if we avoid the categories of opposition and support altogether and see the church as a powerful cultural force that interacted with the secular medical tradition, appropriating and transforming it. Churchmen neither simply repudiated nor simply adopted secular medicine, but put it to use; and to use it was to adapt it to new circumstances, thereby subtly (or, in some respects, radically) altering its character. It is not too strong to claim that within Christendom there was a fusion of secular and religious healing traditions. In its new context, Greco-Roman medicine would have to be accommodated to Christian ideas of divine omnipotence, providence, and miracles. In the radically new institutional setting provided by the monasteries, it was not only nurtured and preserved through a dangerous period in European history, but it was also pressed into service on behalf of Christian ideals of charity (one important outcome of which was the development and spread of hospitals). And eventually, its institutionalization in the universities restored its contact with various branches of philosophy and elevated its status as a science.

One further development of critical importance requires our attention before we leave the early medieval period. The translation of Greek medical works into Arabic began in the eighth century and continued through the tenth. When it was finished, most of the major Greek medical sources were available in Arabic, including Dioscorides' *De materia medica*, many Hippocratic works, and nearly all the works of Galen. The magnitude of the gap between Islamic and Western access to this Greek medical literature can be illustrated by reference to the Galenic corpus: only two or three of Galen's works were available in Latin before the eleventh century, whereas Ḥunayn ibn Isḥāq (808–73) listed 129 Galenic works known to him in Baghdad, 40 of which he claimed to have personally translated into Arabic.

This Greek medical literature served as a foundation on which a sophisticated Islamic medical tradition would be built (see above, chap. 8). Several features of this medical tradition require brief mention. First, Islamic medicine was built on a full mastery of Greek medical literature and an assimilation of many of the aims and much of the content of Greek medicine. Second, central to the medical thought that emerged were Galenic anatomy and physiology and Galenic theories of health, disease (including epidemic disease), diagnosis, and therapy. An important aspect of Galenic influence was the linkage it revealed between medicine and philosophy—a linkage that became characteristic of much Islamic medical thought.

Third, Galenic medical theory did not rigidly constrain medical thought and practice in Islam, but functioned as a framework to be extended,

Fig. 13.3. Arabic surgical instruments from the treatise by Abū al-Qāsim al-Zahrāwī (Abulcasis), *On Surgery and Instruments*. Oxford, Bodleian Library, MS Huntington 156, fol. 85v.

modified, and integrated with other medical and philosophical systems; medicine in Islam was a dynamic, rather than a static, enterprise. Fourth, not only did Greek medical works circulate in translation, but along with them a large native medical literature was produced by Islamic physicians. This original Arabic literature contained a great deal of variety, of course, but particularly prominent was a series of comprehensive, encyclopedic works that surveyed large segments, or even the whole, of medical theory and practice. Three such encyclopedic works that were to have a profound influence on later Western medicine (see above, chap. 8) were the *Almansor* of Rhazes (al-Rāzī, d. ca. 930), the *Pantegni* (or *Universal Art*) of Haly Abbas ('Alī ibn 'Abbās al-Majūsī, d. 994), and the *Canon of Medicine* of Avicenna (Ibn Sīnā, 980–1037). These, along with many other translated works, helped to shape and redirect Western medicine in the later Middle Ages.[12]

THE TRANSFORMATION OF
WESTERN MEDICINE

In the eleventh and twelfth centuries, a number of influences began to impinge on the European medical tradition and alter its character. The political
and economic renewal of the period, accompanied by dramatic population
increase, led to far-reaching social change, including the urbanization and
expansion of educational opportunity. In the new urban schools the curriculum was broadened, as emphasis came to be placed on subjects that had been
of minor significance, or even totally absent, in the monastic setting. Meanwhile, reform movements within monasticism were attempting to diminish
monastic involvement in secular culture (see above, chap. 9). The convergence
of these movements brought about a shift in the location of medical education
from the monasteries to the urban schools, with a corresponding shift toward
professionalization and secularization. At the same time there was a growing
demand among urban elites for the services of skilled medical practitioners,
which contributed to the emergence of medical practice as a lucrative (and
sometimes prestigious) career.

The earliest example of renewed urban medical activity is at Salerno, in
southern Italy, in the tenth century. By the end of the century, Salerno had
acquired a reputation for its numerous and skilled medical practitioners, including clergy and women. There seems to have been no school in any formal
sense, but simply a center (increasingly a famous center) of medical activity,
with ample opportunities for men and women to master the healing arts
through apprenticeship. What flourished at Salerno in the tenth century and
into the eleventh was not medical learning but skill in the healing arts. In the
course of the eleventh century, some of the practitioners at Salerno began to
produce medical writings of a practical sort. Early in the twelfth century, the
literature emanating from Salerno began to broaden and become more theoretical, reflecting the philosophical orientation of the Arabic medical texts
beginning to circulate in Latin translation. Many of the new texts were teaching texts, connected (apparently) with the emergence of organized medical
instruction at Salerno.[13]

The translations from the Arabic that influenced medical activity at
Salerno in the twelfth century soon transformed medical instruction and
medical practice throughout Europe. The earliest translations appear to
have been those of Constantine the African (fl. 1065–85), a Benedictine
monk at the monastery of Monte Cassino in southern Italy, who had close
ties with Salerno (fig 13.4). Constantine, whose knowledge of Arabic was

Fig. 13.4. Constantine the African practicing urinalysis. Oxford, Bodleian Library, MS Rawlinson C.328, fol. 3r (15th c.). For commentary, see Loren C. MacKinney, *Medical Illustrations in Medieval Manuscripts*, pp. 12–13.

undoubtedly connected with his north African origins, translated works of Hippocrates and Galen, the *Pantegni* of Haly Abbas, medical works by Ḥunayn ibn Isḥāq, and other sources. He was followed by other translators over the next 150 years, in southern Italy, Spain, and elsewhere, who little by little rendered from Arabic to Latin much of the corpus of Greco-Arabic medicine. At Toledo, Gerard of Cremona (ca. 1114–87) translated nine Galenic treatises, Rhazes' *Almansor*, and Avicenna's great *Canon of Medicine*. These new texts vastly broadened and deepened Western medical knowledge, giving it a much more philosophical orientation than it had possessed during the early Middle Ages and ultimately shaping the form and content of medical instruction in the newly founded universities.[14]

MEDICAL PRACTITIONERS

Today we generally think of medicine as a learned profession, which can be practiced only by those who have undergone a long period of schooling and acquired appropriate professional credentials. But if we project such a model onto the Middle Ages, we will be sorely misled. A far more useful modern analogue would be carpentry. Carpentry covers a continuum from elementary home maintenance through the professional carpentry of the building trades to civil engineering and architecture. Carpentry of the simplest kind falls into the realm of general knowledge (almost everybody knows, or is willing to learn, something about elementary home repair); the weekend amateur, who (for example) restores antiques for a hobby, may command considerable knowledge and skill; the building trades are staffed by expert professionals who have, for the most part, learned their trade through apprenticeship; and finally, the civil engineer and the architect bring theoretical knowledge to bear on the subject.

So it was with the practice of medicine in the Middle Ages. Simple domestic medicine, practiced in the home, was the property of almost everybody. If more expertise was required, every community had people known to have a knack for treating certain kinds of ailments, and we begin here to move up the ladder of medical expertise and specialization. Most villages would have midwives, bonesetters, and people knowledgeable in herbs and herbal remedies. In the cities, one would find a variety of "empirics" with such specialties as the treatment of wounds, dental problems, and certain kinds of surgery (for example, lancing boils, repair of hernia, or removal of kidney stones). At a higher level of professionalization were apothecaries, trained surgeons, skilled professional medical practitioners educated through apprenticeship, and finally university-educated physicians. This was by no means a static or strictly linear hierarchy, nor was it invariable from place to place; it was also complicated by the existence of both secular and religious practitioners (clerics, for example, who frequently combined conventional medical practice with religious duties) at many of the levels; moreover, the lines of demarcation were rarely clear, because the regulation or licensing of medical practitioners, which would have demanded relatively clear categories, was only slowly instituted in the course of the later medieval period and never became universally effective. But some semblance of this classification scheme was generally characteristic of the medieval medical scene.[15]

We have only the sketchiest data on the numbers of medical practitioners in medieval Europe. We can learn something, however, from the fragments of data in our possession. In 1338, Florence (which was undoubtedly blessed with far more physicians per capita than the average European city) had approximately 60 licensed medical practitioners of all kinds (including surgeons and unlettered "empirics") for a population of 120,000. Twenty years later, after the population had been decimated by the black death, Florence had

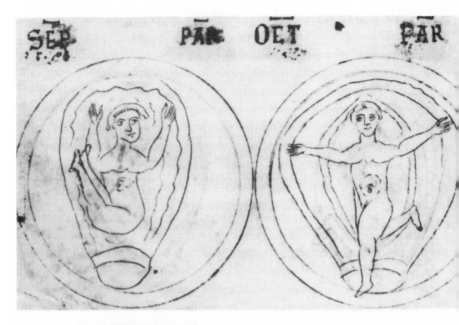

SEP PAR OCT · PAR

Coeliture mliex: Lib' trotule

Fig. 13.5 *(above)*. Fetuses in the womb. Copenhagen, Kongelige Bibliotek, MS Gl. kgl. Saml. 1653 4°, fol. 18r (12th c.).

Fig. 13.6 *(left)*. Trotula, a twelfth-century Salernitan medical practitioner. London, Wellcome Institute Library, MS 544, p. 65 (12th c.).

56 licensed medical practitioners for a population of about 42,000; and this ratio of 12 or 13 physicians for every 10,000 residents held for the remainder of the century.[16] Access in rural areas to a trained physician must have been far less common. Included among medieval medical practitioners were substantial numbers of women, active in obstetrics and gynecology but also in other medical specialties. The most famous of these is Trota or Trotula, from twelfth-century Salerno, who may not have written the gynecological work usually attributed to her but seems to have produced a more general work of practical medical remedies and advice. In certain parts of Europe Jewish medical practitioners were also numerous.[17]

MEDICINE IN THE UNIVERSITIES

The medical practitioners about whom we know the most are those who studied or taught in the formally organized medical schools of medieval Europe. Because these physicians were literate and left written records that have survived, we can learn something about their identities, their studies, and the kind of medical practice in which they engaged.[18]

Formal medical studies seem to have appeared first in the cathedral schools of the tenth and eleventh centuries—not for the purpose of educating professional physicians, but as an aspect of general education. At Chartres, for example, medical studies appeared by about 990, and in the next century medical instruction could be found in similar schools elsewhere.[19] However, it was at Salerno in the twelfth century that the newly translated medical works of the Greco-Arabic tradition were first assimilated, and it was there that medicine began to emerge as a learned profession. The driving force behind these developments was not mere intellectual curiosity or medical altruism (though a measure of both no doubt existed), but the desire for status and professional advancement. Physicians already at the top of the medical hierarchy outlined above, and therefore already literate, perceived the possibility of elevating their status by imitating other learned professions, such as law, in demanding that practitioners acquire appropriate formal credentials. The aim was to elevate the status of medicine from art or craft to science. Developments at Salerno were influential, and in the thirteenth century medical faculties became prominent at the universities of Montpellier, Paris, and Bologna. Medical faculties of lesser significance were created at Padua, Ferrara, Oxford, and elsewhere.

The institutionalization of medicine in the medieval universities was of enormous importance for the course of medical theory and practice. In the

Fig. 13.7. Medical instruction. From a copy of Avicenna's *Canon of Medicine*, Paris, Bibliothèque Nationale, MS Lat. 14023, fol. 769v (14th c.).

first place, it assured the continuation and the continuity of medical studies and the existence, from the Middle Ages to the present, of an influential community of university-educated physicians. Second, the establishment of medical studies in the university (as opposed to some other possible institutional home) created a linkage between medicine and other branches of knowledge that profoundly shaped the development of medicine. Specifically, a degree in the faculty of arts came to be a typical (if not quite universal) prerequisite for medical studies; and this meant that medical students came equipped with the logical and philosophical tools that would transform medicine (for better or for worse) into a rigorous, scholastic enterprise. It also gave medicine access to Aristotelian natural philosophy, which would provide medicine with some of its important principles, and to astrological theory (and its companion, astronomy), which would become a universal part of the physician's diagnostic and therapeutic armory. Let us give the medical curriculum a brief examination.

Teaching, first at Salerno and later in the other medical schools, coalesced, for a time, around a collection of brief treatises known collectively as the *Articella*.

This collection included an introduction to medicine by Ḥunayn ibn Isḥāq (known in the West as Johannitius), several short works from the Hippocratic corpus, and books on urinalysis and diagnosis by pulse. In the fourteenth and fifteenth centuries, these were supplemented by the works of Galen, Rhazes, Haly Abbas, Avicenna, and others. This curriculum had a marked philosophical orientation—medical theory being required to conform to broader principles of natural philosophy. And the teaching methods employed were the typical scholastic ones of commentary on authoritative texts and debate over disputed questions. But that did not mean (as has sometimes been alleged) that university medicine was a purely theoretical, textbook activity. In fact, many university professors of medicine engaged in private practice on the side, and medical students were frequently required to obtain practical experience.[20]

Finally, do we have any idea of the numbers of students involved? We do have scraps of relevant data. During a period of fifteen years early in the fifteenth century, the University of Bologna (one of the foremost medical schools in Europe) granted sixty-five degrees in medicine and one in surgery. During a thirty-six-year period a little later in the same century, the University of Turin (also in northern Italy) awarded a total of thirteen medical doctorates. And during its first sixty years of existence (beginning in 1477) the University of Tübingen awarded medical degrees at the rate of about one every other year. The number of medical students, of course, was far higher than the number of degree recipients, since most students did not complete the course of studies: the ratio 10:1 has been suggested as a possible multiplier. About all we learn from these numbers is that university-trained physicians, and especially physicians with doctorates in medicine, were rare creatures, members of an urban elite and accessible, for the most part, only for the rich and powerful.[21]

DISEASE, DIAGNOSIS, PROGNOSIS, AND THERAPY

The medical theories held and the diagnostic measures and therapies employed by a medieval medical practitioner varied with the practitioner's level of education, specialty, and professional circumstances. We know most, of course, about the views and procedures of learned physicians; but we have reason to believe that their beliefs and practices filtered downward and therefore influenced other kinds of healers. For example, we have ample evidence that Latin medical treatises were translated into vernacular languages, or translated and excerpted, for the benefit of medical practitioners who were literate but could not read Latin.[22] At the same time, it is clear that folk medicine and folk remedies had a tendency to filter upward and influence professional and even (to some extent) learned medicine. We will not be far off, therefore, if we judge the following elements of medical belief and practice to have been present, to varying degrees, in much medieval healing activity.

Fundamental to medieval theories of disease was the idea that every person has a characteristic complexion or temperament, determined by the balance of the four elements and their corresponding qualities (hot, cold, wet, dry) in the person's body. It was understood that complexion was peculiar to the individual; the balance that was normal for one would be abnormal for another. Closely associated with the theory of complexion was the idea, stemming from Galen and the Hippocratics, that the body contains four principal, physiologically significant fluids or humors—blood, phlegm, black bile, and red or yellow bile—and that these humors are the vehicles by which the proper balance of qualities is maintained. It was understood that health is associated with proper balance, illness with imbalance. For example, fever was conceived to be the result of abnormal heat emanating from the heart. Finally, health and disease were thought to be influenced by a set of conditions called the "nonnaturals": the air breathed, food and drink, sleep and wakefulness, activity and rest, retention and elimination (of nutrients), and state of mind.[23]

If sickness is the result of deviation from a person's normal complexion, then therapy must be directed toward the restoration of balance. Various techniques were available for the achievement of this end. The first was dietary; since the humors are the end products of the food consumed, a suitable diet was absolutely essential to the maintenance of health. Drugs, classified according to their predominant qualities, could also be prescribed to help restore balance. And if more heroic treatment seemed called for, it was possible to eliminate excess bodily fluids through purging, "puking," and bloodletting. In order to determine which of these measures to employ, the physician would need to inquire into the patient's lifestyle or regimen (such matters as diet, exercise, sleep, sexual activity, and bathing) in order to ascertain his or her specific complexion and the regimen required to maintain it. Indeed, for maximum effect the physician should closely monitor the patient's activities over an extended period of time—a realistic aim only for a physician (presumably learned) in the employ of a wealthy patron. Having observed his patron-patient over a period of time, the learned physician would (in theory) be in a position to offer the advice needed for the maintenance or recovery of health. The ideal that governed learned medicine (and, to some extent, less

Fig. 13.8. An apothecary shop. London, British Library, MS Sloane 1977, fol. 49v (14th c.). By permission of the British Library.

learned varieties of medical practice) thus portrayed the physician as medical advisor, with a primary responsibility for what we would call preventive medicine, but capable of following up with suitable remedies when preventive measures failed.[24]

The most common form of medical intervention was drug therapy, and the ability to identify and prepare drugs, along with knowledge of their therapeutic properties, was therefore an essential part of the repertory of most medieval healers. Drugs could be simple or compound; the most common ingredients were herbal, but animal and mineral substances were also employed. Many drugs were folk remedies, sanctioned by apparently successful use over many generations. For example, long experience had taught local healers that certain plant substances were effective as laxatives or painkillers. There is no question that some medieval drugs were effective; the majority, however, were simply harmless, while a few may have been dangerous. And some were downright disgusting—for example, the belief that pig manure was an effective cure for nosebleed. In this case, the cure might well seem worse than the ailment.[25]

But if there was a substantial empirical (frequently folk) component in medieval drug therapy, there was also a strong theoretical component emanating from the Greek and Arabic medical traditions. Dioscorides' *De materia medica* (in a revised and augmented version) had a very modest circulation in the West; in the twelfth century, new and more influential collections of medical recipes appeared; and finally, fresh translations of works by Galen, Avicenna, and others supplied the theoretical underpinnings needed to organize and systematize pharmaceutical knowledge. The basic theoretical assumption (borrowed, undoubtedly, from Galen) was that natural substances have therapeutic properties, associated with their primary qualities: hot, cold, wet, dry. To this theory, Avicenna added the idea that medicinal substances may also have a "specific form," independent of their primary qualities, which explains therapeutic effects not readily accounted for by the four primary qualities. It was thus through its specific form that theriac (a drug known since antiquity, made from viper's flesh and other ingredients) acquired the remarkable curative properties assigned to it in the twelfth-century *Antidotarium Nicolai*:

> Theriac . . . is good for the most serious afflictions of the entire human body: against epilepsy, catalepsy, apoplexy, headache, stomach ache, and migraine; for hoarseness of voice and constriction of the chest; against bronchitis, asthma, spitting of blood, jaundice, dropsy, pneumonia, colic, intestinal wounds, nephritis, the stone, and choler; it induces menstruation and expels the dead fetus; it cures leprosy, smallpox, intermittent chills, and other chronic ills; it is especially good against all poisons, and the bites of snakes and reptiles . . . ; it clears up every failing of the senses[?], it strengthens the heart, brain, and liver, and makes and keeps the entire body incorrupt.[26]

Another area of theoretical concern was the problem of determining how the properties of compound medicines depended on the qualities of their simple ingredients. Elaborate theoretical discussions (including mathematical analysis) of this problem were undertaken by both Islamic and European authors. Indeed, the doctrines of the intensification and remission of forms and qualities discussed above (chap. 12) were developed in part because of their applicability to pharmaceutical theory.[27]

We cannot conclude this discussion of medieval disease and treatment without discussing two prominent diagnostic techniques—urinalysis and the examination of pulse. Both had been recommended in antiquity by various writers, including Galen; the further influence of two short treatises, one on pulse and one on urine, contained in the *Articella* collection, as well as longer discussions in Avicenna's *Canon of Medicine*, assured their centrality in later medieval diagnosis. It was held that urinalysis could reveal the state of the liver, while pulse reflected the state of the heart. The critical features of the urine were color, consistency, odor, and clarity. For example, an early thirteenth-century medical writer, Giles of Corbeil, maintained that "thick urine, whitish, milky, or bluish-white, indicates dropsy, colic, the stone, headache, excess of phlegm, rheum in the members, or a flux."[28] Charts revealing the connection between different colors of urine and various ailments were a common feature of medieval medical writing (see fig. 13.9).

In taking a patient's pulse, the physician attempted to determine its strength, duration, regularity, breadth, and so forth (fig. 13.10). Many varieties of pulse were differentiated and various classification schemes developed. An anonymous treatise of the thirteenth century offered the following scheme:

> The varieties of pulses are differentiated by the physician in a number of ways, in particular according to five considerations: (1) motion of the arteries; (2) condition of the artery; (3) duration of diastole and systole; (4) strengthening or weakening of pulsation; (5) regularity or irregularity of the beat. Ten varieties of pulse derive from these considerations.[29]

Failing pulse could be used to foretell the time of death and was therefore useful for prognosis as well as diagnosis.

Thus far we have sidestepped one pervasive element in medical theory and practice, which hovered over and shaped what the medieval healer believed and the therapeutic measures that he or she prescribed. This was medical astrology, which harbored the belief that planetary influence is implicated in both the cause and the cure of disease. There were good reasons for believing in such planetary influence. One was medical authority: several of the Hippocratic works contained passages that could be interpreted as affirmations of celestial influence, and during the later Middle Ages a treatise on astrological medicine circulated under Hippocrates' name. But far more importantly, anybody who had grasped the fundamentals of natural philosophy, as we have seen above, knew that the heavens exercised an influence on the human body and its environment; and there was no reason at all to doubt that this would have an effect on health and the course of disease (fig. 13.11).[30]

Celestial influence began at conception, contributing to the temperament or complexion of the newly conceived embryo. After birth, every human was

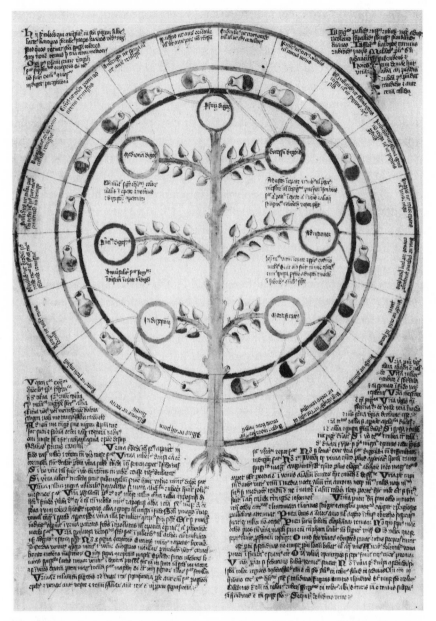

Fig. 13.9. A urine color chart, which connects variations in the color of urine with various stages of digestion. London, Wellcome Institute Library, MS 49, fol. 42r (15th c.). For further commentary, see Nancy G. Siraisi, *Medieval and Early Renaissance Medicine*, p. 126.

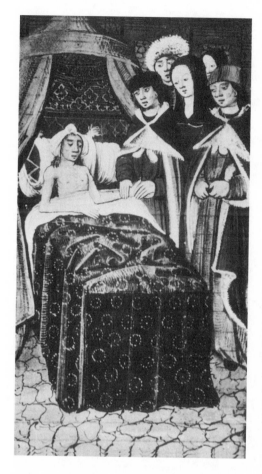

Fig. 13.10. Diagnosis by pulse. Glasgow University Library, MS Hunter 9, fol. 76r (15th c.). For a discussion of this illustration, see Loren C. MacKinney, *Medical Illustrations from Medieval Manuscripts*, pp. 16–17. By permission of the Librarian, Glasgow University Library.

the recipient of a continuous flow of celestial forces, either directly or through the surrounding air, and these influences affected temperament, health, and disease. Indeed, astrological influence was frequently invoked to explain major epidemics, such as the black death of 1347–51. Pressed for an explanation of this particular plague, the medical faculty at the University of Paris concluded that it resulted from corruption of the air caused by a conjunction of Jupiter, Saturn, and Mars in 1345.[31]

If illness struck, the physician needed to take account of the planetary configuration in order to prescribe effective treatment. Preparation and administration of drugs had to be properly timed to coincide with favorable planetary configurations, and proper dosage depended on astrological factors. It was also necessary to determine propitious times for surgical procedures,

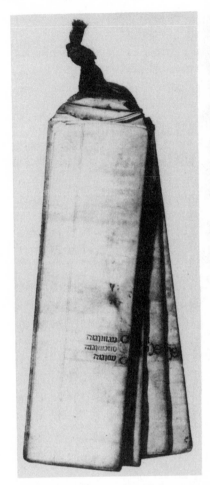
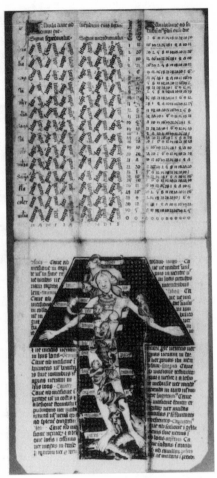

Fig. 13.11. A physician's girdle book. London, Wellcome Institute Library. A handy guide for the physician, meant to be suspended from the belt. The left-hand illustration shows the book in its folded form. The right-hand illustration shows one of its leaves, this one containing astrological information. For a full discussion, see John E. Murdoch, *Album of Science: Antiquity and the Middle Ages*, pp. 318–19.

such as bloodletting. Surgical treatises often contained "bloodletting figures" that instructed the user on the appropriate times for bleeding from specific bleeding points. Finally, the Hippocratic theory of "critical days," which held that the course of acute diseases is marked by crises or turning points, became linked to astrology; among the factors that were believed to determine the outcome of a crisis was its timing—whether or not it occurred on an astrologically favorable day.

ANATOMY AND SURGERY

Medieval healers were no doubt inclined toward moderate forms of medical intervention, such as control of diet and prescription of drugs. But some ailments and medical emergencies demanded more intrusive measures, and Europe always had medical practitioners willing to invade the body surgically. Many kinds of surgeons could be found, with differing specialties and levels of education, from itinerant empirics specializing in a particular surgical procedure to university-educated surgeons in the employ of king or pope. Surgery was customarily viewed as a craft, beneath the dignity of the university-educated physician; however, surgeons did manage to institutionalize their enterprise in southern European universities (Montpellier and Bologna, for example), thereby acquiring intellectual status. A substantial Arabic surgical literature was made available in the West through the translations of the twelfth and thirteenth centuries, and this stimulated a European tradition of surgical writing. Among the most influential of the European treatises were the *Surgery of Roger Frugard* (twelfth century), which frequently circulated in short sections, and the *Chirurgia magna* (or *Great Surgery*) of Guy de Chauliac (ca. 1290–ca. 1370), physician and surgeon to three popes. Guy's work not only circulated widely in Latin, but was also translated into English, French, Provençal, Italian, Dutch, and Hebrew.[32]

Most surgery was not particularly heroic—the setting of a broken bone, reduction of a dislocated joint, dressing of an ulcer or sore, cleaning and suturing of a wound, or lancing of a boil. Bloodletting and cautery (application of hot irons to various parts of the body in order to create ulcers through which unwanted fluids could drain) were also common procedures.[33] Removal of external hemorrhoids may also have been fairly routine. But some medieval surgeons undertook much more ambitious procedures. Operation for removal of cataract, by inserting a sharp instrument through the cornea and forcing the lens of the eye out of its capsule and down to the bottom of the eye, is one example (fig. 13.12). Removal of bladder stone and surgical correction of hernia are others (fig. 13.13). The following text describes removal of a bladder stone:

If there is a stone in the bladder make sure of it as follows: have a strong person sit on a bench, his feet on a stool; the patient sits on his lap, legs bound to his neck with a bandage, or steadied on the shoulders of the assistants. The physician stands before the patient and inserts two fingers of his right hand into the anus, pressing with his left fist over the

Fig. 13.12 *(left)*. Operation for cataract (above) and nasal polyps (below). Oxford, Bodleian Library, MS Ashmole 1462, fol. 10r (12th c.). For commentary on this figure, see Loren C. MacKinney, *Medical Illustrations from Medieval Manuscripts*, pp. 70–71.

Fig. 13.13 *(below)*. Operation for scrotal hernia. Note that the patient is both tied and held down. Montpellier, Bibliothèque Interuniversitaire, Section Médecine, MS H.89, fol. 23r (14th c.). This illustration is discussed by Loren C. MacKinney, *Medical Illustrations from Medieval Manuscripts*, pp. 78–80.

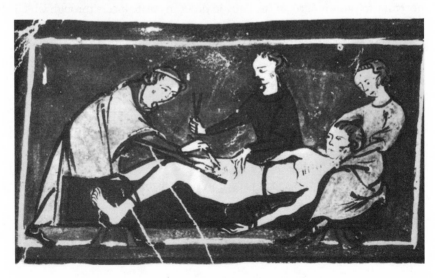

patient's pubes. With his fingers engaging the bladder from above, let him work over all of it. If he finds a hard, firm pellet it is a stone in the bladder. . . . If you want to extract the stone, precede it with light diet and fasting for two days beforehand. On the third day, . . . locate the stone, bring it to the neck of the bladder; there, at the entrance, with two fingers above the anus incise lengthwise with an instrument and extract the stone.34

As an example of dangerous surgery, fracture of the skull sometimes required trephining (the making of small holes in the skull with a saw) in order to reduce pressure and drain blood and pus. All surgeries were performed with only the most modest use of sedatives or anesthetics; if there was anything obviously heroic about medieval surgery, it was the patient.[35]

How much human anatomy did the medieval surgeon or physician know, and what place did anatomical instruction and firsthand anatomical investigation have in the education of medical practitioners? Despite Galen's stress on the importance of anatomical knowledge for the successful treatment of disease, the connection between anatomical knowledge and the clinical side of medical practice remained as tenuous during the Middle Ages as it had been in antiquity. Most medieval practitioners no doubt found that they could get along quite nicely with a minimum of anatomical knowledge, for the advice they dispensed and the dietary and herbal remedies they prescribed rarely, if ever, depended on detailed structural knowledge of the human body. The surgeon's requirements were undoubtedly greater but still modest; much of the required knowledge was common property through such daily experience as animal butchery, and the rest could be obtained by experience in the course of apprenticeship or surgical practice.

Nonetheless, the translations of the twelfth century provoked new interest in anatomical questions. The translation of Galen's anatomical writings and Arabic works based on them (books by Avicenna, Haly Abbas, Rhazes, and later Averroes) brought to the West a body of anatomical literature that demanded attention—not because it promised a large, immediate impact on healing practices, but because it belonged to the body of medical theory that learned physicians considered their intellectual property. The new interest in anatomical knowledge first found expression in the form of actual anatomical dissections in twelfth-century Salerno. The object of dissection in this case was a pig, considered anatomically analogous to humans.

Human dissection appears to have begun in certain Italian universities, especially Bologna, late in the thirteenth century. The picture is murky, but the

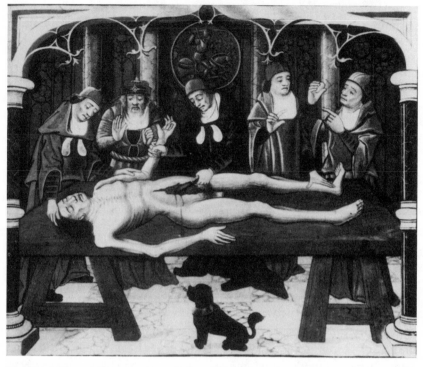

Fig. 13.14. Human dissection. Paris, Bibliothèque Nationale, MS Fr. 218, fol. 56r (late 15th c.).

purpose seems originally to have been legal—autopsies within the law faculty for the purpose of determining the cause of death—the practice spreading subsequently, by steps we know nothing about, to include dissections for medical instruction. By 1316, Mondino dei Luzzi (d. ca. 1326), who taught at Bologna, had become sufficiently skilled in human dissection to write a dissection manual entitled *Anatomia*, which became the standard guide to human dissection for the next two centuries.[36]

In the course of the fourteenth century, dissection became a regular part of medical instruction at Padua, Bologna, and a few other universities. In his *Chirurgia magna*, Guy de Chauliac described the procedures of his master at Bologna, Nicolaus Bertrucius:

Having laid the dead body on the table, he made four lessons on it. In the first the nutritive members [stomach and intestines] were treated, since they decay the soonest. In the second, the spiritual members [heart, lungs, and trachea], in the third the animal members [skull, brain, eyes, and ears], and in the fourth the extremities were treated. And following the commentary on the book of *Sects* [of Galen], in each there are nine things to see: that is, to know the situation, the substance, the constitution, the number, the figure, the relations of connections, the actions and uses, and the diseases which affect them.... We perform anatomies also on bodies dried in the sun, or consumed in the earth, or submerged in running or boiling water. This reveals the anatomy at least of the bones, cartilage, joints, large nerves, tendons, and ligaments.[37]

Such dissections were generally performed on the corpses of criminals, whose execution might be timed to meet the needs of the medical school. They were infrequent, an annual dissection being perhaps the most common pattern. And it is important to understand that the medical student was an observer rather than an experimenter; the function of dissection was to illustrate the Galenic text; this was not research, but pedagogy.[38]

Older histories of medicine have often criticized medieval physicians for adopting a methodology that made texts, rather than cadavers, the primary anatomical authority. The unfortunate result of this methodology, it has been argued, was the continued propagation of a variety of errors in Galen's account of human anatomy. What are we to think of such criticisms? There is no question that medieval physicians found Galenic anatomy an awesome achievement and were therefore inclined to attach great (though not absolute) authority to Galenic texts, but it does not follow that they were fools. Consider a modern parallel: the modern anatomical textbook is also a remarkable achievement, and when a medical student taking the obligatory human anatomy course finds what appears to be a discrepancy between text and cadaver, he or she interprets this discrepancy as a variation in the cadaver rather than a mistake in the textbook. We should not be surprised to see medieval physicians and surgeons behaving similarly. They had every reason to believe that Galen had gotten it right (as, for the most part, he had) and to view the study of Galenic texts as the surest and most efficient, not to mention cleanest, way of acquiring anatomical knowledge.

Despite the secondary importance of anatomical dissection within medical education, we have seen that a tradition of anatomical dissection did develop late in the thirteenth and early in the fourteenth century. It grew in strength and sophistication over the next two hundred years, while maintaining a continuous dialogue with the textual tradition of anatomical knowledge. In the fifteenth century it became allied with the technology of printing, which made possible the cheap production of texts and the faithful reproduction of anatomical drawings. The quality of anatomical drawings was further enhanced

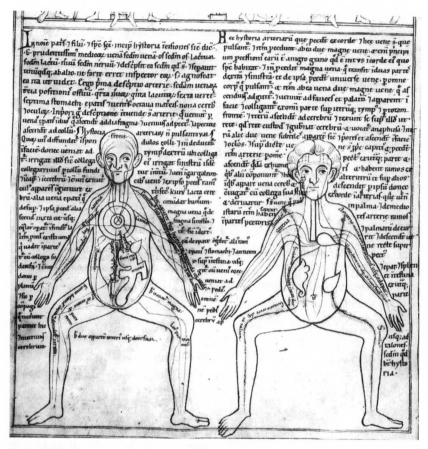

Fig. 13.15. Human anatomy, illustrating the Galenic conception of veins (left) and arteries (right). Munich, Bayerische Staatsbibliothek, CLM 13002, fol. 2v (12th c.). For commentary and additional anatomical drawings, see Siraisi, *Medieval and Early Renaissance Medicine*, pp. 92–95.

by contributions from the growing company of talented artists. And in the sixteenth century, these factors combined with renewed access to the Greek text of Galen to produce the stunning anatomical achievements of Andreas Vesalius (1514–64) and others.

DEVELOPMENT OF THE HOSPITAL

I conclude the discussion of medieval medicine on an institutional note, with a brief account of one of the most celebrated medieval medical achievements— the invention of the hospital. One of the difficulties in tracing the origin of the hospital is deciding what the term means. If, by "hospital," we mean anything called "hospice" or "hospital," then we include many institutions that offered food and shelter to paupers and pilgrims, including the sick, but which provided little or no specialized medical care. If, however, we wish to reserve the term for institutions dedicated to the treatment of the sick, including the provision of skilled medical care, then we are applying a much more stringent criterion. The former sort of hospital, which was common throughout medieval Europe (often maintained by monasteries or communities of lay brethren), will not interest us. It is the latter kind of institution that will be the object of our attention.[39]

Where, then, did the hospital as a medical institution come from? Its origins seem to lie in the Byzantine Empire, where, probably about the fourth century, ideals of Christian charity led to the establishment of hospitals that provided specialized medical care. One of the earliest for which we have hard evidence was the Sampson hospital (named after a saint of the fourth century) in Constantinople; here, early in the seventh century, for example, a church official suffering from a groin infection was hospitalized for surgery and convalescence. Other Byzantine hospitals were organized along the same lines: in the twelfth century, the Pantokrator hospital, also in Constantinople, had space for fifty patients (thirty-eight male and twelve female); to meet their medical and other needs, the hospital employed a staff of forty-seven, including physicians and surgeons.[40]

This Byzantine model became known in both Islam and the West, where it interacted with, and helped to shape, indigenous traditions of health care. In Islam, we find comparable institutions early in the ninth century, perhaps owing to the influence of the Barmak family, which occupied a position of power under the caliph Hārūn al-Rashīd (786–809). No doubt there were many strands in the transmission of the Byzantine model to the West; one of them seems to have come as a by-product of the conquest of Jerusalem in 1099, during the First Crusade. Shortly after the fall of Jerusalem, the lay brothers (subsequently known as "Hospitallers") who operated the hospital of Saint John in Jerusalem reorganized it on the Byzantine model. Because of its prominent location and large size, it became renowned throughout Europe; visitors a century later reported that it housed a thousand patients or more. The Hospitallers eventually established a string of hospitals in Italy and southern France. Through the promulgation of various statutes regulating these hospitals (requiring, in one version, the hiring of four physicians to treat patients), the Jerusalem pattern became familiar in the West, where it influenced the conception of charitable care for the ill and the indigent, encouraging the development of the hospital as a specialized medical institution.[41]

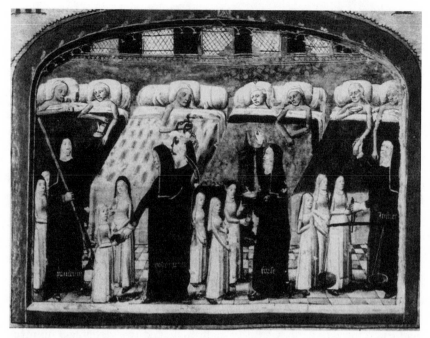

Fig. 13.16. A medieval hospital. From Jean Henry, *Le livre de vie active des religieuses de l'Hôtel-Dieu* (late 15th c.). Paris, Centre de l'Image de l'Assistance Publique. This illustration is discussed by Marie-José Imbault-Huart, *La médecine au moyen âge*, p. 168.

This is, to be sure, a very sketchy picture, with many remaining uncertainties. Whatever the precise details of transmission and assimilation, the model of the hospital as a medical institution spread rapidly in the West in the twelfth and thirteenth centuries, to the point where hospitals could be found in cities and towns throughout Europe. They might be large or small, containing anywhere from hundreds of beds to half a dozen. Their sponsorship could be either religious or secular. Their clientele was principally from the lower classes, though there were exceptions. They were typically staffed by professional physicians who were paid an annual salary for their labors. Considerable thought was given to the needs of the patients—cleanliness and diet, for example. Beds consisted of straw mattresses suspended on ropes from bedposts, designed to hold two, or even three, patients. An account of medical facilities in Milan, written about 1288, is instructive:

> In the city, including the suburbs . . . there are ten hospitals for the sick. . . . The principal one is the Hospital of the Brolo, very rich in

possessions, and founded by Geoffrey de Bussero in 1145. In it . . . are found, particularly in bad times, more than five hundred bed patients and as many more not lying down. All these receive food at the expense of the hospital itself. Besides them, there are 350 babies or more, placed with individual nurses after their birth, under the hospital's care. Every kind of poor person, except lepers, for whom another hospital is reserved, is received there, and kindly and bountifully restored to health, bed as well as food being provided. Also, all the poor needing surgical care are diligently cared for by three surgeons who are assigned to this particular task.[42]

Though surely putting the best face on things, this account reveals the impressive level of care to which a medieval hospital might aspire.

NATURAL HISTORY

Medicine was no doubt the principal repository of biological knowledge during the Middle Ages, but it was not the only one. Aristotelian natural philosophy included a large component of zoological and botanical information. Encyclopedias almost always contained sections on plants and animals. Herbals and bestiaries specialized in the plant and animal kingdoms, respectively. And finally, medieval people had intimate firsthand knowledge of the local flora and fauna. This chapter will conclude with a brief examination of medieval botanical and zoological knowledge.

Medieval botanical knowledge was closely linked with medicine, since the principal use of plants (if we ignore those that formed part of the European diet) was in herbal remedies. If the medicinal use of herbs was to be effective, manuals were needed, which would describe the various herbs and their therapeutic uses. A significant herbal literature thus developed, most of it designed for practical purposes. The model was Dioscorides' *De materia medica*, in its revised Latin translation, which arranged medicinal substances in alphabetical order so as to facilitate use. A typical entry in an herbal would include the name or names of the plant, an account of its identifying features, including habitat, a description of the medicinally significant parts and their therapeutic properties, and instructions regarding preparation and use. The alphabetical arrangement of the herbal reveals that practical aims (the ability to look up a medicinal substance by name) prevailed over classification according to biological type or any other theoretical consideration.[43]

But alongside these practically oriented herbals, there was also a more theoretical or philosophical literature that placed plant life within the context

Fig. 13.17. A page from the *Herbal* of Pseudo-Apuleius, describing and illustrating couchgrass, sword lily, and rosemary. Oxford, Bodleian Library, MS Ashmole 1431, fol. 21r (12th c.). Described in Joan Evans, ed., *The Flowering of the Middle Ages*, pp. 190, 352.

of natural philosophy. Most of this literature descended in one way or another from the book *On Plants*, attributed to Aristotle and believed by medieval scholars to be his, but probably written by Nicholas of Damascus (1st century B.C.). A few commentaries were written on this treatise (perhaps a dozen are known), by far the most impressive of which was *On Vegetables* by Albert the Great (ca. 1200–1280). Albert's *On Vegetables* contains a paraphrase of *On Plants*, accompanied by Albert's own attempt to bring intellectual order to the natural philosophy of plants, and a concluding alphabetical list of herbs and their uses. A reading of this work reveals the extraordinary skill, unmatched by any contemporary, with which Albert observed and described botanical phenomena.[44]

One might expect close parallels between botanical and zoological literature. However, zoological knowledge had few applications in the medical realm and little practical value elsewhere; and consequently, there was no zoological counterpart of that repository of practical botanical knowledge—the herbal. As in the case of botany, there was an underlying Aristotelian textual tradition, for Aristotle had written a series of large and important zoological works. These were rendered into Latin (along with an influential commentary by Avicenna) and attracted considerable attention—not so much for the detailed zoological information they contained as for their bearing on more general issues in natural philosophy. Once again Albert the Great was one of the major figures, producing, in his massive *On Animals* (occupying more than 1,800 pages in modern English translation) and other works, a large body of descriptive and theoretical zoology. Of particular interest are his discussions of nutrition and embryology. His treatment of conception and embryological development, for example, was dependent not only on Aristotle's theories of conception, but to a very substantial degree on his own observations of the reproductive behavior of animals. The history of medieval zoology has yet to be written, but in Albert the Great we undoubtedly see the philosophical side near its zenith.[45]

Besides zoological works in the Aristotelian tradition, there were various other genres of literature on animals—two of which have attracted considerable attention. One of these consists of practical treatises on falconry. The most famous of the genre was written in Sicily by Emperor Frederick II (about the middle of the thirteenth century) and titled *On the Art of Hunting with Birds*. The most famous observation in this most famous treatise on birds is Frederick's experimental determination that vultures locate their food by sight rather than by smell—ascertained by observing their inability to find food when their eyes were covered.[46]

If Frederick's treatise on falconry seems remarkably practical and modern, devoid of the fanciful or metaphysical content that we have come to associate with the Middle Ages, our final example of medieval literature on animals goes to the other extreme. The medieval bestiary is often presented as an example of medieval inability to observe the world objectively and get zoological knowledge straight. Medieval bestiaries are all descended from an anonymous treatise entitled *Physiologus*, emanating from Alexandria and written originally in Greek (perhaps about the year A.D. 200), subsequently translated into Latin and all of the major European vernacular languages. The *Physiologus* and the medieval books inspired by it are collections of animal lore arranged in short entries or chapters under the names of the respective animals—numbering from about forty in the *Physiologus* to more than a hundred in some of the later bestiaries.[47]

The typical entry in a bestiary begins with an etymological explanation of the animal's name. For example, "the vulture is thought to have been named for its slow flight [*a volatu tardo*]." If the animal has distinctive physical characteristics, these will be reported next, followed by an account of unusual or interesting behavior and a description of admirable and regrettable character traits. From this same twelfth-century bestiary, we learn that the hedgehog is covered with spikes and curls itself into a ball for protection; that the fox is a "fraudulent and ingenious animal" that plays dead in order to catch its prey; that cranes move about in military formation; that the serpent called "basilisk" can kill with the power of its glance; that the lynx's urine turns into a precious stone; that lions are compassionate and courageous, and that their eyebrows and manes offer a clue to their disposition. Finally, many (but not all) entries go on to draw a moral or make a theological point on the basis of the animal description. The hedgehog is an example of prudence, the crane of courtesy and responsibility. The fox is employed as a type of the devil, who entices carnal man through fraudulent behavior. And the male lion, breathing life into its stillborn offspring after three days, represents God the Father raising Christ from the dead.[48]

How are we to judge such an odd mixture of fact, fancy, and parable? The bestiary certainly does not read like a modern zoology manual, and on this basis interested parties have sometimes portrayed the people who compiled the bestiaries as incompetent or unsuccessful zoologists. The assumption is that they were trying (or should have been trying) to write modern zoology manuals but could not figure out how to do it; and their most serious deficiency was their apparent inability to distinguish between fact and fancy. But it is, of course, ridiculous to insist that medieval people share our interests and priorities. That medieval scholars were capable of writing something rather like a zoology manual is clear enough from the analogous case of the herbal or from the books on falconry that we have touched upon. And their failure to make the bestiary into a zoological manual must, therefore, derive from the adoption of different aims.

What purpose, then, was the bestiary meant to serve? It was a collection of animal lore and mythology, rich in symbolism and associations, meant to

Fig. 13.18. A page from a medieval bestiary, showing boar, ox, and bull. London, British Library, MS Harley 3244, fol. 47r (early 13th c.). By permission of the British Library.

instruct and entertain. And it surely did not occur, either to the compiler or to the reader, to inquire whether the stories were true in the sense that the claims of Aristotelian natural philosophy were expected to be true. A bestiary succeeded insofar as it effectively brought its reader into a world of traditional mythology, metaphor, and similitude.[49] We have similar mythologies of our own. Consider the lore surrounding the groundhog as a forecaster of the duration of winter, solemnly reported each February (at least in my part of the country) in newspapers and on radio and television. Does anybody believe in the truth of this forecast? Probably not; but to ask the question is to display a woeful misapprehension of the purpose of groundhog lore, which is not the "scientific" communication of meteorological truth, but participation in traditional community ritual, with all of the social and psychological benefits thus entailed.

Most of us become quite skillful at discriminating among different kinds of literary and artistic products in our own culture. We immediately know the difference between a scientific proposition, which must meet a variety of stiff epistemological tests in order to count as truly "scientific," and a Dr. Seuss story or a weather forecast offered to us by Jimmy the groundhog, which have quite different functions and must therefore be measured by different criteria. We need to become equally discerning in our study of medieval people and their achievements, including the various genres of art and literature that they produced. Just as we have seen (chap. 11, above) that the medieval mappamundi generally had purposes quite different from those of a modern world atlas, so must we cease to presume that all medieval books that touch on natural phenomena were meant for philosophical or scientific purposes analogous to ours when we write a scientific textbook, and to understand that they may have been meant to please and inform their readers on a variety of other levels. As we acquire this kind of sophisticated discernment of the products of medieval culture, learning to judge the achievement in the light of the aim, we will be on our way to a fuller appreciation of the character, the achievements, and, yes, the charm of the Middle Ages.

14 ❧ The Legacy of Ancient and Medieval Science

THE CONTINUITY QUESTION

My attempt, in this book, to reconstruct the lives, beliefs, and activities of historical actors from the ancient and medieval past has surely raised more questions than it has been able to answer. I would like to conclude this volume by approaching a cluster of questions that have doubtless occurred to most readers and that are also of great interest to me. What did all of that ancient and medieval scientific activity amount to? Was it really "science"? What difference did it make in the long run? Did it leave a permanent impression on the course or the shape of Western science, or was it an inconsequential cul-de-sac that ultimately led nowhere? Or to pose the question in one of its most common forms, were medieval and early modern science continuous with each other, or discontinuous? This is the celebrated "continuity question," which has been the basis of a persistent, but civilized, feud between medievalists and historians of early modern science. I would like to conclude this volume by cautiously approaching this question—circumspectly examining some of the outstanding scientific achievements of antiquity and the Middle Ages (both Islamic and Christian) in the interest of understanding the degree and shape of ancient and medieval influence on the science of sixteenth- and seventeenth-century Europe.[1]

But the first order of business is to do battle with centuries of entrenched opinion among those who denigrate medieval science, viewing the Middle Ages as a period of unrelieved scientific ignorance and superstition. Such opinions have received ample (if seriously misinformed) scholarly support, and in the mass media the adjective "medieval" has become a synonym for all that is deplorable. An early advocate of this negative opinion was Francis Bacon (1561–1626), who wrote in his *New Organon* (1620) that the ages between antiquity and his own era were "unprosperous" for the sciences, "for neither the Arabians nor the Schoolmen need be mentioned, who in the intermediate times rather crushed the sciences with a multitude of treatises, than increased their weight." A century later, Voltaire (1694–1778) elevated the level of anti-medieval rhetoric, writing of the "general decay and degeneracy" that characterized the Middle Ages, and of the "cunning and simplicity... brutality and artifice," of the medieval mind.[2]

The views of Bacon and Voltaire were sharpened and widely disseminated in the second half of the nineteenth century by the distinguished Swiss historian of the Renaissance, Jacob Burckhardt (1818–97), who argued in his *The Civilization of the Renaissance in Italy* (1860) that "the Middle Ages... spared themselves the trouble of induction and free inquiry." And in its most influential manifestation, Andrew Dickson White used the supposed ignorance and futility of medieval science as a weapon in his widely influential diatribe (1896) against the evils of a Christianity that, in his opinion, interfered with the development of the natural sciences: "The establishment of Christianity," he wrote, "arrested the normal development of the physical sciences for over fifteen hundred years.... There was created an atmosphere in which the germs of physical sciences could hardly grow—an atmosphere in which all seeking in Nature for truth as truth was regarded as futile." Finally, to demonstrate that such views are still alive and well, I quote Charles Freeman, in his *The Closing of the Western Mind: The Rise of Faith and the Fall of Reason* (2003): By the fifth century of the Christian era, he argues, "not only has rational thought been suppressed, but there has been a substitution for it of 'mystery, magic, and authority.'" It is little wonder, given this kind of scholarly backing, that the ignorance and degradation of the Middle Ages has become an article of faith among the general public, achieving the status of invulnerability merely by virtue of endless repetition.[3]

A pro-medieval counterattack was mounted in the early decades of the twentieth century by the French physicist and philosopher Pierre Duhem (1861–1916). While exploring the origins of the science of statics, Duhem encountered the works of fourteenth-century mathematicians at Oxford and Paris who, in his judgment, had laid the foundations for modern science, anticipating some of the most fundamental achievements of Galileo and his contemporaries.[4] Duhem's claims set off the continuity debate, which erupted with a certain regularity throughout the twentieth century. Early support for Duhem's campaign came from the influential medievalists Charles Homer Haskins (1870–1937) and Lynn Thorndike (1882–1965), writing in the 1920s and '30s.[5] The decades after World War II saw a dramatic expansion of historical research on medieval science; increased activity led to improved status and fresh claims about the magnitude and significance of the medieval scientific achievement. One of the leading figures in the postwar movement was Marshall Clagett (1916–2005), who made his mark primarily through the editing and translation of medieval scientific and mathematical texts. Another was Anneliese Maier (1905–71), who produced a series of brilliant studies in which she demonstrated by example how to read the sources more carefully and with closer attention to their philosophical context. While challenging many of Duhem's more extreme claims and offering an analysis of medieval natural philosophy far subtler and more cautious than his, Maier reaffirmed the importance of the medieval contribution, both conceptual and methodological, to the forging of modern science.[6]

Sparring has continued to the present, though to my eye the intensity level of the debate has diminished. For one thing, no informed historian of science would now support the extreme negative opinions of Francis Bacon, Voltaire, Burckhardt, or A. D. White. Alistair Crombie (1915–96) and Alexandre Koyré (1892–1964) traded opinions in the 1950s and 1960s—Crombie pleading that "a systematic theory of experimental science was understood by enough [thirteenth- and fourteenth-century] philosophers . . . to produce the methodological revolution to which modern science owes its origin." Koyré responded by denying the importance of methodology in the abstract for the origins of modern science and questioning whether, in any case, medieval methodologists had actually anticipated seventeenth-century methodology.[7] By now, early in the twenty-first century, both parties have made concessions, and despite an occasional quarrel, it appears that relative peace has broken out. The early modernists no longer question whether important scientific achievements emerged from the Middle Ages; and few medievalists now defend a strong version of the claim for continuity between medieval and early modern science.[8]

CANDIDATES FOR REVOLUTIONARY STATUS

Given the background sketched above, should we accept the construct of a seventeenth-century scientific revolution? And if the answer is affirmative, what was its relationship to the classical tradition of ancient and medieval science? Some observers of early modern science have expressed skepticism, on semantic grounds, about the very possibility of a seventeenth-century scientific revolution because we have no universally accepted definition of either "science" or "revolution." But the same is true of most interesting words. We have no universally accepted definition of "Middle Ages," "Renaissance," "Reformation," "fall of Rome," "art," "music," "religion," "philosophy," and so on. All of these are abstractions with debatable meanings that vary from one linguistic community to another—and from person to person within a given linguistic community. They are labels that we require if we are to communicate with one another, and they are inevitably fuzzy. To quibble about the label "scientific revolution" is thus (in my opinion) a waste of time. We should reserve our quibbling (if we can agree on the meaning of that word) for those occasions for which quibbling is suited—namely, discussion of such things as scientific beliefs and practices, rather than what to name them.

Not to quibble, then, but to inform: in my usage, the term "revolution" represents fundamental change, with no limits on elapsed time. The latter condition is my answer to those who have argued that a revolution that requires a century for its completion can't be a true revolution; revolutions, they believe, must be quick. I am unmoved by this argument; and my unabridged dictionary supports me, defining "revolution" simply as "radical and pervasive change." Call it what you like, do we see what can be regarded as examples of "radical and pervasive agents of change" in seventeenth-century European science—fundamental and with sufficient breadth, depth, and influence to qualify for *revolutionary* status?[9]

Two candidates, it appears to me, currently command significant support among historians of early modern science. The first is an alleged "healing" of an ancient schism separating physics and mathematics, and the creation, in the sixteenth and seventeenth centuries, of the new discipline (or collection of disciplines) that we call "mathematical science." Aristotle, it has been argued, distinguished between mathematics and physics (or natural philosophy) on disciplinary grounds and implicitly forbad the crossing of the disciplinary boundary that separated them. It is claimed, moreover, that ancient and medieval scholars largely (or almost entirely) accepted this prohibition, with the unfortunate consequence that physical science and mathematical science lived separate (and therefore mostly sterile) scientific lives until the early modern period. Medieval scholars could do physics and they could do mathematics, but disciplinary boundaries prohibited them from applying the content and techniques of the one to topics in the other. In short, mathematical modeling was separated by an uncrossable chasm from the exploration of physical reality. A revolution in science resulted, therefore, when Copernicus, Galileo, and other early modern scholars united the two enterprises, thereby creating genuine mathematical physics and setting science (or the physical sciences, at least) on the road to modernity.

We owe this argument largely to the influence of Pierre Duhem, writing in the first decade of the twentieth century, and resurrected by Robert S.

Westman in an influential article published in 1980. Westman argued for the existence of disciplinary communities with rigorously patrolled boundaries, which prohibited pre-Copernican mathematical astronomers or any other kind of mathematical scientist from dealing with questions of physical reality. Westman then used this model to explain (precariously, I believe) certain elements of Andreas Osiander's prefatory letter to Copernicus's *De revolutionibus*.[10]

The only problem with this interpretation is the historical record, which casts serious doubt on its veracity. It is true that Aristotle clearly and frequently distinguished between mathematics and physics in various works, including his *Posterior Analytics*, *Physics*, *Metaphysics*, and *On the Heavens*; he also discussed disciplinary boundaries and principles of subordination. But he consistently *rejected* a prohibition against crossing the boundary between physics and mathematics. He wrote about such matters in his *Physics*:

> The next point to consider is how the mathematician differs from the student of nature [the physicist]; for natural bodies contain surfaces and volumes, lines and points, and these are the subject-matter of mathematics. Further, is astronomy different from natural science or a department of it? It seems absurd that the student of nature should be supposed to know the nature of sun or moon, but not to know any of their essential attributes, particularly as the writers on nature obviously do discuss their shape and whether the earth and the world are spherical or otherwise.[11]

In practice, Aristotle repeatedly and unapologetically applied mathematics to the physical world—most obviously in his analysis of motion, where mathematical proportionalities played a major role, and quantifiable things such as weight or power were considered intrinsic properties and therefore aspects of the natures of the things studied.[12] As for the Middle Ages, the historical record once again comes to our rescue, revealing astronomy, optics, dynamics (theory of motion), and theory of weights as examples of successful sciences that were rooted in both physics and mathematics—traditions adopted and extended by Copernicus, Galileo, Kepler, and many other early modern scientific practitioners.[13]

A convincing paradigmatic case can be made of medieval *perspectiva* (our geometrical optics). I have devoted many decades to research on the major figures in the history of ancient and medieval optics, and except for Euclid (fl. 300 B.C.) and one or two minor figures, have not encountered any evidence that would support the myth (for that is what it has become) that we are exploring.[14] I have no way of knowing what optical scholars thought about the crossing of disciplinary boundaries *in general*, but I do know that all of them regularly crossed the boundary between mathematics and physics without, apparently, the slightest apprehension—concerned equally to discover the physical realities of light, color, reflection, refraction, and vision, and to situate them within a mathematical framework. Mathematical physics was certainly not an invention of the sixteenth or seventeenth century.

The second candidate for early modern revolutionary status is methodological—the invention and practice of the "experimental method" (according to the defenders of this thesis) by such sixteenth- and seventeenth-century scientists as Galileo, William Gilbert, Robert Boyle, and many others. According to defenders of this theory, the sterile scholastic debates and syllogistic demonstrations of ancient and medieval natural philosophy came to an end, replaced by experimental science, with its firsthand observation and manipulation under controlled conditions.

Before we look at examples, I hope for agreement among readers on two matters. First, we need to recognize the gap that separates methodological *theory* from methodological *practice*. What Aristotle *said* about scientific method in abstract methodological treatises and what his followers (and Aristotle himself) *actually did* in their scientific pursuits were very often two different things; and the same claim can be made, in general, for scientists who have troubled themselves to write on scientific methodology. In any case, our concern here will be exclusively with methodological *practice*. Second, we need to define the word "experiment." For present purposes, I am inclined to define it narrowly, by what I take to be its primary epistemological function: an attempt to confirm or disconfirm a theoretical claim about the nature or behavior of the material world by an observation (under controlled conditions if necessary) made for that purpose, or the gathering of data against which future anticipated theoretical claims may be tested.[15]

If these matters may be considered settled, we can go in search of ancient and medieval scientific experiments. They are not hard to find. Ptolemy (and his sources of astronomical data) are primary examples of planetary observation, employing a variety of astronomical instruments, in order to confirm or disconfirm the adequacy of his (and his predecessors') geometrical models for the planets. Similarly in his optics, Ptolemy deployed apparatus in contrived experiments, intended to gather the quantitative data that a successful mathematical theory of the refraction of light would be obliged to predict.[16] In medieval Islam, Ibn al-Haytham (ca. 965–ca.1039) performed experiments designed to prove or disprove the truth of optical theories.[17] Kamāl al-Dīn's fourteenth-century creation of a theory of the rainbow on the basis of experiments with light rays passing through water-filled glass globes is another excellent example—duplicated about the same time in medieval Christendom by the Dominican friar Theodoric of Freiberg.[18] We may also safely infer that the Maragha, Samarkand, and Istanbul observatories developed research programs based on organized observation of the heavens meant to deliver numerical data by which to confirm or disconfirm astronomical theories.[19]

Did experimental efforts continue in the European Middle Ages? Certainly! And, as in antiquity, they were most plentiful in the mathematical sciences, where Ptolemaic influence remained strong. One of the most striking occurred when the sixth-century Alexandrian Platonist John Philoponus simultaneously released two objects of different weights in order to disprove the Aristotelian theory that speed of descent is proportional to the weight of the body. According to Philoponus, "if you let fall from the same height two weights of which one is many times as heavy as the other, you will see that the . . . difference in time [of descent] is a very small one. And so if the difference in the weights is not considerable, that is, if one is, let us say, double the other, there will be no difference or else an imperceptible difference in time."[20]

Experimentation continued through the later Middle Ages, wherever it met a scientific need. Levi ben Gerson (1288–1344) engaged in active astronomical observations, made with the assistance of a variety of instruments, in order to refute aspects of Ptolemy's planetary models. Johannes de Muris, who taught at the Sorbonne in Paris during the first half of the fourteenth century, undertook observations to test and correct existing astronomical data on planetary motions and positions. His solar eclipse observations, for example, discredited certain predictions of the Alfonsine astronomical tables.[21]

Roger Bacon (ca. 1220–ca. 1292) does not deserve the reputation of "founder of experimental science" often bestowed on him. However, he did become an influential propagandist for empirical methodology, advocating the gathering of empirical evidence in all of the sciences. He argued that the first prerogative of experimental science is to verify conclusions drawn from arguments within the other sciences by submitting them to the test of experience; and it is clear from various publications that he practiced what he preached. See, especially, two parts of his *Opus maius*, one entitled *Scientia experimentalis*, where Bacon campaigned for the practice of experimental science, the other entitled *Perspectiva*, where he practiced (whenever possible) what he had been preaching.[22]

Roughly contemporary with Bacon, Peter Peregrinus of Maricourt manipulated magnets in order to gain an understanding of their properties and behavior—discoveries that anticipated many of those that would subsequently be made in the seventeenth century by William Gilbert, often identified as one of the founders of experimental science.[23] And who could deny the status of experimental scientist to the thirteenth-century Franciscan friar Paul of Taranto, who initiated an alchemical tradition characterized methodologically by laboratory manipulation of substances in the attempt to discover the pathway to transmutation?[24] Perhaps this litany of ancient and medieval experiments is overkill; but my purpose has been to make irrefutable the claim that ancient and medieval experiments were not rare exceptions to usual scientific practice but really quite plentiful, made whenever their ability to confirm or disconfirm a scientific claim was recognized.

If all of this is true, what credit is left for Francis Bacon (1561–1626), popularly celebrated as the founder (or *a* founder) of experimental science?[25] This Bacon (no descendant of Roger) argued, in books filled with references to *empiricism* and *experiment*, for the experimental interrogation of nature. However, what he and the Baconian tradition of the seventeenth century gave us was not a new method of experiment, but a new *rhetoric* of experiment, coupled with full exploitation of the possibilities of experiment in programs of scientific investigation.

THE SCIENTIFIC REVOLUTION

Where, then, can we locate this elusive revolution of sixteenth- and seventeenth-century science? I believe that Alexandre Koyré, who, in the 1950s and 1960s, disputed Crombie's focus on experimental science as the revolutionary agent, has put his finger on the right place. The underlying source of revolutionary novelty in the sixteenth and seventeenth centuries, he argued, was metaphysical and cosmological rather than methodological.[26]

Aristotelian and Platonic metaphysics shared a long and complicated relationship, including a certain amount of skirmishing, but in the later Middle Ages Aristotle's teleological metaphysics of nature, matter, form, substance, actuality and potentiality, the four qualities, and the four causes prevailed without serious challenge.[27] A rival metaphysics, Epicurean atomism, became known largely through the long philosophical poem by the Roman, Lucretius (d. ca. 55 B.C.), *On the Nature of Things*, known in Carolingian court circles by the early ninth century but not widely circulated before its early fifteenth-century revival. The atomism found in this treatise was reinforced by Diogenes

Laertius' third-century *Lives and Teachings of the Ancient Philosophers* (which devoted a book to Epicurus) and various writings of Cicero (106–43 B.C.), which advocated a mechanistic universe of lifeless, indivisible atoms moving randomly in an infinite void.[28] Employed and developed in the seventeenth century by Galileo in Italy, René Descartes and Pierre Gassendi in France, Robert Boyle and Isaac Newton in England, and many others, by the end of the century the "mechanical philosophy" (as it has come to be called) had become dominant. The organic universe of medieval metaphysics and cosmology had been routed by the lifeless machinery of the atomists.[29]

The result was a radical conceptual shift, which altered the foundations of natural philosophy as practiced for nearly the preceding two thousand years. Consider some of the consequences. In exchange for the purposeful, organized, organic world of Aristotelian natural philosophy, the new metaphysics offered a mechanical world of lifeless matter, unceasing local motion, and random collisions. It stripped away the sensible qualities so central to Aristotelian natural philosophy, offering them second-class citizenship as secondary qualities, or even reducing them to the status of sensory illusions. In place of the explanatory capabilities of form and matter, it offered the size, shape, and motion of invisible corpuscles—elevating local motion to a position of preeminence within the category of change and reducing all causality to efficient and material causality. As for Aristotelian teleology, which discovered purpose *within* nature, defenders of this new mechanical philosophy substituted the purposes of a creator God, imposed on nature from without.

The metaphysics of the mechanical philosophy reverberated through the scientific disciplines of the seventeenth century, transforming the ways of thinking about all manner of subjects. I do not believe that we can go to the extreme defended by A. Rupert Hall, who argued that the scientific revolution was a wholistic, cultural transformation that "refuses to dissolve into fragments," that was "an unbroken and interlocking series of new discoveries."[30] Surely this is a huge overstatement: the seventeenth-century "scientific revolution" was not a single, all-encompassing event. No doubt there were connections, but surely we can agree that different disciplines have different histories, develop at different rates, pursue different questions, practice different methods, and respond differently to external circumstances. If we limit ourselves to the big, metaphysical picture, overlooking developments at a disciplinary level, we risk missing many of the central realities of seventeenth-century science. If our goal is to enrich and enlarge our understanding of medieval and seventeenth-century scientific change, we cannot limit our gaze to the metaphysician's study; we must look to the laboratory or workplace and the field, one discipline at a time.

Limitations of space do not permit any such venture here. But I conclude this book by returning to the question of continuity. Revolution is coupled in the minds of many people with repudiation of the past, severed connections, a throwing out of the old and bringing in of the new—in short, more or less complete discontinuity. And it is not difficult to find cases in the seventeenth century that appear to follow such a pattern. However, if we look at individual disciplines, we will find that *revolutionary* achievements in many disciplines were built *on medieval foundations* and *out of resources* provided by the classical tradition. Revolution does not demand total rupture with the past.

A sextet of examples will illustrate. Many more could be produced. (1) The individual planetary models in Copernicus's heliocentric model of the planetary system drew their mathematical structures and nearly all of their data from Ptolemy and the Ptolemaic astronomical tradition. Copernicus's contribution was to deploy these resources to build what was ultimately (with further additions by Kepler) a successful heliocentric model.[31] But the pieces and much of the pattern were of ancient vintage. (2) Kepler's new theory of the retinal image (genuinely new and of revolutionary significance for visual theory) emerged not by repudiation of the dominant medieval theory of vision, but by accepting and rigorously applying all of its defining claims—demanding that in a successful theory of vision all rays of light entering the eye must participate in producing an image of the visual field.[32] (3) Galileo's dynamics and kinematics of motion drew substantially from fourteenth-century developments at the Universities of Oxford and Paris. Central to Galileo's early dynamics was the idea that projectile motion is the result of an impressed force, an *impeto*—clearly a sixteenth-century cousin of the *impetus* of the fourteenth and subsequent centuries. And the first two propositions of his mature kinematics of uniformly accelerated motion, presented in his *Two New Sciences*, employ graphing techniques of the fourteenth century to replicate and prove the Merton rule (mean speed theorem).[33]

(4) Nicole Oresme (ca. 1320–82), perhaps the greatest mathematician of the later Middle Ages, devised a predecessor of Cartesian coordinates of the seventeenth century—the graphing techniques referred to just above.[34] (5) Galenic medical theory and practice dominated Western medicine into the seventeenth century and beyond.[35] (6) Finally, even the mechanical philosophy, which I regard as the centerpiece of the scientific revolution, was a replay of the Epicurean atomism of the third century B.C., passed down within the classical tradition and appropriately Christianized.

These examples are not meant to diminish the luster of the scientific revolution or its creators and practitioners. My aim is simply to introduce caution and realism into the picture. No scientist really begins at the beginning, without any expectations, theoretical knowledge, or methodological commitments. Twenty-first-century scientists (even newly minted ones) do not walk into the laboratory with vacant brains, but with minds chock full of knowledge and expectations. The same was true of the scientific leaders of the seventeenth century. The brilliance of the creators of the scientific revolution is revealed not only in their repudiation of the past and creation of theoretical novelties, but also in their ability to re-deploy inherited scientific ideas, theories, assumptions, methodologies, instrumentation, and data, and put them to new theoretical uses. The scientific revolution took place within an ideologically rich human environment; it had ideologically rich historical foundations, and with those foundations came continuities.[36]

Notes

CHAPTER ONE

1. This point has been nicely made by David Pingree, "Hellenophilia versus the History of Science."

2. On ancient and medieval attitudes toward technology, see Elspeth Whitney, *Paradise Restored.*

3. The discussion of oral tradition in this section is heavily indebted to Jack Goody and Ian Watt, "The Consequences of Literacy" (p. 306 for the quoted phrase); Jack Goody, *The Domestication of the Savage Mind*; Jan Vansina, *Oral tradition as History.* See also Bronislaw Malinowski, *Myth in Primitive Psychology.*

4. This is certainly true of prehistoric cultures. Contemporary preliterate communities may have seen or heard about writing by contact with the literate world outside, but until they themselves have learned to write, it is doubtful that they can be said to have grasped the *idea* of writing.

5. Goody and Watt, "Consequences of Literacy," pp. 307–11. On oral tradition as "charter," see Malinowski, *Myth in Primitive Psychology*, pp. 42–44.

6. H. Frankfort and H. A. Frankfort, "Myth and Reality," pp. 24–25.

7. Jan Vansina, *The Children of Woot*, pp. 30–31, 198; Vansina, *Oral Tradition*, pp. 117, 125–29.

8. Vansina, *Oral Tradition*, pp. 130–33.

9. Vansina, *Children of Woot*, pp. 30–31. On origin myths and their relation to worldview, see also Vansina, *Oral Tradition*, pp. 133–37.

10. John A. Wilson, "The Nature of the Universe," p. 63. For a recent and thorough discussion of Egyptian cosmology and cosmogony, see Marshall Clagett, *Ancient Egyptian Science*, vol. 1, pt. 1, pp. 263–372. On Egyptian religion, see James H. Breasted, *Development of Religion and Thought in Ancient Egypt.*

11. On Babylonian creation myths, see Thorkild Jacobsen, "Mesopotamia: The Cosmos as State"; S. G. F. Brandon, *Creation Legends of the Ancient Near East*, chap. 3.

12. On primitive or folk medicine, see Henry E. Sigerist, *A History of Medicine*, vol. 1: *Primitive and Archaic Medicine*; John Scarborough, ed., *Folklore and Folk Medicines.*

13. The idea of "primitive mentality" was developed by Lucien Lévy-Bruhl in his *How Natives Think*; for a critique, see Goody, *Domestication of the Savage Mind*, chap. 1; G. E. R. Lloyd, *Demystifying Mentalities*, introduction.

370 Notes to pages 10–17

14. On "truth," especially "historical truth," see Vansina, *Oral Tradition*, pp. 21–24, 129–33.

15. Goody and Watt, "Consequences of Literacy," pp. 311–19. See also Barry Powell's reconstruction of the invention of Greek alphabetic writing: *Homer and the Origin of the Greek Alphabet*.

16. Paraphrasing Goody, *Domestication of the Savage Mind*, p. 76.

17. Ibid., chap. 3.

18. Ibid., chap. 5.

19. Goody and Watt, "Consequences of Literacy," pp. 319–43; Lloyd, *Demystifying Mentalities*, chap. 1.

20. The answer is 14. On Egyptian mathematics, see Otto Neugebauer, *The Exact Sciences in Antiquity*, chap. 4; B. L. van der Waerden, *Science Awakening: Egyptian, Babylonian, and Greek Mathematics*, chap. 1; G. J. Toomer, "Mathematics and Astronomy"; R. J. Gillings, "The Mathematics of Ancient Egypt"; Carl B. Boyer, *A History of Mathematics*, chap. 2.

21. Richard Parker, "Egyptian Astronomy, Astrology, and Calendrical Reckoning."

22. On Babylonian mathematics, see Neugebauer, *Exact Sciences in Antiquity*, chaps. 2–3; van der Waerden, *Science Awakening: Egyptian, Babylonian, and Greek Mathematics*, chaps. 2–3; van der Waerden, "Mathematics and Astronomy in Mesopotamia"; Boyer, *History of Mathematics*, chap. 2.

23. For an analysis of the question of ancient "algebra," see Sabetai Unguru, "History of Ancient Mathematics: Some Reflections on the State of the Art"; Unguru, "On the Need to Rewrite the History of Greek Mathematics."

24. Here I follow Francesca Rochberg, *The Heavenly Writing: Divination, Horoscopy, and Astronomy in Mesopotamian Culture*, chap. 2. See also Pingree, "Hellenophilia versus the History of Science," p. 556.

25. Rochberg, *Heavenly Writing*. The quoted phrase, representing the content of an actual Babylonian clay tablet, is on p. 55. On Babylonian astronomy, see also Neugebauer, *Exact Sciences in Antiquity*, chap. 5 (pp. 104–8 for zig-zag functions); van der Waerden and Huber, *Science Awakening II: The Birth of Astronomy*, chaps. 2–8; van der Waerden, "Mathematics and Astronomy in Mesopotamia; Asger Aaboe, "On Babylonian Planetary Theories"; and the essays collected in Neugebauer, *Astronomy and History*. For a highly technical account, see Neugebauer, *A History of Ancient Mathematical Astronomy*, 1:347–555. For a popular account, see Stephen Toulmin and June Goodfield, *The Fabric of the Heavens: The Development of Astronomy and Dynamics*, chap. 1. On canine urination as an omen, see Rochberg, *Heavenly Writing*, pp. 51, 55.

26. The contending opinions are nicely summarized by Rochberg, *Heavenly Writing*, chap. 5.

27. This point has been convincingly argued by Alexander Jones, "The Adaptation of Babylonian Methods in Greek Numerical Astronomy." See also below, chap. 5.

28. I borrow these data from Neugebauer, *Exact Sciences in Antiquity*, pp. 105–9; see also Toulmin and Goodfield, *The Fabric of the Heavens*, pp. 48–50.

29. In the Babylonian sexagesimal (base 60) system of counting, there are 360 degrees in a circle, 60 minutes in a degree, 60 seconds in a minute, and 60 thirds in a second. According to this tablet, the increase or decrease in speed (measured in degrees) between one month and the next is on the order of a third of a degree.

30. Sigerist, *History of Medicine*, 1:276. On Egyptian medicine, besides Sigerist, see Paul Ghalioungui, *The House of Life, Per Ankh: Magic and Medical Science in Ancient Egypt*; Ghalioungui, *The Physicians of Pharaonic Egypt*; John R. Harris, "Medicine." On surgery, see Guido Majno, *The Healing Hand*, chap. 3.

31. B. Ebbell, *The Papyrus Ebers*.

32. James H. Breasted, *The Edwin Smith Surgical Papyrus*.

33. On Mesopotamian medicine, see Sigerist, *History of Medicine*, 1, pt. 4; Robert Biggs, "Medicine in Ancient Mesopotamia"; Majno, *Healing Hand*, chap. 2.

CHAPTER TWO

1. Homer, *The Odyssey*, trans. Robert Fagles, pp. 77–78.

2. On the authorship, dating, and historicity of the *Odyssey*, see Bernard Knox's introduction to ibid.

3. Hesiod, *Theogony and Works and Days*, trans. M. L. West, pp. 6–7.

4. *The Poems of Hesiod*, trans. R. M. Frazer, p. 32. See also Robert Graves, *The Greek Myths*, vol. 1; Friedrich Solmsen, *Hesiod and Aeschylus*.

5. See the interesting analysis of this problem by Paul Veyne, *Did the Greeks Believe in Their Myths?*

6. On early Greek philosophy, see especially the following books by G. E. R. Lloyd: *Early Greek Science: Thales to Aristotle*, chap. 1; *Magic, Reason, and Experience*; *The Revolutions of Wisdom*; *Methods and Problems in Greek Science*; and (with Nathan Sivin), *The Way and the Word*. See also G. S. Kirk and J. E. Raven, *The Presocratic Philosophers*; Jonathan Barnes, *Early Greek Philosophy*.

7. On the early Greeks, see Thomas Cahill, *Sailing the Wine-Dark Sea: Why the Greeks Matter*, pp. 9–14.

8. *Metaphysics*, I.3, 983a6–20; portions of trans. borrowed from Kirk and Raven, *Presocratic Philosophers*, p. 87; from Aristotle, *Complete Works*, ed. Jonathan Barnes; and from the trans. of Hugh Tredennick in the Loeb Classical Library. On the Milesians, see Lloyd, *Early Greek Science*, chap. 2, and works cited above in n. 6.

9. Kirk and Raven, *Presocratic Philosophers*, pp. 108–9.

10. Simplicius quoting Theophrastus, ibid., p. 144.

11. G. E. R. Lloyd, *Demystifying Mentalities*, esp. chap. 1; Lloyd, *Early Greek Science*, pp. 10–15.

12. Kirk and Raven, *Presocratic Philosophers*, pp. 1–9; Gregory Vlastos, *Plato's Universe*, pp. 5–10.

13. Plato, *Cratylus*, 402a; Kirk and Raven, *Presocratic Philosophers*, p. 97.

14. On the atomists, see David Furley, *The Greek Cosmologists*, chaps. 9–11; Kirk and Raven, *Presocratic Philosophers*, chap. 17; Jonathan Barnes, *The Presocratic Philosophers*, 2:40–75; Cyril Bailey, *The Greek Atomists and Epicurus*.

372   *Notes to pages 30–43*

15. See the papers in Christoph Lüthy, John E. Murdoch, and William Newman, eds., *Late Medieval and Early Modern Corpuscular Matter Theories*, esp. Newman's article, pp. 291–329.

16. For Thales on the gods, see Lloyd and Sivin, *The Way and the Word*, p. 145.

17. Kirk and Raven, *Presocratic Philosophers*, pp. 328–29; Furley, *Greek Cosmologists*, chap. 7.

18. Aristotle, *Metaphysics*, I.5.985b33–986a2, in Aristotle, *Complete Works*, 2:1559. On the Pythagoreans, see also Kirk and Raven, *Presocratic Philosophers*, chap. 9; Furley, *Greek Cosmologists*, chap. 5; Barnes, *Presocratic Philosophers*, 2:76–94; Lloyd, *Early Greek Science*, chap. 3.

19. Kirk and Raven, *Presocratic Philosophers*, chap. 6; Vlastos, *Plato's Universe*, pp. 6–10.

20. On Parmenides, see Kirk and Raven, *Presocratic Philosophers*, chap. 10; Furley, *Greek Cosmologists*, pp. 36–42; Lloyd, *Early Greek Science*, pp. 37–39; Barnes, *Presocratic Philosophers*, 1: chaps. 10–11.

21. Kirk and Raven, *Presocratic Philosophers*, chap. 11; Barnes, *Presocratic Philosophers*, 1: chaps. 12–13. Aristotle's comment is found in his *Physics*, VI.2.233a22–23. In a second paradox, Zeno describes a race between Achilles (noted for his swiftness) and a tortoise (noted for its slowness): if the tortoise is given a head start, however small, Achilles will never be able to catch it, since by the time Achilles reaches the tortoise's starting point, the tortoise will have moved beyond it to a new position; by the time Achilles reaches this new position, the tortoise will have moved beyond that; and so on, ad infinitum.

22. Lloyd, *Early Greek Science*, chap. 4; Kirk and Raven, *Presocratic Philosophers*, chaps. 14, 17.

23. Kirk and Raven, *Presocratic Philosophers*, p. 422. See also Lloyd, *Early Greek Science*, chap. 4.

24. Kirk and Raven, *Presocratic Philosophers*, pp. 325, 394.

25. The scholarship on Plato is enormous. I have been heavily influenced by Vlastos, *Plato's Universe*, and the translation-commentaries of the various Platonic dialogues by Francis M. Cornford. For brief, recent introductions, see R. M. Hare, *Plato*; David J. Melling, *Understanding Plato*.

26. Plato, *Republic*, bk. VII, 514a–521b.

27. Lloyd, *Early Greek Science*, pp. 68–72; Plato, *Phaedo*, 65b; Plato, *Republic*, bk. VII, 532, trans. Francis M. Cornford, p. 252.

28. Vlastos, *Plato's Universe*, chap. 2. On Plato's cosmology, see also *Plato's Cosmology: The "Timaeus" of Plato*, trans. and commentary by Francis M. Cornford; Richard D. Mohr, *The Platonic Cosmology*.

29. Vlastos, *Plato's Universe*, chap. 3.

30. Ibid., chap. 2

31. The quoted passages are from *Plato's Cosmology*, 30d, p. 40, and 34b, p. 58.

32. Vlastos, *Plato's Universe*, pp. 61–65; Friedrich Solmsen, *Plato's Theology*.

CHAPTER THREE

1. For more on the Lyceum, see below, chap. 4. On Aristotle's relationship to Alexander, see Peter Green, *Alexander of Macedon*, pp. 53–62.

2. There is a great deal of excellent introductory literature on Aristotle; see especially G. E. R. Lloyd, *Aristotle: The Growth and Structure of His Thought*; Jonathan Barnes, *Aristotle*; Abraham Edel, *Aristotle and His Philosophy*.

3. Barnes, *Aristotle*, pp. 32–51; Edel, *Aristotle*, chaps. 3–4; Lloyd, *Aristotle*, chap. 3.

4. The technical name for this Aristotelian doctrine is "hylomorphism"—from *hyle* and *morphe*, the Greek terms for matter and form.

5. Aristotle's epistemology is dealt with in Edel, *Aristotle*, chaps. 12–15; Lloyd, *Aristotle*, chap. 6; Jonathan Lear, *Aristotle: The Desire to Understand*, chap. 4; Marjorie Grene, *A Portrait of Aristotle*, chap. 3.

6. On this subject, see Jonathan Barnes, "Aristotle's Theory of Demonstration"; G. E. R. Lloyd, *Magic, Reason, and Experience*, pp. 200–220.

7. On change, see Edel, *Aristotle*, pp. 54–60; Sarah Waterlow, *Nature, Change, and Agency in Aristotle's "Physics,"* chaps. 1, 3.

8. For Aristotle's conception of "nature," see Sarah Waterlow, *Nature, Change, and Agency*, chaps. 1–2; Edel, *Aristotle*, chap. 5 (p. 71 for the quoted phrase); James A. Weisheipl, "The Concept of Nature."

9. Aristotle was explicit about this in his *Politics*, instructing that "we must look for the intentions of nature in things which retain their nature, and not in things which are corrupted." *Politics*, I.4, 1254a35–37, trans. B. Jowett, in Aristotle, *Complete Works*, 2:1990.

10. See especially Peter Harrison's forthcoming *Adam's Encyclopaedia: The Fall of Man and the Foundations of Science, 1500–1700*; also Waterlow, *Nature, Change, and Agency*, pp. 33–34; Ernan McMullin, "Medieval and Modern Science: Continuity or Discontinuity?" pp. 103–29, esp. 118–19.

11. Edel, *Aristotle*, chap. 5.

12. See especially Friedrich Solmsen, *Aristotle's System of the Physical World*; Lloyd, *Aristotle*, chaps. 7–8.

13. *On the Heavens*, I.4.270b13–16, quoted from Aristotle, *Complete Works*, ed. Barnes, 1:451.

14. Lloyd, *Aristotle*, chap. 7.

15. Ibid., chap. 8. On alchemy, see below, chap. 12.

16. For Aristotle on the void, see Solmsen, *Aristotle's System of the Physical World*, pp. 135–43; David Furley, *Cosmic Problems*, pp. 77–90.

17. Furley, *Cosmic Problems*, chaps. 12–13.

18. Aristotle dealt with the shape of the earth in *On the Heavens*, II.13. See also D. R. Dicks, *Early Greek Astronomy to Aristotle*, pp. 196–98. The myth that ancient and medieval people believed in a flat earth is discussed below in chap. 7.

19. Waterlow, *Nature, Change, and Agency*, pp. 103–4.

20. For a careful analysis of the fine points, see James A. Weisheipl, "The Principle *Omne quod movetur ab alio movetur* in Medieval Physics."

21. On natural motion, see Aristotle's *On the Heavens*, 1.6, and *Physics*, IV.8. On forced motion, see *Physics*, VIII.5. For discussion, see Marshall Clagett, *The Science of Mechanics in the Middle Ages*, pp. 421–33; Clagett, *Greek Science in Antiquity*, pp. 64–68.

22. Lloyd, *Aristotle*, pp. 139–58.

23. There has been a recent burst of interest in Aristotle's biology. See especially three books by G. E. R. Lloyd: *Aristotelian Explorations*; *Aristotle*, chap. 4; and *Early Greek Science*, pp. 115–24. See also Anthony Preus, *Science and Philosophy in Aristotle's Biological Works*; Martha Craven Nussbaum, *Aristotle's "De motu animalium"*; Pierre Pellegrin, *Aristotle's Classification of Animals*; Allan Gotthelf and James G. Lennox, eds., *Philosophical Issues in Aristotle's Biology*.

24. Aristotle, *On the Parts of Animals*, 1.5. See also Lloyd, *Aristotle*, pp. 69–73.

25. Lloyd, *Aristotle*, pp. 76–81, 86–90; Lloyd, *Early Greek Science*, pp. 116–18; Pellegrin, *Aristotle's Classification of Animals*.

26. *History of Animals*, VI.3.561a3–19, in *Complete Works*, 1:883.

27. Lloyd, *Aristotle*, pp. 90–93; D. M. Balme, "The Place of Biology in Aristotle's Philosophy."

28. Aristotle, *De generatione animalium*, II.1.733b25–27, in *Complete Works*, 1:1138.

29. On Aristotle's doctrine of the soul and its faculties, see Lloyd, *Aristotle*, chap. 9; Ross, *Aristotle*, chap. 5; J. L. Ackrill, *Aristotle the Philosopher*, pp. 68–78. On medieval reactions, see chap. 10, below.

30. Aristotle, *De generatione animalium*, II.1.733a34–733b14, in *Complete Works*, ed. Barnes, 1:1138. On biological reproduction, see also Ross, *Aristotle*, pp. 117–22; Preus, *Science and Philosophy in Aristotle's Biological Works*, pp. 48–107.

31. Aristotle, *On the Parts of Animals*, III.6.668b33–669a7. On teleology in Aristotle's biology, see also Ross, *Aristotle*, pp. 122–27; Nussbaum, *Aristotle's "De motu animalium,"* pp. 59–106.

32. On method in Aristotle's biology, see Lloyd, *Aristotle*, pp. 76–81; Lloyd, *Magic, Reason, and Experience*, pp. 211–20; Nussbaum, *Aristotle's "De motu animalium,"* pp. 107–42.

CHAPTER FOUR

1. The classic sources on ancient education, to be used with caution, are H. I. Marrou, *A History of Education in Antiquity*; Werner Jaeger's three-volume *Paideia: The Ideals of Greek Culture*. For more recent scholarship, see John Patrick Lynch, *Aristotle's School*; Robin Barrow, *Greek and Roman Education*.

2. Lynch, *Aristotle's School*, pp. 65–66. On sophistic teaching in general, see pp. 38–54.

3. On Plato's Academy, see Lynch, *Aristotle's School*, pp. 54–63; Harold Cherniss, *The Riddle of the Early Academy*.

4. Cherniss, *Riddle of the Early Academy*, p. 65.

5. On the Lyceum, see Lynch, *Aristotle's School*, chaps. 1, 3; also Felix Grayeff, *Aristotle and His School*.

6. Lynch, *Aristotle's School*, chap. 6.

7. Recent underwater archaeological investigations in Alexandria's harbor have uncovered what appears to be the ancient royal quarter, where the Museum and library would have been located. I urge skepticism regarding the half a million rolls.

8. The best source on the Alexandrian Museum and library, including their social context, is P. M. Fraser, *Ptolemaic Alexandria*, esp. 1:305–35. See also Lynch, *Aristotle's School*, pp. 121–23, 194.

9. On Theophrastus as a natural philosopher, see G. E. R. Lloyd, *Greek Science after Aristotle*, chap. 2; J. B. McDiarmid, "Theophrastus." For more on Aristotle's theory of vision, see David Lindberg, *Theories of Vision from al-Kindi to Kepler*, pp. 6–9.

10. On Theophrastus and the Lyceum, see Lynch, *Aristotle's School*, pp. 97–108. The quotation comes from p. 101, with a minor alteration.

11. Ibid., pp. 101–3, 193.

12. On Strato, see Lloyd, *Greek Science after Aristotle*, pp. 15–20; Marshall Clagett, *Greek Science in Antiquity*, pp. 68–71; H. B. Gottschalk, "Strato of Lampsacus"; David Furley, *Cosmic Problems*, pp. 149–60. On his relationship to the Lyceum, see Lynch, *Aristotle's School*, passim.

13. On the ancient Aristotelian commentators, see the articles collected by Richard Sorabji, ed., *Aristotle Transformed*.

14. For the quotations, see Diogenes Laertius, *Lives of Eminent Philosophers*, trans. R. D. Hicks, 2:649, 667 (with a couple of word changes). On Epicurean philosophy, see A. A. Long, *Hellenistic Philosophy*, 2d ed.; David J. Furley, *Two Studies in the Greek Atomists*; Elizabeth Asmis, *Epicurus' Scientific Method*; Lloyd, *Greek Science after Aristotle*, chap. 3; Cyril Bailey, *The Greek Atomists and Epicurus*; and the sources edited and translated by A. A. Long and D. N. Sedley, *The Hellenistic Philosophers*, 2 vols.

15. Quotations (from Lucretius's *De rerum natura*, II.15 and IV.840) are borrowed from Long and Sedley, *Hellenistic Philosophers*, 1:47; and Lucretius, *De rerum natura*, trans. W. H. D. Rouse and M. F. Smith, rev. 2d ed., p. 343.

16. Asmis, *Epicurus' Scientific Method*, chap. 8.

17. Lloyd, *Greek Science after Aristotle*, pp. 23–24.

18. On Stoic philosophy in general, see Lloyd, *Greek Science after Aristotle*, chap. 3; F. H. Sandbach, *The Stoics*; Long, *Hellenistic Philosophy*; Marcia L. Colish, *The Stoic Tradition from Antiquity to the Early Middle Ages*; the articles collected in Ronald H. Epp, ed., *Recovering the Stoics*; and the sources gathered in Long and Sedley, *The Hellenistic Philosophers*, 2 vols.

19. On Stoic natural philosophy, in addition to the sources cited above, see David E. Hahm, *The Origins of Stoic Cosmology*; S. Sambursky, *Physics of the Stoics*.

20. A. A. Long, "The Stoics on World-Conflagration and Everlasting Recurrence"; Hahm, *Origins of Stoic Cosmology*, chap. 6.

21. Cicero, *On Divination*, 1.125–26, quoted by Long and Sedley, *Hellenistic Philosophers*, 1:337.

CHAPTER FIVE

1. On this question, see Friedrich Solmsen, *Aristotle's System of the Physical World*, pp. 46–48, 259–62; David C. Lindberg, "On the Applicability of Mathematics to Nature"; James A. Weisheipl, *The Development of Physical Theory in the Middle Ages*, pp. 13–17, 48–62.

2. See Aristotle's *Metaphysics*, XI.3. 1061a30–35, but also his *Physics*, II.2. 193b22–31.

3. This is a nice theory, but I doubt that it is anywhere close to the whole story. On Greek mathematics, see B. L. van der Waerden, *Science Awakening: Egyptian, Babylonian, and Greek Mathematics*, chaps. 4–8; Carl B. Boyer, *A History of Mathematics*, chaps. 4–11; Thomas Heath, *A History of Greek Mathematics*. For an overview of recent research, see J. L. Berggren, "History of Greek Mathematics: A Survey of Recent Research."

4. Wilbur Knorr, *The Evolution of the Euclidean Elements*.

5. On Euclid, see Heath, *Greek Mathematics*, chap. 11; Boyer, *History of Mathematics*, chap. 7; also Thomas Heath's translation of Euclid's *Elements*, with lengthy and detailed commentary.

6. E. J. Dijksterhuis, *Archimedes*; T. L. Heath, ed., *The Works of Archimedes*. On Archimedes' medieval influence, see Archimedes, *Archimedes in the Middle Ages*, ed. and trans. Marshall Clagett, 5 vols.

7. For an exhaustive analysis of Apollonius's *Conica*, see Sabetai Unguru and Michael N. Fried, *Apollonius of Perga's "Conica": Text, Context, Subtext*.

8. On early Greek astronomy, see especially Bernard R. Goldstein and Alan C. Bowen, "A New View of Early Greek Astronomy"; D. R. Dicks, *Early Greek Astronomy to Aristotle*; Lloyd, *Early Greek Science*, chap. 7; Thomas Heath, *Aristarchus of Samos, The Ancient Copernicus*. For a highly technical account, see Otto Neugebauer, *A History of Ancient Mathematical Astronomy*, 2:571–776. For a very readable account of intermediate technicality, see James Evans, *The History and Practice of Ancient Astronomy*.

9. For a useful discussion of the basic planetary phenomena and the two-sphere model, see Thomas S. Kuhn, *The Copernican Revolution*, chap. 1; also Michael J. Crowe, *Theories of the World from Antiquity to the Copernican Revolution*, chap. 1.

10. On Plato's astronomical knowledge, see Dicks, *Early Greek Astronomy*, chap. 5.

11. On Eudoxus and the hippopede, see ibid., chap. 6; Otto Neugebauer, "On the 'Hippopede' of Eudoxus"; David Hargreave, "Reconstructing the Planetary Motions of the Eudoxean System."

12. Dicks, *Early Greek Astronomy*, chap. 7.

13. Aristotle, *Metaphysica*, XII.8, 1073b17–1074a16.

14. On Aristotle's Unmoved Mover, see G. E. R. Lloyd, *Aristotle: The Growth and Structure of His Thought*, pp. 140–58. On the role of love in moving the spheres, see Aristotle, *Metaphysica*, XII.7, 1072b3.

15. Ibid., XII.8, 1074a1–5.

16. For further discussion of the debate over the aims of astronomy, see below, chaps. 11, 14.

17. Heath, *Aristarchus of Samos*, pt. 1, chap. 18; Otto Neugebauer, "On the Allegedly Heliocentric Theory of Venus by Heraclides Ponticus"; G. J. Toomer, "Heraclides Ponticus." On the subsequent history of the idea of a sun-centered motion for Mercury and Venus, see Eastwood, "Kepler as Historian of Science: Precursors of Copernican Heliocentrism according to De revolutionibus, I, 10."

18. Heath, *Aristarchus of Samos*, pt. 2; G. E. R. Lloyd, *Greek Science after Aristotle*, pp. 53–61. We have almost no details about Aristarchus's life. Since the island of Samos was under Ptolemaic rule during his lifetime, it is possible that he undertook astronomical and cosmological studies in Alexandria; see P. M. Fraser, *Ptolemaic Alexandria*, 1:397; William H. Stahl, "Aristarchus of Samos."

19. Heath, *Aristarchus of Samos*, pt. 2, chap. 3; G. J. Toomer, "Hipparchus."

20. D. R. Dicks, "Eratosthenes"; Albert Van Helden, *Measuring the Universe*, chap. 2.

21. On Hipparchus and Babylonian astronomy, see G. J. Toomer, "Hipparchus," pp. 207–24; Alexander Jones, "The Adaptation of Babylonian Methods in Greek Numerical Astronomy"; James Evans, *The History and Practice of Ancient Astronomy*, pp. 213–16; Otto Neugebauer, *The Exact Sciences in Antiquity*, chap. 5. On Hellenistic astronomy more generally, see Neugebauer, *History of Ancient Mathematical Astronomy*, 2:779–1058.

22. For introductions to Ptolemaic astronomy, see Lloyd, *Greek Science after Aristotle*, chap. 8; Crowe, *Theories of the World*, chaps. 3–4. For more technical discussions, see G. J. Toomer, "Ptolemy"; James Evans, *The History and Practice of Ancient Astronomy*, passim; Neugebauer, *History of Ancient Mathematical Astronomy*, 1:21–343; Olaf Pedersen, *A Survey of the Almagest*; and Ptolemy's astronomical text in English translation: Ptolemy, *Almagest*, ed. and trans. G. J. Toomer. The name "Ptolemy" offers opportunity for confusion. It does not signify descent from the ruling Ptolemaic dynasty. It may represent a certain geographical sector of the city of Alexandria, used by citizens as a "tribal" name. Its importance for us, in any case, is that it signifies that Claudius Ptolemy was not a recent immigrant, as were many of the early Alexandrian intellectuals, but was descended from a line of Alexandrian citizens.

23. Toomer, "Ptolemy," pp. 192–94.

24. Bernard R. Goldstein, *The Arabic Version of Ptolemy's "Planetary Hypotheses"*; G. E. R. Lloyd, "Saving the Appearances." See also below, chap. 11, pp. 269–70.

25. On ancient theories of vision, see David C. Lindberg, *Theories of Vision from al-Kindi to Kepler*, chap. 1.

26. On the geometrical approach to vision, see A. Mark Smith, "Saving the Appearances of the Appearances"; Albert Lejeune, *Euclide et Ptolémée*; Lindberg, *Theories of Vision*, pp. 11–17.

27. On Ptolemy, see the English translation of his *Optica*, with commentary, by A. Mark Smith, *Ptolemy's Theory of Visual Perception*; Smith's "Ptolemy's Search for a Law of Refraction"; Lindberg, *Theories of Vision*, pp. 15–17; Albert Lejeune, *Recherches sur la catoptrique grecque*; Lejeune, *Euclide et Ptolémée*.

28. For a clear and informative account of Ptolemy's theory of vision, see Smith, *Ptolemy's Theory of Visual Perception*, pp. 21–35.

29. For example, Ptolemy constructed the following table, comparing angles of incidence with their corresponding angles of refraction for light (or visual rays) passing from air to water:

Angle of incidence	10°	20°	30°	40°	50°	60°	70°	80°
Angle of refraction	8°	15½°	22½°	29°	35°	40½°	45½°	50°

Note that the differences between successive angles of refraction form an arithmetic series: 7½ 7, 6½, 6, 5½, 5, 4½. For an analysis of these results, see Smith, *Ptolemy's Theory of Visual Perception*, pp. 152–66.

30. Marshall Clagett, *Science of Mechanics in the Middle Ages*, chap. 1; Joseph E. Brown, "The Science of Weights"; Ernest A. Moody and Marshall Clagett, *The Medieval Science of Weights*.

31. For analyses of Archimedes' work, see the sources cited above, n. 6.

CHAPTER SIX

1. On "primitive" Greek medicine, see Fridolf Kudlien, "Early Greek Primitive Medicine." On varieties of Greek medical practitioners, see Owsei Temkin, "Greek Medicine as Science and Craft"; Lloyd, *Magic, Reason, and Experience*, pp. 37–49.

2. Ludwig Edelstein, "The Distinctive Hellenism of Greek Medicine," reprinted in Edelstein, *Ancient Medicine*, pp. 367–97; for the attitudes of Homer and Hesiod, see pp. 376–78. See also James Longrigg, "Presocratic Philosophy and Hippocratic Medicine." On magic and religion in Greek medicine, see Ludwig Edelstein, "Greek Medicine and Its Relation to Religion and Magic"; G. E. R. Lloyd, *Magic, Reason, and Experience*, chap. 1; Lloyd, *The Revolutions of Wisdom*, chap. 1.

3. Emma J. Edelstein and Ludwig Edelstein, *Asclepius: A Collection and Interpretation of the Testimonies*, 1:235.

4. There is a very large literature on Hippocratic medicine. For recent interpretations, see Wesley D. Smith, *The Hippocratic Tradition*; Owsei Temkin, *Hippocrates in a World of Pagans and Christians*; G. E. R. Lloyd's introduction to his edition of the *Hippocratic Writings*. See also Lloyd, *Early Greek Science*, chap. 5; Lloyd, *Magic, Reason, and Experience*; Longrigg, "Presocratic Philosophy and Hippocratic Medicine"; and the first three articles in Edelstein's *Ancient Medicine*.

5. On the relationship of medicine to philosophy, see Longrigg, "Presocratic Philosophy and Hippocratic Medicine"; Ludwig Edelstein, "The Relation of Ancient Philosophy to Medicine"; Lloyd, *Magic, Reason, and Experience*, pp. 86–98.

6. Quoted (with one change of wording) from the translation of J. Chadwick and W. N. Mann, in Lloyd, ed., *Hippocratic Writings*, pp. 237–38. On medicine and the supernatural in the Hippocratic corpus, see especially Temkin, *Hippocrates in a World of Pagans and Christians*; Lloyd, *Revolutions of Wisdom*, chap. 1; Lloyd, *Magic, Reason, and Experience*, chap. 1; Longrigg, "Presocratic Philosophy and Hippocratic Medicine."

7. *The Nature of Man*, trans. J. Chadwick and W. N. Mann, in Lloyd, ed., *Hippocratic Writings*, p. 262. The theory of four humors did not dominate Hippocratic physiology,

as it would Galenic and subsequent physiologies. Some Hippocratic writers accepted only two humors (usually bile and phlegm), and many discussed no humoral theory at all.

8. *Epidemics*, I.i, trans. J. Chadwick and W. N. Mann, in Lloyd, ed., *Hippocratic Writings*, pp. 87–88, with minor changes of punctuation.

9. Trans. Chadwick and Mann, in Lloyd, ed., *Hippocratic Writings*, p. 79. Despite his skepticism, the author of this treatise advances hypotheses of his own.

10. Ibid., p. 247.

11. Edelstein, "Greek Medicine and Its Relation to Religion and Magic," pp. 241–43; Temkin, *Hippocrates in a World of Pagans and Christians*.

12. See *Hippocrates, with an English Translation*, 4:423, 437.

13. On Hellenistic medicine, in addition to sources cited below, see John Scarborough, *Roman Medicine*; Ralph Jackson, *Doctors and Diseases in the Roman Empire*.

14. For an excellent survey of these developments, consult James Longrigg, "Anatomy in Alexandria in the Third Century B.C."

15. On Herophilus, see the authoritative volume by Heinrich von Staden, *Herophilus*; also Longrigg, "Superlative Achievement," pp. 164–77.

16. On Erasistratus, see James Longrigg, "Erasistratus"; Longrigg, "Superlative Achievement," pp. 177–84; G. E. R. Lloyd, *Greek Science after Aristotle*, pp. 80–85.

17. On the medical sects, see Heinrich von Staden, "Hairesis and Heresy: The Case of the *haireseis iatrikai*"; also Michael Frede, "The Method of the So-Called Methodical School of Medicine"; Ludwig Edelstein, "The Methodists"; Edelstein, "Empiricism and Skepticism in the Teaching of the Greek Empiricist School"; P. M. Fraser, *Ptolemaic Alexandria*, 1:338–76.

18. On Galen's life and times, see Vivian Nutton, "The Chronology of Galen's Early Career"; Nutton, "Galen in the Eyes of His Contemporaries."

19. Fraser, *Ptolemaic Alexandria*, 1:339. On Galen's thought, see Owsei Temkin, *Galenism*; Luis García Ballester, "Galen as a Medical Practitioner: Problems in Diagnosis"; Smith, *Hippocratic Tradition*, chap. 2; John Scarborough, "Galen Redivivus: An Essay Review"; Phillip De Lacy, "Galen's Platonism"; the essays contained in Fridolf Kudlien and Richard J. Durling, eds., *Galen's Method of Healing*; Lloyd, *Greek Science after Aristotle*, chap. 9. See also the introductions to Galen's *On the Usefulness of the Parts of the Body*, ed. and trans. Margaret T. May; Peter Brain, *Galen on Bloodletting*. See also chap. 13, below.

20. Galen, *On the Art of Healing*, 1.2, quoted from García Ballester, "Galen as a Medical Practitioner," with minor improvements in punctuation.

21. I have been strongly tempted to prepare a schematic diagram of Galen's physiological system, but I have reluctantly concluded that this cannot be done without foisting an unacceptable amount of modern anatomical and physiological knowledge onto Galen. For previous attempts to diagram Galen's physiology, see Charles Singer, *A Short History of Anatomy and Physiology from the Greeks to Harvey*, p. 61; Karl E. Rothschuh, *History of Physiology*, p. 19.

22. On (at least) one occasion Galen referred to the possibility of a "natural spirit" or "natural pneuma" in the venous blood; this suggestion was taken up by his followers,

who made it a canonical part of the Galenic system; see Owsei Temkin, "On Galen's Pneumatology."

23. Galen, *On the Natural Faculties*, trans. A. J. Brock, III.15, p. 321, with minor editing (some following Lloyd, *Greek Science after Aristotle*, p. 149).

24. In addition to works on Galen cited above, see Galen, *On Respiration and the Arteries*, ed. and trans. David J. Furley and J. S. Wilkie.

25. On Galen's methodology, in addition to works cited above, see Galen, *Three Treatises on the Nature of Science*.

26. Galen, *On the Usefulness of the Parts of the Body*, III.10, 1:189.

27. Ibid., VII.1, 2:729–31.

28. For example, George Sarton's *Galen of Pergamon*.

29. Temkin, *Galenism*, p. 24.

CHAPTER SEVEN

1. Horace, *Epistles*, II.1.156.

2. On these developments, see especially Elizabeth Rawson, *Intellectual Life in the Late Roman Republic*.

3. On Cicero, see below, pp. 137–39.

4. See especially William H. Stahl, *Roman Science*, pp. 50 (where Stahl refers to the "curse of the popularizer") and 55 (where he labels popularizers as "hacks").

5. On Aratus, see ibid., pp. 36–38. Stahl, though frequently a harsh critic, is also a useful source on the Roman popularization of Greek science.

6. See, for example, Arnold Reymond, *History of the Sciences in Greco-Roman Antiquity*, trans. Ruth Gheury de Bray, p. 92.

7. Quoted by Cicero, *De republica*, I. xviii.30.

8. For attempts to reconstruct the scientific content of Varro's *Disciplines*, see Rawson, *Intellectual Life in the Late Roman Republic*, pp. 158–64; Stephen Gersh, *Middle Platonism and Neoplatonism: The Latin Tradition*, 2:825–40; William H. Stahl, Richard Johnson, and E. L. Burge, *Martianus Capella and the Seven Liberal Arts*, 1:44–53.

9. When we speak of "Platonists" in this period, we always mean members of one or another of the philosophical traditions emanating from Plato and the Academy. Many of these "Platonists" defended doctrines that Plato would have disowned. For a useful analysis of Cicero's philosophy and its relationship to the Platonic tradition, see Gersh, *Middle Platonism and Neoplatonism*, 1:53–154.

10. Lucretius, *De rerum natura*, trans. W. H. D. Rouse and M. F. Smith. For a brief description, see Stahl, *Roman Science*, pp. 80–83.

11. On these authors, see Stahl, *Roman Science*, chap. 6; Gersh, *Middle Platonism and Neoplatonism*, chap. 3; and the relevant articles in the *Dictionary of Scientific Biography*. On Seneca, see also his *Physical Science in the Time of Nero: Being a Translation of the "Quaestiones naturales" of Seneca*, trans. John Clarke. For Celsus, see his *De medicina*, trans. W. G. Spencer.

12. See the articles in Roger French and Frank Greenaway, eds., *Science in the Early Roman Empire: Pliny the Elder, His Sources and Influence*; also Stahl, *Roman Science*,

chap. 7; for an older, but fuller, analysis see Lynn Thorndike, *A History of Magic and Experimental Science*, 1:41–99.

13. On Pliny's method, see A. Locher, "The Structure of Pliny the Elder's Natural History."

14. Pliny the Younger, *Letters*, trans. William Melmoth, revised by W. M. L. Hutchinson, III.5, 1:198.

15. Pliny the Elder, *Natural History*, trans. H. Rackham, W. H. S. Jones, and D. E. Eicholz, II.25–37, II.54, II.86, VII.2, IX.8.

16. Ibid., II.6–22; Olaf Pedersen, "Some Astronomical Topics in Pliny"; Bruce S. Eastwood, "Plinian Astronomy in the Middle Ages and Renaissance."

17. Gersh, *Middle Platonism and Neoplatonism*, chap. 7; Macrobius, *Commentary on the Dream of Scipio*, trans. William H. Stahl; Stahl, *Roman Science*, pp. 153–69.

18. On Martianus, see Stahl et al., *Martianus Capella*, 1:9–20. This work also contains a full translation of *The Marriage of Philology and Mercury*, with accompanying commentary.

19. Ibid., 2:278. For the definition of a perfect number, see just below. Martianus's identification of 9 as a perfect number (in violation of the Euclidean definition) is, according to Stahl, an indication of "the reverence" in which it was held.

20. There has been considerable discussion of the sources of Martianus's astronomical knowledge. See ibid., 1:50–53; Eastwood, "Plinian Astronomy in the Middle Ages and Renaissance," pp. 198–99.

21. For discussion of the ensuing controversy, see Bruce S. Eastwood, "Kepler as Historian of Science: Precursors of Copernican Heliocentrism According to *De revolutionibus* I, 10"; Eastwood, "Johannes Scottus Eriugena, Sun-Centered Planets, and Carolingian Astronomy." On Martianus, see also Eastwood, "Invention and Reform in Latin Planetary Astronomy." The quotation is from Stahl et al.'s translation: *Martianus Capella and the Seven Liberal Arts*, vol. 2, p. 333.

22. For a list of other translators and their translations, see Marshall Clagett, *Greek Science in Antiquity*, pp. 154–56.

23. On this question, see Gersh, *Middle Platonism and Neoplatonism*, pp. 421–34. Gersh also discusses Calcidius's philosophical stance.

24. On Boethius, see Lorenzo Minio-Paluello, "Boethius, Anicius Manlius Severinus"; Gersh, *Middle Platonism and Neoplatonism*, chap. 9; Clagett, *Greek Science in Antiquity*, pp. 150–53.

25. See, for example, Henry Chadwick, *The Early Church*.

26. On this subject, see David C. Lindberg: "Early Christian Attitudes toward Nature"; Lindberg, "Science and the Early Church"; Lindberg, "The Medieval Church Encounters the Classical Tradition: St. Augustine, Roger Bacon, and the Handmaiden Metaphor"; Lindberg, "Science and the Medieval Church."

27. See especially Henry Chadwick, *Early Christian Thought and the Classical Tradition*; Charles N. Cochrane, *Christianity and Classical Culture*; A. H. Armstrong and R. A. Markus, *Christian Faith and Greek Philosophy*.

28. Augustine's use of the feminine gender in this metaphor (handmaiden rather than manservant) had nothing to do with notions of female inferiority, but derived

382 *Notes to pages 150–160*

simply from the gender (grammatically speaking) of the Latin noun *philosophia*. The mistress, *theologia*, was also feminine.

29. On Roman education, see especially Stanley F. Bonner, *Education in Ancient Rome*; also H. I. Marrou, *A History of Education in Antiquity*; N. G. Wilson, *Scholars of Byzantium*, esp. pp. 8–27; Robin Barrow, *Greek and Roman Education*. On early medieval education, see Pierre Riché, *Education and Culture in the Barbarian West, Sixth through Eighth Centuries*; M. L. W. Laistner, *Thought and Letters in Western Europe, A.D. 500–900*, new ed., chaps. 2–3.

30. On monasticism and monastic schools, see Jean Leclercq, O.S.B., *The Love of Learning and the Desire for God: A Study of Monastic Culture*; Riché, *Education and Culture*, chap. 4.

31. The evidence against a proliferation of external monastic schools is persuasively presented in M. M. Hildebrandt, *The External School in Carolingian Society*.

32. Laistner, *Thought and Letters*, chap. 5. The Irish did not "save civilization," as claimed by Thomas Cahill in his popular and engaging book, *How the Irish Saved Civilization*. Irish monasticism made an important contribution, but it was not the only, and not even the major, contributor to European intellectual life in the early Middle Ages.

33. On Cassiodorus and Vivarium, see James J. O'Donnell, *Cassiodorus*.

34. On Isidore, see Evelyn Edson, *Mapping Time and Space: How Medieval Mapmakers Viewed Their World*, pp. 36–50; Jacques Fontaine, *Isidore de Séville et la culture classique dans l'Espagne wisigothique*; J. N. Hillgarth, "Isidore of Seville, St."Two older, but influential, sources that are unfairly and ahistorically critical of Isidore, failing (among other things) to understand that he accepted the sphericity of the earth, are Stahl, *Roman Science*, pp. 218–23; Ernest Brehaut, *An Encyclopedist of the Dark Ages: Isidore of Seville*.

35. On Bede, see Charles W. Jones, "Bede"; Wesley M. Stevens, *Bede's Scientific Achievement*; Edson, *Mapping Time and Space*, pp. 50–52; Peter Hunter Blair, *The World of Bede*, esp. chap. 24; Stephen C. McCluskey, *Astronomies and Cultures in Early Medieval Europe*.

36. For a compact, general history of the Byzantine Empire from about 330 to 1453, see Philip Whitting, ed., *Byzantium: An Introduction*. See also the following sequence of articles (with bibliographies) in the *Dictionary of the Middle Ages*: T. E. Gregory, "Byzantine Empire: History (330–1025)"; Charles M. Brand, "Byzantine Empire: History (1025–1204)"; John W. Barker, "Byzantine Empire: History (1204–1453)." The history of Byzantine science has been woefully neglected. I know of no overview of the subject in a Western language, apart from Anne Tihon's forthcoming "Byzantine Science."

37. On learning in the Byzantine Empire, see N. G. Wilson, *Scholars of Byzantium*; F. E. Peters, *The Harvest of Hellenism*; L. G. Westerink, "Philosophy and Theology, Byzantine."

38. For a convenient discussion of the Byzantine Greek commentators on Aristotle, see Richard Sorabji's general introduction to Christian Wildberg's translation of John Philoponus's *Against Aristotle on the Eternity of the World*, pp. 1–17. On Themis-

tius and Simplicius, see G. Verbeke's articles in the *Dictionary of Scientific Biography*, Ilsetraut Hadot, ed., *Simplicius: sa vie, son oeuvre, sa survie*. On Philoponus, see Richard Sorabji, ed., *Philoponus and the Rejection of Aristotelian Science*.

39. On arithmetic and numerals, see Heath, *History of Greek Mathematics*, I, 26–117; Boyer, *History of Mathematics*, pp. 272–73.

40. Boyer, *History of Mathematics*, pp. 111–31; Tihon, "Byzantine Science." On the achievements of Anthemius and Isidore, see G. L. Huxley, "Anthemius of Tralles"; Ivor Bulmer-Thomas, "Isidorus of Miletus."

41. For more on medieval geography, see chap. 11, below.

42. No educated person could have failed to know that Plato had proclaimed, and Aristotle had demonstrated, the sphericity of the earth. On Cosmas, see Dilke, "Cartography in the Byzantine Empire," pp. 261–63; and the introduction to Winstedt, *The Christian Topography of Cosmas Indicopleustes*. This book also contains the full Greek text of Cosmas' "Christian Topography." The American essayist and novelist Washington Irving launched the myth in the 1820s in his *A History of the Life and Voyages of Christopher Columbus*, 4 vols. (published in Paris, though written in English). For more on the theory of a flat earth, see Russell, *Inventing the Flat Earth*; Christine Garwood, *Flat Earth: The History of an Infamous Idea*.

43. On Theon and Hypatia, see G. J. Toomer, "Theon of Alexandria"; Edna E. Kramer, "Hypatia."

44. Anne Tihon, "Byzantine Science." On the astrolabe, see chap. 11, below.

45. Edward Rosen, "Regiomontanus, Johannes"; J. D. North, *The Norton History of Astronomy and Cosmology*, pp. 253–59. For the quotation, see Michael Shank, "Regiomontanus on Ptolemy, Physical Orbs, and Astronomical Fictionalism," forthcoming.

46. On the subjects treated in this paragraph, see Tihon, "Byzantine Science." On hospitals, see also Michael W. Dols, "Origins of the Islamic Hospital: Myth and Reality," pp. 382–84; Timothy S. Miller, *Birth of the Hospital in the Byzantine Empire*.

47. For elaboration of the argument of this final sentence, see chap. 14, below.

CHAPTER EIGHT

1. For an excellent analysis of the process of cultural diffusion in general, see F. E. Peters, *Allah's Commonwealth*; also Peters, *Aristotle and the Arabs: The Aristotelian Tradition in Islam*; and Peters, *Harvest of Hellenism*. G. W. Bowersock's *Hellenism in Late Antiquity* counters the theory that the spread of Hellenism in Asia was superficial. For an exhaustive analysis of Alexander the Great's military campaigns, see Peter Green, *Alexander of Macedon, 356–323 B.C.*

2. Vastly oversimplified, the Nestorian position endowed Christ with two distinct natures, one human and the other divine, whereas their opponents, Monophysites, declared Christ's divinity and humanity to be joined in a single unified nature. For more, see D. W. Johnson, "Nestorianism"; Johnson, "Monophysitism." A brief, informative account of the struggle is found in W. H. C. Frend, *The Early Church*, chap. 19.

3. Peters, *Aristotle and the Arabs*, chap. 2; Peters, *Allah's Commonwealth*, introduction.

4. De Lacy O'Leary, *How Greek Science Passed to the Arabs*, pp. 150–53; Peters, *Allah's Commonwealth*, pp. 318, 377–78, 383, 529; Peters, *Aristotle and the Arabs*, pp. 44–45, 53, 59; Majid Fakhry, *A History of Islamic Philosophy*, pp. 15–16. This would seem to be a trivial matter were it not for the ubiquity of the legend.

5. For reappraisal of the Gondeshapur legend, see Dols, "The Origins of the Islamic Hospital: Myth and Reality." I am grateful also for correspondence regarding this problem with Vivian Nutton.

6. Among the innumerable books on the early history of Islam, I have found the following particularly useful: Albert Hourani, *A History of the Arab Peoples*; F. E. Peters, *Allah's Commonwealth*. For lavish illustrations accompanied by an excellent text, see Bernard Lewis, ed., *Islam and the Arab World*.

7. Peters, *Allah's Commonwealth*, pp. 143, 400–402.

8. For the argument of this paragraph and the following, I am heavily indebted to Dimitri Gutas, *Greek Thought, Arabic Culture: The Graeco-Arabic Translation Movement in Baghdad and Early ʿAbbāsid Society*—a book that has revolutionized our (or at least my) understanding of the translation movement; see especially the introduction and chaps. 1–4. For the quotation, see p. 16.

9. Ibid., p. 36.

10. Ibid., pp. 61–69.

11. Ibid., pp. 121–36

12. Ibid., pp. 53–60.

13. On Ḥunayn, in addition to Gutas, see Lufti M. Saʿdi, "A Bio-Bibliographical Study of Ḥunayn ibn Isḥāq al-Ibadi (Johannitius)"; Georges C. Anawati, "Ḥunayn ibn Isḥāq"; Albert Z. Iskandar, "Ḥunayn the Translator; Ḥunayn the Physician." On the translations more generally, see Peters, *Allah's Commonwealth*; Peters, *Aristotle and the Arabs*; Jamil Ragep, "Islamic Culture and the Natural Sciences"; also O'Leary, *How Greek Science Passed to the Arabs*; Fakhry, *History of Islamic Philosophy*, pp. 16–31.

14. I am indebted in this section to Jamil Ragep, "Islamic Culture and the Natural Sciences"; A. I. Sabra, "The Appropriation and Subsequent Naturalization of Greek Science in Medieval Islam"; Sabra, "The Scientific Enterprise."

15. Fakhry, *A History of Islamic Philosophy*, deals throughout with the relationship between Hellenistic philosophy and orthodox Islamic theology, but see especially pp. 112–24, 244–61. See also Fazlur Rahman, *Islam*, 2d ed., pp. 120–27; W. Montgomery Watt, *Islamic Philosophy and Theology*; Sabra, "The Appropriation and Subsequent Naturalization of Greek Science in Medieval Islam." For an older view, see G. E. von Grunebaum, "Muslim World View and Muslim Science."

16. Sabra, "Appropriation and Subsequent Naturalization of Greek Science," pp. 236, 237, for the quotations.

17. On Islamic education, see Rahman, *Islam*, pp.181–92; George Makdisi, *The Rise of Colleges*. On education in the sciences, see Françoise Micheau, "The Scientific Institutions in the Medieval Near East," pp. 994–1007; A. I. Sabra, 'Science, Islamic," pp. 84–86.

18. William E. Gohlman, *The Life of Ibn Sina*, pp. 23, 31 for the quotations. On Ibn Sīnā, see also Fakhry, *A History of Islamic Philosophy*, pp. 147–83. For another autodidact, see the study of Ibn Ridwān by Michael W. Dols, trans., *Medieval Islamic Medicine: Ibn Ridwān's Treatise "On the Prevention of Bodily Ills in Egypt."*

19. For wide-ranging accounts of Islamic scientific achievements, see the articles in Roshdi Rashed, ed., *Encyclopedia of the History of Arabic Science*, 3 vols. (now probably the best source); Jan P. Hogendijk and A. I. Sabra, eds., *The Enterprise of Science in Islam: New Perspectives*; M. L. J. Young, J. D. Latham, R. B. Serjeant, eds., *Religion, Learning and Science in the ʿAbbasid Period*; Frans Rosenthal, *The Classical Heritage in Islam*; and articles in David C. Lindberg and Michael H. Shank, eds., *The Middle Ages*, vol. 2 (forthcoming) of *The Cambridge History of Science*. See also A. I. Sabra, "The Scientific Enterprise"; Howard R. Turner, *Science in Medieval Islam: An Illustrated Introduction*. For a discussion of Islamic antecedents of scientific developments in Western Europe, see chaps. 12–13, below.

20. Sensitive to the fact that al-Khwārizmī's *Algebra* contains no equations and looks suspiciously like geometry, historians of mathematics have begun to refer to it as "geometrical" or "rhetorical" algebra; see Sabetai Unguru, "On the Need to Rewrite the History of Greek Mathematics"; Unguru and Michael N. Fried, *Apollonius of Perga's "Conica,"* chap. 1.

The best collection of general sources for Islamic mathematics is in Rashed's *Encyclopedia of the History of Arabic Science*, vol. 2. On arithmetic and algebra, see Ahmad S. Saidan, "Numeration and Arithmetic," pp. 331–48; Roshdi Rashed, "Algebra," pp. 349–75. Also Carl Boyer, *A History of Mathematics*, chaps. 13–14.

21. On geometry and trigonometry, see Boris Rosenfeld and Adolf Youschkevitch, "Geometry"; Marie-Thérèse Debarnot, "Trigonometry." For medieval engineering, technology, and the like, see Donald R. Hill, *Islamic Science and Engineering*.

22. For Islamic astrology, see David Pingree, "Abū Maʿshar," Pingree, "Māshāʾallāh." On Māshāʾallāh, see also E. S. Kennedy and Pingree, *The Astrological History of Māshāʾallāh*.

23. Early Hindu influences on Islamic astronomy were largely displaced by access to Ptolemy's *Almagest* and other Greek sources, as was indigenous folk astronomy.

24. The best and most convenient general sources on the history of Islamic astronomy are Régis Morelon, "General Survey of Arabic Astronomy"; Morelon, "Eastern Arabic Astronomy between the Eighth and the Eleventh Centuries"; George Saliba, "Arabic Planetary Theories after the Eleventh Century A.D."; Saliba, *A History of Arabic Astronomy*.

25. Saliba, *History of Arabic Astronomy*, p. 14.

26. Aydin Sayili, *The Observatory in Islam and Its Place in the General History of the Observatory*, chap. 2; Morelon, "General Survey of Arabic Astronomy," p. 26.

27. Willy Hartner, "Al-Battānī." The quotation is Hartner's translation, p. 508.

28. The Tusi-couple consists of two circles, the larger having twice the radius of the smaller. If the smaller circle is allowed to roll inside the circumference of the larger (without slippage), then a point on the circumference of the smaller circle moves in a back-and-forth straight line along a diameter of the larger circle. See North, *Norton History of Astronomy and Cosmology*, pp. 192–96.

29. For a short account of the Maragha school and its agenda and achievements, see North, *Norton History of Astronomy and Cosmology*, pp. 192–99. For longer analyses, see Saliba, "Arabic Planetary Theories," pp. 86–127; Saliba, *History of Arabic Astronomy*, pp. 245–317.

30. Among historians of early astronomy, it is universally agreed that this cannot have been a coincidence. Two of the leading authorities, Swerdlow and Neugebauer, write in their *Mathematical Astronomy in Copernicus's De Revolutionibus*, 1:47: "The question therefore is not whether, but when, where, and in what form [Copernicus] learned of the Maragha theory." On al-Shāṭir's achievements, see Saliba, "Arabic Planetary Theories," pp. 86–90, 108–14; Saliba, *History of Arabic Astronomy*, pp. 272–78, 300–304.

31. Micheau, "The Scientific Institutions in the Medieval Near East," pp. 992–1006; Sayili, *The Observatory in Islam and Its Place in the General History of the Observatory*, pp. 187–223; Sabra, "Science, Islamic," pp. 84–86. On astronomical instrumentation, see François Charette, "The Locales of Islamic Astronomical Instrumentation."

32. On the Islamic contribution to optics, see below, chap. 12; Roshdi Rashed, "Geometrical Optics"; David C. Lindberg, *Theories of Vision from al-Kindi to Kepler*; Lindberg and Katherine H. Tachau, "The Science of Light and Color: Seeing and Knowing." For Ibn al-Haytham in particular, see Sabra, "Ibn al-Haytham's Revolutionary Project in Optics"; A. Mark Smith, *Alhacen's Theory of Visual Perception*; A. I. Sabra, *The Optics of Ibn al-Haytham* (the last two of which include partial translations, Sabra's from the original Arabic, Smith's from the medieval Latin version).

33. Roshdi Rashed, "Geometrical Optics," pp. 655–60. Snell's law states that when light is refracted as it passes from one transparent medium to another (from air to glass, for example), the ratio between the sine of the angle of incidence and the sine of the angle of refraction has a fixed value, determined by the relative optical densities of the two media.

34. Roshdi Rashed, "Kamāl al-Dīn"; Carl B. Boyer, *The Rainbow: From Myth to Mathematics*, pp. 127–29. It has sometimes been argued (as by Boyer) that the true Islamic creator of this theory was Kamāl al-Dīn's teacher, al-Shīrāzī. Descartes presented the theory of the rainbow in his *Discourse on Method*, "Meteorology," 8th discourse.

35. For excellent short accounts of medieval Islamic medicine, see Emilie Savage-Smith, "Medicine"; Savage-Smith, "Medicine in Medieval Islam." I am indebted to Emilie Savage-Smith for the shape of the following account of Islamic medicine, but not for possible misunderstandings and misstatements.

36. I borrow this list of diseases from Manfred Ullmann, *Islamic Medicine*, p. 1.

37. Savage-Smith, "Medicine in Medieval Islam."

38. On Ḥunayn and his team, see n. 13, above.

39. Shlomo Pines, "Al-Rāzī"; Lutz Richter-Bernburg, "al-Majūsī," *Encyclopedia Iranica*, 1:837–38. On Al-Zahrāwī, see Savage-Smith, "Medicine," pp. 942–48.

40. For Ibn Sīnā, see Savage-Smith, "Medicine," pp. 921–26; Georges C. Anawati and Albert Z. Iskandar, "Ibn Sīnā." For Ibn Sīnā and the Aristotelian tradition, see Dimitri Gutas, *Avicenna and the Aristotelian Tradition: Introduction to Reading Avicenna's Philosophical Works*.

41. On the influence of Islamic medical writings on medieval Europe, see Danielle Jacquart, "The Influence of Arabic Medicine in the Medieval West." On ophthalmology, see Savage-Smith, "Medicine," pp. 948–50.

42. Savage-Smith, "Medicine," pp. 933–36; p. 934 for the quotation.

43. Albert Z. Iskandar, "Ibn al-Nafis" (p. 603 for the quoted paragraph, slightly modified). The first European discovery of the pulmonary transit (typically, but incorrectly, referred to as the "pulmonary circulation") was by Michael Servetus (1511–53). However, it was independently discovered and first made public by Realdo Colombo (1510–59) several years later. Unlike the planetary models of Ibn al-Shāṭir, which could not have been reproduced by Copernicus with such exactitude without knowledge of Ibn al Shāṭir's achievement, there is no such necessity in this case, for discovery of the pulmonary transit in Europe was an inevitability, given the anatomical talent and serious anatomical studies under way at the University of Padua and other European medical schools in the sixteenth century. It is true that portions of Ibn al-Nafis's commentary on Ibn Sīnā's *Canon of Medicine* were translated into Latin early in the fifteenth century. On this question, see P. E. Pormann and Emilie Savage-Smith, *Medieval Islamic Medicine*.

44. Micheau, "The Scientific Institutions in the Medieval Near East"; Savage-Smith, "Medicine."

45. On Umayyad Spain, see especially María Rosa Menocal, *The Ornament of the World: How Muslims, Jews, and Christians Created a Culture of Tolerance in Medieval Spain*.

46. For a sampling of opinion on this subject, see A. I. Sabra, "Science, Islamic," pp. 87–88; Sabra, "The Appropriation and Subsequent Naturalization of Greek Science," pp. 238–40; Gustave E. von Grunebaum, *Medieval Islam: A Study in Cultural Orientation*, p. 339; von Grunebaum, *Classical Islam: A History 600 A.D.–1258 A.D.*, pp. 198–201; Max Meyerhof, "Science and Medicine," pp. 337–42; Muzaffar Iqbal, *Islam and Science*, pp. 125–70; Ignaz Goldziher, "The Attitude of Orthodox Islam toward the 'Ancient Sciences'" (published in 1915 but still of interest); Turner, *Science in Medieval Islam*, pp. 201–7; H. Floris Cohen, *The Scientific Revolution: A Historiographical Inquiry*, pp. 483–88; Lindberg, in the first edition of *The Beginnings of Western Science*, pp. 180–82; and a much-disputed book by Toby E. Huff, *The Rise of Early Modern Science: Islam, China, and the West*, chaps. 1–5.

47. Menocal, *The Ornament of the World*.

48. Rodney Stark, *For the Glory of God: How Monotheism Led to Reformations, Science, Witch-Hunts, and the End of Slavery*, employs the history of medieval science in an ill-informed attempt to prove that only a Christian culture could produce genuine science.

49. This research is just now emerging. See İhsan Fazlıoğlu, "The Mathematical/ Astronomical School of Samarqand as a Background for Ottoman Philosophy and Science." Also F. Jamil Ragep, "Copernicus and His Islamic Predecessors: Some Historical Remarks"; Charette, "The Locales of Islamic Astronomical Instrumentation."

CHAPTER NINE

1. John Marenbon, *Early Medieval Philosophy (480–1150)*, chaps. 4–5; M. L. W. Laistner, *Thought and Letters in Western Europe*, chaps. 3–4; Gillian R Evans, *The Thought of Gregory the Great*, pp. 55–68. Many of the themes covered in this chapter are also treated by Edward Grant, *God and Reason in the Middle Ages*; Grant, *Science and Religion, 400 B.C.–A.D. 1550: From Aristotle to Copernicus*.

2. Quoted by Rosamond McKitterick, "The Carolingian Renaissance," p. 152. This article offers an excellent short account of the educational reforms that emerged from Charlemagne's court. See also her earlier *The Carolingians and the Written Word*.

3. On Alcuin and the Carolingian educational reforms more generally, see Heinrich Fichtenau, *The Carolingian Empire*, chap. 4; John Marenbon, *From the Circle of Alcuin to the School of Auxerre*, chap. 2; Laistner, *Thought and Letters in Western Europe*, chap. 7. For a careful discussion of the exact meaning and significance of the decree to establish monastery schools, see M. M. Hildebrandt, *The External School in Carolingian Society*.

4. For an excellent, comprehensive, study of Carolingian astronomy, see Bruce S. Eastwood's forthcoming book, *Ordering the Heavens: Roman Astronomy in the Carolingian Renaissance*; also the broader account in Stephen McCluskey, *Astronomies and Cultures in Early Medieval Europe*, chaps. 5, 8. Also relevant are Bruce S. Eastwood, *The Revival of Planetary Astronomy in Carolingian and Post-Carolingian Europe*; Eastwood and Gerd Grasshoff, *Planetary Diagrams for Roman Astronomy in Medieval Europe, ca. 800–1500*. For a close look at what computus involved, see Faith Wallis, trans., *Bede: The Reckoning of Time*.

5. An impressive nonmathematical account of Roman astronomy and cosmology (divided equally between Latin text and English translation) appears in Pliny the Elder, *Natural History*, trans. H. Rackham, W. H. S. Jones, and D. E. Eicholz, II.6–22, pp. 170–255.

6. Eastwood, *Ordering the Heavens*, chaps. 1–2.

7. Trans. Bruce Eastwood in *Ordering the Heavens*, chap. 4, from Leiden Universiteitsbibliotheek, MS Voss. lat. F.48, fol. 79v.

8. The former challenge, according to Eastwood, *Ordering the Heavens*, chap. 1, was not satisfactorily met until the eleventh century.

9. Eastwood and Grasshoff, *Planetary Diagrams for Roman Astronomy in Medieval Europe*, fig. II.5, p. 30.

10. On Eriugena and his circle, see John J. O'Meara, *Eriugena*; Marenbon, *Early Medieval Philosophy*, chap. 6; Marenbon, *From the Circle of Alcuin*, chaps. 3–4.

11. See especially Robert Hillenbrand, "Cordoba"; María Rosa Menocal, *Ornament of the World: How Muslims, Jews, and Christians Created a Culture of Tolerance in Medieval Spain*; Philip K. Hitti, *History of the Arabs: From the Earliest Times to the Present Day*, chaps. 34–35.

12. The most important study of Gerbert's Spanish venture, loaded with detail, is a recent paper by Marco Zuccato, "Gerbert of Aurillac and a Tenth-Century Jewish Channel for the Transmission of Arabic Science to the West." For other sources on

Gerbert, see D. J. Struik, "Gerbert"; Harriet Pratt Lattin, ed. and trans., *The Letters of Gerbert with His Papal Privileges as Sylvester II*; Cora E. Lutz, *Schoolmasters of the Tenth Century*, chap. 12.

13. On the technology of this period, see especially Lynn White, Jr., *Medieval Technology and Social Change*; Donald Hill, *A History of Engineering in Classical and Medieval Times*; Jean Gimpel, *The Medieval Machine: The Industrial Revolution of the Middle Ages*. On the water wheel, see Terry S. Reynolds, *Stronger than a Hundred Men: A History of the Vertical Water Wheel*, chap. 2.

14. David Herlihy, "Demography."

15. Nicholas Orme, *English Schools of the Middle Ages*; John J. Contreni, "Schools, Cathedral"; Contreni, *The Cathedral School of Laon from 850 to 930*; Marenbon, *Early Medieval Philosophy*, chap. 10; John W. Baldwin, *The Scholastic Culture of the Middle Ages*, chap. 3; Richard W. Southern, "The Schools of Paris and the School of Chartres"; Southern, "From Schools to University"; Paul F. Grendler, *Schooling in Renaissance Italy*, esp. chap. 1.

16. Southern, "The Schools of Paris and the School of Chartres," pp. 114–18; Jean Leclercq, "The Renewal of Theology," pp. 72–73.

17. See Richard W. Southern, *Medieval Humanism and Other Studies*, chap. 5; the vehement reply by Nikolaus Häring, "Chartres and Paris Revisited"; and Southern's rejoinder in "The Schools of Paris and the School of Chartres."

18. Charles Homer Haskins, *The Renaissance of the Twelfth Century*, chaps. 4, 7.

19. Colin Morris, *The Discovery of the Individual, 1050–1200*, p. 46. On the rationalistic turn of the eleventh and twelfth centuries, see also the ambitious book by Alexander Murray, *Reason and Society in the Middle Ages*.

20. Although his intellectual formation occurred, at least in part, within the monastic tradition—in his late twenties he studied at the monastery of Bec in northern France—Anselm faithfully represents the broader intellectual currents of his day and did much to shape the theological traditions of the twelfth-century schools.

21. Jasper Hopkins, *A Companion to the Study of St. Anselm*; G. R. Evans, *Anselm and a New Generation*; Richard W. Southern, *Saint Anselm*, esp. pp. 123–37; Southern, *Medieval Humanism*, chap. 2. On the distinction between monastic and "scholastic" theology in the twelfth century, see Jean Leclercq, "The Renewal of Theology."

22. Abelard's *Epistolae*, no. 17, in *Patrologia latina*, vol. 178, col. 375. For a brief account of Abelard's life and thought, see David E. Luscombe, *Peter Abelard*; Luscombe, "Peter Abelard."

23. On the idea of a twelfth-century renaissance, see Charles Burnett, "The Twelfth-Century Renaissance" (forthcoming). On twelfth-century Platonism, see M.-D. Chenu, *Nature, Man, and Society in the Twelfth Century*, chap. 2; Tullio Gregory, "The Platonic Inheritance." On twelfth-century natural philosophy in general, see Winthrop Wetherbee, "Philosophy, Cosmology, and the Twelfth-Century Renaissance" and other essays in Dronke, *History of Twelfth-Century Western Philosophy*. Also *Adelard of Bath: Conversations with His Nephew*, trans. Charles Burnett. Older, but still useful, are Charles Homer Haskins, *Studies in the History of Mediaeval Science*; Lynn

Thorndike, *A History of Magic and Experimental Science*, vol. 2, chaps. 35–50. On other specific aspects of twelfth-century philosophy, see the citations below.

24. Nikolaus M. Häring, "The Creation and Creator of the World According to Thierry of Chartres and Clarenbaldus of Arras"; Peter Dronke, "Thierry of Chartres"; J. M. Parent, *La doctrine de la création dans l'école de Chartres*.

25. On the idea of nature, see Tullio Gregory, "La nouvelle idée de nature et de savoir scientifique au XIIe siècle"; and a number of the essays contained in *La filosofia della natura nel medioevo*.

26. On William of Conches, see Tullio Gregory, *Anima mundi: La filosofia di Guglielmo di Conches e la scuola di Chartres*; Dorothy Elford, "William of Conches"; Thorndike, *History of Magic*, vol. 2, chap. 37; Joan Cadden, "Science and Rhetoric in the Middle Ages: The Natural Philosophy of William of Conches." For his *Dragmaticon philosophiae*, see the translation by Italo Ronca and Matthew Curr, *A Dialogue on Natural Philosophy*. On Adelard of Bath, see Charles Burnett, ed., *Adelard of Bath*. For the quoted passages, see William of Conches, *Philosophia mundi*, ed. Gregor Maurach, 1.22, pp. 32–33; Adelard of Bath, *Questiones naturales*, trans. Burnett in *Adelard of Bath, Conversations with His Nephew*, pp. 97, 99 (with one minor change of wording). For useful summaries and analyses of the problem, see William J. Courtenay, "Nature and the Natural in Twelfth Century Thought"; Chenu, *Nature, Man, and Society*.

27. The quoted passages are taken from Tullio Gregory, "The Platonic Inheritance," pp. 65, 57. Cf. similar remarks by Adelard of Bath, *Quaestiones naturales*, trans. Burnett, pp. 97, 99.

28. William J. Courtenay, "Nature and the Natural in Twelfth-Century Thought"; Courtenay, "The Dialectic of Divine Omnipotence."

29. On humanism, see Morris, *Discovery of the Individual*; Southern, *Medieval Humanism*, chap. 4. For a significant qualification, see Caroline Walker Bynum, "Did the Twelfth Century Discover the Individual?"

30. On medieval astrology, see Olaf Pedersen, "Astrology"; the essays contained in Patrick Curry, ed., *Astrology, Science, and Society: Historical Essays*; J. D. North, *The Norton History of Astronomy and Cosmology*, pp. 259–71; North, "Celestial Influence: The Major Premiss of Astrology"; North, "Astrology and the Fortunes of Churches." For further discussion and additional bibliography, see the final section of chap. 11, below.

31. On mathematics in the twelfth century, see Charles Burnett, "Scientific Speculations"; Gillian R. Evans, *Old Arts and New Theology*, pp. 119–36; Evans, "The Influence of Quadrivium Studies in the Eleventh- and Twelfth-Century Schools"; Guy Beaujouan, "The Transformation of the Quadrivium." For the quoted passage, see Häring, "The Creation and Creator of the World According to Thierry of Chartres," p. 196.

32. Charles Burnett, "Translation and Transmission of Greek and Islamic Science to Latin Christendom" (forthcoming); Burnett, "Translation and Translators, Western European"; David C. Lindberg, "The Transmission of Greek and Arabic Learning to the West"; Marie-Thérèse d'Alverny, "Translations and Translators"; Millas-Vallicrosa, "Translations of Oriental Scientific Works"; Jean Jolivet, "The Arabic Inheritance."

33. Michael McVaugh, "Constantine the African."

34. Richard Lemay, "Gerard of Cremona." For a list of Gerard's translations, see Edward Grant, ed., *A Source Book in Medieval Science*, pp. 35–38.

35. For two different opinions, see Lemay, "Gerard of Cremona," pp. 174–75; d'Alverny, "Translations and Translators," pp. 453–54.

36. Lorenzo Minio-Paluello, "Moerbeke, William of."

37. On the importance of astrology in the revival of Aristotle, see Richard Lemay, *Abu Maʿshar and Latin Aristotelianism in the Twelfth Century*.

38. M. B. Hackett, "The University as a Corporate Body," p. 37.

39. Excellent introductions to the history of the universities are to be found in Baldwin, *The Scholastic Culture of the Middle Ages*; Astrik L. Gabriel, "Universities"; Alan B. Cobban, *The Medieval Universities: Their Development and Organization*; Michael H. Shank, "The Social and Institutional Background of Medieval Latin Science." Older classics, still useful, are Charles H. Haskins, *The Rise of Universities*; Hastings Rashdall, *The Universities of Europe in the Middle Ages*, ed. F. M. Powicke and A. B. Emden, 3 vols. For excellent recent work on the English universities, see Catto, *The Early Oxford Schools*, vol. 1 of *History of the University of Oxford*; William J. Courtenay, *Schools and Scholars in Fourteenth-Century England*; Alan B. Cobban, *The Medieval English Universities: Oxford and Cambridge to c. 1500*. On Paris, see Stephen C. Ferruolo, *The Origins of the University: The Schools of Paris and Their Critics, 1100–1215*. For a narrowly focused social portrait, see William J. Courtenay, *Parisian Scholars in the Early Fourteenth Century: A Social Portrait*.

40. On patronage and privileges, see Pearl Kibre, *Scholarly Privileges in the Middle Ages*; Guy Fitch Lytle, "Patronage Patterns and Oxford Colleges, c. 1300–c. 1530."

41. I owe these estimates to my colleague, William J. Courtenay.

42. For the data, see James H. Overfield, "University Studies and the Clergy in Pre-Reformation Germany," pp. 277–86.

43. For actual data on student mortality, see Guy Fitch Lytle, "The Careers of Oxford Students in the Later Middle Ages," p. 221.

44. Baldwin, *Scholastic Culture*; James A. Weisheipl, "Curriculum of the Faculty of Arts at Oxford in the Fourteenth Century"; Weisheipl, "Developments in the Arts Curriculum at Oxford in the Early Fourteenth Century"; and the relevant articles in Catto, *The Early Oxford Schools*.

45. On science in the medieval curriculum, in addition to the works of Baldwin and Weisheipl cited above, see Pearl Kibre, "The Quadrivium in the Thirteenth Century Universities (with Special Reference to Paris)"; Guy Beaujouan, "Motives and Opportunities for Science in the Medieval Universities"; Edward Grant, "Science and the Medieval University"; James A. Weisheipl, "Science in the Thirteenth Century"; Edith Dudley Sylla, "Science for Undergraduates in Medieval Universities."

46. It needs to be stressed that learning in the Middle Ages was conceived as the mastery of a set of standard texts. This is in contrast to the modern view, which views education as the mastery of certain subjects and considers the choice of specific texts to be incidental.

CHAPTER TEN

1. On the earliest dissemination of Aristotle's works in the West, see Aleksander Birkenmajer, "Le rôle joué par les médecins et les naturalistes dans la réception d'Aristote au XIIe et XIIIe siècles"; Richard Lemay, *Abu Maʿshar and Latin Aristotelianism in the Twelfth Century.* On the reception of Aristotle in the universities, see the excellent discussion by Fernand Van Steenberghen, *Aristotle in the West*; a parallel analysis can be found in Van Steenberghen's *The Philosophical Movement in the Thirteenth Century.* For an excellent survey of Aristotelianism in the West, see William A. Wallace, "Aristotle in the Middle Ages." On Oxford, see Van Steenberghen, *Aristotle in the West*, chap. 6; D. A Callus, "Introduction of Aristotelian Learning to Oxford." Many of the themes of this chapter are also covered by Edward Grant, *God and Reason in the Middle Ages*; Grant, *Science and Religion, 400 B.C.–A.D.1550: From Aristotle to Copernicus.*

2. On Aristotelianism at Paris, see Van Steenberghen, *Aristotle in the West*, chaps. 4–5; John W. Baldwin, *Masters, Princes, and Merchants: The Social Views of Peter the Chanter and His Circle*, 1:104–7; Richard C. Dales, *The Intellectual Life of Western Europe in the Middle Ages*, pp. 243–46. For a translation of documents bearing on the events in Paris, see Lynn Thorndike, *University Records and Life in the Middle Ages*, pp. 26–40; reprinted, with additional notes, in Edward Grant, ed., *A Source Book in Medieval Science*, pp. 42–44.

3. For the Latin text, see Henricus Denifle and Aemilio Chatelain, *Chartularium Universitatis Parisiensis*, 1:138, 143. For another English translation, which includes more of the text, see Thorndike, *University Records*, p. 40.

4. Van Steenberghen, *Aristotle in the West*, pp. 89–110; David C. Lindberg, ed. and trans., *Roger Bacon's Philosophy of Nature*, pp. xvi–xvii.

5. Van Steenberghen, *Aristotle in the West*, pp. 17–18, 64–66, 127–28. A brief account of Avicenna's philosophy can be found in Majid Fakhry, *A History of Islamic Philosophy*, pp. 147–83; Georges C. Anawati and Albert Z. Iskandar, "Ibn Sīnā." See also chap. 8, above.

6. Van Steenberghen, *Aristotle in the West*, pp. 18–20, 89–93. The most important translator of Averroes was Michael Scot (d. ca. 1235), beginning in 1217 and continuing into the 1230s, but there is no evidence that his translations were used at Paris until after 1230; see ibid., pp. 89–94; Lorenzo Minio-Paluello, "Michael Scot." On Averroes' philosophy, see Fakhry, *History of Islamic Philosophy*, pp. 302–25; Roger Arnaldez and Albert Z. Iskandar, "Ibn Rushd."

7. See, for example, Aristotle, *On the Heavens*, 1.10–11. For a discussion of Aristotle's doctrine, see Friedrich Solmsen, *Aristotle's System of the Physical World*, pp. 51, 266–74, 288, 422–24.

8. See, for example, Thomas Aquinas, Siger of Brabant, and Bonaventure, *On the Eternity of The World*, trans. Cyril Vollert et al.; Boethius of Dacia, *On the Supreme Good, On the Eternity of the World, On Dreams*; Richard C. Dales, "Time and Eternity in the Thirteenth Century." For a full account of medieval discussions, see Dales, *Medieval Discussions of the Eternity of the World.*

9. For a short account of determinism and indeterminism in Aristotle, see Abraham Edel, *Aristotle and His Philosophy*, pp. 95, 389–401. For a full analysis, see Richard Sorabji, *Necessity, Cause, and Blame*. For an excellent analysis of the Islamic attack on this problem, see Barry S. Kogan, *Averroes and the Metaphysis of Causation*.

10. For the biblical doctrine, see Matthew 10:29–31.

11. On Aristotle's theory of the soul, see G. E. R. Lloyd, *Aristotle*, chap. 9. On the Christian response, see Fernand Van Steenberghen, *Thomas Aquinas and Radical Aristotelianism*, pp. 29–70; Knowles, *Evolution of Medieval Thought*, pp. 206–18, 292–96.

12. For an extended account of Averroistic monopsychism and the Western response, see Van Steenberghen, *Thomas Aquinas and Radical Aristotelianism*, pp. 29–74.

13. For a full account, see William J. Courtenay, *Teaching Careers at the University of Paris in the Thirteenth and Fourteenth Centuries*.

14. On Grosseteste and his scholarly career, see the excellent study by James McEvoy, *The Philosophy of Robert Grosseteste*; for the dating of Grosseteste's commentary on the *Posterior Analytics*, see pp. 512–14. On Grosseteste's life and work, see also Richard W. Southern, *Robert Grosseteste*; D. A. Callus, ed., *Robert Grosseteste, Scholar and Bishop*. On Grosseteste's investigation of Aristotle's logic and its effect on his scientific methodology, see the seriously exaggerated claims by A. C. Crombie, *Robert Grosseteste and the Origins of Experimental Science, 1100–1700*, chaps. 3–4 (no informed scholar of my acquaintance accepts Crombie's claim that Grosseteste was the founder of experimental science; nevertheless, he remains an important figure, and Crombie provides a good account of some of his achievements).

15. On Grosseteste's cosmogony, see below, chap. 11, and the accompanying notes.

16. On Bacon's scientific career, see Stewart C. Easton, *Roger Bacon and His Search for a Universal Science*; Theodore Crowley, *Roger Bacon: The Problem of the Soul in His Philosophical Commentaries*. For a convenient biographical sketch, see Lindberg, *Roger Bacon's Philosophy of Nature*, pp. xv–xxvi. On Bacon's scientific method and scientific achievements, see Lindberg, *Roger Bacon and the Origins of "Perspectiva" in the Middle Ages*, pp. xxii–xciv. The widespread mythology that portrays Bacon as a lone defender of something akin to modern science against theological opposition is a considerable misrepresentation.

17. On the term "handmaiden" and its lack of gender implications, see above, chap. 7, n. 28.

18. Trans. Lindberg, from John H. Bridges, ed., *Opus majus of Roger Bacon*, 3:36. On Bacon's defense of the new philosophy, see David C. Lindberg, "Science as Handmaiden: Roger Bacon and the Patristic Tradition."

19. Bonaventure's position in relation to the various philosophical traditions of the thirteenth century has been much disputed. For an account of the alternatives and an attempt to adjudicate among them, see Van Steenberghen, *Aristotle in the West*, pp. 147–62; Knowles, *Evolution of Medieval Thought*, pp. 236–48; John Francis Quinn, *The Historical Constitution of St. Bonaventure's Philosophy*, especially pp. 841–96. These works will lead the reader to additional sources.

20. On Albert's life and works, see James A. Weisheipl, "The Life and Works of St. Albert the Great," in Weisheipl, ed., *Albertus Magnus and the Sciences*, pp. 13–51; also appendix 1 to the same volume, pp. 565–77.

21. Quoted by Benedict M. Ashley, "St. Albert and the Nature of Natural Science," p. 78. On Albert's thought, besides the essays contained in Weisheipl, *Albertus Magnus and the Sciences*, see Van Steenberghen, *Aristotle in the West*, pp. 167–81; Francis J. Kovach and Robert W. Shahan, eds., *Albert the Great: Commemorative Essays*.

22. On Albert's sources, see the various essays in Weisheipl, *Albertus Magnus and the Sciences*.

23. Anybody who doubts Albertus's achievements as a field biologist should read his massive, two-volume *On Animals*, trans. Kenneth F. Kitchell Jr. and Irven Michael Resnick. See also Karen Reeds, "Albert on the Natural Philosophy of Plant Life"; see also excellent essays in Weisheipl, *Albertus Magnus and the Sciences*, on Albert as an observer of flora, fauna, and minerals. Finally, see Reeds, *Botany in Medieval and Renaissance Universities*; and, with Tomomi Kinukawa, "Natural History" (forthcoming).

24. Albert's theory of the soul is discussed by Anton C. Pegis, *St. Thomas and the Problem of the Soul in the Thirteenth Century*, chap. 3; Katharine Park, "Albert's Influence on Medieval Psychology." For Albert's views on the eternity of the world, see the introduction to Thomas Aquinas, Siger of Brabant, and Bonaventure, *On the Eternity of the World*, trans. Vollert et al., p. 13.

25. On Albert's naturalist program and the question of Noah's flood, see Albert's *De causis proprietatibus elementorum*, 1.2.9, in Albert the Great, *Opera omnia*, ed. Augustus Borgnet, 9:618–19. Cf. Lynn Thorndike, *History of Magic and Experimental Science*, 2:535.

26. The literature on Thomas Aquinas is enormous. On his life, see James A. Weisheipl, *Friar Thomas d'Aquino: His Life, Thought, and Works*. Useful summaries of his scholarly achievement (here arranged in order of ascending length) are Knowles, *Evolution of Medieval Thought*, chap. 21; Ralph McInerny, *St. Thomas Aquinas*; M.-D. Chenu, *Toward Understanding St. Thomas*; Etienne Gilson, *The Christian Philosophy of St. Thomas Aquinas*. Most discussions of Thomas's philosophy (including all of the above) have been written by modern-day Thomists, committed to Thomas's philosophy and not averse to extolling its virtues. Some of these works are marred, therefore, by a tendency to see Thomas (because he was "right") as the glorious culmination of medieval thought. For a brief account that manages to capture the essentials of Thomas's achievement while avoiding value judgments, see Julius Weinberg, *A Short History of Medieval Philosophy*, chap. 9.

27. Thomas Aquinas, *Faith, Reason, and Theology: Questions I–IV of His Commentary on the De Trinitate of Boethius*, trans. Armand Maurer, p. 48. The first two of these four questions are devoted to the legitimacy of employing philosophy in matters of faith.

28. Ibid., pp. 48–49.

29. For an excellent analysis of Thomas's position on the eternity of the world and the nature of the soul, see Van Steenberghen, *Thomas Aquinas and Radical Aristotelianism*, chaps. 1–2.

30. For a discussion of radical Aristotelianism and its consequences, see the excellent survey by Edward Grant, "Science and Theology in the Middle Ages."

31. In the judgment of Siger's foremost modern interpreter, this was not a matter of caving in to theology, but of being led by the force of Thomas's philosophical arguments to rethink and correct his own philosophical position. See Fernand Van Steenberghen, *Les oeuvres et la doctrine de Siger de Brabant*; Van Steenberghen, *Aristotle in the West*, pp. 209–29; Van Steenberghen, *Thomas Aquinas and Radical Aristotelianism*, pp. 6–8, 35–43, 89–95. It is inconceivable to me that Siger's philosophical purity would not have been compromised in some measure by the need to arrive at a theologically orthodox conclusion.

32. Boethius of Dacia, *On the Supreme Good, On the Eternity of the World, On Dreams*, pp. 36–67, quoting from p. 47 (with modest improvements).

33. For a short account of the condemnations, see Van Steenberghen, *Aristotle in the West*, chap. 9; John F. Wippel, "The Condemnations of 1270 and 1277 at Paris"; Edward Grant, "The Condemnation of 1277, God's Absolute Power, and Physical Thought in the Late Middle Ages"; and for important context, J. M. M. H. Thijssen, "What Really Happened on 7 March 1277?" For other analyses, see Pierre Duhem, *Le système du monde*, vol. 6; Roland Hissette, *Enquête sur les 219 articles condamnés à Paris le 7 mars 1277*. For a translation of the decree of 1277 and the condemned propositions, see Ralph Lerner and Muhsin Mahdi, eds., *Medieval Political Philosophy: A Sourcebook*, pp. 335–54; a selection of propositions relevant to natural philosophy appears, with introduction and running commentary, in Edward Grant, *A Source Book in Medieval Science*, pp. 45–50.

34. Pierre Duhem, *Etudes sur Léonard de Vinci*, 2:412; Anneliese Maier, *Zwischen Philosophie und Mechanik*, pp. 122–24; Edward Grant, "The Condemnation of 1277, God's Absolute Power, and Physical Thought in the Late Middle Ages," pp. 226–31; Hissette, *Enquête sur les 219 articles*, pp. 118–20.

35. J. M. M. H. Thijssen, "What Really Happened on 7 March 1277?"

36. Duhem, *Etudes sur Léonard de Vinci*, 2:412; Duhem, *Système du monde*, 6:66. For survival of the Duhem claim, qualified and weakened but still recognizable, see Edward Grant, "Late Medieval Thought, Copernicus, and the Scientific Revolution"; Grant, "Condemnation of 1277."

37. On the question of void space, see Edward Grant, *Much Ado about Nothing: Theories of Space and Vacuum from the Middle Ages to the Scientific Revolution*; also Grant, "Condemnation of 1277," pp. 232–34.

38. For an excellent historical analysis of the question of divine omnipotence, see Francis Oakley, *Omnipotence, Covenant, and Order*.

39. Transubstantiation is the process by which, according to Catholic doctrine, the eucharistic bread and wine are transformed into the body and blood of Christ.

40. See Grant, "Science and Theology in the Middle Ages," pp. 54–70; Grant, *Nicole Oresme and the Kinematics of Circular Motion*.

41. For the impact of the condemnations on natural philosophy, see Grant, "Condemnation of 1277."

42. William A. Wallace, "Thomism and Its Opponents"; Knowles, *Evolution of Medieval Thought*, chap. 24; Grant, "Condemnation of 1277." The quoted passages

come, respectively, from Marshall Clagett, *The Science of Mechanics in the Middle Ages*, p. 536 (with minor modifications); Nicole Oresme, *Le livre du ciel et du monde*, ed. and trans. A. D. Menut and A. J. Denomy, p. 369.

43. On late medieval and Renaissance Aristotelianism, see John Herman Randall, Jr., *The School of Padua and the Emergence of Modern Science*; Charles B. Schmitt, *Aristotle and the Renaissance*. The quotation (paraphrased) is taken from William J. Courtenay and Katherine H. Tachau, "Ockham, Ockhamists, and the English-German Nation at Paris, 1339–1341," p. 61.

44. On the epistemological discussions of the late thirteenth and fourteenth centuries, see Marilyn McCord Adams, *William Ockham*, 1:551–629; Eileen Serene, "Demonstrative Science." On Ockham, see also William J. Courtenay, "Ockham, William of."

45. Oakley, *Omnipotence, Covenant, and Order*, chap. 3; William J. Courtenay, "The Critique on Natural Causality in the Mutakallimun and Nominalism." For a full discussion of divine omnipotence and its implications for natural philosophy, see Courtenay's *Capacity and Volition: A History of the Distinction of Absolute and Ordained Power*; Amos Funkenstein, *Theology and the Scientific Imagination from the Middle Ages to the Seventeenth Century*, pp. 117–201.

46. And it was generally held that those exceptions were built into the universe from the moment of creation; see above, chap. 9.

47. See the essays in Courtenay, *Covenant and Causality*, esp. chap. 4: "The Dialectic of Divine Omnipotence"; and chap. 5: "The Critique on Natural Causality in the Mutakallimun and Nominalism."

48. Experimental science was not invented in the seventeenth century. Examples of experimentation as an agent of exploration or confirmation can be found in all of the cultures covered by this book. What was new in the seventeenth century was the creation of a rhetoric of experimentation and full exploitation of the possibilities of experiment in programs of scientific investigation.

49. This thesis is persuasively argued in a forthcoming book by Peter Harrison, *Adam's Encyclopaedia: The Fall of Man and the Foundations of Science*.

CHAPTER ELEVEN

1. I have decided not to employ theoretical classification schemes ("divisions of the sciences") developed during the Middle Ages, because I believe that catering to the conceptual framework of my audience for pedagogical reasons is, in this case, more important than rigid historical purity. People interested in these classification schemes should read James A. Weisheipl, "Classification of the Sciences in Medieval Thought"; Weisheipl, "The Nature, Scope, and Classification of the Sciences."

2. For a good example of twelfth-century cosmology, see Winthrop Wetherbee, trans., *The Cosmographia of Bernardus Silvestris*, with introduction and notes by Wetherbee. See also chap. 9, above. On medieval cosmology as portrayed in literary contexts, see C. S. Lewis, *The Discarded Image*.

3. For bibliography on Grosseteste, see chap. 10, n. 14.

4. Grosseteste refers to this form as "first form" or "corporeal form." For more on corporeal form, see below, chap. 12.

5. On Grosseteste's cosmology, see the excellent study by James McEvoy, *The Philosophy of Robert Grosseteste*, pp. 149–88, 369–441. For a short version, see David C. Lindberg, "The Genesis of Kepler's Theory of Light: Light Metaphysics from Plotinus to Kepler," pp. 14–17.

6. Pierre Duhem, *Le système du monde*, 10 vols. Excerpts from these ten volumes have been translated into English in Pierre Duhem, *Medieval Cosmology: Theories of Infinity, Place, Time, Void, and the Plurality of Worlds*, ed. and trans. Roger Ariew. For definitive modern treatments, see Edward Grant, *Planets, Stars, and Orbs: The Medieval Cosmos, 1200–1687*; Grant, *Much Ado about Nothing: Theories of Space and Vacuum from the Middle Ages to the Scientific Revolution*. I am also indebted, in the account that follows, to the excellent summary of medieval cosmology in Edward Grant, "Cosmology," in David C. Lindberg, ed., *Science in the Middle Ages*, and the articles collected in Grant's *Studies in Medieval Science and Natural Philosophy*. Thomas Aquinas's cosmology is nicely treated in the Blackfriars edition of his *Summa Theologiae*, vol. 10: *Cosmogony*, ed. and trans. William A. Wallace.

7. Edward Grant, "Medieval and Seventeenth-Century Conceptions of an Infinite Void Space beyond the Cosmos"; Grant, *Much Ado about Nothing*, esp. chaps. 5–6.

8. Edward Grant, "The Medieval Doctrine of Place: Some Fundamental Problems and Solutions," esp. pp. 72–79.

9. Edward Grant, "Cosmology," in David C. Lindberg, ed., *Science in the Middle Ages*, pp. 275–79; Grant, "Celestial Orbs in the Latin Middle Ages," pp. 159–62; Grant, "Science and Theology in the Middle Ages," pp. 63–64.

10. For representative medieval texts, see those included in Lynn Thorndike, ed. and trans., *The Sphere of Sacrobosco and Its Commentators*, p. 206. For discussion, see Edward Grant, "Celestial Matter: A Medieval and Galilean Cosmological Problem"; Grant, "Celestial Orbs," pp. 167–72; Grant, "Cosmology," in Lindberg, ed., *Science in the Middle Ages*, pp. 286–88.

11. James A. Weisheipl, "The Celestial Movers in Medieval Physics"; Edward Grant, *Planets, Stars, and Orbs*; Grant, "Cosmology," in Lindberg, ed., *Science in the Middle Ages*, pp. 284–86.

12. For these data, see Grant, "Cosmology," in Lindberg, ed., *Science in the Middle Ages*, p. 292; Francis S. Benjamin and G. J. Toomer, eds. and trans., *Campanus of Novara and Medieval Planetary Theory: "Theorica planetarum,"* pp. 356–63. Campanus defines a mile as the equivalent of 4,000 cubits and gives the circumference of the earth as 20,400 miles (Benjamin and Toomer, p. 147). For more on ideas of cosmic size, see Bernard R. Goldstein and Noel Swerdlow, "Planetary Distances and Sizes in an Anonymous Arabic Treatise Preserved in Bodleian MS Marsh 621"; Albert Van Helden, *Measuring the Universe: Cosmic Dimensions from Aristarchus to Halley*.

13. Pierre Duhem, *To Save the Phenomena: An Essay on the Idea of Physical Theory from Plato to Galileo* (1969); first published in French in 1908. For the ancient origins of the realism/instrumentalism distinction, see also G. E. R. Lloyd, "Saving the Appearances."

14. G. E. R. Lloyd, "Saving the Appearances."

15. Grant, "Cosmology," in Lindberg, ed., *Science in the Middle Ages*, pp. 265–68.

16. On early medieval astronomy, see the first few items in Bruce S. Eastwood, *Astronomy and Optics from Pliny to Descartes*; Eastwood, "Plinian Astronomical Diagrams in the Early Middle Ages"; Stephen C. McCluskey, *Astronomies and Cultures in Early Medieval Europe*; McCluskey, "Gregory of Tours, Monastic Timekeeping, and Early Christian Attitudes to Astronomy." See also above, chap. 9, on the Carolingians.

17. The best introductory explanation of the astrolabe is by J. D. North, "The Astrolabe."

18. On the reception of Arabic astronomy in the West, see Henri Hugonnard-Roche, "The Influence of Arabic Astronomy in the Medieval West."

19. Thorndike, *Sphere of Sacrobosco*, supplies the Latin text of this treatise, an English translation, and a very useful introduction. On medieval European astronomy, see John North, *The Norton History of Astronomy and Cosmology*, chaps. 9–10; North, "Astronomy and Astrology" (forthcoming); Olaf Pedersen, "Astronomy"; Pedersen, "Corpus Astronomicum and the Traditions of Mediaeval Latin Astronomy." My organization of this material owes much to Pedersen.

20. For the remaining planets, see Pedersen, "Astronomy," pp. 316–18; also Pedersen's translation of the *Theorica* in Edward Grant, ed., *A Source Book in Medieval Science*, pp. 451–65.

21. On the *Toledan Tables*, see G. J. Toomer, "A Survey of the Toledan Tables." On the *Alfonsine Tables*, see North, *Norton History of Astronomy and Cosmology*, pp. 217–23. Extracts from the *Alphonsine Tables* have been translated and annotated by Victor E. Thoren, in Grant, *Source Book*, pp. 465–87. Thoren and North each provide a sample calculation.

22. North, *Norton History of Astronomy and Cosmology*, pp. 234–41.

23. On Bacon, see Pierre Duhem, *Un fragment inédit de l'Opus tertium de Roger Bacon, précédé d'une étude sur ce fragment*, pp. 128–37. For a detailed analysis of Guido's theory, see Michael H. Shank, "Rings in a Fluid Heaven: The Equatorium-Driven Physical Astronomy of Guido de Marchia (fl. 1291–1310)." See also Claudia Kren, "Bernard of Verdun"; Kren, "Homocentric Astronomy in the Latin West: The De reprobatione ecentricorum et epiciclorum of Henry of Hesse."

24. On medieval astrology, see North, *Norton History of Astronomy and Cosmology*; North, *Horoscopes and History*; North, "Celestial Influence: The Major Premiss of Astrology"; North, "Astrology and the Fortunes of Churches"; Olaf Pedersen, "Astrology"; Edward Grant, "Medieval and Renaissance Scholastic Conceptions of the Influence of the Celestial Region on the Terrestrial"; Lewis, *Discarded Image*, pp. 102–10; and the papers contained in Patrick Curry, ed., *Astrology, Science, and Society: Historical Essays* (especially that of Richard Lemay, "The True Place of Astrology in Medieval Science and Philosophy").

25. On Mesopotamian astrology, see Francesca Rochberg, *The Heavenly Writing: Divination, Horoscopy, and Astronomy in Mesopotamian Culture*; B. L. van der Waerden and Peter Huber, *Science Awakening II: The Birth of Astronomy*, chap. 5; Richard Olson,

Science Deified and Science Defied: The Historical Significance of Science in Western Culture, pp. 34–56.

26. Aristotle is quoted from *Meteorologica*, 1.2, trans. E. W. Webster, in *The Complete Works of Aristotle*, ed. Jonathan Barnes, p. 555.

27. Ptolemy, *Tetrabiblos*, 1.2, ed. and trans. F. E. Robbins, pp. 5–13 (with one change of wording). On Ptolemy's astrology, see also Jim Tester, *History of Western Astrology*, chap. 4; Long, "Astrology: Arguments Pro and Contra," pp. 178–83.

28. Augustine, *City of God*, V.6, 2:157. On Augustine's attitude toward astrology see also his *Confessions*, IV.3 and VII.6; Theodore Otto Wedel, *The Mediaeval Attitude toward Astrology*, pp. 20–24; Joshua D. Lipton, *The Rational Evaluation of Astrology in the Period of Arabo-Latin Translation, ca. 1126–1187 A.D.*, pp. 133–35; Tester, *History of Western Astrology*, chap. 5.

29. Wedel, *Mediaeval Attitude toward Astrology*, chap. 2.

30. *The Didascalicon of Hugh of St. Victor: A Medieval Guide to the Arts*, trans. Jerome Taylor, p. 68. For the Latin text of the second quotation, see Charles S. F. Burnett, "What Is the Experimentarius of Bernardus Silvestris? A Preliminary Survey of the Material." For the third quotation (possibly from William of Conches), see Lipton, *Rational Evaluation of Astrology*, p. 145. Lipton's study contains a very useful analysis of twelfth-century astrology; see also Wedel, *Mediaeval Attitude toward Astrology*, pp. 60–63.

31. Tester, *History of Western Astrology*, pp. 152–53.

32. Lemay, *Abu Ma'shar*, pp. 41–132; David Pingree, "Abū Maʿshar al-Balkhī."

33. See, for example, Nancy G. Siraisi, *Taddeo Alderotti and His Pupils: Two Generations of Italian Medical Learning*, pp. 140–45.

34. G. W. Coopland, *Nicole Oresme and the Astrologers*, pp. 53–57. On Oresme, see also Stefano Caroti, "Nicole Oresme's Polemic against Astrology in His 'Quodlibeta.'"

35. The secondary rainbow is produced in the same manner, except that the radiation responsible for it undergoes two internal reflections. For a translation of Theodoric's treatise *On the Rainbow*, see Edward Grant, ed., *A Source Book in Medieval Science*, pp. 435–41. For analysis, see Carl B. Boyer, *The Rainbow: From Myth to Mathematics*, chaps. 3–5. Medieval meteorology is treated in John Kirtland Wright, *The Geographical Lore of the Time of the Crusades: A Study in the History of Medieval Science and Tradition in Western Europe*, pp. 166–81; Nicholas H. Steneck, *Science and Creation in the Middle Ages: Henry of Langenstein (d. 1397) on Genesis*, pp. 84–87.

36. On the origin of the "flat earth" myth, see above, chap. 7, n. 42.

37. For a sketch of medieval geography, see Lewis, *Discarded Image*, pp. 139–46, from whom this point and the terminology for expressing it have been borrowed. For a full-length survey, see Wright, *Geographical Lore*. On cartography in particular, see David Woodward, "Medieval Mappaemundi."

38. William H. Stahl, *Roman Science*, pp. 115–19, 221–22.

39. On types of medieval maps and their functions, see the articles in *History of Cartography*, ed. J. B. Harley and David Woodward, vol. 1. On the two maps mentioned here, see Woodward "Medieval Mappaemundi," pp. 290, 310.

40. On *mappaemundi*, see the thorough study by Woodward, "Medieval Mappaemundi."

41. On portolan charts and Ptolemy's mapping techniques, see Tony Campbell, "Portolan Charts from the Late Thirteenth Century to 1500"; O. A. W. Dilke, "The Culmination of Greek Cartography in Ptolemy."

42. Marshall Clagett, *The Science of Mechanics in the Middle Ages*, pp. 594–99.

43. Nicole Oresme, *Le livre du ciel et du monde*, pp. 525, 531, with a variety of improvements and corrections. For analysis, see Marshall Clagett, *The Science of Mechanics in the Middle Ages*, pp. 583–88; Edward Grant, *Physical Science in the Middle Ages*, pp. 63–70.

44. Oresme, *Livre du ciel*, p. 537 (with modified punctuation).

45. Ibid., pp. 537–39.

CHAPTER TWELVE

1. On Aristotelian natural philosophy, see chap. 3, above, and the citations provided there. On subsequent developments within the Aristotelian tradition, see Helen S. Lang, *Aristotle's Physics and Its Medieval Varieties*; Harry Austryn Wolfson, *Crescas' Critique of Aristotle: Problems of Aristotle's "Physics" in Jewish and Arabic Philosophy*, Leclerc, *Nature of Physical Existence*; Norma E. Emerton, *The Scientific Reinterpretation of Form*, chaps. 2–3.

2. On the Aristotelian conception of "nature," see Sarah Waterlow, *Nature, Change, and Agency in Aristotle's "Physics"*; Lang, *Aristotle's Physics and Its Medieval Varieties*, chap. 1.

3. G. E. R. Lloyd, *Aristotle*, pp. 164–75; Anneliese Maier, "The Theory of the Elements and the Problem of Their Participation in Compounds."

4. Chap. 11, above. See also Wolfson, *Crescas' Critique*, pp. 580–90; Arthur Hyman, "Aristotle's 'First Matter' and Avicenna's and Averroes' 'Corporeal Form.'" On the significance of the idea of corporeal form within medieval Christian thought, see D. E. Sharp, *Franciscan Philosophy at Oxford in the Thirteenth Century*, pp. 186–89.

5. For a challenging discussion of Greek and medieval conceptions of matter, see the articles collected in Ernan McMullin, ed., *The Concept of Matter in Greek and Medieval Philosophy*; also Leclerc, *Nature of Physical Existence*, chaps. 8–9.

6. On the Aristotelian doctrine of mixtio, see Friedrich Solmsen, *Aristotle's System of the Physical World*, chap. 19; Waterlow, *Nature, Change, and Agency*, pp. 82–85; Emerton, *Scientific Reinterpretation of Form*, chap. 3.

7. On the medieval doctrine of mixtio, see E. J. Dijksterhuis, *The Mechanization of the World Picture*, pp. 200–204; Emerton, *Scientific Reinterpretation of Form*, pp. 77–85; and most usefully Anneliese Maier, *An der Grenze von Scholastik und Naturwissenschaft*, 2d ed., pp. 3–140, the introductory portion of which appears in English as "Theory of the Elements."

8. On minima, see Dijksterhuis, *Mechanization*, pp. 205–9; Emerton, *Scientific Reinterpretation of Form*, pp. 85–93.

9. The recent history of medieval alchemy owes a great deal to the research and publications of William Newman, who has drawn the many fragments of medieval alchemical theory into a coherent and convincing story. What I wish to make clear is that, because he is almost my only source, I am in the awkward position of offering

a miniature version of *his* story in *my* words—reducing a book-length argument to a few paragraphs. For Newman's writings on the subject, see under his name in the bibliography. For other useful sources, see Robert Halleux, *Les textes alchimiques*; Halleux, "Alchemy"; Manfred Ullmann, "Al-Kīmiyā'"; Georges C. Anawati, "Arabic Alchemy"; E. J. Holmyard, *Alchemy*.

10. The sulphur and mercury in question were not the common minerals by those names, but the pure essences thought to provide the various qualities needed to produce metals, sometimes referred to as "philosophical sulphur" and "philosophical mercury." Philosophical sulphur was frequently identified as the active, spiritual principle; philosophical mercury as the passive, material principle. On the origins of the sulphur-mercury theory, see E. J. Holmyard, *Alchemy*, pp. 75, 145–46. On its medieval use, see Pearl Kibre, "Albertus Magnus on Alchemy."

11. Newman, "Medieval Alchemy."

12. Newman, "Medieval Alchemy." On alchemical apparatus and processes, see also Holmyard, *Alchemy*, chap. 4. On chemical/alchemical apparatus, see R. G. W. Anderson, "The Archaeology of Chemistry"; Newman, "Alchemy, Assaying, and Experiment."

13. The identification of Paul of Taranto as author of the *Summa perfectionis* (a long-standing puzzle) is the achievement of William Newman, "The *Summa perfectionis* and Late Medieval Alchemy." Also Newman, "Experimental Corpuscular Theory in Aristotelian Alchemy: From Geber to Sennert"; Newman, *Atoms and Alchemy*.

14. But see Newman's defense, *Atoms and Alchemy*, pp. 67–68.

15. Quotation from Newman, *Atoms and Alchemy*, p. 41. The same point is made in many other Newman publications.

16. Ibid., pp. 23–24. On later alchemy see Newman, ibid.; Allen G. Debus, *Man and Nature in the Renaissance*, chap. 2; Debus, *The Chemical Philosophy: Paracelsian Science and Medicine in the Sixteenth and Seventeenth Centuries*.

17. The word "local" in "local motion" does not bear its usual modern connotation of "nearby" or "in the vicinity." Rather, it borrows its meaning from its Latin root, "locus," simply meaning "place." Thus "local motion" is change of place.

18. *Physics*, III.1, 200b14–15. For a recent survey of medieval theories of motion, see Walter Roy Laird, "Change and Motion" (forthcoming).

19. On the nature of motion, see John E. Murdoch and Edith D. Sylla, "The Science of Motion," pp. 213–22. See also the works of Anneliese Maier: *Zwischen Philosophie und Mechanik*, chaps. 1–3; *Die Vorläufer Galileis im 14. Jahrhundert*, 2d ed., chap. 1; and the English translation of the latter, appearing in *On the Threshold of Exact Science* as "The Nature of Motion."

20. John E. Murdoch, "The Development of a Critical Temper: New Approaches and Modes of Analysis in Fourteenth-Century Philosophy, Science, and Theology," pp. 60–61; Murdoch and Sylla, "Science of Motion," pp. 216–17; Maier, "The Nature of Motion," pp. 30–31.

21. Murdoch and Sylla, "Science of Motion," pp. 217–18; Maier, "The Nature of Motion," pp. 33–38; Maier, *Zwischen Philosophie und Mechanik*, pp. 121–31.

22. For more on Aristotle's mathematization of nature, see Edward Hussey, "Aristotle's Mathematical Physics: A Reconstruction."

23. Marshall Clagett, *The Science of Mechanics in the Middle Ages*, pp. 163–86.

24. Any description of motion that includes the concept of "force" (as in Newton's second law of motion: $F = ma$) is a "dynamic" claim. On Gerard, see ibid., pp. 184–97; Clagett, "The *Liber de motu* of Gerard of Brussels and the Origins of Kinematics in the West"; Murdoch and Sylla, "Science of Motion," pp. 222–23; Wilbur R. Knorr, "John of Tynemouth alias John of London: Emerging Portrait of a Singular Medieval Mathematician," pp. 312–22.

25. But velocity was treated as a scalar rather than a vector quantity. That is, it had magnitude, but was indifferent as to direction.

26. Clagett, *Science of Mechanics*, chap. 4.

27. Two theories competed to explain in physical terms how intensification and remission occur. The "addition/subtraction" theory claimed that a form is intensified by addition of a new part of form, remitted by subtraction of a part of the original form. The "replacement" theory held that the original form is annihilated and replaced by a new form of greater or lesser intensity. For more on this problem, see Edith D. Sylla, "Medieval Concepts of the Latitude of Forms: The Oxford Calculators," pp. 230–33; Murdoch and Sylla, "Science of Motion," pp. 231–33. On the intensification and remission of qualities in general, see also Clagett, *Science of Mechanics*, pp. 205–6, 212–15; Murdoch and Sylla, "Science of Motion," pp. 233–37.

28. This notion goes back at least to Galen; see Clagett, *Giovanni Marliani and Late Medieval Physics*, pp. 34–36.

29. Clagett, *Science of Mechanics*, pp. 212–13.

30. If the equality of the triangle and the rectangle is not evident by inspection, draw a diagonal line from B to D, thus dividing rectangle BCDE into two equal triangles. Note then that both rectangle ACDF and triangle ACG have been subdivided into four small, equal triangles.

31. On the geometrical representation of qualities, see Marshall Clagett, ed. and trans., *Nicole Oresme and the Medieval Geometry of Qualities and Motions*, pp. 50–121; Clagett, *Science of Mechanics*, chap. 6; Murdoch and Sylla, "Science of Motion," pp. 237–41. On the Merton rule, see Clagett, *Science of Mechanics*, chap. 5.

32. Galileo, *Two New Sciences*, trans. Stillman Drake, p. 165.

33. On this exceedingly technical question, see Richard Sorabji, *Matter, Space, and Motion: Theories in Antiquity and Their Sequel*, chap. 13; James A. Weisheipl, *Nature and Motion in the Middle Ages*, chaps. 4–5. For the Aristotelian texts, see Aristotle's *Physics*, II.1, VII.1, and VIII.4.

34. On Philoponus, see Clagett, *Science of Mechanics*, pp. 508–10. For more recent studies, which do full justice to the radical Neoplatonic character of Philoponus's attack on Aristotelian dynamics, see Michael Wolff, "Philoponus and the Rise of Preclassical Dynamics"; Lang, *Aristotle's Physics and Its Medieval Varieties*, chap. 5; Sorabji, *Matter, Space, and Motion*, chap. 14.

35. For the latest word on this subject, see Fritz Zimmermann, "Philoponus' Impetus Theory in the Arabic Tradition"; Sorabji, *Matter, Space, and Motion*, pp. 237–38. See also Clagett, *Science of Mechanics*, pp. 510–17.

36. On impetus theory, see Clagett, *Science of Mechanics*, pp. 521–25 (quotation from p. 524, with one change of wording); Lang, *Aristotle's Physics and Its Medieval Varieties*, pp. 164–72; Anneliese Maier, "The Significance of the Theory of Impetus for Scholastic Natural Philosophy." Unknown to Buridan, Philoponus had anticipated his suggestion that impetus or impressed force could be used to explain celestial motion; see Sorabji, *Matter, Space, and Motion*, p. 237.

37. Morris R. Cohen and I. E. Drabkin, *A Source Book in Greek Science*, p. 220, with several emendations. See also Clagett, *Science of Mechanics*, pp. 433–35, 546–47.

38. Whether Galileo ever performed the famous experiment from the tower of Pisa is a matter of scholarly dispute. (He never claimed to have done so.) See Lane Cooper, *Aristotle, Galileo, and the Tower of Pisa.*

39. For a good dose of serious medieval mathematical physics, see H. Lamar Crosby, Jr., ed. and trans., *Thomas of Bradwardine, His "Tractatus de Proportionibus": Its Significance for the Development of Mathematical Physics*, pp. 32–36, on the refutation of these three relationships.

40. Crosby, *Thomas of Bradwardine*, pp. 38–45. The classic analyses of Bradwardine's "function" and his predecessors are by Maier, *Die Vorläufer Galileis*, pp. 81–110 (partially translated in Maier's *On the Threshold of Exact Science*, pp. 61–75); Ernest A. Moody, "Galileo and Avempace: The Dynamics of the Leaning Tower Experiment"; Clagett, *Science of Mechanics*, chap. 7.

41. A. G. Molland, "The Geometrical Background to the 'Merton School,'" esp. pp. 116–21 (p. 120, for the quotation); Murdoch and Sylla, "Science of Motion," pp. 225–26; Edith D. Sylla, "Compounding Ratios: Bradwardine, Oresme, and the First Edition of Newton's *Principia.*"

42. Murdoch and Sylla, "Science of Motion," pp. 227–30; Clagett, *Marliani*, chap. 6; Clagett, *Science of Mechanics*, p. 443. On Swineshead's work, see John E. Murdoch and Edith D. Sylla, "Swineshead, Richard." On Oresme, see Nicole Oresme, *"De proportionibus proportionum" and "Ad pauca respicientes."*

43. On medieval optics in general, see David C. Lindberg, *Theories of Vision from al-Kindi to Kepler*; Lindberg, "The Science of Optics"; Lindberg, "Optics, Western European"; the papers collected in Lindberg's *Studies in the History of Medieval Optics*; Lindberg and Katherine H. Tachau, "The Science of Light and Color: Seeing and Knowing"; the optical papers contained in Bruce S. Eastwood, *Astronomy and Optics from Pliny to Descartes*; A. Mark Smith, "Getting the Big Picture in Perspectivist Optics." For the Greek and Islamic background, see chaps. 5, 8, above.

44. "Alhacen" is the correct medieval spelling of a name that has been misspelled ("Alhazen") since Friedrich Risner invented that misspelling in the sixteenth century. The full twentieth-century transliteration of the name is Abū ʿAlī al-Ḥasan ibn al-Ḥasan ibn al-Haytham; and Alhacen (with a soft "c") is the medieval Latin transliteration of the middle segments: al-Ḥasan. All extant medieval manuscripts of his great *Optics* spell the name with a "c." See David C. Lindberg, *Roger Bacon and the Origins of "Perspectiva" in the Middle Ages*, p. xxiii, n. 75.

45. It could be argued that extramission of rays was a necessary feature of mathematical theories of vision, for it was the conical emanation of rays from the eye

that defined the visual cone, which in turn made the mathematical analysis of vision possible.

46. Aristotle's *Meteorology*, III.4–5; Lindberg, *Theories of Vision*, p. 217, n. 39.

47. On Alhacen's optical achievement, see the definitive translation and commentary by A. I. Sabra, ed. and trans., *The Optics of Ibn al-Haytham: Books I–III, On Direct Vision*. For shorter accounts, see Sabra, "Ibn al-Haytham"; Sabra, "Form in Ibn al-Haytham's Theory of Vision"; Lindberg, *Theories of Vision*, chap. 4.

48. Above, chap. 5.

49. For the Latin text of the translation of the first three books of Alhacen's *Optics*, accompanied by an English translation and excellent notes and commentary, see A. Mark Smith, *Alhacen's Theory of Visual Perception*, 2 vols. (more forthcoming). On the Western reception of Greek and Islamic optics, see (in addition to sources already cited) David C. Lindberg, "Alhazen's Theory of Vision and Its Reception in the West"; Lindberg, "Roger Bacon and the Origins of Perspectiva in the West."

50. On Bacon's optics, see David C. Lindberg, ed. and trans., *Roger Bacon's Philosophy of Nature: A Critical Edition, with English Translation, Introduction, and Notes, of "De multiplicatione specierum" and "De speculis comburentibus"*; Lindberg, *Theories of Vision*, chap. 6; Lindberg, *Roger Bacon and the Origins of "Perspectiva" in the Middle Ages*.

51. On Bacon's Neoplatonism, see David C. Lindberg, "Roger Bacon on Light, Vision, and the Universal Emanation of Force"; Lindberg, "The Genesis of Kepler's Theory of Light: Light Metaphysics from Plotinus to Kepler," pp. 12–23; Lindberg, *Roger Bacon's Philosophy of Nature*, pp. liii–lxxi.

52. Lindberg and Tachau, "The Science of Light and Color: Seeing and Knowing"; Lindberg, *Theories of Vision*, chap. 9; Katherine H. Tachau, *Vision and Certitude in the Age of Ockham: Optics, Epistemology, and the Foundations of Semantics, 1250–1345*. Pecham's *Perspectiva communis* is available in David C. Lindberg, ed. and trans., *John Pecham and the Science of Optics*. A project to translate Witelo's massive *Perspectiva* is under way. Three volumes have been published: *Witelonis Perspectivae liber primus* and *Witelonis Perspectivae liber secundus et liber tertius*, ed. and trans. Sabetai Unguru; *Witelonis Perspectivae liber quintus*, ed. and trans. A. Mark Smith.

CHAPTER THIRTEEN

1. For the basic framework of this chapter, I am indebted to Nancy G. Siraisi, *Medieval and Early Renaissance Medicine: An Introduction to Knowledge and Practice*; Michael McVaugh, "Medicine, History of." Other important sources include McVaugh, *Medicine before the Plague: Practitioners and Their Patients in the Crown of Aragon 1285–1345*; Faye Getz, *Medicine in the English Middle Ages*; Danielle Jacquart and Claude Thomasset, *Sexuality and Medicine in the Middle Ages*; Katharine Park, "Medicine and Society in Medieval Europe, 500–1500." For an excellent set of translated medical texts (selected, annotated, and in some cases translated by Michael McVaugh), see Edward Grant, ed., *A Source Book in Medieval Science*, pp. 700–808. For medical illustrations, see Loren C. MacKinney, *Medical Illustrations in Medieval Manuscripts*;

Peter M. Jones, *Medieval Medical Miniatures*; Marie-Jose Imbault-Huart, *La médecine au moyen âge à travers les manuscrits de la Bibliothèque Nationale.*

2. On early medieval medicine, see especially Vivian Nutton, "Early Medieval Medicine and Natural Science"; John M. Riddle, "Theory and Practice in Medieval Medicine"; Henry E. Sigerist, "The Latin Medical Literature of the Early Middle Ages"; Edward Kealey, *Medieval Medicus: A Social History of Anglo-Norman Medicine*; Linda E. Voigts, "Anglo-Saxon Plant Remedies and the Anglo-Saxons"; M. L. Cameron, "The Sources of Medical Knowledge in Anglo-Saxon England"; Siraisi, *Medieval and Early Renaissance Medicine*, pp. 5–13; and (old but still useful) Loren C. MacKinney, *Early Medieval Medicine, with Special Reference to France and Chartres.*

3. See William D. Sharpe, ed. and trans., *Isidore of Seville: The Medical Writings*; Celsus, *De medicina, with an English Translation.*

4. On Dioscorides, see John M. Riddle, *Dioscorides on Pharmacy and Medicine*; Riddle, "Dioscorides." See the latter, pp. 125–33, on *Ex herbis femininis* (a work not restricted to remedies for feminine ailments). On medical recipes, see also Voigts, "Anglo-Saxon Plant Remedies"; Sigerist, "Latin Medical Literature," pp. 136–41; MacKinney, *Early Medieval Medicine*, pp. 31–38.

5. MacKinney, *Early Medieval Medicine*, pp. 47–49, 61–73.

6. The relevant passage from Cassiodorus's *Institutiones* is quoted by MacKinney, *Early Medieval Medicine*, p. 51. On monastic medicine more generally, see ibid., pp. 50–58.

7. See especially Darrel Amundsen and Gary B. Ferngren, "The Early Christian Tradition"; Amundsen, "The Medieval Catholic Tradition"; Amundsen, "Medicine and Faith in Early Christianity"; Siraisi, *Medieval and Early Renaissance Medicine*, pp. 7–9.

8. Amundsen, "Medieval Catholic Tradition," p. 79; Grant, *Source Book*, pp. 773–74.

9. Quoted by Siraisi, *Medieval and Early Renaissance Medicine*, p. 14, from *Bernard of Clairvaux, Letters*, no. 388, trans. Bruno Scott James (Chicago: Regnery, 1953), pp. 458–59.

10. Amundsen, "Medicine and Faith in Early Christianity," pp. 333–49 (p. 338 for the quotation from Basil). On Tertullian, see *De corona*, 8, and *Ad nationes*, 11.5, in Alexander Roberts and James Donaldson, eds., *The Ante-Nicene Fathers*, rev. by A. Cleveland Coxe (Grand Rapids: Eerdmans, 1986), 3:97, 134. See also Siraisi, *Medieval and Early Renaissance Medicine*, p. 9.

11. Bede, *Ecclesiastical History of the English People*, IV.31, in Bede, *Baedae opera historica*, trans. J. E. King, 2:191–93 (with changes). On the cult of saints, see the brilliant study by Peter Brown, *The Cult of Saints: Its Rise and Function in Latin Christianity*; also Amundsen, "Medieval Catholic Tradition," pp. 79–82. On miraculous cures, see Ronald C. Finucane, *Miracles and Pilgrims: Popular Beliefs in Medieval England*, esp. chaps. 4–5.

12. On Islamic medicine, see Emily Savage-Smith, "Medicine in Medieval Islam"; Michael W. Dols, *Medieval Islamic Medicine: Ibn Ridwān's Treatise "On the Prevention of Bodily Ills in Egypt"*; Manfred Ullmann, *Islamic Medicine*; Franz Rosenthal, "The Physician in Medieval Muslim Society"; articles by Max Meyerhof, collected in his

Studies in Medieval Arabic Medicine: Theory and Practice; Siraisi, *Medieval and Early Renaissance Medicine*, pp. 11–13. An older reference source, still useful, is Lucien Leclerc, *Histoire de la médecine arabe*.

13. The classic work on Salerno is Paul Oskar Kristeller, "The School of Salerno: Its Development and Its Contribution to the History of Learning." See also McVaugh, "Medicine," pp. 247–49; Morris Harold Saffron, *Maurus of Salerno: Twelfth-Century "Optimus Physicus" with His "Commentary on the Prognostics of Hippocrates."*

14. Danielle Jacquart, "The Influence of Arabic Medicine in the Medieval West"; Michael McVaugh, "Constantine the African"; McVaugh, "Medicine," pp. 248–49; also above, chap. 9.

15. Katharine Park, "Medical Practice"; Park, *Doctors and Medicine in Early Renaissance Florence*, pp. 58–76; Siraisi, *Medieval and Early Renaissance Medicine*, pp. 17–21; Kealey, *Medieval Medicus*, chap. 2.

16. Park, *Doctors and Medicine*, pp. 54–58.

17. On women healers, Monica H. Green, ed. and trans., *The Trotula: A Medieval Compendium of Women's Medicine*; Green, *Women's Healthcare in the Medieval West: Texts and Contexts*; Green, "Women's Medical Practice and Medical Care in Medieval Europe"; John Benton, "Trotula, Women's Problems, and the Professionalization of Medicine in the Middle Ages"; Siraisi, *Medieval and Early Renaissance Medicine*, pp. 27, 34, 45–46; Edward J. Kealey, "England's Earliest Women Doctors." On Jewish practitioners, see Elliot N. Dorff, "The Jewish Tradition"; Luis García Ballester, Lola Ferre, and Edward Feliu, "Jewish Appreciation of Fourteenth-Century Scholastic Medicine."

18. On medicine in the universities, see Siraisi, *Medieval and Early Renaissance Medicine*, chap. 3; McVaugh, "Medicine," pp. 249–52; Vern L. Bullough, *The Development of Medicine as a Profession: The Contribution of the Medieval University to Modern Medicine*, esp. chap. 3; Faye M. Getz, "The Faculty of Medicine before 1500."

19. McVaugh, "Medicine," p. 247.

20. On the curriculum, see Siraisi, *Medieval and Early Renaissance Medicine*, pp. 65–77; Siraisi, *Taddeo Alderotti and His Pupils: Two Generations of Italian Medical Learning*, chaps. 4–5; Siraisi, *Avicenna in Renaissance Italy: The "Canon" and Medical Teaching in Italian Universities after 1500*, chap. 3; Getz, "Faculty of Medicine"; McVaugh, "Medicine," pp. 248–52. A good idea of the content of the *Articella* collection can be obtained from the annotated translation of its most basic component, the *Isagoge* of Ḥunayn ibn Isḥāq (Johannitius), contained in Grant, *Source Book*, pp. 705–15.

21. The numerical data presented here have been drawn from Siraisi, *Medieval and Early Renaissance Medicine*, pp. 63–64. For quantitative data from Oxford, see Getz, "Faculty of Medicine."

22. Faye M. Getz, "Charity, Translation, and the Language of Medical Learning in Medieval England"; Getz, *Healing and Society in Medieval England*.

23. Siraisi, *Medieval and Early Renaissance Medicine*, pp. 101–6; Danielle Jacquart, "Anatomy, Physiology, and Medical Theory"; Grant, *Source Book*, pp. 705–9; L. J. Rather, "The 'Six Things Non-natural': A Note on the Origins and Fate of a Doctrine and a Phrase."

24. On the treatment of disease, see Siraisi, *Medieval and Early Renaissance Medicine*, chap. 5; Jacquart, "Anatomy, Physiology, and Medical Theory"; Grant, *Source Book*, pp. 775–91.

25. On drug therapy, see Siraisi, *Medieval and Early Renaissance Medicine*, pp. 141–49 (p. 148 on the application of pig dung for nosebleed); Jones, *Medieval Medical Miniatures*, chap. 4.

26. Translated by Michael McVaugh in Grant, *Source Book*, p. 788. This list of curative properties is followed by the recipe for theriac. On theriac, see also McVaugh, "Theriac at Montpellier."

27. See, for example, Michael McVaugh, "Arnald of Villanova and Bradwardine's Law"; McVaugh, "Quantified Medical Theory and Practice at Fourteenth-Century Montpellier." Also McVaugh's introduction to *Arnald de Villanova, Opera medica omnia*, vol. 2: *Aphorismi de gradibus*.

28. Translated by Michael McVaugh, in Grant, *Source Book*, p. 749. On urinalysis, see MacKinney, *Medical Illustrations*, pp. 9–14; Jones, *Medieval Medical Miniatures*, pp. 58–60.

29. Translated by Michael McVaugh in Grant, *Source Book*, p. 746. On pulse, see also MacKinney, *Medical Illustrations*, pp. 15–19.

30. For medical astrology, see Siraisi, *Taddeo Alderotti*, pp. 140–45; Siraisi, *Medieval and Early Renaissance Medicine*, pp. 68, 111–12, 123, 128–29, 134–36, 149–52; Jones, *Medieval Medical Miniatures*, pp. 69–74.

31. On the black death, see McVaugh, "Medicine," p. 253; Siraisi, *Medieval and Early Renaissance Medicine*, pp. 128–29; Grant, *Source Book*, pp. 773–74; Daniel Williman, ed., *The Black Death: The Impact of the Fourteenth-Century Plague*, pp. 9–22; David Herlihy, *The Black Death and the Transformation of the West*; Philip Ziegler, *The Black Death*.

32. Siraisi, *Medieval and Early Renaissance Medicine*, chap. 6; MacKinney, *Medical Illustrations*, chap. 8. On Roger Frugard, see Siraisi, *Medieval and Early Renaissance Medicine*, pp. 162–66; MacKinney, *Medical Illustrations*, passim (under the name "Rogerius"). On Guy de Chauliac, see Vern L. Bullough, "Chauliac, Guy de."

33. Linda E. Voigts and Michael R. McVaugh, *A Latin Technical Phlebotomy and Its Middle English Translation*; MacKinney, *Medical Illustrations*, pp. 55–61.

34. MacKinney, *Medical Illustrations*, pp. 80–81.

35. Stupefactives capable of putting the patient to sleep were available, but it is not clear how widely they were used; see Linda E. Voigts and Robert P. Hudson, "'A drynke that men callen dwale to make a man to slepe whyle men kerven him': A Surgical Anesthetic from Late Medieval England."

36. Vern L. Bullough, "Mondino de' Luzzi"; Bullough, *Development of Medicine as a Profession*, pp. 61–65; Siraisi, *Medieval and Early Renaissance Medicine*, pp. 86–97; Jacquart, "Anatomy, Physiology, and Medical Theory."

37. Quoted from Bullough, *Development of Medicine as a Profession*, p. 64, with several changes of wording.

38. For an excellent account of social attitudes toward dissection in the Renaissance (attitudes that also have resonances in the later Middle Ages), see Katharine

Park, "The Criminal and the Saintly Body: Autopsy and Dissection in Renaissance Italy."

39. On the origins of the hospital in the narrower sense, see especially Timothy S. Miller, *The Birth of the Hospital in the Byzantine Empire*; Miller, "The Knights of Saint John and the Hospitals of the Latin West"; Michael W. Dols, "The Origins of the Islamic Hospital: Myth and Reality"; Kealey, *Medieval Medicus*, chaps. 4–5.

40. Miller, "Knights of Saint John," pp. 723–25.

41. It seems doubtful that the last word has been said on this complex subject. I have followed Dols, "Origins of the Islamic Hospital," pp. 382–84; Miller, "Knights of Saint John," pp. 717–23, 726–33. On the Barmak family, see above, chap. 8.

42. On hospitals in the West, see Katharine Park, "Medical Practice." The quotation is from C. H. Talbot, *Medicine in Medieval England*, pp. 177–78.

43. For a good introduction to medieval botanical knowledge, herbals in particular, see Jerry Stannard, "Medieval Herbals and Their Development"; Stannard, "Natural History," pp. 443–49; Karen Reeds and Tomomi Kinukawa, "Natural History."

44. On Albert's botanical knowledge, see Karen Reeds, "Albert on the Natural Philosophy of Plant Life"; Jerry Stannard, "Albertus Magnus and Medieval Herbalism." On Albert's biological studies, see also above, chap. 10.

45. On medieval zoology, see Stannard, "Natural History," pp. 432–43. On Albert's contributions, see Joan Cadden, "Albertus Magnus' Universal Physiology: The Example of Nutrition"; Luke Demaitre and Anthony A. Travill, "Human Embryology and Development in the Works of Albertus Magnus"; Robin S. Oggins, "Albertus Magnus on Falcons and Hawks." The full text of *De animalibus* has been translated by Kenneth F. Kitchell, Jr., and Irven Michael Resnick, *Albertus Magnus On Animals: A Medieval "Summa Zoologica,"* 2 vols. Selections from Albert's *De animalibus* appear in Albert the Great, *Man and the Beasts, "De animalibus,"* books 22–26, trans. James J. Scanlan.

46. Charles Homer Haskins, "Science at the Court of the Emperor Frederick II"; Haskins, "The De arte venandi cum avibus of Frederick II."

47. On medieval bestiaries and the *Physiologus*, see the introduction to Michael J. Curley, trans., *Physiologus*, pp. ix–xxxviii; Stannard, "Natural History," pp. 430–43; C. S. Lewis, *The Discarded Image*, pp. 146–52; Willene B. Clark and Meradith T. McMunn, eds., *Beasts and Birds of the Middle Ages*.

48. All of the examples in this paragraph are from T. H. White, trans., *The Bestiary: A Book of Beasts*; quotations from pp. 108–10, 52.

49. See the wonderfully illuminating discussion of sixteenth-century zoological literature (relevant also to the medieval scene) by William B. Ashworth, Jr., "Natural History and the Emblematic World View," pp. 304–6.

CHAPTER FOURTEEN

1. For broad discussions of the continuity debate, see David C. Lindberg, "Conceptions of the Scientific Revolution from Bacon to Butterfield"; Bruce S. Eastwood, "On the Continuity of Western Science from the Middle Ages"; Edward Grant, *The Foundations of Modern Science in the Middle Ages*, pp. 168–206.

2. Francis Bacon, *New Organon*, 4:77; François Marie Arouet de Voltaire, *Works*, trans. T. Smollett, T. Francklin, et al., 39 vols. (London: J. Newbery et al. 1761–74), 1:82.

3. Jacob Burckhardt, *The Civilization of the Renaissance in Italy*, pp. 371; Andrew Dickson White, *History of the Warfare of Science with Theology in Christendom*, 1:375; Charles Freeman, *The Closing of the Western Mind: The Rise of Faith and the Fall of Reason*, p. xviii.

4. Pierre Duhem, *Etudes sur Léonard de Vinci*; Duhem, *Le système du monde*; Duhem, *Les origines de la statique*. On Duhem see also R. N. D. Martin, "The Genesis of a Mediaeval Historian"; Stanley Jaki, *Uneasy Genius: The Life and Work of Pierre Duhem*.

5. Charles Homer Haskins, *Studies in the History of Mediaeval Science*; Haskins, *The Renaissance of the Twelfth Century*; Lynn Thorndike, *A History of Magic and Experimental Science*; Thorndike, *Science and Thought in the Fifteenth Century*.

6. On Clagett, see the bibliography, below, and the list of his publications supplied in an appendix to Edward Grant and John E. Murdoch, eds., *Mathematics and Its Applications to Science and Natural Philosophy in the Middle Ages*, pp. 325–28. On Maier, see Anneliese Maier, "The Achievements of Late Scholastic Natural Philosophy"; Steven D. Sargent, introduction to Maier, *On the Threshold of Exact Science*, pp. 11–16; John E. Murdoch and Edith D. Sylla, "Anneliese Maier and the History of Medieval Science."

7. A. C. Crombie, *Augustine to Galileo: The History of Science A.D. 400–1650*, p. 273. This book went through various revisions—also a change of name, when it appeared in 1959 as *Medieval and Early Modern Science*. Also Crombie, *Robert Grosseteste and the Origins of Experimental Science 1100–1700*, especially pp. 9–10. For an assessment of Crombie's achievement, see Eastwood, "On the Continuity of Western Science." For Koyré, see "The Origins of Modern Science: A New Interpretation" and "Galileo and Plato," in Koyré's *Metaphysics and Measurement: Essays in the Scientific Revolution*. For another voice in the debate, see A. Rupert Hall, "On the Historical Singularity of the Scientific Revolution of the Seventeenth Century."

8. But Edward Grant comes close in various writings, including *The Foundations of Modern Science in the Middle Ages*. See also William R. Newman for the restricted realms of alchemy and experimentalism, in his *Atoms and Alchemy: Chymistry and the Experimental Origins of the Scientific Revolution*.

9. On the subject of scientific revolutions, see Thomas S. Kuhn, *The Structure of Scientific Revolutions*; Paul Thagard, *Conceptual Revolutions*; and I. Bernard Cohen, *Revolution in Science*. On possible meanings of the word "revolution," see the latter, chap. 18.

10. Pierre Duhem, *To Save the Phenomena: An Essay on the Idea of Physical Theory from Plato to Galileo*; Robert S. Westman, "The Astronomer's Role in the Sixteenth Century: A Preliminary Study." See also G. E. R. Lloyd, "Saving the Appearances."

11. Aristotle, *Physics*, II.2, 19b23–31, in Aristotle, *Complete Works*, vol. 1, p. 332. See also Wallace, *Causality and Scientific Explanation*, 1:16–21. To gain a fuller sense of the complexity of the relationship of astronomy to physics among Aristotle and his commentators, see Thomas Heath, *Mathematics in Aristotle*, pp. 11–12, 98–100, 211–14.

12. See Edward Hussey, "Aristotle's Mathematical Physics: A Reconstruction." For Ptolemy on the mingling of physics and mathematics, see Olaf Pedersen, *A Survey of the Almagest*, pp. 26–46; and Ptolemy's *Almagest*, ed. and trans. G. J. Toomer, passim. Also see Edward Grant, *A Source Book in Medieval Science*, the index under "Ptolemy."

13. See also chap. 13, above.

14. See Lindberg, *Catalogue of Medieval and Renaissance Optical Manuscripts*, which catalogs more than six hundred medieval and Renaissance optical manuscripts. See also Lindberg, *Theories of Vision from Al-Kindi to Kepler*, passim. I believe that an equally convincing case can be made for astronomy; I chose *perspectiva* because that is my home ground.

15. Broader definitions of "experiment" are also defensible, but I do not wish to enter into that particular debate here, except to say that I see no compelling reason for the requirement, frequently imposed by scientists and the general public, that an "experiment," to qualify for that title, requires the use of artificial means—usually hardware of some sort. Why does it matter, for example, whether the theory that heavy and light bodies fall with the same speed is confirmed by timing the bodies' descent with a mechanical clock or simply noticing that when a large rock and a small stone are dropped in unison, they strike the ground simultaneously? And do we really want to insist that in order to confirm Alhacen's theories about binocular vision, we must manufacture an apparatus that situates two vertical pegs before the eyes in such a way as to demonstrate parallax phenomena, rather than simply making use of two conveniently situated tree stumps? Surely the decisive criterion should not be the presence of hardware, but the epistemological purpose of the observation—namely, to confirm or disconfirm theoretical claims. On experiment in ancient science, see G. E. R. Lloyd, "Experiment in Early Greek Philosophy and Medicine."

16. On Ptolemy's astronomical methods, see Olaf Pedersen, *A Survey of the Almagest*; and Grant, *Source Book in Medieval Science*, the index under "Ptolemy." On Ptolemy's refraction experiments, see A. Mark Smith, *Ptolemy's Theory of Visual Perception*, pp. 229–39

17. For experiment in Ibn al-Haytham's optics, see A. I. Sabra, ed., *The Optics of Ibn al-Haytham, Books I–III*, vol. 2, index, under "experiment, experimental examination." See also Sabra, "Ibn al-Haytham's Revolutionary Project in Optics: The Achievement and the Obstacle."

18. On the theory of the rainbow: for Kamāl al-Dīn see Carl B. Boyer, *The Rainbow: From Myth to Mathematics*, pp. 125–30; for Theodoric of Freiberg, see above, pp. 277–78 and fig. 11.11; also William A. Wallace, *The Scientific Methodology of Theodoric of Freiberg*, pp. 174–224.

19. Aydin Sayili, *The Observatory in Islam and Its Place in the General History of the Observatory*, pp. 187–222, 259–89; also above, chap. 8.

20. Morris R. Cohen and I. E. Drabkin, eds., *A Source Book in Greek Science*, pp. 217–20; also above, chap. 12.

21. Julio Samsó, "Levi ben Gerson"; Emmanuel Poulle, "John of Murs."

22. On Bacon's contribution to scientific experiment, see David Lindberg, *Roger Bacon and the Origins of "Perspectiva" in the Middle Ages*, pp. lii–lxvi; on Bacon's

predecessors, see ibid., pp. xxxvii–xli; Jeremiah Hackett, "Roger Bacon on 'Scientia Experimentalis,'" in Hackett, *Roger Bacon on the Sciences*, pp. 277–315.

23. "The Letter of Peter Peregrinus on the Magnet," in Grant, *A Source Book in Medieval Science*, pp. 368–76.

24. William Newman, *Atoms and Alchemy: Chymistry and the Experimental Origins of the Scientific Revolution*.

25. For an expression of the popular view, see Marie Boas Hall, *The Scientific Renaissance, 1450–1630*, pp. 248–60.

26. See this chapter, n. 7, above.

27. See above, chaps. 2–3.

28. See above, chap. 4. On medieval and Renaissance knowledge of Lucretius, see David Ganz, "The Leiden Manuscripts and Their Carolingian Readers," pp. 92, 95; David J. Furley, "Lucretius."

29. For a succinct discussion of seventeenth-century mechanical philosophy, see Margaret J. Osler, "Mechanical Philosophy"; John Henry, "Atomism"; and their up-to-date bibliographies. See also Richard S. Westfall, "The Scientific Revolution of the Seventeenth Century: The Construction of a New World View."

30. A. Rupert Hall, "On the Historical Singularity of the Scientific Revolution," pp. 210–11.

31. This story has been told many times. Thomas S. Kuhn, *The Copernican Revolution* would be a good place to begin.

32. See David C. Lindberg, *Theories of Vision from al-Kindi to Kepler*, pp. 193 ff.

33. Galileo, *Two New Sciences*, pp. 165–70. See also above, chap. 12.

34. See above, chap. 12; see also Marshall Clagett, "Oresme, Nicole."

35. See Lawrence I. Conrad et al., *The Western Medical Tradition, 800 B.C. to A.D. 1800*, index under "Galen." See also Nancy Siraisi, *Medieval and Renaissance Medicine*.

36. Readers may wonder about the role of institutions, culture, social and economic factors, and religion in the scientific revolution—all topics of great interest, each of which would require a separate book, or perhaps books. I have dealt throughout this book with the relationship between science and religion (a topic on which I have published elsewhere) within the chronological limits of the book. The relationship between science and *religion* in the seventeenth century was exceedingly complicated and the source of much dispute. (See the publications of Peter Harrison for new light on the subject.) One fundamentally important claim that almost all students of the subject would insist upon is that, although seventeenth-century science was shaped by Christian doctrine in many ways, the scientific revolution did not depend on Christian influence for its existence, but *only* for aspects of its shape and character. For entry into the massive literature bearing on this subject, see the "Guide to Further Reading" in David C. Lindberg and Ronald L. Numbers, eds., *When Science and Christianity Meet*; also the articles and bibliographies contained in Lindberg and Numbers' older volume, *God and Nature*; and Gary B. Ferngren, *Science and Religion: A Historical Introduction*.

Bibliography

Aaboe, Asger. "On Babylonian Planetary Theories." *Centaurus*, 5 (1958): 209–77.

Abū Maʿshar. *The Abbreviation of "The Introduction to Astrology": Together with the Medieval Latin Translation of Adelard of Bath*, ed. Charles Burnett, Keiji Yamamoto, and Michio Yano. Leiden: Brill, 1994.

Ackrill, J. L. *Aristotle the Philosopher*. Oxford: Clarendon Press, 1981.

Adams, Marilyn McCord. *William Ockham*, 2 vols. Notre Dame, Ind.: University of Notre Dame Press, 1987.

Adelard of Bath. *Conversations with His Nephew: On the Same and the Different, Questions on Natural Science, and On Birds*, ed. and trans. Charles Burnett. Cambridge: Cambridge University Press, 1998.

Albert the Great. *Man and the Beasts, "De animalibus (books 22–26),"* trans. James J. Scanlan. Binghamton: Medieval and Renaissance Texts and Studies, Center for Medieval and Early Renaissance Studies, 1987.

_____. *Opera omnia*, ed. Augustus Borgnet, 38 vols. Paris: Vives, 1890–99.

_____. See also Kitchell, Kenneth F., Jr., and Resnick, Irven Michael.

Amundsen, Darrel W. "Medicine and Faith in Early Christianity." *Bulletin of the History of Medicine*, 56 (1982): 326–50.

_____. *Medicine, Society, and Faith in the Ancient and Medieval Worlds*. Baltimore: Johns Hopkins University Press, 1996.

_____. "Medieval Canon Law on Medical and Surgical Practice by the Clergy." *Bulletin of the History of Medicine*, 52 (1978): 22–44.

_____. "The Medieval Catholic Tradition." In Numbers, Ronald L., and Amundsen, Darrel W., eds., *Caring and Curing: Health and Medicine in the Western Religious Traditions*, pp. 65–107. New York: Macmillan, 1986.

Amundsen, Darrel W., and Ferngren, Gary B. "The Early Christian Tradition." In Numbers, Ronald L., and Amundsen, Darrel W., eds., *Caring and Curing: Health and Medicine in the Western Religious Traditions*, pp. 40–64. New York: Macmillan, 1986.

Anawati, Georges C. "Arabic Alchemy." *Encyclopedia of the History of Arabic Science*, 3:853–85.

_____. "Ḥunayn ibn Isḥāq." *Dictionary of Scientific Biography*, 15:230–34.

Anawati, Georges C., and Iskandar, Albert Z. "Ibn Sīnā." *Dictionary of Scientific Biography*, 15:494–501.

Anderson, R. G. W. "The Archaeology of Chemistry." In Holmes, Frederic L., and Levere, Trevor H., eds., *Instruments and Experimentation in the History of Chemistry*, pp. 5–34. Cambridge, Mass.: MIT Press, 2000.

Apollonius of Perga. *Conica*. See Fried and Unguru.

Archimedes. *Archimedes in the Middle Ages*, ed. and trans. Marshall Clagett, 5 vols. Madison: University of Wisconsin Press, 1964; Philadelphia: American Philosophical Society, 1976–84.

———. *The Works of Archimedes: Edited in Modern Notation, with Introductory Chapters*, ed. Thomas L. Heath, 2d ed. Cambridge: Cambridge University Press, 1912.

Aristotle. *Complete Works*, ed. Jonathan Barnes, 2 vols. Princeton: Princeton University Press, 1984.

———. *Metaphysics*, trans. Hugh Tredennick, 2 vols. London: Heinemann, 1935.

———. *Physics*, trans. P. H. Wicksteed and F. M. Cornford. 2 vols. London: Heinemann, 1929.

———. See also Nussbaum, Martha Craven.

Armstrong, A. H., ed. *The Cambridge History of Later Greek and Early Medieval Philosophy*. Cambridge: Cambridge University Press, 1970.

Armstrong, A. H., and Markus, R. A. *Christian Faith and Greek Philosophy*. London: Darton, Longman, & Todd, 1960.

Arnald de Villanova. See McVaugh, Michael.

Arnaldez, Roger, and Iskandar, Albert Z. "Ibn Rushd." *Dictionary of Scientific Biography*, 12:1–9.

Arts libéraux et philosophie au moyen âge: Actes du quatrième congrès international de philosophie médiévale, Université de Montréal, 27 August–2 September 1967. Montréal: Institut d'études médiévales, 1969.

Ashley, Benedict M. "St. Albert and the Nature of Natural Science." In Weisheipl, James A., ed., *Albertus Magnus and the Sciences: Commemorative Essays 1980*, pp. 73–102. Toronto: Pontifical Institute of Mediaeval Studies, 1980.

Ashworth, E. J. "Logic." In Lindberg, David C., and Shank, Michael H., eds., *The Cambridge History of Science*, vol. 2: *The Middle Ages*. Cambridge: Cambridge University Press, forthcoming.

Ashworth, William B., Jr. "Natural History and the Emblematic World View." In Lindberg, David C., and Westman, Robert S., eds., *Reappraisals of the Scientific Revolution*, pp. 303–32. Cambridge: Cambridge University Press, 1990.

Asmis, Elizabeth. *Epicurus' Scientific Method*. Ithaca: Cornell University Press, 1984.

Augustine. *The City of God*, trans. William H. Green. London: Heinemann, 1963.

Bacon, Francis. *New Organon*, in *Works*, trans. James Spedding, Robert Ellis, and Douglas Heath, new ed., 15 vols. New York: Hurd & Houghton, 1870–72.

Bacon, Roger. See Bridges, John H; see also Lindberg, David C.

Bailey, Cyril. *The Greek Atomists and Epicurus*. Oxford: Clarendon Press, 1928.

Baldwin, John W. *Masters, Princes, and Merchants: The Social Views of Peter the Chanter and His Circle*, 2 vols. Princeton: Princeton University Press, 1970.

———. *The Scholastic Culture of the Middle Ages*. Lexington, Mass.: D. C. Heath, 1971.

Ballester. See García Ballester.

Balme, D. M. "The Place of Biology in Aristotle's Philosophy." In Gotthelf, Allan, and Lennox, James G., eds., *Philosophical Issues in Aristotle's Biology*, pp. 9–20. Cambridge: Cambridge University Press, 1987.

Barker, John W. "Byzantine Empire: History (1204–1453)." *Dictionary of the Middle Ages*, 2:498–505.

Barnes, Jonathan. *Aristotle*. Oxford: Oxford University Press, 1982.

———. "Aristotle's Theory of Demonstration." In Barnes, Jonathan; Schofield, Malcolm; and Sorabji, Richard, eds., *Articles on Aristotle*, I: *Science*, pp. 65–87. London: Duckworth, 1975.

———. *Early Greek Philosophy*. Harmondsworth: Penguin, 1987.

———. *The Presocratic Philosophers*, 2 vols. London: Routledge & Kegan Paul, 1979.

Barnes, Jonathan; Brunschwig, Jacques; Burnyeat, Myles; and Schofield, Malcolm, eds. *Science and Speculation: Studies in Hellenistic Theory and Practice*. Cambridge: Cambridge University Press, 1982.

Barnes, Jonathan; Schofield, Malcolm; and Sorabji, Richard, eds. *Articles on Aristotle*, I: *Science*. London: Duckworth, 1975.

Barrow, Robin. *Greek and Roman Education*. London: Macmillan, 1967.

Basalla, George, ed. *The Rise of Modern Science: Internal or External Factors?* Lexington, Mass.: D. C. Heath, 1968.

Beaujouan, Guy. "Motives and Opportunities for Science in the Medieval Universities." In Crombie, A. C., ed., *Scientific Change*, pp. 219–36. London: Heinemann, 1963.

———. "The Transformation of the Quadrivium." In Benson, Robert L., and Constable, Giles, eds., *Renaissance and Renewal in the Twelfth Century*, pp. 463–87. Cambridge, Mass.: Harvard University Press, 1982.

Bede. *Baedae opera historica*, trans. J. E. King, 2 vols. London: Heinemann, 1930.

Beller, Eliyahu. "Ancient Jewish Mathematical Astronomy." *Archive for History of Exact Sciences*, 38 (1988): 51–66.

Benjamin, Francis S., and Toomer, G. J., eds. and trans. *Campanus of Novara and Medieval Planetary Theory: "Theorica planetarum."* Madison: University of Wisconsin Press, 1971.

Benson, Robert L., and Constable, Giles, eds. *Renaissance and Renewal in the Twelfth Century*. Cambridge, Mass.: Harvard University Press, 1982.

Benton, John. "Trotula, Women's Problems, and the Professionalization of Medicine in the Middle Ages." *Bulletin of the History of Medicine*, 59 (1985): 30–53.

Berggren, J. L. "History of Greek Mathematics: A Survey of Recent Research." *Historia Mathematica*, 11 (1984): 394–410.

———. "Islamic Mathematics." In Lindberg, David C., and Shank, Michael H., eds., *The Cambridge History of Science*, vol. 2: *The Middle Ages*. Cambridge: Cambridge University Press, forthcoming.

Berman, Harold J. *Law and Revolution: The Formation of the Western Legal Tradition*. Cambridge, Mass.: Harvard University Press, 1983.

Biggs, Robert. "Medicine in Ancient Mesopotamia." *History of Science*, 8 (1969): 94–105.

Birkenmajer, Aleksander. "Le rôle joué par les médecins et les naturalistes dans la réception d'Aristote au XIIe et XIIIe siècles." In Birkenmajer *Etudes d'histoire des sciences et de philosophie du moyen âge*, pp. 73–87. Studia Copernicana, no. 1. Wrocław: Ossolineum, 1970.

Al-Biṭrūjī. *On the Principles of Astronomy*, ed. and trans. Bernard R. Goldstein, 2 vols. New Haven: Yale University Press, 1971.

Blair, Peter Hunter. *The World of Bede*. Cambridge: Cambridge University Press, 1970.

Boethius of Dacia. *On the Supreme Good, On the Eternity of the World, On Dreams*, trans. John F. Wippel. Mediaeval Sources in Translation, no. 30. Toronto: Pontifical Institute of Mediaeval Studies, 1987.

Bonner, Stanley F. *Education in Ancient Rome: From the Elder Cato to the Younger Pliny*. Berkeley and Los Angeles: University of California Press, 1977.

Bowersock, G. W. *Hellenism in Late Antiquity*. Ann Arbor: University of Michigan Press, 1996.

Boyer, Carl B. *A History of Mathematics*. New York: John Wiley, 1968.

———. *The Rainbow: From Myth to Mathematics*. New York: Yoseloff, 1959.

Bradwardine, Thomas. See Crosby, H. Lamar, Jr.

Brain, Peter. *Galen on Bloodletting: A Study of the Origins, Development, and Validity of His Opinions, with a Translation of the Three Works*. Cambridge: Cambridge University Press, 1986.

Brand, Charles M. "Byzantine Empire: History (1025–1204)." *Dictionary of the Middle Ages*, 2:491–98.

Brandon, S. G. F. *Creation Legends of the Ancient Near East*. London: Hodder and Stoughton, 1963.

Breasted, James Henry. *Development of Religion and Thought in Ancient Egypt*. New York: Scribner's, 1912.

———. *The Edwin Smith Surgical Papyrus*, 2 vols. University of Chicago, Oriental Institute Publications, 3–4. Chicago: University of Chicago Press, 1930.

Brehaut, Ernest. *An Encyclopedist of the Dark Ages: Isidore of Seville*. New York: Columbia University Press, 1912.

Bridges, John H., ed. *Opus majus of Roger Bacon*, 3 vols. London: Williams and Norgate, 1900.

Brown, Joseph E. "The Science of Weights." In Lindberg, David C., ed., *Science in the Middle Ages*, pp. 179–205. Chicago: University of Chicago Press, 1978.

Brown, Peter. *Augustine of Hippo: A Biography*. Berkeley and Los Angeles: University of California Press, 1969.

———. *The Cult of Saints: Its Rise and Function in Latin Christianity*. Chicago: University of Chicago Press, 1981.

Brundell, Barry. *Pierre Gassendi: From Aristotelianism to a New Natural Philosophy*. Dordrecht: Reidel, 1987.

Brunschwig, Jacques, and Lloyd, Geoffrey E. R., eds. *The Greek Pursuit of Knowledge*. Cambridge, Mass.: Harvard University Press, 2003.

———. *A Guide to Greek Thought: Major Figures and Trends*. Cambridge, Mass.: Harvard University Press, 2003.

Bullough, Vern L. "Chauliac, Guy de." *Dictionary of Scientific Biography*, 3:218–19.

———. *The Development of Medicine as a Profession: The Contribution of the Medieval University to Modern Medicine.* Basel: Karger, 1966.

———. "Mondino de' Luzzi." *Dictionary of Scientific Biography*, 9:467–69.

Bulmer-Thomas, Ivor. "Isidorus of Miletus." *Dictionary of Scientific Biography*, 7:28–30.

Burckhardt, Jacob. *The Civilization of the Renaissance in Italy*, trans. S. G. C. Middlemore. New York: Modern Library, 1954.

Burnett, Charles S. F., ed. *Adelard of Bath: An English Scientist and Arabist of the Early Twelfth Century.* Warburg Institute Surveys and Texts, no. 14. London: Warburg Institute, 1987.

———. "Scientific Speculations." In Dronke, Peter, ed., *A History of Twelfth-Century Western Philosophy*, pp. 155–66. Cambridge: Cambridge University Press, 1988.

———. "Translation and Translators, Western European." *Dictionary of the Middle Ages*, 12:136–42.

———. "Translation and Transmission of Greek and Islamic Science to Latin Christendom." In Lindberg, David C., and Shank, Michael H., eds., *The Cambridge History of Science*, vol. 2: *The Middle Ages.* Cambridge: Cambridge University Press, forthcoming.

———. "The Transmission of Arabic Astronomy via Antioch and Pisa in the Second Quarter of the Twelfth Century." In Hogendijk, Jan, and Sabra, Abdelhamid I., eds. *The Enterprise of Science in Islam: New Perspectives*, pp. 23–51. Cambridge, Mass.: MIT Press, 2003.

———. "The Twelfth-Century Renaissance." In Lindberg, David C., and Shank, Michael H., eds., *The Cambridge History of Science*, vol. 2: *The Middle Ages.* Cambridge: Cambridge University Press, forthcoming.

———. "What Is the Experimentarius of Bernardus Silvestris? A Preliminary Survey of the Material." *Archives d'histoire doctrinale et littéraire du moyen âge*, 44 (1977): 79–125.

Butterfield, Herbert. *The Origins of Modern Science 1300–1800.* London: G. Bell, 1949.

Bynum, Caroline Walker. "Did the Twelfth Century Discover the Individual?" *Journal of Ecclesiastical History*, 31 (1980): 1–17.

Cadden, Joan. "Albertus Magnus' Universal Physiology: The Example of Nutrition." In Weisheipl, James A., ed., *Albertus Magnus and the Sciences: Commemorative Essays 1980*, pp. 321–29. Toronto: Pontifical Institute of Mediaeval Studies, 1980.

———. *Meanings of Sex Differences in the Middle Ages: Medicine, Science, and Culture.* Cambridge: Cambridge University Press, 1993.

———. "The Organization of Knowledge: Disciplines and Practices." In Lindberg, David C., and Shank, Michael H., eds., *The Cambridge History of Science*, vol. 2: *The Middle Ages.* Cambridge: Cambridge University Press, forthcoming.

———. "Science and Rhetoric in the Middle Ages: The Natural Philosophy of William of Conches." *Journal of the History of Ideas*, 56 (1995): 1–24.

Cahill, Thomas. *How the Irish Saved Civilization.* New York: Doubleday, 1995.

————. *Sailing the Wine-Dark Sea: Why the Greeks Matter*. New York: Random House/Doubleday, 2003.

Callus, D. A. "Introduction of Aristotelian Learning to Oxford." *Proceedings of the British Academy*, 29 (1943): 229–81.

————, ed. *Robert Grosseteste, Scholar and Bishop: Essays in Commemoration of the Seventh Centenary of His Death*. Oxford: Clarendon Press, 1955.

Cambridge History of Science, 8 vols., ed. David C. Lindberg and Ronald L. Numbers. Cambridge: Cambridge University Press, 2003–.

Cameron, M. L. "The Sources of Medical Knowledge in Anglo-Saxon England." *Anglo-Saxon England*, 11 (1983): 135–52.

Campbell, Tony. "Portolan Charts from the Late Thirteenth Century to 1500." In Harley, J. B., and Woodward, David, eds., *The History of Cartography*, 1:317–463. Chicago: University of Chicago Press, 1987.

Caroti, Stefano. "Nicole Oresme's Polemic against Astrology in His 'Quodlibeta.'" In Curry, Patrick, ed., *Astrology, Science, and Society: Historical Essays*, pp. 75–93. Woodbridge, Suffolk: Boydell, 1987.

Carré, Meyrick H. *Realists and Nominalists*. Oxford: Clarendon Press, 1946.

Catto, J. I., ed. *The Early Oxford Schools*. Vol. 1 of Aston, T. H., general ed., *The History of the University of Oxford*. Oxford: Clarendon Press, 1984.

Celsus, Aulus Cornelius. *De medicina, with an English Translation*, trans. W. G. Spencer, 3 vols. London: Heinemann, 1935–38.

Chadwick, Henry. *Early Christian Thought and the Classical Tradition: Studies in Justin, Clement, and Origen*. New York: Oxford University Press, 1966.

————. *The Early Church*. Harmondsworth: Penguin, 1967.

Charette, François. "The Locales of Islamic Astronomical Instrumentation." *History of Science*, 46 (2006): 123–38.

Chenu, M.-D. *Nature, Man, and Society in the Twelfth Century: Essays on New Theological Perspectives in the Latin West*, trans. Jerome Taylor and Lester K. Little. Chicago: University of Chicago Press, 1968.

————. *Toward Understanding St. Thomas*, ed. and trans. A.-M. Landry and D. Hughes. Chicago: Henry Regnery, 1964.

Cherniss, Harold. *The Riddle of the Early Academy*. Berkeley and Los Angeles: University of California Press, 1945.

Cicero. *De republica*, trans. Clinton Walker Keyes. London: Heinemann, 1928.

Cisne, John L. "How Science Survived: Medieval Manuscripts' 'Demography' and Classic Texts' Extinction." *Science*, 307 (25 Feb. 2005): 1305–7.

Clagett, Marshall. *Ancient Egyptian Science: A Source Book*, 3 vols. Philadelphia: American Philosophical Society, 1989–.

————, ed. *Critical Problems in the History of Science*. Madison: University of Wisconsin Press, 1962.

————. *Giovanni Marliani and Late Medieval Physics*. New York: Columbia University Press, 1941.

————. *Greek Science in Antiquity*. London: Abelard-Schuman, 1957.

_____. "The *Liber de motu* of Gerard of Brussels and the Origins of Kinematics in the West." *Osiris*, 12 (1956): 73–175.

_____, ed. and trans. *Nicole Oresme and the Medieval Geometry of Qualities and Motions.* Madison: University of Wisconsin Press, 1968.

_____. "Oresme, Nicole." *Dictionary of Scientific Biography*, 10:223–30.

_____. *The Science of Mechanics in the Middle Ages.* Madison: University of Wisconsin Press, 1959.

_____. "Some Novel Trends in the Science of the Fourteenth Century." In Singleton, Charles S., ed., *Art, Science, and History in the Renaissance*, pp. 275–303. Baltimore: Johns Hopkins University Press, 1968.

_____. *Studies in Medieval Physics and Mathematics.* London: Variorum, 1979.

Clark, Willene B., and McMunn, Meradith T., eds. *Beasts and Birds of the Middle Ages: The Bestiary and Its Legacy.* Philadelphia: University of Pennsylvania Press, 1989.

Clarke, M. L. *Higher Education in the Ancient World.* London: Routledge & Kegan Paul, 1971.

Cobban, Alan B. *The Medieval English Universities: Oxford and Cambridge to c. 1500.* Aldershot: Scolar Press, 1988.

_____. *The Medieval Universities: Their Development and Organization.* London: Methuen, 1975.

Cochrane, Charles N. *Christianity and Classical Culture: A Study of Thought and Action from Augustus to Augustine.* Oxford: Clarendon Press, 1940.

Cohen, Morris R., and Drabkin, I. E., eds. *A Source Book in Greek Science.* Cambridge, Mass.: Harvard University Press, 1958.

Cohen, H. Floris. *The Scientific Revolution: A Historiographical Inquiry.* Chicago: University of Chicago Press, 1994.

Cohen, I. Bernard. *Revolution in Science.* Cambridge, Mass.: Belknap Press, 1985.

Colish, Marcia L. *The Stoic Tradition from Antiquity to the Early Middle Ages*, 2 vols. Leiden: Brill, 1985.

Collingwood, R G. *The Idea of Nature.* Oxford: Clarendon Press, 1945.

Conrad, Lawrence I.; Neve, Michael; Nutton, Vivian; Porter, Roy; and Wear, Andrew. *The Western Medical Tradition: 800 BC to AD 1800.* Cambridge: Cambridge University Press, 1995.

Contreni, John J. *The Cathedral School of Laon from 850 to 930: Its Manuscripts and Masters.* Münchener Beiträge zur Mediävistik und Renaissance-Forschung, vol. 29. Munich: Arbeo-Gesellschaft, 1978.

_____. "Schools, Cathedral." *Dictionary of the Middle Ages*, 11:59–63.

Cooper, Lane. *Aristotle, Galileo, and the Tower of Pisa.* Ithaca: Cornell University Press, 1935.

Coopland, G. W. *Nicole Oresme and the Astrologers: A Study of His Livre de divinacions.* Cambridge, Mass.: Harvard University Press, 1952.

Copenhaver, Brian P. "Natural Magic, Hermetism, and Occultism in Early Modern Science." In Lindberg, David C., and Westman, Robert S., eds., *Reappraisals of the Scientific Revolution*, pp. 261–301. Cambridge: Cambridge University Press, 1990.

Courtenay, William J. *Capacity and Volition: A History of the Distinction of Absolute and Ordained Power.* Quodlibet: Ricerche e strumenti di filosofia medievale, no. 8. Bergamo: Pierluigi Lubrina, 1990.

————. *Covenant and Causality in Medieval Thought.* London: Variorum, 1984.

————. "The Critique on Natural Causality in the Mutakallimun and Nominalism." *Harvard Theological Review,* 66 (1973): 77–94. Reprinted in Courtenay, *Covenant and Causality in Medieval Thought,* chap. 5.

————. "The Dialectic of Divine Omnipotence." In Courtenay, *Covenant and Causality in Medieval Thought,* chap. 4.

————. "Nature and the Natural in Twelfth-Century Thought." In Courtenay, *Covenant and Causality in Medieval Thought,* chap. 3.

————. "Ockham, William of." *Dictionary of the Middle Ages,* 9:209–14.

————. *Parisian Scholars in the Early Fourteenth Century: A Social Portrait.* Cambridge: Cambridge University Press, 1999.

————. *Schools and Scholars in Fourteenth-Century England.* Princeton: Princeton University Press, 1987.

————. *Teaching Careers at the University of Paris in the Thirteenth and Fourteenth Centuries.* Texts and Studies in the History of Mediaeval Education, no. 18. Notre Dame, Ind.: United States Subcommission for the History of Universities, University of Notre Dame, 1988.

Courtenay, William J., and Tachau, Katherine H. "Ockham, Ockhamists, and the English-German Nation at Paris, 1339–1341." *History of Universities,* 2 (1982): 53–96.

Crombie, A. C. *Augustine to Galileo: The History of Science A.D. 400–1650.* London: Falcon, 1952. Reissued as *Medieval and Early Modern Science,* 2 vols. Garden City: Doubleday Anchor, 1959.

————. *Robert Grosseteste and the Origins of Experimental Science, 1100–1700.* Oxford: Clarendon Press, 1953.

————. *Science, Optics, and Music in Medieval and Early Modern Thought.* London: Hambledon, 1990.

Crosby, Alfred W. *The Measure of Reality: Quantification and Western Society, 1250–1600.* Cambridge: Cambridge University Press, 1997.

Crosby, H. Lamar, Jr., ed. and trans. *Thomas of Bradwardine, His "Tractatus de Proportionibus": Its Significance for the Development of Mathematical Physics.* Madison: University of Wisconsin Press, 1961.

Crowe, Michael J. *Theories of the World from Antiquity to the Copernican Revolution.* New York: Dover, 1990.

Crowley, Theodore. *Roger Bacon: The Problem of the Soul in His Philosophical Commentaries.* Dublin: James Duffy; Louvain: Editions de l'Institut Supérieur de Philosophie, 1950.

Cumont, Franz. *Astrology and Religion among the Greeks and Romans.* New York: Putnam's Sons, 1912.

Curley, Michael J., trans. *Physiologus.* Austin: University of Texas Press, 1979.

Curry, Patrick, ed. *Astrology, Science, and Society: Historical Essays.* Woodbridge, Suffolk: Boydell, 1987.

Dales, Richard C. *The Intellectual Life of Western Europe in the Middle Ages.* Washington, D.C.: University Press of America, 1980.

————. "Marius 'On the Elements' and the Twelfth-Century Science of Matter." *Viator,* 3 (1972): 191–218.

————. *Medieval Discussions of the Eternity of the World.* Leiden: Brill, 1990.

————. "Time and Eternity in the Thirteenth Century." *Journal of the History of Ideas,* 49 (1988): 27–45.

d'Alverny, Marie-Thérèse. "Translations and Translators." In Benson, Robert L., and Constable, Giles, eds., *Renaissance and Renewal in the Twelfth Century,* pp. 421–62. Cambridge, Mass.: Harvard University Press, 1982.

Daston, Lorraine, and Park, Katharine. *Wonders and the Order of Nature, 1150–1750.* Cambridge, Mass.: Zone Books, 1998.

Dear, Peter. *Discipline and Experience: The Mathematical Way in the Scientific Revolution.* Chicago: University of Chicago Press, 1995.

————. "Jesuit Mathematical Science and the Reconstitution of Experience in the Early 17th Century." *Studies in History and Philosophy of Science,* 18 (1987): 133–75.

————. *Revolutionizing the Sciences: European Knowledge and Its Ambitions.* Princeton: Princeton University Press, 2001.

Debarnot, Marie-Thérèse. "Trigonometry." *Encyclopedia of the History of Arabic Science,* 2:495–538.

Debus, Allen G. *The Chemical Philosophy: Paracelsian Science and Medicine in the Sixteenth and Seventeenth Centuries,* 2 vols. New York: Science History Publications, 1977.

————. *Man and Nature in the Renaissance.* Cambridge: Cambridge University Press, 1978.

De Lacy, Phillip. "Galen's Platonism." *American Journal of Philology,* 93 (1972): 27–39.

Demaitre, Luke E., and Travill, Anthony A. "Human Embryology and Development in the Works of Albertus Magnus." In Weisheipl, James A., ed., *Albertus Magnus and the Sciences: Commemorative Essays 1980,* pp. 405–40. Toronto: Pontifical Institute of Mediaeval Studies, 1980.

Denifle, Henricus, and Chatelain, Aemilio. *Chartularium Universitatis Parisiensis,* 4 vols. Paris: Delalain, 1889–97.

de Santillana, Giorgio. *The Origins of Scientific Thought: From Anaximander to Proclus, 600 B.C. to A.D. 500.* Chicago: University of Chicago Press, 1961.

de Vaux, Carra. "Astronomy and Mathematics." In Arnold, Thomas, and Guillaume, Alfred, eds., *The Legacy of Islam,* pp. 376–97. London: Oxford University Press, 1931.

Dicks, D. R. *Early Greek Astronomy to Aristotle.* Ithaca: Cornell University Press, 1970.

————. "Eratosthenes." *Dictionary of Scientific Biography,* 4:388–93.

Dictionary of Scientific Biography, 16 vols. New York: Scribner's, 1970–80.

Dijksterhuis, E. J. *Archimedes*, trans. C. Dikshoorn. Copenhagen: Munksgaard, 1956.

————. *The Mechanization of the World Picture*, trans. C. Dikshoorn. Oxford: Clarendon Press, 1961.

Dilke, O. A. W. "Cartography in the Byzantine Empire." In Harley, J. B., and Woodward, David, eds., *The History of Cartography*, 2:258–75. Chicago: University of Chicago Press, 1992.

————. "The Culmination of Greek Cartography in Ptolemy." In Harley, J. B., and Woodward, David, eds., *The History of Cartography*, 1:177–200. Chicago: University of Chicago Press, 1987.

Diogenes Laertius. *Lives of Eminent Philosophers*, trans. R D. Hicks, 2 vols. London: Heinemann, 1925.

Dodge, Bayard. *Muslim Education in Medieval Times*. Washington, D.C.: Middle East Institute, 1962.

Dols, Michael W., trans. *Medieval Islamic Medicine: Ibn Ridwān's Treatise "On the Prevention of Bodily Ills in Egypt,"* with an Arabic text edited by Adil S. Gamal. Berkeley and Los Angeles: University of California Press, 1984.

————. "The Origins of the Islamic Hospital: Myth and Reality." *Bulletin of the History of Medicine*, 61 (1987): 367–90.

Dorff, Elliot N. "The Jewish Tradition." In Numbers, Ronald L., and Amundsen, Darrel W., eds., *Caring and Curing: Health and Medicine in the Western Religious Traditions*, pp. 5–39. New York: Macmillan, 1986.

Dorn, Harold. *The Geography of Science*. Baltimore: Johns Hopkins University Press, 1991,

Drake, Stillman. "The Uniform Motion Equivalent of a Uniformly Accelerated Motion from Rest." *Isis*, 63 (1972): 28–38.

Dreyer, J. L. E. *History of the Planetary Systems from Thales to Kepler*. Cambridge: Cambridge University Press, 1906. Reissued as *A History of Astronomy from Thales to Kepler*, ed. W. H. Stahl. New York: Dover, 1953.

Dronke, Peter, ed. *A History of Twelfth-Century Western Philosophy*. Cambridge: Cambridge University Press, 1988.

————. "Thierry of Chartres." In Dronke, Peter, ed., *A History of Twelfth-Century Western Philosophy*, pp. 358–85.

Duhem, Pierre. *Etudes sur Léonard de Vinci*, 3 vols. Paris: Hermann, 1906–13.

————, ed. *Un fragment inédit de l'Opus tertium de Roger Bacon, précédé d'une étude sur ce fragment*. Quaracchi: Collegium S. Bonaventurae, 1909.

————. *Medieval Cosmology: Theories of Infinity, Place, Time, Void, and the Plurality of Worlds*, ed. and trans. Roger Ariew. Chicago: University of Chicago Press, 1985.

————. *Les orgines de la statique*, 2 vols. Paris: Hermann, 1905–6.

————. *Le système du monde*, 10 vols. Paris: Hermann, 1913–59.

————. *To Save the Phenomena: An Essay on the Idea of Physical Theory from Plato to Galileo*, trans. Edmund Doland and Chaninah Maschler. Chicago: University of Chicago Press, 1969.

Düring, Ingemar. "The Impact of Aristotle's Scientific Ideas in the Middle Ages." *Archiv für Geschichte der Philosophie,* 50 (1968): 115–33.

Easton, Stewart C. *Roger Bacon and His Search for a Universal Science.* Oxford: Basil Blackwell, 1952.

Eastwood, Bruce S. "Astronomical Images and Planetary Theory in Carolingian Studies of Martianus Capella." *Journal for the History of Astronomy,* 31 (2000): 1–28.

———. *Astronomy and Optics from Pliny to Descartes.* London: Variorum, 1989.

———. "The Astronomy of Macrobius in Carolingian Europe: Dungal's Letter of 811 to Charles the Great." *Early Medieval Europe,* 3 (1994): 117–34.

———. "Chalcidius's Commentary on Plato's *Timaeus* in Latin Astronomy of the Ninth to Eleventh Centuries." In Nauta, L., and Vanderjagt, A., eds., *Between Demonstration and Imagination: Essays in the History of Science and Philosophy,* pp. 171–209. Leiden: Brill, 1999.

———. "Cosmology, Astronomy, and Mathematics." In Lindberg, David C., and Shank, Michael H., eds., *The Cambridge History of Science,* vol. 2: *The Middle Ages.* Cambridge: Cambridge University Press, forthcoming.

———. "Invention and Reform in Latin Planetary Astronomy." In Herren, Michael W.; McDonough, C. J.; and Arthur, Ross G., eds., *Latin Culture in the Eleventh Century,* pp. 264–97. Turnhout, Belgium: Brepols, 2002.

———. "Johannes Scottus Eriugena, Sun-Centered Planets, and Carolingian Astronomy." *Journal for the History of Astronomy,* 32 (2001): 281–324.

———. "Kepler as Historian of Science: Precursors of Copernican Heliocentrism According to De revolutionibus, I, 10." *Proceedings of the American Philosophical Society,* 126 (1982): 367–94.

———. "On the Continuity of Western Science from the Middle Ages: A.C. Crombie's 'Augustine to Galileo.'" *Isis,* 83 (1992): 84–99.

———. *Ordering the Heavens: Roman Astronomy and Cosmology in the Carolingian Renaissance.* Leiden: Brill, in press.

———. "Plinian Astronomical Diagrams in the Early Middle Ages." In Grant, Edward, and Murdoch, John E., eds., *Mathematics and Its Applications to Science and Natural Philosophy in the Middle Ages: Essays in Honor of Marshall Clagett,* pp. 141–72. Cambridge: Cambridge University Press, 1987.

———. "Plinian Astronomy in the Middle Ages and Renaissance." In French, Roger, and Greenaway, Frank, eds., *Science in the Early Roman Empire: Pliny the Elder, His Sources and Influence,* pp. 197–251. Totawa, New Jersey: Barnes & Noble, 1986.

———. *The Revival of Planetary Astronomy in Carolingian and Post-Carolingian Europe.* Ashgate: Variorum, 2002.

Eastwood, Bruce, and Grasshoff, Gerd. *Planetary Diagrams for Roman Astronomy in Medieval Europe, ca. 800–1500.* Transactions of the American Philosophical Society, vol. 94, pt. 3. Philadelphia: American Philosophical Society, 2004.

Ebbell, B. *The Papyrus Ebers, the Greatest Egyptian Medical Document.* Copenhagen: Munksgaard, 1939.

Edel, Abraham. *Aristotle and His Philosophy.* Chapel Hill: University of North Carolina Press, 1982.

Edelstein, Emma J., and Edelstein, Ludwig. *Asclepius: A Collection and Interpretation of the Testimonies*, 2 vols. Baltimore: Johns Hopkins University Press, 1945.

Ede, Andrew, and Cormack, Lesley B. *A History of Science in Society: From Philosophy to Utility*. Peterborough, Ontario: Broadview Press, 2004.

Edelstein, Ludwig. *Ancient Medicine: Selected Papers of Ludwig Edelstein*, ed. Owsei Temkin and C. Lilian Temkin. Baltimore: Johns Hopkins University Press, 1967.

——. "The Distinctive Hellenism of Greek Medicine." *Bulletin of the History of Medicine*, 40 (1966): 197–255. Reprinted in Edelstein, *Ancient Medicine*, pp. 367–97. Baltimore: Johns Hopkins University Press, 1967.

——. "Empiricism and Skepticism in the Teaching of the Greek Empiricist School." In Edelstein, *Ancient Medicine: Selected Papers of Ludwig Edelstein*, pp. 195–203. Baltimore: Johns Hopkins University Press, 1967.

——. "Greek Medicine and Its Relation to Religion and Magic." *Bulletin of the Institute of the History of Medicine*, 5 (1937): 201–46.

——. "The Methodists." In Edelstein, *Ancient Medicine*, pp. 173–91. Baltimore: Johns Hopkins University Press, 1967.

——. "The Relation of Ancient Philosophy to Medicine." *Bulletin of the History of Medicine*, 26 (1952): 299–316. Reprinted in Edelstein, *Ancient Medicine*, pp. 349–66. Baltimore: Johns Hopkins University Press, 1967.

Edson, Evelyn. *Mapping Time and Space: How Medieval Mapmakers Viewed Their World*. London: British Library, 1997.

Edson, Evelyn, and Savage-Smith, Emilie. *Medieval Views of the Cosmos*. Oxford: Bodleian Library, 2004.

Elford, Dorothy. "William of Conches." In Dronke, Peter, ed., *A History of Twelfth-Century Western Philosophy*, pp. 308–27. Cambridge: Cambridge University Press, 1988.

Emerton, Norma E. *The Scientific Reinterpretation of Form*. Ithaca: Cornell University Press, 1984.

Encyclopedia of the History of Arabic Science. See Rashed, Roshdi.

Endress, Gerhard. "Mathematics and Philosophy in Medieval Islam." In Hogendijk, Jan, and Sabra, Abdelhamid I., eds., *The Enterprise of Science in Islam: New Perspectives*, pp. 121–176. Cambridge, Mass.: MIT Press, 2003.

Epp, Ronald H., ed. *Recovering the Stoics*. Supplement to the *Southern Journal of Philosophy*, vol. 23 (1985).

Euclid. *The Elements*, trans. Thomas Heath, 3 vols. Cambridge: Cambridge University Press, 1908.

Evans, Gillian R. *Anselm and a New Generation*. Oxford: Clarendon Press, 1980.

——. "The Influence of Quadrivium Studies in the Eleventh- and Twelfth-Century Schools." *Journal of Medieval History*, 1 (1975): 151–64.

——. *Old Arts and New Theology: The Beginnings of Theology as an Academic Discpline*. Oxford: Clarendon Press, 1980.

——. *The Thought of Gregory the Great*. Cambridge: Cambridge University Press, 1986.

Evans, James. *The History and Practice of Ancient Astronomy.* Oxford: Oxford University Press, 1998.

Evans, Joan, ed. *The Flowering of the Middle Ages.* London: Thames and Hudson, 1966.

Fakhry, Majid. *A History of Islamic Philosophy.* New York: Columbia University Press, 1970.

Farrington, Benjamin. *Greek Science,* rev. ed. Harmondsworth: Penguin, 1961.

Fazlioğlu, İhsan. "The Mathematical/Astronomical School of Samarqand as a Background for Ottoman Philosophy and Science." *Journal for the History of Arabic Science,* vol. 14 (2007): forthcoming.

Feingold, Mordechai. *The Mathematicians' Apprenticeship: Science, Universities, and Society in England, 1560–1640.* Cambridge: Cambridge University Press, 1984.

Ferguson, Wallace K. *The Renaissance in Historical Thought.* Boston: Houghton Mifflin, 1948.

Ferngren, Gary B., ed. *Science and Religion: A Historical Introduction.* Baltimore: Johns Hopkins University Press, 2002.

Ferruolo, Stephen C. *The Origins of the University: The Schools of Paris and Their Critics, 1100–1215.* Stanford: Stanford University Press, 1985.

Fichtenau, Heinrich. *The Carolingian Empire,* trans. Peter Munz. Oxford: Basil Blackwell, 1957.

La filosofia della natura nel medioevo: Atti del Terzo Congresso Internazionale di Filosofia Medioevale, 31 August–5 September 1964. Milan: Società Editrice Vita e Pensiero, 1966.

Finley, M. I. The *World of Odysseus,* rev. ed. New York: Viking Press, 1965.

Finucane, Ronald C. *Miracles and Pilgrims: Popular Beliefs in Medieval England.* Totowa, N.J.: Rowman and Littlefield, 1977.

Flint, Valerie I. J. *The Rise of Magic in Early Medieval Europe.* Princeton: Princeton University Press, 1991.

Fontaine, Jacques. *Isidore de Séville et la culture classique dans l'Espagne wisigothique,* 2d ed., 3 vols. Paris: Etudes Augustiniennes, 1983.

Francesco of Capuano. *Sphera mundi . . . cum commentariis.* Venice: Giunta, 1518.

Frankfort, H., and Frankfort, H. A. "Myth and Reality." In Frankfort, H.; Frankfort, H. A.; Wilson, John A.; and Jacobsen, Thorkild, *Before Philosophy: The Intellectual Adventure of Ancient Man,* pp. 11–36. Baltimore: Penguin, 1951.

Frankfort, H.; Frankfort, H. A.; Wilson, John A.; and Jacobsen, Thorkild. *Before Philosophy: The Intellectual Adventure of Ancient Man.* Baltimore: Penguin, 1951.

Fraser, P. M. *Ptolemaic Alexandria,* 3 vols. Oxford: Clarendon Press, 1972.

Frede, Michael. "The Method of the So-Called Methodical School of Medicine." In Barnes, Jonathan; Brunschwig, Jacques; Burnyeat, Myles; and Schofield, Malcolm, eds., *Science and Speculation: Studies in Hellenistic Theory and Practice,* pp. 1–23. Cambridge: Cambridge University Press, 1982.

Freeman, Charles. *The Closing of the Western Mind: The Rise of Faith and the Fall of Reason.* New York: Knopf, 2003.

French, Roger. *Ancient Natural History.* London: Routledge, 1994.

French, Roger, and Greenaway, Frank, eds. *Science in the Early Roman Empire: Pliny the Elder, His Sources and Influence.* Totawa, N.J.: Barnes & Noble, 1986.

Frend, W. H. C. *The Early Church.* Philadelphia: Fortress Press, 1982.

Freudenthal, Gad. *Science in the Medieval Hebrew and Arabic Traditions.* Aldershot, England: Variorum, 2005.

Fried, Michael N, and Unguru, Sabetai. *Apollonius of Perga's "Conica": Text, Context, Subtext.* Leiden: Brill, 2001.

Funkenstein, Amos. *Theology and the Scientific Imagination from the Middle Ages to the Seventeenth Century.* Princeton: Princeton University Press, 1986.

Furley, David J. *Cosmic Problems: Essays on Greek and Roman Philosophy of Nature.* Cambridge: Cambridge University Press, 1989.

———. *The Greek Cosmologists,* vol. 1: *The Formation of the Atomic Theory and Its Earliest Critics.* Cambridge: Cambridge University Press, 1987.

———. "Lucretius." *Dictionary of Scientific Biography,* 8:536–39.

———. *Two Studies in the Greek Atomists.* Princeton: Princeton University Press, 1967.

Gabriel, Astrik L. "Universities." *Dictionary of the Middle Ages,* 12:282–300.

Galen. *On Respiration and the Arteries,* ed. and trans. David J. Furley and J. S. Wilkie. Princeton: Princeton University Press, 1984.

———. *On the Natural Faculties,* trans. A. J. Brock. London: Heinemann, 1963.

———. *On the Usefulness of the Parts of the Body,* ed. and trans. Margaret T. May, 2 vols. Ithaca: Cornell University Press, 1968.

———. *Three Treatises on the Nature of Science,* ed. and trans. Richard Walzer and Michael Frede. Indianapolis: Hackett, 1985.

Galileo. *Two New Sciences,* trans. Stillman Drake. Madison: University of Wisconsin Press, 1974.

Ganz, David. "The Leiden Manuscripts and Their Carolingian Readers." In *Medieval Manuscripts of the Latin Classics: Production and Use* (Proceedings of the Seminar in the History of the Book to 1500, 1993). Leiden, 1996.

García Ballester, Luis. "Galen as a Medical Practitioner: Problems in Diagnosis." In Nutton, Vivian, ed., *Galen: Problems and Prospects,* pp. 13–46. London: Wellcome Institute for the History of Medicine, 1981.

García Ballester, Luis; Ferre, Lola; and Feliu, Edward. "Jewish Appreciation of Fourteenth-Century Scholastic Medicine." *Osiris,* n.s. 6 (1990): 85–117.

Garsoïan, Nina G. "Nisibis." *Dictionary of the Middle Ages,* 9:141–42.

Garwood, Christine. *Flat Earth: The History of an Infamous Idea.* London: Macmillan, 2007.

Gascoigne, John. "A Reappraisal of the Role of the Universities in the Scientific Revolution." In Lindberg, David C., and Westman, Robert S., eds., *Reappraisals of the Scientific Revolution,* pp. 207–60. Cambridge: Cambridge University Press, 1990.

Gersh, Stephen. *Middle Platonism and Neoplatonism: The Latin Tradition,* 2 vols. Notre Dame, Ind.: University of Notre Dame Press, 1986.

Getz, Faye M. "Charity, Translation, and the Language of Medical Learning in Medieval England." *Bulletin of the History of Medicine,* 64 (1990): 1–17.

———. "The Faculty of Medicine before 1500." In Catto, J. I., and Evans, Ralph, eds., vol. 2 of *The History of the University of Oxford*, pp 373–405. Oxford: Clarendon Press, 1992.

———. *Healing and Society in Medieval England*. Madison: University of Wisconsin Press, 1991.

———. *Medicine in the English Middle Ages*. Princeton: Princeton University Press, 1998.

———. "Western Medieval Medicine." *Trends in History*, 4, nos. 2–3 (1988): 37–54.

Ghalioungui, Paul. *The House of Life, Per Ankh: Magic and Medical Science in Ancient Egypt*, 2d ed. Amsterdam: B. M. Israel, 1973.

———. *The Physicians of Pharaonic Egypt*. Cairo: Al-Ahram Center for Scientific Translations, 1983.

Gillings, R J. "The Mathematics of Ancient Egypt." *Dictionary of Scientific Biography*, 15:681–705.

Gilson, Etienne. *The Christian Philosophy of St. Thomas Aquinas*, trans. L. K. Shook. New York: Random House, 1956.

Gimpel, Jean. *The Medieval Machine: The Industrial Revolution of the Middle Ages*. New York: Holt, Rinehart and Winston, 1976.

Gingerich, Owen. "Islamic Astronomy." *Scientific American*, 254, no. 4 (April 1986): 74–83.

Gohlman, William E., ed. and trans. *The Life of Ibn Sina: A Critical Edition and Annotated Translation* (Albany: State University of New York Press, 1974).

Goldstein, Bernard R. *The Arabic Version of Ptolemy's "Planetary Hypotheses."* Transactions of the American Philosophical Society, n.s., vol. 57, pt. 4. Philadelphia: American Philosophical Society, 1967.

———. *Al-Biṭrūjī On the Principles of Astronomy*, 2 vols. New Haven: Yale University Press, 1971.

———. *The Astronomy of Levi ben Gerson (1288–1344)*. New York: Springer, 1985.

———. "John of Murs." *Dictionary of Scientific Biography*, 7:128–33.

———. "Theory and Observation in Ancient and Medieval Astronomy." *Isis*, 63 (1972): 39–47.

———. *Theory amd Observation in Ancient and Medieval Astronomy*. London: Variorum Reprints, 1985.

Goldstein, Bernard R., and Bowen, Alan C. "A New View of Early Greek Astronomy." *Isis*, 74 (1983): 330–40.

Goldstein, Bernard R., and Swerdlow, Noel. "Planetary Distances and Sizes in an Anonymous Arabic Treatise Preserved in Bodleian MS Marsh 621." *Centaurus*, 15 (1970): 135–70.

Goldziher, Ignaz. "The Attitude of Orthodox Islam toward the 'Ancient Sciences.'" Trans. from German original (1915). In Swartz, Merlin L., ed., *Studies on Islam*, pp. 185–215. Oxford: Oxford University Press, 1981.

———. "Catholic Tendencies and Particularism in Islam." Trans. from German original (1914). In Swartz, Merlin L., ed., *Studies on Islam*, pp. 123–39. Oxford: Oxford University Press, 1981.

Goody, Jack. *The Domestication of the Savage Mind*. Cambridge: Cambridge University Press, 1977.

Goody, Jack, and Watt, Ian. "The Consequences of Literacy." *Comparative Studies in Society and History*, 5 (1962–63): 304–45.

Gottfried, Robert S. *Doctors and Medicine in Medieval England 1340–1530*. Princeton: Princeton University Press, 1986.

Gotthelf, Allan, and Lennox, James G., eds. *Philosophical Issues in Aristotle's Biology*. Cambridge: Cambridge University Press, 1987.

Gottschalk, H. B. "Strato of Lampsacus." *Dictionary of Scientific Biography*, 13:91–95.

Grant, Edward. "Aristotelianism and the Longevity of the Medieval World View." *History of Science*, 16 (1978): 93–106.

———. *Celestial Matter: A Medieval and Galilean Cosmological Problem." Journal of Medieval and Renaissance Studies*, 13 (1983): 157–86.

———. "Celestial Orbs in the Latin Middle Ages." *Isis*, 78 (1987): 153–73.

———. "The Condemnation of 1277, God's Absolute Power, and Physical Thought in the Late Middle Ages." *Viator*, 10 (1979): 211–44.

———. "Cosmology." In Lindberg, David C., ed., *Science in the Middle Ages*, pp. 265–302. Chicago: University of Chicago Press, 1978.

———. "Cosmology." In Lindberg, David C., and Shank, Michael H., eds., *The Cambridge History of Science*, vol. 2: *The Middle Ages*. Cambridge: Cambridge University Press, forthcoming.

———. *The Foundations of Modern Science in the Middle Ages: Their Religious, Institutional, and Intellectual Contexts*. Cambridge: Cambridge University Press, 1996.

———. *God and Reason in the Middle Ages*. Cambridge: Cambridge University Press, 2001.

———. "Late Medieval Thought, Copernicus, and the Scientific Revolution." *Journal of the History of Ideas*, 23 (1962): 197–220.

———. "Medieval and Renaissance Scholastic Conceptions of the Influence of the Celestial Region on the Terrestrial." *Journal of Medieval and Renaissance Studies*, 17 (1987): 1–23.

———. "Medieval and Seventeenth-Century Conceptions of an Infinite Void Space beyond the Cosmos." *Isis*, 60 (1969): 39–60.

———. "The Medieval Doctrine of Place: Some Fundamental Problems and Solutions." In Maierù, A., and Paravicini Bagliani, A., eds., *Studi sul XIV secolo in memoria di Anneliese Maier*, pp. 57–79. Storia e Letteratura, Raccolta di studi e testi, no. 151. Rome: Edizioni di Storia e Letteratura, 1981.

———. *Much Ado about Nothing: Theories of Space and Vacuum from the Middle Ages to the Scientific Revolution*. Cambridge: Cambridge University Press, 1981.

———, ed. and trans. *Nicole Oresme and the Kinematics of Circular Motion: Tractatus de commensurabilitate vel incommensurabilitate motuum celi*. Madison: University of Wisconsin Press, 1971.

———. *Physical Science in the Middle Ages*. New York: Wiley, 1971.

———. *Planets, Stars, and Orbs: The Medieval Cosmos, 1200–1687*. Cambridge: Cambridge University Press, 1994.

————. *Science and Religion, 400 B.C.–A.D. 1550: From Aristotle to Copernicus.* Westport, Conn.: Greenwood, 2004.

————. "Science and the Medieval University." In Kittelson, James M., and Transue, Pamela J., eds., *Rebirth, Reform, and Resilience: Universities in Transition 1300–1700,* pp. 68–102. Columbus: Ohio State University Press, 1984.

————. "Science and Theology in the Middle Ages." In Lindberg, David C., and Numbers, Ronald L., eds., *God and Nature: Historical Essays on the Encounter between Christianity and Science,* pp. 49–75. Berkeley and Los Angeles: University of California Press, 1986.

————, ed. *A Source Book in Medieval Science.* Cambridge, Mass.: Harvard University Press, 1974.

————. *Studies in Medieval Science and Natural Philosophy.* London: Variorum, 1981.

Grant, Edward, and Murdoch, John E., eds. *Mathematics and Its Applications to Science and Natural Philosophy in the Middle Ages: Essays in Honor of Marshall Clagett.* Cambridge: Cambridge University Press, 1987.

Grant, Robert M. *Miracle and Natural Law in Graeco-Roman and Early Christian Thought.* Amsterdam: North-Holland, 1952.

Graves, Robert. *The Greek Myths.* 2 vols. Harmondsworth: Penguin, 1955.

Grayeff, Felix. *Aristotle and His School.* London: Duckworth, 1974.

Green, Monica H., ed. and trans. *The Trotula: A Medieval Compendium of Women's Medicine.* Philadelphia: University of Pennsylvania Press, 2001.

————. *Women's Healthcare in the Medieval West: Texts and Contexts.* Aldershot: Ashgate, 2000.

————. "Women's Medical Practice and Medical Care in Medieval Europe." *Signs,* 14 (1989): 434–73.

Green, Peter. *Alexander of Macedon, 356–323 B.C.: A Historical Biography.* Berkeley and Los Angeles: University of California Press, 1991.

Greene, Mott. *Natural Knowledge in Preclassical Antiquity.* Baltimore: Johns Hopkins University Press, 1992.

Gregory, T. E. "Byzantine Empire: History (330–1025)." *Dictionary of the Middle Ages,* 2:481–91.

Gregory, Tullio. *Anima mundi: La filosofia di Guglielmo di Conches e la scuola di Chartres.* Florence: G. C. Sansoni, 1955.

————. "La nouvelle idée de nature et de savoir scientifique au XIIe siècle." In Murdoch, John E., and Sylla, Edith D., eds., *The Cultural Context of Medieval Learning,* pp. 193–212. Boston Studies in the Philosophy of Science, 26. Dordrecht: Reidel, 1975.

————. "The Platonic Inheritance." In Dronke, Peter, ed., *A History of Twelfth-Century Western Philosophy,* pp. 54–80. Cambridge: Cambridge University Press, 1988.

Grendler, Paul F. *Schooling in Renaissance Italy: Literacy and Learning, 1300–1600.* Baltimore: Johns Hopkins University Press, 1989.

Grene, Marjorie. *A Portrait of Aristotle.* Chicago: University of Chicago Press, 1963.

Griffin, Jasper. *Homer.* Oxford: Oxford University Press, 1980.

Gunther, R. T. *Early Science in Oxford,* vol. 2. Oxford: Oxford University Press, 1923.

Gutas, Dimitri. *Avicenna and the Aristotelian Tradition: Introduction to Reading Avicenna's Philosophical Works.* Leiden: Brill, 1988.

———. *Greek Thought, Arabic Culture: The Graeco-Arabic Translation Movement in Baghdad and Early ʿAbbāsid Society (2nd–4th/8th–10th centuries).* London: Routledge, 1998.

Hackett, Jeremiah, ed. *Roger Bacon and the Sciences: Commemorative Essays.* Leiden: Brill, 1997.

Hackett, M. B. "The University as a Corporate Body." In Catto, J. I., ed., *The Early Oxford Schools*, vol. 1 of *The History of the University of Oxford*, general ed. T. H. Aston, pp. 37–95. Oxford: Clarendon Press, 1984.

Hadot, Ilsetraut, ed. *Simplicius: sa vie, son oeuvre, sa survie.* In *Actes du Colloque international de Paris*, 28 Sept.–1 Oct. 1985. Berlin: Walter de Gruyter, 1987.

Hahm, David E. *The Origins of Stoic Cosmology.* Columbus: Ohio State University Press, 1977.

Hall, A. Rupert. "Merton Revisited; or, Science and Society in the Seventeenth Century." *History of Science*, 2 (1963): 1–16.

———. "On the Historical Singularity of the Scientific Revolution of the Seventeenth Century." In Elliott, J. H., and Koenigsberger, H. G., eds., *The Diversity of History: Essays in Honour of Sir Herbert Butterfield*, pp. 199–221. London: Routledge & Kegan Paul, 1970.

———. *The Revolution in Science 1500–1750.* London: Longman, 1983.

———. *The Scientific Revolution 1500–1800.* London: Longmans, Green, 1954.

Hall, Marie Boas. *The Scientific Renaissance.* New York: Harper, 1962.

Halleux, Robert. "Alchemy." *Dictionary of the Middle Ages*, 1:134–40.

———. *Les textes alchimiques.* Typologie des sources du moyen âge occidental, no. 32. Turnhout: Brepols, 1979.

Hamarneh, Sami. "Al-Majūsī." *Dictionary of Scientific Biography*, 9:40–42.

———. "Al-Zahrāwī." *Dictionary of Scientific Biography*, 14:584–85.

Hamilton, Edith. *Mythology.* Boston: Little, Brown, 1942.

Hansen, Bert. *Nicole Oresme and the Marvels of Nature: A Study of His "De causis mirabilium" with Critical Edition, Translation, and Commentary.* Toronto: Pontifical Institute of Mediaeval Studies, 1985.

Haq, S. Nomanul. "Greek Alchemy or Shīʿī Metaphysics? A Preliminary Statement concerning Jābir ibn Hayyān's Zāhir and Bātin." *Bulletin of the Royal Institute for Inter-faith Studies* 4, no. 2 (Autumn/Winter 2002): 19–32.

Hare, R. M. *Plato.* Oxford: Oxford University Press, 1982.

Hargreave, David. "Reconstructing the Planetary Motions of the Eudoxean System." *Scripta Mathematica*, 28 (1970): 335–45.

Häring, Nikolaus. "Chartres and Paris Revisited." In O'Donnell, J. Reginald, ed., *Essays in Honour of Anton Charles Pegis*, pp. 268–329. Toronto: Pontifical Institute of Mediaeval Studies, 1974.

———. "The Creation and Creator of the World According to Thierry of Chartres and Clarenbaldus of Arras." *Archives d'histoire doctrinale et littéraire du moyen âge*, 22 (1955): 137–216.

Harley, J. B., and Woodward, David, eds. *The History of Cartography*, vol. 1: *Cartography in Prehistoric, Ancient, and Medieval Europe and the Mediterranean*. Chicago: University of Chicago Press, 1987.

Harris, John R. "Medicine." In Harris, John R., ed., *The Legacy of Egypt*, 2d ed., pp. 112–37. Oxford: Clarendon Press, 1971.

Harris, William V. *Ancient Literacy*. Cambridge, Mass.: Harvard University Press, 1989.

Harrison, Peter. *Adam's Encyclopaedia: The Fall of Man and the Foundations of Science*. Cambridge: Cambridge University Press, forthcoming.

———. *The Bible, Protestantism, and the Rise of Natural Science*. Cambridge: Cambridge University Press, 1998.

———. "Curiosity, Forbidden Knowledge, and the Reformation of Natural Philosophy in Early-Modern England." *Isis*, 92 (2001): 265–90.

———. "Original Sin and the Problem of Knowledge in Early Modern Europe." *Journal of the History of Ideas*, 63 (2002): 239–59.

———. "Voluntarism and Early Modern Science." *History of Science*, 40 (2002): 63–89.

Hartner, Willy. "Al-Battānī." *Dictionary of Scientific Biography*, 1:507–16.

Haskins, Charles Homer. "The De arte venandi cum avibus of Frederick II." *English Historical Review*, 36 (1921): 334–55. Reprinted in Haskins, *Studies in the History of Mediaeval Science*, pp. 299–326.

———. *The Renaissance of the Twelfth Century*. Cambridge, Mass.: Harvard University Press, 1927.

———. *The Rise of Universities*. Providence: Brown University Press, 1923.

———. "Science at the Court of the Emperor Frederick II." *American Historical Review*, 27 (1922): 669–94. Reprinted in Haskins, *Studies in the History of Mediaeval Science*, pp. 242–71.

———. *Studies in the History of Mediaeval Science*. Cambridge, Mass.: Harvard University Press, 1924.

Heath, Thomas L. *Aristarchus of Samos, The Ancient Copernicus: A History of Greek Astronomy to Aristarchus*. Oxford: Clarendon Press, 1913.

———. *A History of Greek Mathematics*, 2 vols. Oxford: Clarendon Press, 1921.

———. *Mathematics in Aristotle*. Oxford: Clarendon Press, 1949.

Helton, Tinsley, ed. *The Renaissance: A Reconsideration of the Theories and Interpretations of the Age*. Madison: University of Wisconsin Press, 1961.

Henry, John. "Atomism." In Hessenbruch, Arne, ed., *Reader's Guide to the History of Science*, pp. 56–59. London: Fitzroy Dearborn, 2000.

Herlihy, David. *The Black Death and the Transformation of the West*, ed. Samuel K. Cohn, Jr. Cambridge, Mass.: Harvard University Press, 1997.

———. "Demography." *Dictionary of the Middle Ages*, 4:136–48.

Hesiod. *The Poems of Hesiod*, trans. R. M. Frazer. Norman: University of Oklahoma Press, 1983.

———. *Theogony and Works and Days*, trans., with introduction and notes, by M. L. West. Oxford: Oxford University Press, 1988.

Hildebrandt, M. M. *The External School in Carolingian Society*. Leiden: Brill, 1991.

Hill, Donald R. *A History of Engineering in Classical and Medieval Times*. London: Croom Helm, 1984.

———. *Islamic Science and Engineering*. Edinburgh: Edinburgh University Press, 1993.

Hillenbrand, Robert. "Cordoba." In *Dictionary of the Middle Ages*, 3:597–601.

Hillgarth, J. N. "Isidore of Seville, St." *Dictionary of the Middle Ages*, 6:563–66.

Hippocrates, with an English Translation, trans. W. H. S. Jones, E. T. Withington, and Paul Potter, 6 vols. London: Heinemann, 1923–88.

Hissette, Roland. *Enquête sur les 219 articles condamnés à Paris le 7 mars 1277*. Philosophes médiévaux, no. 22. Louvain: Publications universitaires, 1977.

Hitti, Philip K. *History of the Arabs: From the Earliest Times to the Present Day*. London: Macmillan, 1937.

Hogendijk, Jan P., and Sabra, Abdelhamid I., eds. *The Enterprise of Science in Islam: New Perspectives*. Cambridge, Mass.: MIT Press, 2003.

Holmes, Frederic L., and Levere, Trevor H., eds. *Experiments and Experimentation in the History of Chemistry*. Cambridge, Mass.: MIT Press, 2000.

Holmyard, E. J. *Alchemy*. Harmondsworth: Penguin, 1957.

Hopkins, Jasper. *A Companion to the Study of St. Anselm*. Minneapolis: University of Minnesota Press, 1972.

Hoskin, Michael, and Molland, A. G. "Swineshead on Falling Bodies: An Example of Fourteenth-Century Physics." *British Journal for the History of Science*, 3 (1966): 150–82.

Hourani, Albert. *A History of the Arab Peoples*. Cambridge, Mass.: Harvard University Press, 1991.

Høyrup, Jens. *In Measure, Number, and Weight: Studies in Mathematics and Culture*. Albany: State University of New York Press, 1994.

Huff, Toby E. *The Rise of Early Modern Science: Islam, China, and the West*. Cambridge: Cambridge University Press, 1993.

Hugh of St. Victor. *The Didascalicon of Hugh of St. Victor: A Medieval Guide to the Arts*, ed. and trans. Jerome Taylor. New York: Columbia University Press, 1961.

Hugonnard-Roche, Henri. "The Influence of Arabic Astronomy in the Medieval West." *Encyclopedia of the History of Arabic Science*, 1:284–305.

Hussey. Edward. "Aristotle's Mathematical Physics: A Reconstruction." In Judson, Lindsay, ed., *Aristotle's Physics: A Collection of Essays*, chap. 9: 213–42. Oxford: Clarendon Press, 1991.

Huxley, G. L. "Anthemius of Tralles." *Dictionary of Scientific Biography*, 1:169–70.

Hyman, Arthur. "Aristotle's 'First Matter' and Avicenna's and Averroes' 'Corporeal Form.'" In *Harry Austryn Wolfson Jubilee Volume*, 1:385–406. Jerusalem: American Academy for Jewish Research, 1965.

Imbault-Huart, Marie-José. *La médecine au moyen âge à travers les manuscrits de la Bibliothèque Nationale*. Paris: Editions de la Porte Verte, 1983.

Iqbal, Muzaffar. *Islam and Science*. Burlington, Vt.: Ashgate, 2002.

Irving, Washington. *A History of the Life and Voyages of Christopher Columbus*, 4 vols. Paris: A. and W. Galignani, 1828.

Isidore of Seville. See Sharpe, William D.

Iskandar, Albert Z. "Ḥunayn the Translator; Ḥunayn the Physician." *Dictionary of Scientific Biography*, 15:234–39.

———. "Ibn al-Nafis." *Dictionary of Scientific Biography*, 9:602–6.

Jackson, Ralph. *Doctors and Diseases in the Roman Empire*. Norman: University of Oklahoma Press, 1988.

Jacob, Margaret C. *The Cultural Meaning of the Scientific Revolution*. New York: Knopf, 1988.

Jacobsen, Thorkild. "Mesopotamia: The Cosmos as State." In Frankfort, H.; Frankfort, H. A.; Wilson, John A.; and Jacobsen, Thorkild, *Before Philosophy: The Intellectual Adventure of Ancient Man*, pp. 137–99. Baltimore: Penguin, 1951.

Jacquart, Danielle. "Anatomy, Physiology, and Medical Theory." In Lindberg, David C., and Shank, Michael H., eds., *The Cambridge History of Science*, vol. 2: *The Middle Ages*. Cambridge: Cambridge University Press, forthcoming.

———. "The Influence of Arabic Medicine in the Medieval West." 3:963–84.

Jacquart, Danielle, and Thomasset, Claude. *Sexuality and Medicine in the Middle Ages*, trans. Matthew Adamson. Princeton: Princeton University Press, 1988.

Jaeger, Werner. *Paideia: The Ideals of Greek Culture*, 3 vols., trans. Gilbert Highet. Oxford: Oxford University Press, 1939.

Jaki, Stanley. *Uneasy Genius: The Life and Work of Pierre Duhem*. The Hague: Nijhoff, 1984.

Johnson, D. W. "Monophysitism." *Dictionary of the Middle Ages*, 8:476–79.

———. "Nestorianism." *Dictionary of the Middle Ages*, 9:104–8.

Jolivet, Jean. "The Arabic Inheritance." In Dronke, Peter, ed., *A History of Twelfth-Century Western Philosophy*, pp. 113–48. Cambridge: Cambridge University Press, 1988.

Jones, Alexander. "The Adaptation of Babylonian Methods in Greek Numerical Astronomy." *Isis*, 82 (1991): 441–53. Reprinted in Shank, Michael H., ed, *The Scientific Enterprise in Antiquity and the Middle Ages*, pp. 96–109. Chicago: University of Chicago Press, 2000.

Jones, Charles W. "Bede." *Dictionary of the Middle Ages*, 2:153–56.

Jones, Peter M. *Medieval Medical Miniatures*. London: British Library in association with the Wellcome Institute for the History of Medicine, 1984.

Judson, Lindsay, ed. *Aristotle's Physics: A Collection of Essays*. Oxford: Clarendon Press, 1991.

Kahn, Charles H. *Anaximander and the Origins of Greek Cosmology*. New York: Columbia University Press, 1960.

Kaiser, Christopher. *Creation and the History of Science*. Grand Rapids: Eerdmans, 1991.

Kargon, Robert Hugh. *Atomism in England from Hariot to Newton*. Oxford: Clarendon Press, 1966.

Kari-Niazov, T. N. "Ulugh Beg." *Dictionary of Scientific Biography*, 13:535–37.

Kealey, Edward J. "England's Earliest Women Doctors." *Journal of the History of Medicine*, 40 (1985): 473–77.

———. *Medieval Medicus: A Social History of Anglo-Norman Medicine*. Baltimore: Johns Hopkins University Press, 1981.

Kennedy, E. S. "The Arabic Heritage in the Exact Sciences." *Al-Abhath: A Quarterly Journal for Arab Studies*, 23 (1970): 327–44.

———. "The Exact Sciences." In Frye, R. N., ed., *The Cambridge History of Iran*, vol. 4: *The Period from the Arab Invasion to the Saljuqs*, pp. 378–95. Cambridge: Cambridge University Press, 1975.

———. "The History of Trigonometry: An Overview." In Kennedy, E. S., et al., *Studies in the Islamic Exact Sciences*, pp. 3–29.

Kennedy, E. S., and David Pingree. *The Astrological History of Māshā'allāh*. Cambridge, Mass.: Harvard University Press, 1971.

Kennedy, E. S., with colleagues and former students. *Studies in the Islamic Exact Sciences*. Beirut: American University of Beirut, 1983.

Kheirandish, Elaheh. "The Mathematical Sciences in Islam." In Lindberg, David C., and Shank, Michael H., eds., *The Cambridge History of Science*, vol. 2: *The Middle Ages*. Cambridge: Cambridge University Press, forthcoming.

Kibre, Pearl. "Albertus Magnus on Alchemy." In Weisheipl, James A., ed., *Albertus Magnus and the Sciences: Commemorative Essays 1980*, pp. 187–202. Toronto: Pontifical Institute of Mediaeval Studies, 1980.

———. "'Astronomia' or 'Astrologia Ypocratis.'" In Hilfstein, Erna; Czartoryski, Paweł; and Grande, Frank D., eds., *Science and History: Studies in Honor of Edward Rosen*, pp. 133–56. Studia Copernicana, no. 16. Wrocław: Ossolineum, 1978.

———. "The Quadrivium in the Thirteenth Century Universities (with Special Reference to Paris)." In *Arts libéraux et philosophie au moyen âge: Actes du quatrième congrès international de philosophie médiévale*, Université de Montréal, 27 August–2 September 1967, pp. 175–91. Montréal: Institut d'études médiévales, 1969.

———. *Scholarly Privileges in the Middle Ages*. Cambridge, Mass.: Mediaeval Academy of America, 1962.

———. *Studies in Medieval Science: Alchemy, Astrology, Mathematics, and Medicine*. London: Hambledon Press, 1984.

Kibre, Pearl, and Siraisi, Nancy G. "The Institutional Setting: The Universities." In Lindberg, David C., ed., *Science in the Middle Ages*, pp. 120–44. Chicago: University of Chicago Press, 1978.

Kieckhefer, Richard. *Magic in the Middle Ages*. Cambridge: Cambridge University Press, 1990.

King, David A. *Islamic Astronomical Instruments*. London: Variorum, 1987.

———. *Islamic Mathematical Astronomy*. London: Variorum, 1986.

Kirk, G. S., and Raven, J. E. *The Presocratic Philosophers: A Critical History with a Selection of Texts*. Cambridge: Cambridge University Press, 1960.

Kitchell, Kenneth F., Jr., and Resnick, Irven Michael, trans. *Albertus Magnus On Animals: A Medieval "Summa Zoologica,"* 2 vols. Baltimore: Johns Hopkins University Press, 1999.

Knorr, Wilbur. "Archimedes and the Pseudo-Euclidean Catoptrics: Early Stages in the Ancient Geometric Theory of Mirrors." *Archives internationales d'histoire des sciences*, 35 (1985): 28–105.

———. *The Evolution of the Euclidean Elements: A Study of the Theory of Incommensurable Magnitudes and Its Significance for Early Greek Geometry*. Dordrecht: D. Reidel, 1975.

———. "John of Tynemouth alias John of London: Emerging Portrait of a Singular Medieval Mathematician." *British Journal for the History of Science*, 23 (1990): 293–330.

Knowles, David. *The Evolution of Medieval Thought*. New York: Vintage, 1964.

Kogan, Barry S. *Averroes and the Metaphysics of Causation*. Albany: State University of New York Press, 1985.

Kovach, Francis J., and Shahan, Robert W., eds. *Albert the Great: Commemorative Essays*. Norman: University of Oklahoma Press, 1980.

Koyré, Alexandre. *The Astronomical Revolution: Copernicus, Kepler, Borelli*, trans. R. E. W. Maddison. Paris: Hermann, 1973.

———. *From the Closed World to the Infinite Universe*. Baltimore: Johns Hopkins University Press, 1957.

———. *Galileo Studies*, trans. John Mepham. Atlantic Highlands, N.J.: Humanities Press, 1978.

———. *Metaphysics and Measurement: Essays in the Scientific Revolution*. London: Chapman & Hall, 1968.

———. *Newtonian Studies*. London: Chapman & Hall, 1965.

———. "The Origins of Modern Science: A New Interpretation." *Diogenes*, 16 (Winter 1956): 1–22.

Kramer, Edna E. "Hypatia." *Dictionary of Scientific Biography*, 6:615–16.

Kren, Claudia. *Alchemy in Europe: A Guide to Research*. New York: Garland, 1990.

———. "Astronomy." In Wagner, David L., ed., *The Seven Liberal Arts in the Middle Ages*, pp. 218–47. Bloomington: Indiana University Press, 1983.

———. "Bernard of Verdun." *Dictionary of Scientific Biography*, 2:23–24.

———. "Homocentric Astronomy in the Latin West: The De reprobatione ecentricorum et epiciclorum of Henry of Hesse." *Isis*, 59 (1968): 269–81.

———. *Medieval Science and Technology: A Selected, Annotated Bibliography*. New York: Garland, 1985.

Kretzmann, Norman, ed. *Infinity and Continuity in Ancient and Medieval Thought*. Ithaca: Cornell University Press, 1982.

Kretzmann, Norman; Kenny, Anthony; and Pinborg, Jan, eds. *The Cambridge History of Later Medieval Philosophy*. Cambridge: Cambridge University Press, 1982.

Kristeller, Paul Oskar. "The School of Salerno: Its Development and Its Contribution to the History of Learning." *Bulletin of the History of Medicine*, 17 (1945): 138–94.

Kudlien, Fridolf. "Early Greek Primitive Medicine." *Clio Medica*, 3 (1968): 305–36.

Kudlien, Fridolf, and Durling, Richard J., eds. *Galen's Method of Healing*. Leiden: Brill, 1991.

Kuhn, Thomas S. *The Copernican Revolution: Planetary Astronomy in the Development of Western Thought*. Cambridge, Mass.: Harvard University Press, 1957.

_____. "Mathematical versus Experimental Traditions in the Development of Physical Science." *Journal of Interdisciplinary History*, 7 (1976): 1–31. Reprinted in Kuhn, *The Essential Tension: Selected Studies in Scientific Tradition and Change*, pp. 31–65. Chicago: University of Chicago Press, 1977.

_____. *The Structure of Scientific Revolutions*. Chicago: University of Chicago Press, 1962.

Kunitzsch, Paul. *Stars and Numbers: Astronomy and Mathematics in the Medieval Arab and Western Worlds*. Aldershot: Ashgate, 2004.

Laird, Walter Roy. "Change and Motion." In Lindberg, David C., and Shank, Michael H., eds., *The Cambridge History of Science*, vol. 2: *The Middle Ages*. Cambridge: Cambridge University Press, forthcoming.

Laistner, M. L. W. *Christianity and Pagan Culture in the Later Roman Empire*. Ithaca: Cornell University Press, 1951.

_____. *Thought and Letters in Western Europe*, A.D. *500–900*, new ed. London: Methuen, 1957.

Lang, Helen S. *Aristotle's Physics and Its Medieval Varieties*. Albany: State University of New York Press, 1992.

Langermann, Y. Tzvi. *The Jews and the Sciences in the Middle Ages*. London: Variorum, 1999.

_____. "Science in the Jewish Communities." In Lindberg, David C., and Shank, Michael H., eds., *The Cambridge History of Science*, vol. 2: *The Middle Ages*. Cambridge: Cambridge University Press, forthcoming.

Lassner, Jacob. *The Shaping of ʿAbbāsid Rule*. Princeton: Princeton University Press, 1980.

Lattin, Harriet Pratt, ed. and trans. *The Letters of Gerbert with His Papal Privileges as Sylvester II*. New York: Columbia University Press, 1961.

Lear, Jonathan. *Aristotle: The Desire to Understand*. Cambridge: Cambridge University Press, 1988.

Leclerc, Ivor. *The Nature of Physical Existence*. London: George Allen & Unwin, 1972.

Leclerc, Lucien. *Histoire de la médecine arabe*, 2 vols. Paris: Ernest Leroux, 1876.

Leclercq, Jean, O.S.B. *The Love of Learning and the Desire for God: A Study of Monastic Culture*, trans. Catherine Misrahi. New York: Fordham University Press, 1961.

_____. "The Renewal of Theology." In Benson, Robert L., and Constable, Giles, eds., *Renaissance and Renewal in the Twelfth Century*, pp. 68–87. Cambridge, Mass.: Harvard University Press, 1982.

Lejeune, Albert. *Euclide et Ptolémée: Deux stades de l'optique géométrique grecque*. Louvain: Bibliothèque de l'Université, 1948.

_____. *Recherches sur la catoptrique grecque*. Brussels: Palais des Académies, 1957.

Lemay, Richard. *Abū Maʿshar and Latin Aristotelianism in the Twelfth Century: The Recovery of Aristotle's Natural Philosophy through Arabic Astrology*. Beirut: American University of Beirut, 1962.

———. "Gerard of Cremona." *Dictionary of Scientific Biography*, 15:173–92.

———. "The True Place of Astrology in Medieval Science and Philosophy." In Curry, Patrick, ed., *Astrology, Science, and Society: Historical Essays*, pp. 57–73. Woodbridge, Suffolk: Boydell, 1987.

Lerner, Ralph, and Mahdi, Muhsin, eds. *Medieval Political Philosophy: A Sourcebook*. New York: Free Press of Glencoe, 1963.

Lettinck, P. *Aristotle's* Physics *and Its Reception in the Arabic World: With an Edition of the Unpublished Parts of Ibn Bājja's Commentary on the Physics*. Leiden: Brill, 1994.

Levy-Bruhl, Lucien. *How Natives Think*, trans. Lilian A. Clare. London: George Allen & Unwin, 1926.

Lewis, Bernard, ed. *Islam and the Arab World: Faith, People, Culture*. New York: Knopf, 1976.

Lewis, C. S. *The Discarded Image: An Introduction to Medieval and Renaissance Literature*. Cambridge: Cambridge University Press, 1964.

Lewis, Christopher. *The Merton Tradition and Kinematics in Late Sixteenth and Early Seventeenth Century Italy*. Padua: Antenore, 1980.

Liebeschutz, H. "Boethius and the Legacy of Antiquity." In Armstrong, A. H., ed., *The Cambridge History of Later Greek and Early Medieval Philosophy*, pp. 538–64. Cambridge: Cambridge University Press, 1970.

Lindberg, David C. "Alhazen's Theory of Vision and Its Reception in the West." *Isis*, 58 (1967): 321–41.

———. *A Catalogue of Medieval and Renaissance Optical Manuscripts*. Subsidia Mediaevalia, IV. Toronto: Pontifical Institute of Mediaeval Studies, 1975.

———. "Conceptions of the Scientific Revolution from Bacon to Butterfield: A Preliminary Sketch." In Lindberg, David C., and Westman, Robert S., eds., *Reappraisals of the Scientific Revolution*, pp. 1–26.

———. "Continuity and Discontinuity in the History of Optics: Kepler and the Medieval Tradition." *History and Technology*, 4 (1987): 423–40.

———. "Early Christian Attitudes toward Nature." In Ferngren, Gary B., ed., *The History of Science and Religion in the Western Tradition*, pp. 243–47. Garland: New York, 2000. Reprinted in Ferngren, Gary B., ed., *Science and Religion: A Historical Introduction*, pp. 47–56. Baltimore: Johns Hopkins University Press, 2002.

———. "The Genesis of Kepler's Theory of Light: Light Metaphysics from Plotinus to Kepler." *Osiris*, n.s. 2 (1986): 5–42.

———, ed. and trans. *John Pecham and the Science of Optics: Perspectiva communis*," with an introduction and critical notes. Madison: University of Wisconsin Press, 1970.

———. "Laying the Foundations of Geometrical Optics: Maurolico, Kepler, and the Medieval Tradition." In Lindberg, David C., and Cantor, Geoffrey, *The Discourse of Light from the Middle Ages to the Enlightenment*, pp. 1–65. Los Angeles: William Andrews Clark Memorial Library, 1985.

———. "The Medieval Church Encounters the Classical Tradition: Saint Augustine, Roger Bacon, and the Handmaiden Metaphor," in Lindberg, David C., and Numbers, Ronald L., eds., *When Science and Christianity Meet*, pp. 7–32. Chicago: University of Chicago Press, 2003.

———. "Medieval Science and Its Religious Context." In Thackray, Arnold, ed., *Constructing Knowledge in the History of Science. Osiris*, n.s. 10 (1995): 61–79.

———. "Medieval Science and Religion." In Ferngren, Gary B., ed., *The History of Science and Religion in the Western Tradition*, pp. 59–67. Garland: New York, 2000. Reprinted in Ferngren, Gary B., ed. *Science and Religion: A Historical Introduction*, pp. 57–72. Baltimore: Johns Hopkins University Press, 2002.

———. "On the Applicability of Mathematics to Nature: Roger Bacon and His Predecessors." *British Journal for the History of Science*, 15 (1982): 3–25.

———. "Optics, Western European." *Dictionary of the Middle Ages*, 9:247–53.

———. "A Reconsideration of Roger Bacon's Theory of Pinhole Images." *Archive for History of Exact Sciences*, 6 (1970): 214–23.

———. *Roger Bacon and the Origins of "Perspectiva" in the Middle Ages: A Critical Edition and English Translation of Bacon's "Perspectiva," with Introduction and Notes*. Oxford: Clarendon Press, 1996.

———. "Roger Bacon and the Origins of Perspectiva in the West." In Grant, Edward, and Murdoch, John E., eds., *Mathematics and Its Applications to Science and Natural Philosophy in the Middle Ages: Essays in Honor of Marshall Clagett*, pp. 249–68. Cambridge: Cambridge University Press, 1987.

———. "Roger Bacon on Light, Vision, and the Universal Emanation of Force." In Hackett, Jeremiah, ed., *Roger Bacon and the Sciences: Commemorative Essays*, pp. 243–75. Leiden: Brill, 1997.

———, ed. and trans. *Roger Bacon's Philosophy of Nature: A Critical Edition, with English Translation, Introduction, and Notes, of "De multiplicatione specierum" and "De speculis comburentibus."* Oxford: Clarendon Press, 1983.

———. "Science and the Early Church." In Lindberg, David C., and Numbers, Ronald L., eds., *God and Nature: Historical Essays on the Encounter between Christianity and Science*, pp. 19–48.

———. "Science and the Medieval Church." In Lindberg, David C., and Shank, Michael H., eds., *The Cambridge History of Science*, vol. 2: *The Middle Ages*. Cambridge: Cambridge University Press, forthcoming.

———. "Science as Handmaiden: Roger Bacon and the Patristic Tradition." *Isis*, 78 (1987): 518–36.

———, ed. *Science in the Middle Ages*. Chicago: University of Chicago Press, 1978.

———. "The Science of Optics." In Lindberg, ed., *Science in the Middle Ages*, pp. 338–68.

———. *Studies in the History of Medieval Optics*. London: Variorum, 1983.

———. *Theories of Vision from al-Kindi to Kepler*. Chicago: University of Chicago Press, 1976.

———. "The Theory of Pinhole Images from Antiquity to the Thirteenth Century." *Archive for History of Exact Sciences*, 5 (1968): 154–76.

———. "The Theory of Pinhole Images in the Fourteenth Century." *Archive for History of Exact Sciences*, 6 (1970): 299–325.

———. "The Transmission of Greek and Arabic Learning to the West." In Lindberg, David C., ed., *Science in the Middle Ages*, pp. 52–90.

Lindberg, David C., and Numbers, Ronald L., eds. *God and Nature: Historical Essays on the Encounter between Christianity and Science*. Berkeley and Los Angeles: University of California Press, 1986.

————, eds. *When Science and Christianity Meet*. Chicago: University of Chicago Press, 2003.

Lindberg, David C., and Shank, Michael H., eds., *The Cambridge History of Science*, vol. 2: *Medieval Science* (forthcoming).

Lindberg, David C., and Tachau, Katherine H. "The Science of Light and Color: Seeing and Knowing." In Lindberg, David C., and Shank, Michael H., eds., *The Cambridge History of Science*, vol. 2: *The Middle Ages*. Cambridge: Cambridge University Press, forthcoming.

Lindberg, David C., and Westman, Robert S., eds. *Reappraisals of the Scientific Revolution*. Cambridge: Cambridge University Press, 1990.

Lipton, Joshua D. *The Rational Evaluation of Astrology in the Period of Arabo-Latin Translation, ca. 1126–1187 A.D.* Ph.D. dissertation, University of California-Los Angeles.

Little, A. G., ed. *Roger Bacon Essays*. Oxford: Clarendon Press, 1914.

Livesey, Steven J. *Theology and Science in the Fourteenth Century: Three Questions on the Unity and Subalternation of the Sciences from John of Reading's Commentary on the Sentences*. Studien und Texte zur Geistesgeschichte des Mittelalters, vol. 25. Leiden: Brill, 1989.

Lloyd, G. E. R. *Adversaries and Authorities: Investigations into Ancient Greek and Chinese Science*. Cambridge: Cambridge University Press, 1996.

————. *Aristotelian Explorations*. Cambridge: Cambridge University Press, 1996.

————. *Aristotle: The Growth and Structure of His Thought*. Cambridge: Cambridge University Press, 1968.

————. *Demystifying Mentalities*. Cambridge: Cambridge University Press, 1990.

————. "Experiment in Early Greek Philosophy and Medicine." In Lloyd, *Methods and Problems in Greek Science*, chap. 4. Cambridge: Cambridge University Press, 1991.

————. *Early Greek Science: Thales to Aristotle*. London: Chatto & Windus, 1970.

————. *Greek Science after Aristotle*. London: Chatto & Windus, 1973.

————, ed. *Hippocratic Writings*. Harmondsworth: Penguin, 1978.

————. *Magic, Reason, and Experience: Studies in the Origins and Development of Greek Science*. Cambridge: Cambridge University Press, 1979.

————. *Methods and Problems in Greek Science: Selected Papers*. Cambridge: Cambridge University Press, 1991.

————. *The Revolutions of Wisdom: Studies in the Claims and Practice of Ancient Greek Science*. Berkeley and Los Angeles: University of California Press, 1987.

————. "Saving the Appearances." *Classical Quarterly*, 28 (1978): 202–22.

————. *Science, Folklore, and Ideology: Studies in the Life Sciences in Ancient Greece*. Cambridge: Cambridge University Press, 1983.

Lloyd, G. E. R., and Sivin, Nathan, *The Way and the Word: Science and Medicine in Early China and Greece*. New Haven: Yale University Press, 2002.

Locher, A. "The Structure of Pliny the Elder's Natural History." In French, Roger, and Greenaway, Frank, eds., *Science in the Early Roman Empire*, pp. 20–29. Totawa, N.J.: Barnes & Noble, 1986.

Long, A. A. "Astrology: Arguments Pro and Contra." In Barnes, Jonathan; Brunschwig, Jacques; Burnyeat, Myles; and Schofield, Malcolm, eds., *Science and Speculation: Studies in Hellenistic Theory and Practice*, pp. 165–92. Cambridge: Cambridge University Press, 1982.

———. *Hellenistic Philosophy: Stoics, Epicureans, Sceptics*, 2d ed. London: Duckworth, 1974.

———. "The Stoics on World-Conflagration and Everlasting Recurrence." In Epp, Ronald H., ed., *Recovering the Stoics*, pp. 13–37. Supplement to *The Southern Journal of Philosophy*, vol. 23 (1985).

Long, A. A., and Sedley, D. N. *The Hellenistic Philosophers*, 2 vols. Cambridge: Cambridge University Press, 1987.

Long, Pamela O. *Openness, Secrecy, Authorship: Technical Arts and the Culture of Knowledge from Antiquity to the Renaissance*. Baltimore: Johns Hopkins University Press, 2001.

Longrigg, James. "Anatomy in Alexandria in the Third Century B.C." *British Journal for the History of Science*, 21 (1988): 455–88.

———. "Erasistratus." *Dictionary of Scientific Biography*, 4:382–86.

———. "Presocratic Philosophy and Hippocratic Medicine." *History of Science*, 27 (1989): 1–39.

———. "Superlative Achievement and Comparative Neglect: Alexandrian Medical Science and Modern Historical Research." *History of Science*, 19 (1981): 155–200.

Lucretius. *De rerum natura*, trans. W. H. D. Rouse and M. F. Smith, rev. 2d ed. London: Heinemann, 1982.

Luscombe, David E. *Peter Abelard*. London: Historical Association, 1979.

———. "Peter Abelard." In Dronke, Peter, ed., *A History of Twelfth-Century Western Philosophy*, pp. 279–307. Cambridge: Cambridge University Press, 1988.

Lüthy, Christoph; Murdoch, John E.; and Newman, William R., eds., *Late Medieval and Early Modern Corpuscular Matter Theories*. Leiden: Brill, 2001.

Lutz, Cora E. *Schoolmasters of the Tenth Century*. Hamden, Conn.: Archon, 1977.

Lynch, John Patrick. *Aristotle's School: A Study of a Greek Educational Institution*. Berkeley and Los Angeles: University of California Press, 1972.

Lytle, Guy Fitch. "The Careers of Oxford Students in the Later Middle Ages." In Kittelson, James M., and Transue, Pamela J., eds., *Rebirth, Reform, and Resilience: Universities in Transition 1300–1700*, pp. 213–53. Columbus: Ohio State University Press, 1984.

———. "Patronage Patterns and Oxford Colleges, c. 1300–c. 1530." In Stone, Lawrence, ed., *The University in Society*, 1:111–49. Princeton: Princeton University Press, 1974.

Machamer, Peter. *The Cambridge Companion to Galileo*. Cambridge: Cambridge University Press, 1998.

MacKinney, Loren C. *Early Medieval Medicine, with Special Reference to France and Chartres*. Baltimore: Johns Hopkins University Press, 1937.

———. *Medical Illustrations in Medieval Manuscripts*. London: Wellcome Historical Medical Library, 1965.

Macrobius. *Commentary on the Dream of Scipio*, trans. with introduction and notes by William H. Stahl. New York: Columbia University Press, 1952.

Mahoney, Michael S. "Another Look at Greek Geometrical Analysis." *Archive for History of Exact Sciences*, 5 (1968): 318–48.

———. "Mathematics." In Lindberg, David C., ed., *Science in the Middle Ages*, pp. 145–78. Chicago: University of Chicago Press, 1978.

Maier, Anneliese. "The Achievements of Late Scholastic Natural Philosophy." In Maier, *On the Threshold of Exact Science*, pp. 143–70. Philadelphia: University of Pennsylvania Press, 1982.

———. *An der Grenze von Scholastik und Naturwissenschaft*, 2d ed. Rome: Edizioni di Storia e Letteratura, 1952.

———. *Ausgehendes Mittelalter: Gesammelte Aufsätze zur Geistesgeschichte des 14. Jahrhunderts*, 3 vols. Rome: Edizioni di Storia e Letteratura, 1964–77.

———. *Metaphysische Hintergründe der spätscholastischen Naturphilosophie*. Rome: Edizioni di Storia e Letteratura, 1955.

———. "The Nature of Motion." In Maier, *On the Threshold of Exact Science*, pp. 21–39. Philadelphia: University of Pennsylvania Press, 1982.

———. *On the Threshold of Exact Science: Selected Writings of Anneliese Maier on Late Medieval Natural Philosophy*, trans. Steven D. Sargent. Philadelphia: University of Pennsylvania Press, 1982.

———. "The Significance of the Theory of Impetus for Scholastic Natural Philosophy." In Maier, *On the Threshold of Exact Science*, pp. 76–102. Philadelphia: University of Pennsylvania Press, 1982.

———. "The Theory of the Elements and the Problem of Their Participation in Compounds." In Maier, *On the Threshold of Exact Science*, pp. 124–42. Philadelphia: University of Pennsylvania Press, 1982.

———. *Die Vorläufer Galileis im 14. Jahrhundert*, 2d ed. Rome: Edizioni di Storia e Letteratura, 1966.

———. *Zwischen Philosophie und Mechanik*. Rome: Edizioni di Storia e Letteratura, 1958.

Maierù, A., and Paravicini Bagliani, A., eds. *Studi sul XIV secolo in memoria di Anneliese Maier*. Storia e Letteratura, Raccolta di studi e testi, no. 151. Rome: Edizioni di Storia e Letteratura, 1981.

Majno, Guido. *The Healing Hand: Man and Wound in the Ancient World*. Cambridge, Mass.: Harvard University Press, 1975.

Makdisi, George. *The Rise of Colleges: Institutions of Learning in Islam and the West*. Edinburgh: Edinburgh University Press, 1981.

Malinowski, Bronislaw. *Myth in Primitive Psychology*. New York: W. W. Norton, 1926.

Marenbon, John. *Early Medieval Philosophy (480–1150): An Introduction*. London: Routledge & Kegan Paul, 1983.

————. *From the Circle of Alcuin to the School of Auxerre: Logic, Theology, and Philosophy in the Early Middle Ages*. Cambridge: Cambridge University Press, 1981.

Marrou, H. I. *A History of Education in Antiquity*, trans. George Lamb. New York: Sheed and Ward, 1956.

Martin, R. N. D. "The Genesis of a Mediaeval Historian: Pierre Duhem and the Origins of Statics." *Annals of Science*, 33 (1976): 119–29.

McCluskey, Stephen C. *Astronomies and Cultures in Early Medieval Europe*. Cambridge: Cambridge University Press, 1998.

————. "Gregory of Tours, Monastic Timekeeping, and Early Christian Attitudes to Astronomy." *Isis*, 81 (1990): 9–22.

————. "Natural Knowledge in the Early Middle Ages." In Lindberg, David C., and Shank, Michael H., eds., *The Cambridge History of Science*, vol. 2: *The Middle Ages*. Cambridge: Cambridge University Press, forthcoming.

McColley, Grant. "The Theory of the Diurnal Rotation of the Earth." *Isis*, 26 (1937): 392–402.

McDiarmid, J. B. "Theophrastus." *Dictionary of Scientific Biography*, 13:328–34.

McEvoy, James. *The Philosophy of Robert Grosseteste*. Oxford: Clarendon Press, 1982.

McInerny, Ralph. *St. Thomas Aquinas*. Notre Dame, Ind.: University of Notre Dame Press, 1982.

McKitterick, Rosamond. "The Carolingian Renaissance." In Story, Joanna, ed., *Charlemagne: Empire and Society*, pp. 151–66. Manchester, England: Manchester University Press, 2005.

————. *The Carolingians and the Written Word*. Cambridge: Cambridge University Press, 1989.

McMullin, Ernan, ed. *The Concept of Matter in Greek and Medieval Philosophy*. Notre Dame, Ind.: University of Notre Dame Press, 1963.

————. "Conceptions of Science in the Scientific Revolution." In Lindberg, David C., and Westman, Robert S., eds., *Reappraisals of the Scientific Revolution*, pp. 27–86. Cambridge: Cambridge University Press, 1990.

————. "Empiricism and the Scientific Revolution." In Singleton, Charles, ed., *Art, Science, and History in the Renaissance*, pp. 331–69. Baltimore: Johns Hopkins University Press, 1967.

————. "Galileo on Science and Scripture." In Machamer, Peter, ed., *Cambridge Companion to Galileo*, pp. 271–347. Cambridge: Cambridge University Press, 1998.

————. "Medieval and Modern Science: Continuity or Discontinuity?" *International Philosophical Quarterly*, 5 (1965): 103–29.

McVaugh, Michael. "Arnald of Villanova and Bradwardine's Law." *Isis*, 58 (1967): 56–64.

————, ed. *Arnald de Villanova, Opera medica omnia*, vol. 2: *Aphorismi de gradibus*. Granada: Seminarium historiae medicae Granatensis, 1975.

————. "Constantine the African." *Dictionary of Scientific Biography*, 3:393–95.

————. "The Experimenta of Arnald of Villanova." *Journal of Medieval and Renaissance Studies*, 1 (1971): 107–18.

———. *Medicine before the Plague: Practitioners and Their Patients in the Crown of Aragon, 1285–1345.* Cambridge: Cambridge University Press, 1993.

———. "Medicine, History of." *Dictionary of the Middle Ages,* 8:247–54.

———. "The Nature and Limits of Medical Certitude." *Osiris,* n.s. 6 (1990): 62–84.

———. "Quantified Medical Theory and Practice at Fourteenth-Century Montpellier." *Bulletin of the History of Medicine,* 43 (1969): 397–413.

———. "Theriac at Montpellier." *Sudhoffs Archiv,* 56 (1972): 113–44.

Melling, David J. *Understanding Plato.* Oxford: Oxford University Press, 1987.

Menocal, María Rosa. *Ornament of the World: How Muslims, Jews, and Christians Created a Culture of Tolerance in Medieval Spain.* Boston: Little, Brown and Co., 2002.

Merton, Robert K. *Science, Technology, and Society in Seventeenth Century England.* Originally published in *Osiris,* 4 (1938): 360–632. Reissued, with a new introduction, New York: Harper and Row, 1970.

Meyerhof, Max. "Science and Medicine." In Arnold, Thomas, and Guillaume, Alfred, eds., *The Legacy of Islam,* pp. 311–55. London: Oxford University Press, 1931.

———. *Studies in Medieval Arabic Medicine: Theory and Practice,* ed. Penelope Johnstone. London: Variorum, 1984.

Micheau, Françoise. "The Scientific Institutions in the Medieval Near East." *Encyclopedia of the History of Arabic Science,* 3:985–1007.

Migne, J.-P., ed. *Patrologia latina,* 221 vols. in 223. Paris: J.-P. Migne, 1844–91.

Millas-Vallicrosa, J. M. "Translations of Oriental Scientific Works." In Métraux, Guy S., and Crouzet, François, eds., *The Evolution of Science,* pp. 128–67. New York: Mentor, 1963.

Miller, Timothy S. *The Birth of the Hospital in the Byzantine Empire.* Baltimore: Johns Hopkins University Press, 1985.

———. "The Knights of Saint John and the Hospitals of the Latin West." *Speculum,* 53 (1978): 709–33.

Minio-Paluello, Lorenzo. "Boethius, Anicius Manlius Severinus." *Dictionary of Scientific Biography,* 2:228–36.

———. "Michael Scot." *Dictionary of Scientific Biography,* 9:361–65.

———. "Moerbeke, William of." *Dictionary of Scientific Biography,* 9:434–40.

Mohr, Richard D. *The Platonic Cosmology.* Leiden: Brill, 1985.

Molland, A. G. "Aristotelian Holism and Medieval Mathematical Physics." In Caroti, Stefano, ed., *Studies in Medieval Natural Philosophy,* pp. 227–35. Florence: Olschki, 1989.

———. "Continuity and Measure in Medieval Natural Philosophy." *Miscellanea Mediaevalia,* 16 (1983): 132–44.

———. "An Examination of Bradwardine's Geometry." *Archive for History of Exact Sciences,* 19 (1978): 113–75.

———. "The Geometrical Background to the 'Merton School.'" *British Journal for the History of Science,* 4 (1968–69): 108–25.

———. "Mathematics." In Lindberg, David C., and Shank, Michael H., eds., *The Cambridge History of Science,* vol. 2: *The Middle Ages.* Cambridge: Cambridge University Press, forthcoming.

_____. "Nicole Oresme and Scientific Progress." *Miscellanea Mediaevalia*, 9 (1974): 206–20.

Montgomery, Scott L. *Science in Translation: Movements of Knowledge through Cultures and Time*. Chicago: University of Chicago Press, 2000.

Moody, Ernest A. "Galileo and Avempace: The Dynamics of the Leaning Tower Experiment." *Journal of the History of Ideas*, 12 (1951): 163–93, 375–422.

_____. *Studies in Medieval Philosophy, Science, and Logic: Collected Papers 1933–1969*. Berkeley and Los Angeles: University of California Press, 1975.

Moody, Ernest A., and Clagett, Marshall, eds. and trans. *The Medieval Science of Weights*. Madison: University of Wisconsin Press, 1960.

Morelon, Régis. "Eastern Arabic Astronomy between the Eighth and Eleventh Centuries." *Encyclopedia of the History of Arabic Science*, 1:20–57.

_____. "General Survey of Arabic Astronomy." *Encyclopedia of the History of Arabic Science*, 1:1–19.

Morris, Colin. *The Discovery of the Individual, 1050–1200*. New York: Harper and Row, 1972.

Murdoch, John E. *Album of Science: Antiquity and the Middle Ages*. New York: Scribner's, 1984.

_____. "The Development of a Critical Temper: New Approaches and Modes of Analysis in Fourteenth-Century Philosophy, Science, and Theology." *Medieval and Renaissance Studies*, 7 (1978): 51–79.

_____. "From Social into Intellectual Factors: An Aspect of the Unitary Character of Late Medieval Learning." In Murdoch, John E., and Sylla, Edith D., eds., *The Cultural Context of Medieval Learning*, pp. 271–348. Boston Studies in the Philosophy of Science, no. 26. Dordrecht: D. Reidel, 1975.

_____. "Mathesis in philosophiam scholasticam introducta: The Rise and Development of the Application of Mathematics in Fourteenth Century Philosophy and Theology." In *Arts libéraux et philosophie au moyen âge: Actes du quatrième congrès international de philosophie médiévale*, Université de Montréal, 27 August–2 September 1967, pp. 215–54. Montréal: Institut d'études médiévales, 1969.

_____. "Philosophy and the Enterprise of Science in the Later Middle Ages." In Elkana, Yehuda, ed., *The Interaction between Science and Philosophy*, pp. 51–74. Atlantic Highlands, N.J.: Humanities Press, 1974.

Murdoch, John E., and Sylla, Edith D. "Anneliese Maier and the History of Medieval Science." In Maierù, A., and Paravicini Bagliani, A., eds., *Studi sul XIV secolo in memoria di Anneliese Maier*, pp. 7–13. Storia e letteratura: Raccolta di studi e testi, no. 151. Rome: Edizioni Storia e Letteratura, 1981.

_____, eds. *The Cultural Context of Medieval Learning*. Boston Studies in the Philosophy of Science, no. 26. Dordrecht: D. Reidel, 1975.

_____. "The Science of Motion." In Lindberg, David C., ed., *Science in the Middle Ages*, pp. 206–64. Chicago: University of Chicago Press, 1978.

_____. "Swineshead, Richard." *Dictionary of Scientific Biography*, 13:184–213.

Murray, Alexander. *Reason and Society in the Middle Ages*. Oxford: Clarendon Press, 1978.

Nakosteen, Mehdi. *History of Islamic Origins of Western Education, A.D. 800–1359, with an Introduction to Medieval Muslim Education*. Boulder: University of Colorado Press, 1964.

Nasr, Seyyed Hossein. *An Introduction to Islamic Cosmological Doctrines*. Cambridge, Mass.: Belknap Press, 1964.

———. *Science and Civilization in Islam*. Cambridge, Mass.: Harvard University Press, 1968.

Neugebauer, Otto. "Apollonius' Planetary Theory." *Communications on Pure and Applied Mathematics*, 8 (1955): 641–48.

———. *Astronomy and History: Selected Essays*. New York: Springer, 1983.

———. *The Exact Sciences in Antiquity*. Princeton: Princeton University Press, 1952.

———. *A History of Ancient Mathematical Astronomy*, 3 pts. New York: Springer, 1975.

———. "On the Allegedly Heliocentric Theory of Venus by Heraclides Ponticus." *American Journal of Philology*, 93 (1972): 600–601.

———. "On the 'Hippopede' of Eudoxus." *Scripta Mathematica*, 19 (1953): 225–29.

Neugebauer, Otto, and Sachs, A., eds. *Mathematical Cuneiform Texts*. American Oriental Series, vol. 29. New Haven: American Oriental Society, 1945.

Newman, William R. "Alchemy, Assaying, and Experiment." In Holmes, Frederic L., and Levere, Trevor H., eds., *Instruments and Experimentation in the History of Chemistry*, pp. 35–54. Cambridge, Mass.: MIT Press, 2000.

———. *Atoms and Alchemy: Chymistry and the Experimental Origins of the Scientific Revolution*. Chicago: University of Chicago Press, 2006.

———. "Experimental Corpuscular Theory in Aristotelian Alchemy: From Geber to Sennert." In Lüthy, Christoph; Murdoch, John E.; and Newman, William R., eds., *Late Medieval and Early Modern Corpuscular Matter Theories*, pp. 291–329. Leiden: Brill, 2001.

———. "The Genesis of the *Summa perfectionis*." *Archives internationales d'histoire des sciences*, 35 (1985): 240–302.

———. "Medieval Alchemy." In Lindberg, David C., and Shank, Michael H., eds., *The Cambridge History of Science*, vol. 2: *The Middle Ages*. Cambridge: Cambridge University Press, forthcoming.

———. *Promethean Ambitions: Alchemy and the Quest to Perfect Nature*. Chicago: University of Chicago Press, 2004.

———. "The *Summa perfectionis* and Late Medieval Alchemy: A Study of Chemical Traditions, Techniques, and Theories in Thirteenth Century Italy," 4 vols. Ph.D. dissertation, Harvard University, 1986.

———. *The "Summa perfectionis" of Pseudo-Geber: A Critical Edition, Translation, and Study*. Leiden: Brill, 1991.

———. "Technology and Chemical Debate in the Late Middle Ages." *Isis*, 80 (1989): 423–45.

North, J. D. "The Alphonsine Tables in England." In North, *Stars, Minds, and Fate*, pp. 327–59. London: Hambledon, 1989.

———. "The Astrolabe." *Scientific American*, 230, no. 1 (January 1974): 96–106.

———. "Astrology." In *Oxford Dictionary of the Middle Ages*, ed. Robert Bjork, Oxford University Press, forthcoming.

———. "Astrology and the Fortunes of Churches." *Centaurus*, 24 (1980): 181–211.

———. "Astronomy and Astrology." In Lindberg, David C., and Shank, Michael H., eds., *The Cambridge History of Science*, vol. 2: *The Middle Ages*. Cambridge: Cambridge University Press, forthcoming.

———. "Celestial Influence: The Major Premiss of Astrology." In Zambelli, P., ed., *Astrologi hallucinati*, pp. 45–100. Berlin: Walter de Gruyter, 1986.

———. *Chaucer's Universe*. Oxford: Clarendon Press, 1988.

———. *Horoscopes and History*. Warburg Institute Surveys and Texts, XIII. London: Warburg Institute, 1986.

———. *The Norton History of Astronomy and Cosmology*. New York: W. W. Norton, 1994.

———, ed. and trans. *Richard of Wallingford: An Edition of His Writings with Introductions, English Translation, and Commentary*, 3 vols. Oxford: Clarendon Press, 1976.

———. *Stars, Minds, and Fate: Essays in Ancient and Medieval Cosmology*. London: Hambledon, 1989.

———. *The Universal Frame: Historical Essays in Astronomy, Natural Philosophy, and Scientific Method*. London: Hambledon, 1989.

Numbers, Ronald L., and Amundsen, Darrel W., eds. *Caring and Curing: Health and Medicine in the Western Religious Traditions*. New York: Macmillan, 1986.

Nussbaum, Martha Craven. *Aristotle's "De motu animalium": Text with Translation, Commentary, and Interpretive Essays*. Princeton: Princeton University Press, 1978.

Nutton, Vivian. "The Chronology of Galen's Early Career." *Classical Quarterly*, 23 (1973): 158–71.

———. "Early Medieval Medicine and Natural Science." In Lindberg, David C., and Shank, Michael H., eds., *The Cambridge History of Science*, vol. 2: *The Middle Ages*. Cambridge: Cambridge University Press, forthcoming.

———. *From Democedes to Harvey: Studies in the History of Medicine*. London: Variorum, 1988.

———. "Galen in the Eyes of His Contemporaries." *Bulletin of the History of Medicine*, 58 (1984): 315–24.

Oakley, Francis. *Omnipotence, Covenant, and Order: An Excursion in the History of Ideas from Abelard to Leibniz*. Ithaca: Cornell University Press, 1984.

Obrist, Barbara. *La cosmologie médiévale: Textes et images, vol. 1: Les fondements antiques*. Florence: Edizioni del Galluzzo, 2004.

O'Donnell, James J. *Cassiodorus*. Berkeley and Los Angeles: University of California Press, 1979.

Oggins, Robin S. "Albertus Magnus on Falcons and Hawks." In Weisheipl, James A., ed., *Albertus Magnus and the Sciences: Commemorative Essays 1980*, pp. 441–62. Toronto: Pontifical Institute of Mediaeval Studies, 1980.

O'Leary, De Lacy. *How Greek Science Passed to the Arabs*. London: Routledge & Kegan Paul, 1949.

Olson, Richard. *Science Deified and Science Defied: The Historical Significance of Science in Western Culture from the Bronze Age to the Beginnings of the Modern Era, ca. 3500 B.C. to ca. A.D. 1640.* Berkeley and Los Angeles: University of California Press, 1982.

O'Meara, Dominic J. *Pythagoras Revived: Mathematics and Philosophy in Late Antiquity.* Oxford: Clarendon Press, 1989.

O'Meara, John J. *Eriugena.* Oxford: Clarendon Press, 1988.

Oresme, Nicole. *"De proportionibus proportionum" and "Ad pauca respicientes,"* ed. and trans. Edward Grant. Madison: University of Wisconsin Press, 1966.

———. *Le livre du ciel et du monde,* ed. and trans. A. D. Menut and A. J. Denomy. Madison: University of Wisconsin Press, 1968.

Orme, Nicholas. *English Schools of the Middle Ages.* London: Methuen, 1973.

Osler, Margaret J. "Mechanical Philosophy." In Ferngren, Gary B., ed., *Science and Religion: A Historical Introduction,* pp. 143–152. Baltimore: Johns Hopkins University Press, 2002.

———, ed. *Rethinking the Scientific Revolution.* Cambridge: Cambridge University Press, 2000.

Overfield, James H. "University Studies and the Clergy in Pre-Reformation Germany." In Kittelson, James M., and Transue, Pamela J., eds., *Rebirth, Reform, and Resilience: Universities in Transition 1300–1700,* pp. 254–92. Columbus: Ohio State University Press, 1984.

Ovitt, George. "Technology." In Lindberg, David C., and Shank, Michael H., eds., *The Cambridge History of Science,* vol. 2: *The Middle Ages.* Cambridge: Cambridge University Press, forthcoming.

Owen, G. E. L. *Logic, Science, and Dialectic: Collected Papers in Greek Philosophy,* ed. Martha Nussbaum. Ithaca: Cornell University Press, 1986.

Parent, J. M. *La doctrine de la création dans l'école de Chartres.* Paris: J. Vrin, 1938.

Park, Katharine. "Albert's Influence on Medieval Psychology." In Weisheipl, James A., ed., *Albertus Magnus and the Sciences: Commemorative Essays 1980,* pp. 501–35. Toronto: Pontifical Institute of Mediaeval Studies, 1980.

———. *Doctors and Medicine in Early Renaissance Florence.* Princeton: Princeton University Press, 1985.

———. "Medical Practice." In Lindberg, David C., and Shank, Michael H., eds., *The Cambridge History of Science,* vol. 2: *The Middle Ages.* Cambridge: Cambridge University Press, forthcoming.

———. "Medicine and Society in Medieval Europe, 500–1500." In Wear, Andrew, ed., *Medicine in Society: Historical Essays,* pp. 59–90. Cambridge: Cambridge University Press, 1992.

———. "The Criminal and the Saintly Body: Autopsy and Dissection in Renaissance Italy." *Renaissance Quarterly,* 47 (1994): 1–33.

Parker, Richard. "Egyptian Astronomy, Astrology, and Calendrical Reckoning." *Dictionary of Scientific Biography,* 15:706–27.

Pecham, John. See Lindberg, David C.

Pedersen, Olaf. "Astrology." *Dictionary of the Middle Ages,* 1:604–10.

———. "Astronomy." In Lindberg, David C., ed., *Science in the Middle Ages*, pp. 303–36. Chicago: University of Chicago Press, 1978.

———. "The Corpus Astronomicum and the Traditions of Mediaeval Latin Astronomy: A Tentative Interpretation." In *Astronomy of Copernicus and Its Background*, pp. 57–96. Wrocław: Ossolineum, 1975.

———. "The Development of Natural Philosophy 1250–1350." *Classica et Medievalia*, 14 (1953): 86–155.

———. "Some Astronomical Topics in Pliny." In French, Roger, and Greenaway, Frank, eds., *Science in the Early Roman Empire: Pliny the Elder, His Sources and Influence*, 162–96. Totawa, N.J.: Barnes & Noble, 1986.

———. *A Survey of the Almagest*. Acta Historica Scientiarum Naturalium et Medicinalium, vol. 30. Odense: Odense University Press, 1974.

Pedersen, Olaf, and Pihl, Mogens. *Early Physics and Astronomy: A Historical Introduction*. New York: Science History Publications, 1974.

Pegis, Anton C. *St. Thomas and the Problem of the Soul in the Thirteenth Century*. Toronto: Pontifical Institute of Mediaeval Studies, 1934.

Pellegrin, Pierre. *Aristotle's Classification of Animals: Biology and the Conceptual Unity of the Aristotelian Corpus*, trans. Anthony Preus. Berkeley and Los Angeles: University of California Press, 1986.

Peters, F. E. *Allah's Commonwealth: A History of Islam in the Near East, 600–1100 AD*. New York: Simon and Schuster, 1973.

———. *Aristotle and the Arabs: The Aristotelian Tradition in Islam*. New York: New York University Press, 1968.

———. *The Harvest of Hellenism: A History of the Near East from Alexander the Great to the Triumph of Christianity*. New York: Simon and Schuster, 1970.

Phillips, E. D. *Greek Medicine*. London: Thames and Hudson, 1973.

Philoponus, John. *Against Aristotle on the Eternity of the World*, trans. Christian Wildberg. Ithaca: Cornell University Press, 1987.

Pines, Shlomo. "Al-Rāzī." *Dictionary of Scientific Biography*, 11:323–26.

Pingree, David. "Abū Maʿshar al-Balkhī." *Dictionary of Scientific Biography*, 1:32–39.

———. "Hellenophilia versus the History of Science." *Isis*, 83 (1992): 554–63.

———. "Māshāʾallāh." *Dictionary of Scientific Biography*, 9:159–62.

The Planispheric Astrolabe. Greenwich: National Maritime Museum, 1976.

Plato. *Phaedo, The Dialogues of Plato*, trans. B. Jowett, vol. 1, pp. 363–477. Boston: Jefferson Press, n.d.

———. *Plato, with an English Translation*, 10 vols. London: Loeb, 1914–29.

———. *Plato's Cosmology: The "Timaeus" of Plato*, trans. with commentary by Francis M. Cornford. London: Routledge & Kegan Paul, 1957.

———. *Plato's Theory of Knowledge: The "Theaetetus" and the "Sophist" of Plato*, trans. with commentary by Francis M. Cornford. London: Routledge & Kegan Paul, 1935.

———. *The "Republic" of Plato*, trans. Francis M. Cornford. Oxford: Oxford University Press, 1941.

Pliny the Elder. *Natural History,* trans. H. Rackham, W. H. S. Jones, and D. E. Eicholz, 10 vols. London: Heinemann, 1938–62.

Pliny the Younger. *Letters,* with an English translation by William Melmoth, revised by W. M. L. Hutchinson, 2 vols. London: Heinemann, 1961.

Pormann, P. E., and Savage-Smith, Emilie. *Medieval Islamic Medicine.* Edinburgh: University of Edinburgh Press, forthcoming 2007.

Poulle, Emmanuel. "John of Murs." *Dictionary of Scientific Biography,* 7:128–33.

Powell, Barry. *Homer and the Origin of the Greek Alphabet.* Cambridge: Cambridge University Press, 1991.

Preus, Anthony. *Science and Philosophy in Aristotle's Biological Works.* Hildesheim: Georg Olms, 1975.

Ptolemy, Claudius. *L'Optique de Claude Ptolémée,* ed. and trans. Albert Lejeune. Leiden: Brill, 1989.

————. *Ptolemy's Almagest,* ed. and trans. G. J. Toomer. New York: Springer, 1984.

————. *Tetrabiblos,* ed. and trans. F. E. Robbins. London: Heinemann, 1948.

————. See also Smith, A. Mark.

Quinn, John Francis. *The Historical Constitution of St. Bonaventure's Philosophy.* Toronto: Pontifical Institute of Mediaeval Studies, 1973.

Ragep, F. Jamil. "Copernicus and His Islamic Predecessors: Some Historical Remarks." *Filozofski vestnik* 15 (2004): 125–42.

————. "Islamic Culture and the Natural Sciences." In Lindberg, David C., and Shank, Michael H., eds., *The Cambridge History of Science,* vol. 2: *The Middle Ages.* Cambridge: Cambridge University Press, forthcoming.

————. "Tūsī and Copernicus: The Earth's Motion in Context." *Science in Context* 14 (2001): 145–63.

Rahman, Fazlur. *Islam,* 2d ed. Chicago: University of Chicago Press, 1979.

Randall, John Herman, Jr. *The School of Padua and the Emergence of Modern Science.* Padua: Antenore, 1961.

Rashdall, Hastings. *The Universities of Europe in the Middle Ages,* ed. F. M. Powicke and A. B. Emden, 3 vols. Oxford: Clarendon Press, 1936.

Rashed, Roshdi. "Algebra." *Encyclopedia of the History of Arabic Science,* 2:349–75.

————, ed. *Encyclopedia of the History of Arabic Science,* 3 vols. London: Routledge, 1996.

————. "Geometrical Optics." *Encyclopedia of the History of Arabic Science,* 2:643–71.

————. "Kamāl al-Dīn." *Dictionary of Scientific Biography,* 7:212–19.

————. *Les mathématiques infinitésimales du IXe au XIe siècle,* 4: *Ibn al-Haytham, méthodes géométriques, transformations ponctuelles et philosophie des mathématiques.* London: Al-Furqān Islamic Heritage Foundation, 2002.

————. *Oeuvres philosophiques et scientifiques d'al-Kindī,* vol. 1: *L'optique et la catoptrique.* Leiden: Brill, 1997.

Rather, L. J. "The 'Six Things Non-Natural': A Note on the Origins and Fate of a Doctrine and a Phrase." *Clio Medica,* 3 (1968): 337–47.

Rawson, Elizabeth. *Intellectual Life in the Late Roman Republic.* Baltimore: Johns Hopkins University Press, 1985.

Reeds, Karen. "Albert on the Natural Philosophy of Plant Life." In Weisheipl, James A., ed., *Albertus Magnus and the Sciences: Commemorative Essays 1980*, pp. 341–54. Toronto: Pontifical Institute of Mediaeval Studies, 1980.

———. *Botany in Medieval and Renaissance Universities*. New York: Garland, 1991.

Reeds, Karen, and Kinukawa, Tomomi. "Natural History." In Lindberg, David C., and Shank, Michael H., eds., *The Cambridge History of Science*, vol. 2: *The Middle Ages*. Cambridge: Cambridge University Press, forthcoming.

Reymond, Arnold. *History of the Sciences in Greco-Roman Antiquity*, trans. Ruth Gheury de Bray. London: Methuen, 1927.

Reynolds, Terry S. *Stronger than a Hundred Men: A History of the Vertical Water Wheel*. Baltimore: Johns Hopkins University Press, 1983.

Richard of Wallingford. See North, J. D.

Riché, Pierre. *Education and Culture in the Barbarian West, Sixth through Eighth Centuries*, trans. John J. Contreni. Columbia: University of South Carolina Press, 1976.

Richter-Bernburg, Lutz. "Al-Majūsī." *Encyclopedia Iranica*, 1:837–38.

Riddle, John M. "Dioscorides." In Cranz, F. Edward, and Kristeller, Paul O., eds., *Catalogus translationum et commentariorum: Mediaeval and Renaissance Latin Translations and Commentaries. Annotated Lists and Guides*, vol. 6, pp. 1–143. Washington, D.C.: Catholic University of America Press, 1980.

———. *Dioscorides on Pharmacy and Medicine*. Austin: University of Texas Press, 1985.

———. "Theory and Practice in Medieval Medicine." *Viator*, 5 (1974): 157–70.

Roberts, Alexander, and Donaldson, James, eds. *The Ante-Nicene Fathers*, rev. by A. Cleveland Coxe. 10 vols. Grand Rapids: Eerdmans, 1986.

Rochberg, Francesca. *The Heavenly Writing: Divination, Horoscopy, and Astronomy in Mesopotamian Culture*. Cambridge: Cambridge University Press, 2004.

———. "Mesopotamian Cosmology." In Heatherington, Norriss S., ed., *Cosmology: Historical, Literary, Philosophical, Religious, and Scientific Perspectives*, pp. 37–52. New York: Garland, 1993.

Rosen, Edward. "Regiomontanus, Johannes." *Dictionary of Scientific Biography*, 11:348–52.

———. "Renaissance Science as Seen by Burckhardt and His Successors." In Helton, Tinsley, ed., *The Renaissance: A Reconsideration of the Theories and Interpretations of the Age*, pp. 77–103. Madison: University of Wisconsin Press, 1961.

Rosenfeld, Boris A., and Grigorian, A. T. "Thābit ibn Qurra." *Dictionary of Scientific Biography*, 13:288–95.

Rosenfeld, Boris A., and Youschkevitch, Adolf. "Geometry." *Encyclopedia of the History of Arabic Science*, 2:447–94.

Rosenthal, Franz. *The Classical Heritage in Islam*, trans. Emile and Jenny Marmorstein. London: Routledge and Kegan Paul, 1975.

———. "The Physician in Medieval Muslim Society." *Bulletin of the History of Medicine*, 52 (1978): 475–91.

Ross, W. D. *Aristotle: A Complete Exposition of His Works and Thought*, 5th ed. Cleveland: Meridian, 1959.

Rothschuh, Karl E. *History of Physiology*, trans. Guenter B. Risse. Huntington, N.Y.: Krieger, 1973.

Russell, Bertrand. *A History of Western Philosophy*, 2d ed. London: George Allen & Unwin, 1961.

Russell, Jeffrey B. *Inventing the Flat Earth: Columbus and Modern Historians*. Westport, Conn.: Praeger, 1991.

Sabra, Abdelhamid I. "The Andalusian Revolt against Ptolemaic Astronomy: Averroes and al-Biṭrūjī." In Mendelsohn, Everett, ed., *Transformation and Tradition in the Sciences: Essays in Honor of I. Bernard Cohen*, pp. 133–53. Cambridge: Cambridge University Press, 1984.

———. "The Appropriation and Subsequent Naturalization of Greek Science in Medieval Islam: A Preliminary Statement." *History of Science*, 25 (1987): 223–43.

———. "An Eleventh-Century Refutation of Ptolemy's Planetary Theory." In Hilfstein, Erna; Czartoryski, Paweł; and Grande, Frank D., eds., *Science and History: Studies in Honor of Edward Rosen*, pp. 117–31. Studia Copernicana, no. 16. Wrocław: Ossolineum, 1978.

———. "Al-Farghānī." *Dictionary of Scientific Biography*, 4:541–45.

———. "Form in Ibn al-Haytham's Theory of Vision." *Zeitschrift für Geschichte der arabisch-islamischen Wissenschaften*, 5 (1989): 115–40.

———. "Ibn al-Haytham." *Dictionary of Scientific Biography*, 6:189–210.

———. "Ibn al-Haytham's Revolutionary Project in Optics: The Achievement and the Obstacle." In Hogendijk, Jan P., and Sabra, Abdelhamid I., eds., *The Enterprise of Science in Islam: New Perspectives*, pp. 85–118. Cambridge, Mass.: MIT Press, 2003.

———, ed. and trans. *The Optics of Ibn al-Haytham: Books I–III, On Direct Vision*, 2 vols. London: Warburg Institute, 1989.

———. "Science, Islamic." *Dictionary of the Middle Ages*, 11:81–88.

———. "The Scientific Enterprise." In Lewis, Bernard, ed., *Islam and the Arab World*, pp. 181–92. New York: Knopf, 1976.

Sa'di, Lufti M. "A Bio-Bibliographical Study of Hunayn ibn Is-haq al-Ibadi (Johannitius)." *Bulletin of the Institute of the History of Medicine*, 2 (1934): 409–46.

Saffron, Morris Harold. *Maurus of Salerno: Twelfth-century "Optimus Physicus" with His "Commentary on the Prognostics of Hippocrates."* Transactions of the American Philosophical Society, vol. 62, pt. 1. Philadelphia: American Philosophical Society, 1972.

Saidan, Ahmad S. "Numeration and Arithmetic." *Encyclopedia of the History of Arabic Science*, 2:331–48.

Saliba, George. "Arabic Planetary Theories after the Eleventh Century A.D." *Encyclopedia of the History of Arabic Science*, 3:58–127.

———. "Astrology/Astronomy, Islamic." *Dictionary of the Middle Ages*, 1:616–24.

———. "The Development of Astronomy in Medieval Islamic Society." *Arab Studies Quarterly*, 4 (1982): 211–25.

———. *A History of Arabic Astronomy: Planetary Theories during the Golden Age of Islam*. New York: New York University Press, 1994.

Sambursky, S. *The Physical World of Late Antiquity*. London: Routledge & Kegan Paul, 1962.

————. *The Physical World of the Greeks*, trans. Merton Dagut. London: Routledge & Kegan Paul, 1956.

————. *Physics of the Stoics*. London: Routledge & Kegan Paul, 1959.

Samsó, Julio. "Levi ben Gerson." *Dictionary of Scientific Biography*, 8:279–82.

Sandbach, F. H. *The Stoics*. London: Chatto & Windus, 1975.

Sarton, George. *Galen of Pergamon*. Lawrence: University of Kansas Press, 1954.

————. *Introduction to the History of Science*, 2 vols. Washington, D.C.: Williams and Wilkins, 1927–48.

Savage-Smith, Emilie. "Medicine." *Encyclopedia of the History of Arabic Science*, 3:903–62.

————. "Medicine in Medieval Islam." In Lindberg, David C., and Shank, Michael H., eds., *The Cambridge History of Science*, vol. 2: *The Middle Ages*. Cambridge: Cambridge University Press, forthcoming.

Sayili, Aydin. *The Observatory in Islam and Its Place in the General History of the Observatory*. Publications of the Turkish Historical Society, series 7, no. 38. Ankara: Türk Tarih Kurumu Basimevi, 1960.

Scarborough, John, ed. *Folklore and Folk Medicines*. Madison, Wis.: American Institute of the History of Pharmacy, 1987.

————. "Galen Redivivus: An Essay Review." *Journal of the History of Medicine*, 43 (1988): 313–21.

————. *Roman Medicine*. Ithaca: Cornell University Press, 1969.

Schmitt, Charles B. *The Aristotelian Tradition and Renaissance Universities*. London: Variorum, 1984.

————. *Aristotle and the Renaissance*. Cambridge, Mass.: Harvard University Press, 1983.

————. *Reappraisals in Renaissance Thought*. London: Variorum, 1989.

————. *Studies in Renaissance Philosophy and Science*. London: Variorum, 1981.

Schulman, N. M. "Husband, Father, Bishop? Grosseteste in Paris." *Speculum*, 72 (1997): 330–46.

Seneca, Lucius Annaeus. *Physical Science in the Time of Nero: Being a Translation of the "Quaestiones naturales" of Seneca*, trans. John Clarke, notes by Archibald Giekie. London: Macmillan, 1910.

Serene, Eileen. "Demonstrative Science." In Kretzmann, Norman; Kenny, Anthony; and Pinborg, Jan, eds., *The Cambridge History of Later Medieval Philosophy*, pp. 496–517. Cambridge: Cambridge University Press, 1982.

Shank, Michael H. "Regiomontanus on Ptolemy, Physical Orbs, and Astronomical Fictionalism: Goldsteinian Themes in the 'Defense of Theon against George of Trebizond'." *Perspectives on Science*, 10, no. 2 (2002): 179–207.

————. "Rings in a Fluid Heaven: The Equatorium-Driven Physical Astronomy of Guido de Marchia (fl. 1292–1310)." *Centaurus*, 45 (2003): 175–203.

————. "Science in the Fifteenth Century." In Lindberg, David C., and Shank, Michael H., eds., *The Cambridge History of Science*, vol. 2: *The Middle Ages*. Cambridge: Cambridge University Press, forthcoming.

_____, ed. *The Scientific Enterprise in Antiquity and the Middle Ages.* Chicago: University of Chicago Press, 2000.

_____. "The Social and Institutional Background of Medieval Latin Science." In Lindberg, David C., and Shank, Michael H., eds., *The Cambridge History of Science*, vol. 2: *The Middle Ages.* Cambridge: Cambridge University Press, forthcoming.

_____. *"Unless You Believe, You Shall Not Understand": Logic, University, and Society in Late Medieval Vienna.* Princeton: Princeton University Press, 1988.

Shapin, Steven. *The Scientific Revolution.* Chicago: University of Chicago Press, 1996.

Shapin, Steven, and Schaffer, Simon. *Leviathan and the Air-Pump: Hobbes, Boyle, and the Experimental Life.* Princeton: Princeton University Press, 1985.

Sharp, D. E. *Franciscan Philosophy at Oxford in the Thirteenth Century.* Oxford: Clarendon Press, 1930.

Sharpe, William D., ed. and trans. *Isidore of Seville: The Medical Writings.* Transactions of the American Philosophical Society, vol. 54, pt. 2. Philadelphia: American Philosophical Society, 1964.

Sigerist, Henry E. *A History of Medicine*, 2 vols. Oxford: Oxford University Press, 1951–61.

_____. "The Latin Medical Literature of the Early Middle Ages." *Journal of the History of Medicine*, 13 (1958): 127–46.

Singer, Charles. *A Short History of Anatomy and Physiology from the Greeks to Harvey.* New York: Dover, 1957.

Singleton, Charles S., ed. *Art, Science, and History in the Renaissance.* Baltimore: Johns Hopkins University Press, 1968.

Siraisi, Nancy G. *Arts and Sciences at Padua: The Studium of Padua before 1350.* Toronto: Pontifical Institute of Mediaeval Studies, 1973.

_____. *Avicenna in Renaissance Italy: The "Canon" and Medical Teaching in Italian Universities after 1500.* Princeton: Princeton University Press, 1987.

_____. "Introduction." In Williman, Daniel, ed., *The Black Death: The Impact of the Fourteenth-Century Plague*, pp. 9–22. Binghamton: Center for Medieval and Early Renaissance Studies, 1982.

_____. *Medieval and Early Renaissance Medicine: An Introduction to Knowledge and Practice.* Chicago: University of Chicago Press, 1990.

_____. *Taddeo Alderotti and His Pupils: Two Generations of Italian Medical Learning.* Princeton: Princeton University Press, 1981.

Smalley, Beryl. *The Study of the Bible in the Middle Ages.* Oxford: Basil Blackwell, 1952.

Smith, A. Mark, ed. and trans. *Alhacen's Theory of Visual Perception: A Critical Edition, with English Translation and Commentary, of the First Three Books of Alhacen's De aspectibus, the Medieval Latin Version of Ibn al-Haytham's Kitāb al-Manāzir*, 2 vols. Transactions of the American Philosophical Society, vol. 91, pts. 4–5. Philadelphia: American Philosophical Society, 2001.

_____. "Getting the Big Picture in Perspectivist Optics." *Isis*, 72 (1981): 568–89.

————. "Ptolemy's Search for a Law of Refraction: A Case-Study in the Classical Methodology of 'Saving the Appearances' and Its Limitations." *Archive for History of Exact Sciences*, 26 (1982): 221–40.

————, trans. *Ptolemy's Theory of Visual Perception: An English Translation of the "Optics" with Introduction and Commentary.* Transactions of the American Philosophical Society, vol. 86, pt. 2. Philadelphia: American Philosophical Society, 1996.

————. "Saving the Appearances of the Appearances: The Foundations of Classical Geometrical Optics." *Archive for History of Exact Sciences*, 24 (1981): 73–100.

Smith, Wesley D. *The Hippocratic Tradition.* Ithaca: Cornell University Press, 1979.

Solmsen, Friedrich. *Aristotle's System of the Physical World: A Comparison with His Predecessors.* Ithaca: Cornell University Press, 1960.

————. *Hesiod and Aeschylus.* Ithaca: Cornell University Press, 1949.

————. *Plato's Theology.* Ithaca: Cornell University Press, 1942.

Sorabji, Richard, ed. *Aristotle Transformed: The Ancient Commentators and Their Influence.* Ithaca: Cornell University Press, 1990.

————, *Matter, Space, and Motion: Theories in Antiquity and Their Sequel.* Ithaca: Cornell University Press, 1988.

————, ed. *Philoponus and the Rejection of Aristotelian Science.* London: Duckworth, 1987.

————. *Necessity, Cause, and Blame: Perspectives on Aristotle's Theory.* Ithaca: Cornell University Press, 1980.

Southern, Richard W. "From Schools to University." In Catto, J. I., ed., *The Early Oxford Schools*, vol. 1 of *The History of the University of Oxford*, general ed. T. H. Aston, pp. 1–36. Oxford: Clarendon Press, 1984.

————. *Medieval Humanism and Other Studies.* New York: Harper Torchbooks, 1970.

————. *Robert Grosseteste: The Growth of an English Mind in Medieval Europe.* Oxford: Clarendon Press, 1986.

————. *Saint Anselm: A Portrait in a Landscape.* Cambridge: Cambridge University Press, 1990.

————. "The Schools of Paris and the School of Chartres." In Benson, Robert L., and Constable, Giles, eds., *Renaissance and Renewal in the Twelfth Century*, pp. 113–37. Cambridge, Mass.: Harvard University Press, 1982.

Stahl, William H. "Aristarchus of Samos." *Dictionary of Scientific Biography*, 1:246.

————. *Roman Science: Origins, Development and Influence to the Later Middle Ages.* Madison: University of Wisconsin Press, 1962.

Stahl, William H.; Johnson, Richard; and Burge, E. L. *Martianus Capella and the Seven Liberal Arts*, 2 vols. New York: Columbia University Press, 1971–77.

Stannard, Jerry. "Albertus Magnus and Medieval Herbalism." In Weisheipl, James A., ed., *Albertus Magnus and the Sciences: Commemorative Essays 1980*, pp. 355–77. Toronto: Pontifical Institute of Mediaeval Studies, 1980.

————. "Medieval Herbals and Their Development." *Clio Medica*, 9 (1974): 23–33.

————. "Natural History." In Lindberg, David C., ed., *Science in the Middle Ages*, pp. 429–60. Chicago: University of Chicago Press, 1978.

Stark, Rodney. *For the Glory of God: How Monotheism Led to Reformations, Science, Witch-Hunts, and the End of Slavery.* Princeton: Princeton University Press, 2003.

Steneck, Nicholas H. *Science and Creation in the Middle Ages: Henry of Langenstein (d. 1397) on Genesis.* Notre Dame, Ind.: University of Notre Dame Press, 1976.

Stevens, Wesley M. *Bede's Scientific Achievement.* Jarrow upon Tyne: Parish of Jarrow, 1986.

Stock, Brian. *The Implications of Literacy: Written Language and Models of Interpretation in the Eleventh and Twelfth Centuries.* Princeton: Princeton University Press, 1983.

———. *Myth and Science in the Twelfth Century: A Study of Bernard Silvester.* Princeton: Princeton University Press, 1972.

———. "Science, Technology, and Economic Progress." In Lindberg, David C., ed., *Science in the Middle Ages,* pp. 1–51. Chicago: University of Chicago Press, 1978.

Struik, D. J. "Gerbert." *Dictionary of Scientific Biography,* 5:364–66.

Swartz, Merlin L., ed. and trans. *Studies on Islam.* Oxford: Oxford University Press, 1981.

Swerdlow, Noel M. *The Babylonian Theory of the Planets.* Princeton: Princeton University Press, 1998.

———, ed. *Ancient Astronomy and Celestial Divination.* Cambridge, Mass.: MIT Press, 1999.

Swerdlow, Noel M., and Neugebauer, Otto. *Mathematical Astronomy in Copernicus's De Revolutionibus,* 2 pts. New York: Springer, 1984.

Sylla, Edith Dudley. "Compounding Ratios: Bradwardine, Oresme, and the First Edition of Newton's *Principia.*" In Mendelsohn, Everett, ed., *Transformation and Tradition in the Sciences: Essays in Honor of I. Bernard Cohen,* pp. 11–43. Cambridge: Cambridge University Press, 1984.

———. "Galileo and the Oxford Calculatores: Analytical Languages and the Mean Speed Theorem for Accelerated Motion." In Wallace, William A., ed., *Reinterpreting Galileo,* pp. 53–108. Washington, D.C.: Catholic University of America Press, 1986.

———. "Medieval Concepts of the Latitude of Forms: The Oxford Calculators." *Archives d'histoire doctrinale et littéraire du moyen âge,* 40 (1973): 225–83.

———. "Medieval Quantifications of Qualities: The 'Merton School.'" *Archive for History of Exact Sciences,* 8 (1971): 9–39.

———. "Science for Undergraduates in Medieval Universities." In Long, Pamela O., ed., *Science and Technology in Medieval Society,* pp. 171–86. Annals of the New York Academy of Sciences, vol. 441. New York: New York Academy of Sciences, 1985.

Sylla, Edith Dudley, and McVaugh, Michael, eds. *Texts and Contexts in Ancient and Medieval Science: Studies on the Occasion of John E. Murdoch's Seventieth Birthday.* Leiden: Brill, 1997.

Symonds, John Addington. *Renaissance in Italy,* 2 parts. New York: Henry Holt, 1888.

Tachau, Katherine H. *Vision and Certitude in the Age of Ockham: Optics, Epistemology, and the Foundations of Semantics, 1250–1345.* Leiden: Brill, 1988.

Talbot, Charles H. *Medicine in Medieval England*. London: Oldbourne, 1967.

Taylor, F. Sherwood. *The Alchemists*. New York: Henry Schuman, 1949.

Temkin, Owsei. *The Double Face of Janus and Other Essays in the History of Medicine*. Baltimore: Johns Hopkins University Press, 1977.

————. *Galenism: Rise and Decline of a Medical Philosophy*. Ithaca: Cornell University Press, 1973.

————. "Greek Medicine as Science and Craft." *Isis*, 44 (1953): 213–25.

————. *Hippocrates in a World of Pagans and Christians*. Baltimore: Johns Hopkins University Press, 1991.

————. "On Galen's Pneumatology." *Gesnerus*, 8 (1951): 180–89.

Tester, Jim. *A History of Western Astrology*. Woodbridge, Suffolk: Boydell, 1987.

Thagard, Paul. *Conceptual Revolutions*. Princeton: Princeton University Press, 1992.

Thijssen, J. M. M. H. "What Really Happened on 7 March 1277?" In Sylla, Edith, and McVaugh, Michael, eds., *Texts and Contexts in Ancient and Medieval Science: Studies on the Occasion of John E. Murdoch's Seventieth Birthday*, pp. 84–114. Leiden: Brill, 1997.

Thomas Aquinas. *Faith, Reason, and Theology: Questions I–IV of His Commentary on the De Trinitate of Boethius*, trans. Armand Maurer. Toronto: Pontifical Institute of Mediaeval Studies, 1987.

————. *Summa Theologiae* (Blackfriars ed.), vol. 10: *Cosmogony*, ed. and trans. William A. Wallace. New York: McGraw-Hill, 1967.

Thomas Aquinas, Siger of Brabant, and Bonaventure. *On the Eternity of The World*, trans. Cyril Vollert, Lottie H. Kendzierski, and Paul M. Byrne. Mediaeval Philosophical Texts in Translation, no. 16. Milwaukee: Marquette University Press, 1964.

Thorndike, Lynn. *A History of Magic and Experimental Science*, 8 vols. New York: Columbia University Press, 1923–58.

————. *Michael Scot*. London: Nelson, 1965.

————. *Science and Thought in the Fifteenth Century*. New York: Columbia University Press, 1929.

————, ed. and trans. *The Sphere of Sacrobosco and Its Commentators*. Chicago: University of Chicago Press, 1949.

————. *University Records and Life in the Middle Ages*. New York: Columbia University Press, 1944.

Tihon, Anne. "Byzantine Science." In Lindberg, David C., and Shank, Michael H., eds., *The Cambridge History of Science*, vol. 2: *The Middle Ages*. Cambridge: Cambridge University Press, forthcoming.

Toomer, G. J. "Heraclides Ponticus." *Dictionary of Scientific Biography*, 15:202- 5.

————. "Hipparchus." *Dictionary of Scientific Biography*, 15:205–24.

————. "Mathematics and Astronomy." In Harris, John R., ed., *The Legacy of Egypt*, 2d ed., pp. 27–54. Oxford: Clarendon Press, 1971.

————. "Ptolemy." *Dictionary of Scientific Biography*, 11:186–206.

————. "A Survey of the Toledan Tables." *Osiris*, 15 (1968): 1–174.

————. "Theon of Alexandria." *Dictionary of Scientific Biography*, 13:321–24.

Toulmin, Stephen, and Goodfield, June. *The Fabric of the Heavens: The Development of Astronomy and Dynamics*. Chicago: University of Chicago Press, 1999.

Turner, Howard R. *Science in Medieval Islam: An Illustrated Introduction*. Austin: University of Texas Press, 1995.

Ullmann, Manfred. "Al-Kīmiyāʾ." *The Encyclopaedia of Islam*, new ed., vol. 5, fasc. 79–80, pp. 110–15.

———. *Islamic Medicine*, trans. Jean Watt. Edinburgh: Edinburgh University Press, 1978.

Unguru, Sabetai. "History of Ancient Mathematics: Some Reflections on the State of the Art." *Isis*, 70 (1979): 555–65.

———. "On the Need to Rewrite the History of Greek Mathematics." *Archive for History of Exact Sciences*, 15 (1975): 67–114.

van der Waerden, B. L. "Mathematics and Astronomy in Mesopotamia." *Dictionary of Scientific Biography*, 15:667–80.

———. *Science Awakening: Egyptian, Babylonian, and Greek Mathematics*, trans. Arnold Dresden. New York: John Wiley, 1963.

van der Waerden, B. L., and Huber, Peter. *Science Awakening II: The Birth of Astronomy*. Leyden: Noordhoff, 1974.

Van Helden, Albert. *The Invention of the Telescope*. Transactions of the American Philosophical Society, vol. 67, pt. 4. Philadelphia: American Philosophical Society, 1977.

———. *Measuring the Universe: Cosmic Dimensions from Aristarchus to Halley*. Chicago: University of Chicago Press, 1985.

Vansina, Jan. *The Children of Woot: A History of the Kuba Peoples*. Madison: University of Wisconsin Press, 1978.

———. *Oral Tradition as History*. Madison: University of Wisconsin Press, 1985.

Van Steenberghen, Fernand. *Aristotle in the West*, trans. Leonard Johnston. Louvain: Nauwelaerts, 1955.

———. *Les oeuvres et la doctrine de Siger de Brabant*. Paris: Palais des Académies, 1938.

———. *The Philosophical Movement in the Thirteenth Century*. London: Nelson, 1955.

———. *Thomas Aquinas and Radical Aristotelianism*. Washington, D.C.: Catholic University of America Press, 1980.

Verbeke, G. "Simplicius." *Dictionary of Scientific Biography*, 12:440–43.

———. "Themistius." *Dictionary of Scientific Biography*, 13:307–9.

Veyne, Paul. *Did the Greeks Believe in Their Myths?*, trans. Paula Wissing. Chicago: University of Chicago Press, 1988.

Vickers, Brian, ed. *Occult and Scientific Mentalities in the Renaissance*. Cambridge: Cambridge University Press, 1984.

Vlastos, Gregory. *Plato's Universe*. Seattle: University of Washington Press, 1975.

Voigts, Linda E. "Anglo-Saxon Plant Remedies and the Anglo-Saxons." *Isis*, 70 (1979): 250–68.

Voigts, Linda E., and Hudson, Robert P. "'A drynke that men callen dwale to make a man to slepe whyle men kerven hem': A Surgical Anesthetic from Late

Medieval England." In Campbell, Sheila; Hall, Bert; and Klausner, David, eds., *Health, Disease, and Healing in Medieval Culture.* New York: St. Martin's Press, 1992.

Voigts, Linda E., and McVaugh, Michael R. *A Latin Technical Phlebotomy and Its Middle English Translation.* Transactions of the American Philosophical Society, vol. 74, pt. 2. Philadelphia: American Philosophical Society, 1984.

Voltaire, François Marie Arouet de. *Works,* trans. T. Smollett, T. Francklin, et al., 39 vols. London: J. Newbery et al. 1761–74.

von Grunebaum, G. E. *Classical Islam: A History 600 A.D.–1258 A.D.,* trans. Katherine Watson. Chicago: Aldine, 1970.

———. *Medieval Islam: A Study in Cultural Orientation,* 2d ed. Chicago: University of Chicago Press, 1953.

———. "Muslim World View and Muslim Science." In von Grunebaum, *Islam: Essays in the Nature and Growth of a Cultural Tradition,* 2d ed., pp. 111–26. London: Routledge & Kegan Paul, 1961.

von Staden, Heinrich. "Hairesis and Heresy: The Case of the haireseis iatrikai." In Meyer, Ben F., and Sanders, E. P., eds., *Jewish and Christian Self-Definition,* vol. 3: *Self-Definition in the Graeco-Roman World,* pp. 76–100, 199–206. London: SCM Press, 1982.

———. *Herophilus: The Art of Medicine in Early Alexandria.* Cambridge: Cambridge University Press, 1989.

Vööbus, Arthur. *History of the School of Nisibis.* Corpus Scriptorum Christianorum Orientalium, vol. 266. Louvain: Secrétariat du Corpus SCO, 1965.

Wagner, David L., ed. *The Seven Liberal Arts in the Middle Ages.* Bloomington: Indiana University Press, 1983.

Wallace, Willam A. "Aristotle in the Middle Ages." *Dictionary of the Middle Ages,* 1:456–69.

———. *Causality and Scientific Explanation,* 2 vols. Ann Arbor: University of Michigan Press, 1972–74.

———. *Galileo and His Sources: The Heritage of the Collegio Romano in Galileo's Science.* Princeton: Princeton University Press, 1984.

———. "The Philosophical Setting of Medieval Science." In Lindberg, David C., ed., *Science in the Middle Ages,* pp. 91–119. Chicago: University of Chicago Press, 1978.

———. *Prelude to Galileo: Essays on Medieval and Sixteenth-Century Sources of Galileo's Thought.* Boston Studies in the Philosophy of Science, vol. 62. Dordrecht: Reidel, 1981.

———, ed. *Reinterpreting Galileo.* Studies in Philosophy and the History of Science, no. 15. Washington, D.C.: Catholic University of America Press, 1986.

———. *The Scientific Methodology of Theodoric of Freiberg.* Fribourg: Fribourg University Press, 1959.

———. "Thomism and Its Opponents." *Dictionary of the Middle Ages,* 12:38–45.

Wallis, Faith. *Bede: The Reckoning of Time,* trans., with intro. and commentary. Liverpool: Liverpool University Press, 1999.

Walzer, Richard. "Arabic Transmission of Greek Thought to Medieval Europe." *Bulletin of the John Rylands Library*, 29 (1945–46): 160–83.

Waterlow, Sarah. *Nature, Change, and Agency in Aristotle's "Physics": A Philosophical Study*. Oxford: Clarendon Press, 1982.

Watt, W. Montgomery. *Islamic Philosophy and Theology*, 2d ed. Edinburgh: Edinburgh University Press, 1985.

Wedel, Theodore Otto. *The Mediaeval Attitude toward Astrology, Particularly in England*. New Haven: Yale University Press, 1920.

Weinberg, Julius. *A Short History of Medieval Philosophy*. Princeton: Princeton University Press, 1964.

Weisheipl, James A., ed. *Albertus Magnus and the Sciences: Commemorative Essays 1980*. Toronto: Pontifical Institute of Mediaeval Studies, 1980.

———. "The Celestial Movers in Medieval Physics." *Thomist*, 24 (1961): 286–326.

———. "Classification of the Sciences in Medieval Thought." *Mediaeval Studies*, 27 (1965): 54–90.

———. "The Concept of Nature." *New Scholasticism*, 28 (1954): 377–408.

———. "Curriculum of the Faculty of Arts at Oxford in the Fourteenth Century." *Mediaeval Studies*, 26 (1964): 143–85.

———. *The Development of Physical Theory in the Middle Ages*. New York: Sheed and Ward, 1959.

———. "Developments in the Arts Curriculum at Oxford in the Early Fourteenth Century." *Mediaeval Studies*, 28 (1966): 151–75.

———. *Friar Thomas d'Aquino: His Life, Thought, and Works*. Garden City: Doubleday, 1974.

———. "The Life and Works of St. Albert the Great." In Weisheipl, James A., ed., *Albertus Magnus and the Sciences: Commemorative Essays 1980*, pp. 13–51. Toronto: Pontifical Institute of Mediaeval Studies, 1980.

———. *Nature and Motion in the Middle Ages*, ed. William E. Carroll. Washington, D.C.: Catholic University of America Press, 1985.

———. "The Nature, Scope, and Classification of the Sciences." In Lindberg, David C., ed., *Science in the Middle Ages*, pp. 461–82. Chicago: University of Chicago Press, 1978.

———. "The Principle *Omne quod movetur ab alio movetur* in Medieval Physics." *Isis*, 56 (1965): 26–45. Reprinted in Weisheipl, *Nature and Motion in the Middle Ages*, pp. 75–97.

———. "Science in the Thirteenth Century." In Catto, J. I., ed., *The Early Oxford Schools*, vol. 1 of *The History of the University of Oxford*, general ed. T. H. Aston, pp. 435–69. Oxford: Clarendon Press, 1984.

Welch, Alford T. "Koran." *Dictionary of the Middle Ages*, 7:293–98.

Westerink. L. G. "Philosophy and Theology, Byzantine." *Dictionary of the Middle Ages*, 9:560–67.

Westfall, Richard S. *The Construction of Modern Science: Mechanisms and Mechanics*. New York: John Wiley, 1971.

————. "The Scientific Revolution of the Seventeenth Century: The Construction of a New World View." In Torrance, John, ed., *The Concept of Nature: The Herbert Spencer Lectures*, pp. 63–93. Oxford: Clarendon Press, 1992.

Westman, Robert S. "The Astronomer's Role in the Sixteenth Century: A Preliminary Study." *History of Science*, 18 (1980): 105–47.

Wetherbee, Winthrop, trans. *The Cosmographia of Bernardus Silvestris*, with introduction and notes by Wetherbee. New York: Columbia University Press, 1973.

————. "Philosophy, Cosmology, and the Twelfth-Century Renaissance." In Dronke, Peter, ed., *A History of Twelfth-Century Western Philosophy*, pp. 21–53. Cambridge: Cambridge University Press, 1988.

White, Andrew Dickson. *History of the Warfare of Science with Theology in Christendom*. 2 vols. New York: Appleton, 1896.

White, Lynn, Jr. *Medieval Technology and Social Change*. Oxford: Oxford University Press, 1962.

White, T. H., trans. *The Bestiary: A Book of Beasts*. New York: G. P. Putnam's Sons, 1954.

Whitney, Elspeth. *Paradise Restored: The Mechanical Arts from Antiquity through the Thirteenth Century*. Transactions of the American Philosophical Society, vol. 80, pt. 1. Philadelphia: American Philosophical Society, 1990.

Whitting, Philip, ed. *Byzantium: An Introduction*. New York: Harper & Row, 1973.

William of Conches. *A Dialogue on Natural Philosophy (Dragmaticon Philosophiae)*, trans. Italo Ronca and Matthew Curr. Notre Dame, Ind.: University of Notre Dame Press, 1997.

————. *Philosophia mundi*, book 1, ed. Gregor Maurach. Pretoria: University of South Africa, 1974.

Williams, Steven J. "*The Secret of Secrets*": *The Scholarly Career of a Pseudo-Aristotelian Text in the Latin Middle Ages*. Ann Arbor: University of Michigan Press, 2003.

Williman, Daniel, ed. *The Black Death: The Impact of the Fourteenth-Century Plague*. Binghamton: Center for Medieval and Early Renaissance Studies, 1982.

Wilson, Curtis. *William Heytesbury: Medieval Logic and the Rise of Mathematical Physics*. Madison: University of Wisconsin Press, 1960.

Wilson, John A. "The Nature of the Universe." In Frankfort, H.; Frankfort, H. A.; Wilson, John A.; and Jacobsen, Thorkild. *Before Philosophy: The Intellectual Adventure of Ancient Man*, pp. 39–70. Baltimore: Penguin, 1951.

Wilson, N. G. *Scholars of Byzantium*. Baltimore: Johns Hopkins University Press, 1983.

Winstedt, E. O., ed. *The Christian Topography of Cosmas Indicopleustes*. Cambridge: Cambridge University Press, 1909.

Wippel, John F. "The Condemnations of 1270 and 1277 at Paris." *Journal of Medieval and Renaissance Studies*, 7 (1977): 169–201.

Witelo. *Witelonis Perspectivae liber primus: Book I of Witelo's Perspectiva: An English Translation with Introduction and Commentary and Latin Edition of the Mathematical Book of Witelo's "Perspectiva,"* ed. and trans. Sabetai Unguru. Studia Copernicana, no. 15. Wrocław: Ossolineum, 1977.

————. *Witelonis Perspectivae liber tertius: Books II and III of Witelo's Perspectiva: An English Translation with Introduction, Notes, and Commentaries,* ed. and trans. Sabetai Unguru. Wrocław: Ossolineum, 1991.

————. *Witelonis Perspectivae liber quintus: Book V of Witelo's Perspectiva: An English Translation with Introduction and Commentary and Latin Edition of the First Catoptrical Book of Witelo's Perspectiva,* ed. and trans. A. Mark Smith. Studia Copernicana, no. 23. Wrocław: Ossolineum, 1983.

Wolff, Michael. "Philoponus and the Rise of Preclassical Dynamics." In Sorabji, Richard, ed., *Philoponus and the Rejection of Aristotelian Science,* pp. 84–120. London: Duckworth, 1987.

Wolfson, Harry Austryn. *Crescas' Critique of Aristotle: Problems of Aristotle's "Physics" in Jewish and Arabic Philosophy.* Cambridge, Mass.: Harvard University Press, 1929.

Woodward, David. "Geography." In Lindberg, David C., and Shank, Michael H., eds., *The Cambridge History of Science,* vol. 2: *The Middle Ages.* Cambridge: Cambridge University Press, forthcoming.

————. "Medieval Mappaemundi." In Harley, J. B., and Woodward, David, eds., *The History of Cartography,* vol. 1: *Cartography in Prehistoric, Ancient, and Medieval Europe and the Mediterranean,* pp. 286–370. Chicago: University of Chicago Press, 1987.

Wright, John Kirtland. *The Geographical Lore of the Time of the Crusades: A Study in the History of Medieval Science and Tradition in Western Europe.* New York: American Geographical Society, 1925.

Yates, Frances A. *Giordano Bruno and the Hermetic Tradition.* London: Routledge & Kegan Paul, 1964.

————. "The Hermetic Tradition in Renaissance Science." In Singleton, Charles S., ed., *Art, Science, and History in the Renaissance,* pp. 255–74. Baltimore: Johns Hopkins University Press, 1968.

Young, M. J. L.; Latham, J. D.; and Serjeant, R. B., eds. *Religion, Learning, and Science in the ʿAbbāsid Period.* Cambridge: Cambridge University Press, 1990.

Ziegler, Philip. *The Black Death.* New York: Harper and Row, 1969.

Zimmermann, Fritz. "Philoponus' Impetus Theory in the Arabic Tradition." In Sorabji, Richard, ed., *Philoponus and the Rejection of Aristotelian Science,* pp. 121–29. London: Duckworth, 1987.

Zinner, Ernst. "Die Tafeln von Toledo." *Osiris,* 1 (1936): 747–74.

Zuccato, Marco. "Gerbert of Aurillac and a Tenth-Century Jewish Channel for the Transmission of Arabic Science to the West." *Speculum,* 80 (2005): 742–63.

Zupko, Jack. *John Buridan: Portrait of a Fourteenth-Century Arts Master.* Notre Dame, Ind.: University of Notre Dame Press, 2003.

Index

abacus, 203

ʿAbbās, al-, 168

ʿAbbāsids, 168, 170–71, 177, 189; fragmentation of empire of, 173, 190–91; physicians to, 165, 168, 171, 185

ʿAbd al-Rahmān, 215

Abelard, Peter, 207–9, 218

Abulcasis (Abū al-Qāsim al-Zahrāwī), 187, 328

Abū Maʿshar (Albumasar), 274, 275

Academy in Athens: migration to Persia from, 164; of Plato, 35, 45, 69–71, 78, 95, 223

accelerated motion, 300, 303–6, 308, 366. *See also* falling bodies

accidental form, 287, 292

accidental qualities, 247, 249

Achilles, 297

active principles: in Stoic thought, 79–80

actuality and potentiality, 49–50, 62, 64, 233, 295

Adelard of Bath, 210, 212

aether. *See* quintessence

Africa: oral cultures of, 6, 7–8

agriculture: European innovations in, 203–4; prehistoric understanding of nature and, 4

air: Strato on, 74–75; as underlying matter, 28–29. *See also* elements

Albategni (al-Battānī), 178–79, 260

Albert the Great, 237–41, 243–44; on motion, 297; *On Animals*, 353; *On Vegetables*, 352–53

Albumasar (Abū Maʿshar), 274, 275

alchemy, 54, 290–95, 364, 400n9

Alcuin, 195, 196

Alexander of Aphrodisias, 75

Alexander of Tralles, 322

Alexander the Great: Aristotle and, 45; empire of, 67, 71, 132, 163, 165, 166

Alexandria: anatomy and physiology in, 119–22; Byzantine, 159, 160, 161; as center of learning, 67, 71–72, 163; library of, 72, 96, 375n7; medicine in, 125, 126; Museum of, 72, 99, 375n7; as name of many cities, 163; Ptolemy and, 99

Alfonsine Tables, 267, 363

Alfonso X of Castile, 267

algebra: Babylonian mathematics and, 15

Algebra (al-Khwārizmī), 177, 216, 226, 385n20

Alhacen. *See* Ibn al-Haytham (Alhacen)

ʿAlī ibn ʿAbbās al-Majūsī (Haly Abbas), 187, 188, 328, 330, 335, 345

Allah, 166

Almagest. See Ptolemy's *Almagest*

alphabetic writing, 10, 11–12

alteration, 295, 296

anatomy, human: dissection and, 119–
21, 126, 345–47; Galen and, 125,
126–27, 129–30, 314, 317, 345, 347,
348; Hippocratic, 119; Islamic, 186,
188; medieval, 345–48; sixteenth-
century anatomists and, 348
Anaxagoras, 27, 34
Anaximander, 27, 28
Anaximenes, 28
Andalus, al-, 201
Andrew of St. Victor, 212
Andronicus of Rhodes, 74, 75
anesthetics, 345
angels, 259, 260
animals: Aristotle's classification of, 61.
   See also zoology
animism, Platonic, 42
Anselm of Bec, 206–8, 209, 389n20
Anthemius of Tralles, 160
Anticrates of Cnidus, 113
Antidotarium Nicolai, 338
Antiochus of Ascalon, 137
antipodeans, 197, 279
Antoninus Pius, 72
apeiron, 28
Apollonius of Perga, 86, 99, 160
apologetics, Christian, 149
apsides, planetary, 197
Aquinas, Thomas, 241–44; Albert the
   Great and, 237, 241, 243–44; con-
   demnations of 1270 and 1277 and,
   246; elevated to sainthood, 250; on
   motion, 307, 308
Arab: defined, 168–69
Arabian peninsula, 166, 169
Arabic language, 169. *See also* transla-
   tion from Arabic to Latin; transla-
   tion from Greek to Arabic
Arabic numerals, 160, 177, 203
Aramaic, 165
Aratus of Soli, 98, 136, 139
Archimedes, 85–86, 110, 160, 173,
   217
Aristarchus of Samos, 95–96, 377n18

Aristotelian logic: Boethius and, 148,
   194, 199, 206; Christian theology
   and, 206; early Latin translations of,
   225; Gerbert and, 199; in medieval
   scholarship, 233; in medieval universities, 224, 227; Nestorian higher
   education and, 164, 165; Persian
   interest in, 170; Porphyry's introduction to, 176; in twelfth-century
   schools, 206
Aristotelian tradition: Albert the Great
   and, 237–39, 240–41; Aquinas and,
   237, 241, 242–43; astrology and,
   230, 272, 274, 275; Avicenna and,
   187; Boethius and, 148, 194, 199,
   206; in Byzantine culture, 159–60,
   162; early Christians and, 149;
   eternity of universe and, 53, 229,
   237, 240, 243, 244, 246; explanatory
   power of, 66; fourteenth-century
   developments in, 251; Galen's
   medical system and, 125, 128, 129;
   immortality of the soul and, 63, 230,
   237, 240, 244, 246; Islamic culture
   and, 173, 186, 191; Lyceum after
   Aristotle and, 73–76; in medieval cosmology, 255–59, 261; in
   medieval medicine, 334; in medieval natural history, 351, 352, 353;
   in medieval physics, 286–90, 299,
   306–9; medieval points of conflict
   with, 228–33; medieval reconciliation of with theology, 233–43,
   248–51, 253; Nestorians in Persia
   and, 164; Ptolemaic astronomy
   and, 267, 269; radicals of thirteenth
   century, 244–46, 248, 250, 251; in
   Roman Empire, 135; Stoic and Epicurean alternatives to, 81; in twelfth
   century, 209, 210; in university curriculum, 226–28
Aristotle, 45–66; achievement of,
   65–66; astronomical model of, 53,
   59–60, 66, 92–95, 105; as biologist,

60–65, 70, 119, 295; on causation, 50, 51–52, 60, 62–63, 65, 230; on change, 47, 48, 49–50, 51–52, 65, 295–96; on combination and mixture, 288–89, 291; on cosmology, 52–56, 229; deductive demonstration and, 48, 84; epistemology of, 48–49; life of, 45; logic of (*see* Aristotelian logic); mathematics and, 82–83, 95, 105, 360, 361; metaphysics of, 46–48, 63, 70, 249, 364, 365; on Milesian philosophers, 28; on motion, 50, 54–55, 56–60, 82, 295–96, 306, 309–10, 361, 401n17; on motion, critiques of, 160, 249, 310–11; on nature, 27, 373n9; on the nature of things, 50–51, 287, 373n9; on place, 56, 75, 257, 258, 295, 298; at Plato's Academy, 45, 70; on Pythagoreans, 31; on rotating earth, 282; Theophrastus and, 45, 73, 74; on vision, 66, 73, 105, 106, 182–83, 313, 315, 317–18; on Zeno's stadium paradox, 33
Aristotle's works, 45; *Ethics*, 237; *History of Animals*, 61; *Metaphysics*, 92, 172, 176, 234, 261, 361; *Meteorology*, 66, 137, 160, 216, 234; *On Animals*, 234; *On Generation and Corruption*, 160, 172, 216, 234; *On the Generation of Animals*, 62; *On the Heavens*, 53, 57, 66, 160, 216, 234, 261, 361; *On the Parts of Animals*, 61, 129; *On Sense and the Sensible*, 234; *On the Soul*, 160, 172, 234; *Physics*, 57–58, 65–66, 160, 172, 216, 234, 238–39, 286, 295, 298, 299, 306–8, 361; *Posterior Analytics*, 234, 361; *Topics*, 171
arithmetic. *See* numbers
arithmetic series: in Babylonian astronomy, 16–17
armillary sphere, 201–2, 215
arteries, 120–22, 126, 127, 128–29

Articella, 334–35, 339
Asclepiades of Bithynia, 124
Asclepius, 112–13, 118, 131
astrolabe, 98, 161, 177, 215, 263–64, 269
astrology: ancient religions and, 271; Aristotelian philosophy and, 230, 272, 274, 275; Augustine's opposition to, 214, 273–74; Babylonian, 11, 16, 20, 98; Bacon's defense of, 236; Bonaventure on determinism of, 237; condemnation of 1277 and, 247, 249; horoscopic, 16, 271, 274; Isidore of Seville's attack on, 157; Islamic, 178, 183, 214, 272, 273, 274, 275; Islamic opposition to, 273; macrocosm-microcosm analogy in, 139, 214, 272; medicine and, 265, 275, 334, 339, 341–42; medieval, 214, 270–71, 274–77; Plato's cosmology and, 271–72; Ptolemy on, 272–73, 274–75; as respectable natural philosophy, 271; as stimulus to astronomy, 265; translation from Arabic to Latin, 216, 217; translation to Arabic or Syriac, 171, 172; in university education, 223
astronomical instruments, 263–65. *See also* armillary sphere; astrolabe; quadrant
astronomical observatories, 175–78, 182, 191, 363
astronomical tables, 161, 178, 179, 180, 265, 267, 363
astronomy: Aristotle on, 53, 59–60, 66, 92–95, 105; Babylonian (Mesopotamian), 11, 15–17, 98, 99, 271; Byzantine, 161–62; Carolingian, 196–98; Cicero on, 139; cosmology vs., 261–62; distance calculations in, 260; early Greek, 15, 28, 59–60, 86–95; Hellenistic, 98–105; Islamic (*see* Islamic astronomy); Martianus Capella on, 145, 262; mathematical,

astronomy (*cont.*)
11, 15–17, 261–70, 361; Plato on,
37, 38, 41, 42–43, 86–89; Pliny on,
142, 262; realist vs. instrumentalist
models in, 261; translation from
Arabic to Latin, 216, 217; in university education, 223, 266–67. *See
also* planetary astronomy; Ptolemaic
astronomy
Athens: educational institutions of,
69–71, 72, 73, 74
atomism: Asclepiades' embrace of, 124;
of Epicureans, 77–78, 79, 139, 364,
366–367; medieval views of, 30,
290; Plato's geometrical version of,
40–41; al-Rāzī's defense of, 186;
scientific revolution and, 364–65;
seventeenth-century return of, 31.
See also corpuscularism
atomists, Greek, 27, 29–31, 33, 34;
Aristotle's arguments against,
54–55, 61; motion and, 295; vision
and, 105, 313, 315
Atto, 201
augmentation and diminution, 295, 296
Augustine: anti-astrological sentiments
of, 214, 273–74; Bacon's citation of,
236; Bonaventure's thought and,
236; on classical learning, 149, 150,
155, 381n28; thirteenth-century
scholarship and, 236, 237; twelfth-
century scholarship and, 209, 210
Aurillac. *See* Gerbert of Aurillac
Autolycus of Pitane, 299
Avempace (Ibn Bājja), 311
Averroes. *See* Ibn Rushd (Averroes)
Avicenna. *See* Ibn Sīnā (Avicenna)

Babylonians: astronomy, 11, 15–17, 98,
99, 271; calendar, 13, 16; creation
myths, 8, 12; logographic writing,
11; mathematics, 12, 13–15,
371n29; medicine, 18, 20
Bacon, Francis, 357, 359, 364

Bacon, Roger: Aristotle's natural
philosophy and, 228; defense of the
new learning, 234, 236–37, 239; experimental science and, 363; hostility
toward Albert the Great, 240; on
motion, 308; mythology about,
393n16; on optics, 236, 239, 318,
320; and physical-sphere Ptolemaic
model, 269, 270; on rainbow, 277
Bactria, 68, 163, 168
Baghdad: astronomical observations
at, 178; as center of translation,
170–73; founding of, 168; Gondeshapur physicians and, 165, 168,
185; hospitals in, 188; mathematics
of al-Khwārizmī at, 177; origin of
scientific movement in, 189; paper
technology brought to, 169; sacked
by Mongols, 191. *See also* 'Abbāsids
balance beam, 109–10
Barmak family, 168, 349
Basil of Caesarea, 325
Battānī, al- (Albategni), 178–79, 260
Beatus map, 281
Bede, 157–58, 196, 325–26
Benedict (St.) of Nursia, 152, 154, 241
Berbers, 169
Bernard of Chartres, 206
Bernard of Clairvaux, 208, 324–25
Bernard of Verdun, 270
Bernard Sylvester, 210–11
Bessarion, Johannes, 161
bestiaries, 162, 353–56
biology: of Albert the Great, 239,
352–53; of Aristotle, 60–65, 70,
119, 295. *See also* anatomy, human;
botany; medicine; natural history;
physiology; zoology
black death of 1347–51, 341. *See also*
plague
bladder stone, 343, 345
blood, 116, 120–22, 125, 127–29, 336
bloodletting, 116, 122, 185, 336, 342, 343
Bobbio, 199

Boethius, 147–48, 199, 203; Aristotelian logic and, 148, 194, 199, 206; *Arithmetic*, 200; translations from Greek to Latin, 148, 194, 217; twelfth-century scholars and, 209, 214

Boethius of Dacia, 244–46

Boethius of Sidon, 75

Bologna: as educational center, 205–6, 218; University of, 219, 221, 333, 335, 343, 345–47

Bonaventure, 236–37, 243, 246

botany: Byzantine, 162; medieval, 351–53; Theophrastus on, 73. *See also* herbs; plants

Boyle, Robert, 294–95, 362, 365

Bradwardine, Thomas, 257, 300, 311–13

brain, 61, 115, 120, 126, 127, 128–29

Bukhtīshūʿ family, 168, 172, 185

Burckhardt, Jacob, 358, 359

Buridan, John: on motion, 260, 298, 308–9, 403n36; on possibility of rotating earth, 282–83; and theological authority, 250–51, 260

Byzantine Empire: Athenian Academy closed in, 164; hospitals in, 162, 164, 188, 189, 349; Islamic appropriation of texts from, 171; Islamic conquest of, 166; learning in, 159–62; map, 167; origin of, 147; Persian culture and, 170; theological disputes in, 164; Western appropriation of texts from, 215, 217

Cahill, Thomas, 382n32

Cairo, 189

Calcidius, 147, 197, 206, 209, 254

calendar: Babylonian, 13, 16; Bede on, 158; early medieval knowledge of, 262; Egyptian, 13; Greek, 13, 86; monastic education and, 155; religious, 196, 223, 236

Callippus of Cyzicus, 92

Campanus of Novara, 260

Canterbury: Anselm of, 206–7; archbishops of, 250, 260

Carolingian period, 194–203, 205, 364

Carthage, 149

Cassiodorus, 155, 157, 322

Catalonia, 201

cataracts, 188, 343, 344

cathedral schools, 205–6, 333

causation: Aristotle on, 50, 51–52, 60, 62–63, 65, 230; astrological, 272; Roger Bacon on, 318, 320; condemnation of 1277 and, 246; dynamics of motion and, 300, 306–7, 309, 310, 311; early Greek philosophers and, 27, 29; Epicurus's atomism and, 77, 78; Greek mythology and, 25; mechanical philosophy and, 365; natural philosophy and, 3; preliterate ideas of, 6, 7, 9; Stoic philosophy and, 81; thirteenth-century naturalism and, 231, 237; twelfth-century naturalism and, 212

cautery, 185, 343

cave, Plato's allegory of, 36–37, 38

celestial region, 53, 92, 256

celestial spheres: Aristotle on, 53, 59–60, 92–95; Buridan and, 250, 308; Callippus on, 92; Eudoxus on, 59–60, 86–92, 94; Grosseteste on, 255; Isidore of Seville on, 157; magnetic compass and, 271; Martianus Capella on, 145; medieval account of, 257–60; Plato on, 41, 86–89; Ptolemaic astronomy and, 267, 269–70; two-sphere model and, 86–88

Celsus, 139, 321

change: Aristotle on, 47, 48, 49–50, 51–52, 65, 295–96; early Greek philosophers on, 26, 28, 29, 31, 32–33, 34, 43, 372n21; medieval Aristotelians on, 295–96, 299; Plato on, 37, 39, 40

Charlemagne, 194–96
Charles the Bald, 198, 199
Chartres: cathedral school of, 205, 333;
    Thierry of, 210, 215, 255; William
    of Conches and, 212
chemical combination, 66, 288–89, 291
Christianity: anti-astrological senti-
    ment in, 214, 247, 249, 273–74;
    on Arabian peninsula, 166; in
    Byzantine Empire, 159, 161, 162,
    164; classical learning and early
    church, 148–50, 152; comparison
    of, to Islam as patron of science,
    191; Galen's teleology and, 129;
    hospitals and, 349; medicine in
    Middle Ages and, 322–27, 331;
    Nestorian, 164–65, 168, 171, 172,
    185, 186, 383n2; origin of, 148;
    relationship of, to medieval science,
    358; in Spain, 190, 191, 201, 216;
    in western Asia, 163–65. *See also*
    monasteries
Christian theology: Aristotelian
    foundations in, 251; Aristotelian
    naturalism in conflict with, 228–29,
    231, 237, 240–41, 246; medieval
    cosmology and, 259; medieval
    physics and, 286, 298; in medieval
    universities, 218, 219, 221, 223,
    224, 226–28, 231–32, 233; radical
    Aristotelianism and, 244–46, 248;
    reconciliation of, with Aristote-
    lianism, 233–43, 248, 251–53;
    seventeenth-century science and,
    412n36; thirteenth-century as-
    similation of new knowledge and,
    225–26; twelfth-century natural-
    ism and, 212–13. *See also* faith and
    reason; handmaiden of theology,
    science as; omnipotence, divine
Chrysippus of Soli, 78
Cicero, 134, 137–39; on determinism,
    81; Gerbert and, 199, 203; Macrobi-
    us's commentary on, 143; mechanistic

universe of, 365; twelfth-century
    schools and, 206, 209
Clagett, Marshall, 358–59
Clarembald of Arras, 211
Cleanthes of Assos, 78
climatic zones, 279, 280
Clovis, 322
cohesion: Stoic theory of, 79–80
Cologne: Dominican school at, 237
Colombo, Realdo, 189, 387n43
Columbus, Christopher, 137, 161
combination: chemical, 66, 288–89,
    291; mixture, 287, 289, 291
comets, 66, 277
Commentator, 228
Commodus, 125
compass, magnetic, 271
complexion, 274, 336, 339
computus, 196
condemnations of 1270 and 1277,
    246–51, 252, 253, 257, 298
conic sections, 86, 160
Constantine the African, 215, 239,
    329–30
Constantinople: hospitals in, 349;
    Nestorian Christians in, 164. *See
    also* Byzantine Empire
continents, 278, 280
contingency of nature, 253
continuity question, 357–59,
    366–67
contraries, 53–54, 55, 233
Copernicus, Nicolaus: Aristarchus's
    anticipation of, 95–96; community
    of astronomers and, 267; Islamic
    astronomy and, 179, 181, 386n30;
    Martianus Capella cited by, 145;
    mathematical physics and, 360–61;
    planetary distances and, 260; Ptol-
    emaic astronomy and, 102, 103, 162,
    270, 366
Cordoba, 189, 190, 201, 215
Corinth, 124–25, 132
corporeal form, 288, 397n4

corpuscularism: alchemical, 292, 294–95; of Erasistratus, 121; seventeenth-century, 295; of Strato, 75, 121. *See also* atomism
Cosmas Indicopleustes, 161
cosmogony: defined, 12; of Grosseteste, 234; of Hesiod, 22. *See also* creation
cosmology: of Aristotle, 52–56, 229; in Aristotle's time, 95–98; Carolingian, 197; condemnation of 1277 and, 248, 249; cyclic, 81, 139; defined, 12; of early Greek philosophers, 26, 27, 28, 30, 35; of Epicurus, 77; of Grosseteste, 255, 397n4; of Isidore of Seville, 157; Islamic, 166, 178; late medieval, 254–61; vs. mathematical astronomy, 261–62; in oral traditions, 5–7; of Plato, 36, 38–43; Pliny on, 142; Pythagorean, 95; scientific revolution and, 364, 365; of Stoics, 79, 80, 81; twelfth-century schools and, 206, 209–12. *See also* eternity of the universe
crafts, 1, 3, 4
creation: Aristotle's eternal universe and, 229, 240, 243, 244; condemnation of 1277 and, 247; late medieval cosmology and, 258–59, 260, 262; omnipotence of God and, 252, 396n46; Platonic cosmology and, 210, 255; radical Aristotelians on, 244, 245; twelfth-century scholarship on, 210, 215, 255. *See also* cosmogony
creation myths, 7–8, 9, 12
critical days, 342
Crombie, Alistair, 359, 364
Crusades, 203, 216, 349
Cuthbert, 326

Damascus, 166, 168, 180
dark ages, 193
decimal system, 13, 160, 177
deduction, 48, 84
deferent. *See* epicycle-on-deferent model of planetary motion

Demetrius Phaleron, 72
Demiurge, 36, 39, 41–42; astrology and, 214, 271–72; Christian use of, 149; Galen's borrowing of, 130
Democritus, 27, 29, 33, 34, 77
dentistry, 331
Descartes, René: mechanical philosophy and, 365; on the rainbow, 185
design: Galen on, 131
determinism: Aristotle's natural philosophy and, 229–30; astrological, 237, 273, 274, 275; condemnation of 1277 and, 246, 248; Epicurus's avoidance of, 77, 78; of Stoic universe, 81. *See also* free will
diet: as medical therapy, 116, 336. *See also* nutrition
Digest, 206
digestion, 41, 121
Diogenes Laertius, 364–65
diopter, 98
Dioscorides, 162, 186, 322, 327, 338, 351
disease: early Greek medicine and, 111, 113, 115, 118, 119; Egyptian beliefs on, 18; Greek gods and, 111; Hellenistic medicine and, 122, 124, 125–26, 131; medical astrology and, 339; medieval theories of, 336; Mesopotamian beliefs on, 20; Plato on, 41; preliterate beliefs on, 9
dissection, animal: by Aristotle, 61; by Galen, 126
dissection, human, 119–20, 126, 345–47
dogmatist physicians, 122
Dominican order, 232, 233; Albert the Great and, 237, 238; Aquinas and, 241
doxographic tradition, 73, 138, 147
drugs. *See* pharmacological therapy

Duhem, Pierre: on applicability of mathematics to physics, 360; on condemnation of 1277, 247–48; defense of medieval science, 358, 359; on medieval cosmology, 256–57; on realist vs. instrumentalist models, 261

Dunāsh ibn Tamīm al-Qarawī, 201–2, 203

Duns Scotus, John, 251

dynamics, 306–9; defined, 300, 402n24; of Galileo, 366; modern, 308–9; quantification of, 309–13, 361

earth, the: in Aristarchus's heliocentric system, 95–96; Aristotle on, 55–56; possible rotation on axis, 95, 282–85

earth as element. *See* elements

earthquakes, 26, 66

earth's circumference: Eratosthenes on, 96–98, 142, 144; medieval agreement on, 278; Posidonius on, 137

earth's sphericity: Aristotle on, 55–56, 361, 383n42; Carolingian knowledge of, 197; Isidore of Seville on, 157; Martianus Capella on, 41; medieval agreement on, 161, 278; Plato on, 41, 383n42

Ebers papyrus, 18

Ebstorf map, 279

eccentric planetary model, 100, 102, 103–4; Carolingian sources on, 197; in medieval astronomy, 261, 266–67

eclipses: Aristotle on, 55; Babylonian calculations of, 16; Carolingian sources on, 197; early Greeks and, 26, 27, 28; Hipparchus's studies of, 96, 98; Isidore of Seville on, 157; Johannes de Sacrobosco on, 266; medieval comparison of tables with, 363; Pliny on, 142

ecliptic: Carolingian sources on, 197; defined in Greek astronomy, 41, 87–89, 92, 93, 94

education: astronomical, Dunāsh's program for, 201–2, 203; Byzantine, 159; Carolingian, 194–96, 198; eleventh- and twelfth-century schools, 203–9, 218; Greek, 25, 35, 45, 67–76, 78, 137; as mastery of standard texts, 391n46; in medieval Islam, 174–76; monasticism and, 152, 154–55, 198–99, 204–5; in Muslim Spain, 201; natural philosophy in, twelfth-century, 209–15; Roman, 150–57; urbanization and, 204–5, 218, 329. *See also* medical education; universities

Edwin Smith papyrus, 18

efficient cause, 51, 52, 60, 62–63, 365

Egypt: Babylonian astronomy and, 15; calendar, 13; creation myths, 8, 9; eleventh-century science in, 189; Greek influence in, 163; Islamic astronomy in, 179; mathematics, 12–13, 15; medicine, 18, 111; Muslim rule in, 166, 169, 173; writing systems, 10, 11

Einhard, 194

Elea, 32, 133

elements: Aristotle on mixture and, 288–89, 291; Aristotle's form and matter theory of, 53–54, 55, 256, 287–88; corpuscularism of Paul of Tarantino and, 292, 294; disease and, 336; Empedocles on, 31, 33, 40; Heraclitus on, 29; Isidore of Seville on, 157; Islamic commentators on, 288; motion of, 57; Plato on, 38, 40–41, 82, 287; spheres of, 277; Stoics' version of, 80, 81; Theophrastus on, 73; Thierry of Chartres on, 210; in twelfth-century natural philosophy, 213–14. *See also* quintessence; underlying reality

Elements. See Euclid's *Elements*

emanationism, Neoplatonic, 234

Empedocles, 27, 31, 33, 34, 53

empiricism: of Bacon, Francis, 364; of Theophrastus, 73. *See also* experimentation; observation; sense perception

empiricist physicians, Hellenistic, 122, 124

empirics, in medieval medical practice, 331, 343

empyreum, 259

encyclopedic works: of Isidore of Seville, 157; Roman, 137, 139–42, 148

Ennius, 136

Ephesus, 27, 29

Epicurean philosophy, 70–72, 76–78; atomism in, 77–78, 79, 139, 364–65; Cicero's study of, 137; compared to Stoics, 78–79, 81; conflict of Christianity with, 149; Lucretius's expression of, 139; Romans and, 136

Epicurus, 70, 75, 76–78; Garden of, 70, 223

epicycle-on-deferent model of planetary motion, 100–104, 267, 269; Carolingian sources on, 197; of Ibn al-Shāṭir, 180; in medieval astronomy, 261, 266–67; physical-sphere version of, 269–70; Sacrobosco on, 266

Epidaurus, 113, 114

epistemology: fourteenth-century, 251; of Greek philosophers, 34, 37, 48–49; Platonic influence on Cicero, 134; scientific method and, 1. *See also* knowledge

equant model, 102–4; Ibn al-Shāṭir's substitute for, 180; in *Theorica planetarum*, 267; Tusi-couple and, 179

equinoxes, 87, 92, 98

Erasistratus of Ceos, 120–22; Galen and, 125–29

Eratosthenes, 96–98; Martianus Capella on, 144, 145, 197

Eriugena, John Scotus, 198–99, 209, 217

eternity of the universe: Aristotle on, 53, 229; Bonaventure's rejection of, 237, 243; medieval Aristotelians and, 53, 229, 240, 243, 244, 246

ethics: of Epicureans and Stoics, 76–77, 78, 136

Euclid: Albert the Great and, 239; deductive system of, 84; on law of the lever, 110; optics of, 105–7, 108, 217, 314, 317, 318, 361

Euclid's *Elements*, 84–85; assimilated in the West, 226; in Byzantine education, 160; in Islamic culture, 172, 175, 176, 177; Martianus Capella on, 144; translations of, 148, 216, 217

Eudoxus: method of exhaustion and, 85; on planetary motions, 59–60, 86–92, 94, 95, 136; Ptolemy compared to, 99

Eutocius of Ascalon, 160

evaporation, 54, 277

exercise: as medical therapy, 116, 336

exhaustion, method of, 85

Ex herbis femininis, 322

experience: Aristotle on, 48; Hippocratic medicine and, 118. *See also* empiricism; experimentation; observation; sense perception

experimentation: alchemical, 290, 292, 294–95, 364; Roger Bacon on, 236; continuity debate and, 359, 362–64, 410n15; criticism of Aristotle for lack of, 51; divine omnipotence and, 253; factors in emergence of, 253; in Islamic optics, 183–85; as measure of genuine science, 1, 2; Philoponus's falling bodies, 310, 363; in Ptolemy's optics, 107, 108–9, 362, 378n29; seventeenth-century exploitation of, 396n48

extramission theory of vision, 314–15, 316, 318, 403n45

eye: anatomy of, 120, 126, 314, 317. *See also* vision

faith and reason, 206–9, 230–31, 233, 241, 243, 251. *See also* Christian theology; rationalism
falconry, 353
falling bodies: Aristotle on, 54–55, 57, 58, 75, 309; Buridan's impetus and, 308; Philoponus on, 160, 310–11, 363; Strato on, 75
Fallopian tubes, 120
Farghani, al-, 260, 265, 266
Fatimids, 189, 201
Ferrara: medical faculty at, 333
fever, 122, 126, 336
final cause, 51, 52, 60, 62, 63, 64; in nested sphere model, 93; Unmoved Mover as, 260. *See also* teleology (purpose)
finger reckoning, 160, 177
fire: Aristotle on, 53, 55; Cicero on, 138–39; Heraclitus on, 27, 29; Plato's theory of vision and, 41, 105; Stoics on, 80, 81, 259; Strato on, 74–75; Theophrastus on, 73. *See also* elements
firmament, 258–59
first form. *See* corporeal form
fixed stars, 42, 59, 258, 259, 272
flat earth: affirmed, 161; denied, 56, 161
Florence: medical practitioners in, 331, 333
fluxus formae, 297, 298
force: Aristotle on, 57, 58, 295, 306; dynamics and, 306, 307–8, 309, 311–12, 402n24; Neoplatonism and, 318, 320
form, Aristotelian: accidental, 287, 292; Albert the Great and, 240; Aquinas and, 243; in Aristotle's analysis of change, 49, 50, 51, 52; in Aristotle's biology, 62, 63, 64; in Aristotle's

metaphysics, 47, 48; in Aristotle's theory of elements, 287–88, 289; divisibility of substance and, 290; mechanical philosophy and, 365; medieval cosmology and, 259; as mover, 307
forma fluens, 297–98
formal cause, 51, 52, 62–63
forms: intensification of, 301, 402n27; Plato's theory of, 36–37, 38, 39, 46, 48
fracture of skull, 345
Franciscan order, 232, 233, 234, 236, 270
Franciscus de Marchia, 308
Frankfort, H., 6
Frankfort, H. A., 6
Frederick II, 353
freedom, intellectual: in Islamic culture, 173–74; in medieval universities, 224
Freeman, Charles, 358
free will: astrology and, 214, 230, 273; Epicurus's analysis of, 78; prohibition on denial of, 246. *See also* determinism
functional explanation: Aristotle and, 52, 62, 64
fundamental reality. *See* underlying reality

Gaia, 22
Galen, 124–31; Albert the Great and, 239; on design, 131; four humors and, 336; and Greco-Roman culture, 132; influence beyond seventeenth century, 366; Islamic physicians and, 186, 187, 188, 327–28; and medieval medicine, 335, 338, 345, 347; Nestorian physicians and, 185; physiological system of, 127–29, 379n21; translated from Arabic to Latin, 215, 216, 330; translated from Greek to Arabic, 172, 185, 327; translated from

Greek to Latin, 322, 345; translated from Greek to Syriac, 172; Vesalius and, 348; on vision, 314, 316, 317

Galileo Galilei: continuity question and, 366; Duhem on medieval anticipation of, 358; experimentation by, 310, 362, 403n38; and mathematical physics, 360, 361; and mechanical philosophy, 365; and Merton rule, 306, 366; and relativity of motion, 283

Gassendi, Pierre, 365

Geber (Jābir ibn Hayyān), 291–92

gender: Aristotle on, 63, 64

generation: Aristotle's theory of, 63–64

generation and corruption: Aristotle on, 160, 172, 216, 234; astrology and, 272, 275; medieval cosmology and, 256; as type of change, 295, 296

Genesis, 210, 229. *See also* creation

Geoffrey de Bussero, 351

Geoffrey Plantagenet, 212

geographical knowledge: Byzantine, 160–61, 162; medieval, 278–81. *See also* maps

geometry: Aristotle's view of, 83; in Byzantine Empire, 160; Egyptian, 13; Gerbert's knowledge of, 203; Greek, 28, 83–86; Greek planetary astronomy and, 86–92, 98, 99; Islamic, 177–78; law of the lever and, 110; Martianus Capella on, 144; in medieval analysis of qualities, 301–6; Plato and, 37, 39–41, 53, 82; Ptolemy's planetary astronomy and, 104–5; in university education, 223. *See also* Euclid's *Elements*

Gerard of Brussels, 299–300

Gerard of Cremona, 187, 216, 265, 330

Gerbert of Aurillac, 199–200, 201, 202–3, 205, 215, 263

German universities, 221

Gilbert, William, 362, 364

Giles of Corbeil, 339

Giovanni di Casali, 301

God or gods: Aristotle's deity, 60, 65; Cicero on, 138–39; Egyptian medicine and, 18; in Galen's worldview, 130, 131; Greek (*see* Greek gods); in Islam, 166; in mechanical philosophy, 365; in oral traditions, 6, 7, 8; Plato's Demiurge, 36, 39, 41–42, 130, 149, 214, 271–72; Plato's planetary deities, 42–43, 214, 260, 271–72; signs or omens from, 15–16, 27, 271; in Stoics' universe, 80, 81. *See also* Christian theology

Gondeshapur (Jundishapur), 164–65, 168, 185

Goody, Jack, 11

Greek East, 147, 158–62

Greek empire, 67, 132

Greek gods, 21–25, 26–27, 29, 31; medicine and, 111, 112–13, 115, 118, 119

Greek language. *See* translation from Greek to Arabic; translation from Greek to Latin; translation from Greek to Syriac

Gregory IX (pope), 227

Gregory the Great, 194

Grosseteste, Robert, 234, 237, 255; on corporeal form, 288; cosmology of, 255, 397n4; on optics, 318; on rainbow, 277

Guido de Marchia, 270

guilds: universities as, 218–19

Gutas, Dimitri, 169

Guy de Chauliac, 343, 346–47

gymnastike, 68

gynecology, 333

Ḥakam, al-, 189

Hall, A. Rupert, 365

Haly Abbas (Alī ibn 'Abbās al-Majūsī), 187, 188, 328, 330, 335, 345

handmaiden of theology, science as:
  Albert the Great and, 239, 243;
  Aquinas and, 242, 243; Augustine
  on, 149, 150, 381n28; Roger Bacon
  on, 236; Bonaventure on, 237; condemnations of 1270 and 1277 and,
  250; radical Aristotelians and, 246
Hārūn al-Rashīd, 168, 171, 349
Harvey, William, 188
Ḥasdāy, Abū Yusuf, 201, 203
Haskins, Charles Homer, 358
heart, 61, 64, 120–21, 127–28, 188
heat. *See* vital heat
heaviness and lightness, 55
heliocentric system: of Aristarchus,
  95–96; of Copernicus, 162, 366
Hellenistic period: anatomy and
  physiology of, 119–22, 124, 125,
  128; anti-astrological sentiment
  in, 273; astronomical constants
  calculated in, 96; Babylonian as-
  trology/astronomy of, 16; defined,
  67; educational institutions in,
  70–72; lever in, 109–10; mathematicians of, 85–86; medicine of, 111,
  122, 124–31, 322; mythological
  literature of, 22; natural philosophy of, 67; optics in (*see* Ptolemy,
  Claudius: optics of); philosophical
  developments in, 76–81; planetary
  astronomy of, 98–105 (*see also*
  Ptolemaic astronomy)
Hellenization of Islamic culture, 168,
  169
hemorrhoids, 343
Henry II of England, 212
Heraclides of Pontus, 95
Heraclitus, 27, 29, 32
herbals, 322, 351, 352; Dioscorides's *De
  materia medica*, 162, 186, 322, 327,
  338, 351; *Ex herbis femininis*, 322
herbs: Albert the Great on, 353; in
  Hippocratic period, 117, 118; in
  medieval healing, 322, 325, 331,

337, 351; prehistoric understanding
  of, 4, 8
Hermann the Dalmation, 216
Herodotus, 12, 27
Herod the Great, 75
Hero of Alexandria, 107, 108, 160
Herophilus of Chalcedon, 120, 121,
  122, 125, 126, 314
Hesiod, 22, 25, 27, 29, 111
hexis, 80
Heytesbury, William, 300
hieroglyphics, 10
Hindu-Arabic numerals, 160, 177, 203
Hipparchus, 96, 98–99, 105, 136, 142,
  145, 262
Hippocrates of Cos, 113
Hippocratic Oath, 115, 118
Hippocratic writings, 113, 115–19,
  120; influence on Galen, 125, 126;
  medieval medicine and, 335, 336,
  339, 342; translations of, 172, 215,
  322, 327, 330
hippopede, 90–91
Homer, 21–22, 25, 29; healing and,
  111, 112
Honorius of Autun, 210
Horace, 132–33, 206
Hospitallers, 349
hospitals, 162, 164–65, 175–76, 188,
  189, 348–51
house of wisdom, 171
Hugh of Santalla, 216
Hugh of St. Victor, 208, 274
Hulagu Khan, 191
humanism, twelfth-century, 214
humoral theories: Asclepiades' repudiation of, 124; Galen's version of,
  125–26; in Hippocratic medicine,
  115–16, 125, 378n7; in medieval
  medicine, 336
Ḥunayn ibn Isḥāq, 171–72, 185–86,
  188, 318, 327, 330, 335
hylomorphism, 373n4
Hypatia, 161

Ibn al-Haytham (Alhacen): correct name of, 403n44; on optics, 183, 226, 314–18, 320, 362; on planetary models, 179, 267, 269

Ibn al-Nadīm, 179

Ibn al-Nafīs, 188–89, 387n43

Ibn al-Shāṭir, 180

Ibn Bājja (Avempace), 311

Ibn ʿĪsā, Aḥmad, 183

Ibn Māsawayh, Yūḥannā, 172, 186

Ibn Rushd (Averroes): Albert the Great and, 239; attack on Ptolemaic astronomy, 269; celestial spheres and, 261; commentaries on Aristotle, 217, 227, 228; on elements, 288; Galen's anatomical writings and, 345; monopsychism of, 230, 243, 244; on motion, 297, 307, 311; on optics, 317–18; translated by Michael Scot, 392n6; University of Paris and, 251

Ibn Sahl, Abū Saʿd, 183

Ibn Sīnā (Avicenna), 187; Albert the Great and, 239, 240; autobiography of, 176; *Canon of Medicine*, 187, 188, 216, 226, 328, 330, 334, 335, 339, 387n43; commentaries on Aristotle, 187, 217, 226, 228, 229, 297, 307, 353; critique of alchemy, 291; on elements, 288; on optics, 317–18; on the rainbow, 183

immortality of the soul: Aristotelian denial of, 63, 230, 237, 240, 244, 246; Plato's defense of, 149

impetus, 260, 308–9, 366

incommensurable magnitudes, 83, 85, 144, 249

India: Babylonian astronomy and, 15; Greek influence in, 163, 168; influence on Islamic culture, 168, 177, 178; mathematics in, 177, 178

induction, 48

inertia, 309

instantaneous velocity, 300

instruments. *See* astronomical instruments

intensity of a quality, 300–303, 338, 402n27

intromission theory of vision, 313, 314, 315–16, 318

Ionia, 27–28, 29, 34

Irish scholarship: Alcuin and, 196; monasticism and, 155, 158, 198–99, 382n32

Irving, Washington, 161

Isidore of Miletus, 160

Isidore of Seville, 157, 158; on astronomy, 262; on geography, 279, 280; on medicine, 321

Islam: birth of, 166; expansion of, 166–68, 169; observatories in, 175–82, 191, 363; terminology for, 168–69. *See also* Islamic culture

Islamic alchemy, 291–92

Islamic astrology, 178, 214, 272–75

Islamic astronomy, 178–83, 189, 191; Byzantine borrowing from, 161; religious uses of, 171, 174, 192; schooling in, 175; on size of lunar sphere, 260; translation for purposes of, 171, 172, 173; transmission to northern Europe, 201–3, 262–63

Islamic culture: Gerbert's contact with, 199–201; Greek science and, 173–77, 178, 185; Hellenization of, 165, 168, 169–73; terminology for, 168. *See also* Islam; Islamic science

Islamic empire: decline of, 190–91; expansion of, 166–68, 169

Islamic mathematics, 177–78, 189; transmission to West, 199, 201, 202–3, 215

Islamic medicine, 185–89, 191, 192, 327–28; Avicenna's *Canon of Medicine*, 187, 188, 216, 226, 328, 330, 334, 335, 339, 387n43; Galen's writings and, 129, 185, 186, 187, 188, 327–28; hospitals and, 188, 189, 349

Islamic optics, 182–85, 226, 313, 314–18, 320

Islamic science: elements in, 288; fate of, 189–92; in medieval universities, 224. *See also* Islamic culture; *and specific sciences*
ius ubique docendi (right of teaching anywhere), 224

Jābir ibn Hayyān (Geber), 291–92
James of Venice, 217
Jerusalem: hospital of Saint John in, 349
Jews: on Arabian peninsula, 166; expulsion from Spain, 191; in medieval medical practice, 333; in multireligious Spain, 201, 216; in Tunisia, 201
Johannes de Muris, 363
Johannes de Sacrobosco (John of Holywood), 262, 266
Johannitius. *See* Ḥunayn ibn Isḥāq
John of Dumbleton, 300
John of Seville, 216, 265
John Philoponus. *See* Philoponus, John
John XXII (pope), 250
Joseph the Spaniard, 203
Julius Caesar, 132
Jundishapur (Gondeshapur), 164–65, 168, 185
Jurjīs ibn Bukhtīshūʿ, 168, 185
Justinian, 164

Kamāl al-Dīn al-Fārisī, 183–85, 277, 362–63
Kepler, Johannes, 179, 320, 361, 366
Khusraw I, 164
Khusraw II, 164
Khwārizmī, al-, 177, 216, 226, 385n20
kidney stones, 331
Kilwardby, Robert, 250, 260
Kindī, al-, 183, 239, 315, 317
kinematics, 300, 304–6, 366
knowledge: Aristotle on, 62, 65, 82; Cicero on, 138; early Greek philosophers on, 25, 33–34, 43; invention

of writing and, 10; Plato on, 37–38; prehistoric conceptions of, 9. *See also* epistemology
Koran, 166, 169, 175, 176
Koyré, Alexandre, 359, 364
Kuba people of Africa, 7–8, 9

Laon: cathedral school of, 205, 207
Latin: scholarly discourse in, 135, 137; in schools, 151, 206, 209, 218, 220. *See also* translation from Arabic to Latin; translation from Greek to Latin; translation from Latin to vernacular languages
Latin West, 147–48, 149, 162, 193
law: Aristotelian foundations in, 251; education in, 206, 218, 219, 221, 223
laws of nature: miracles and, 213; preliterate ideas of causation and, 6; Stoic cosmos and, 81. *See also* natural law
Lesbos, 45, 60, 73
Leucippus, 27, 29, 31, 33, 77
lever, 109–10
Levi ben Gerson, 363
Leyden papyrus, 18
liberal arts: Boethius's handbooks on, 148; in Byzantine education, 159; Charlemagne and, 194; higher education in, 218, 219–20, 221, 223; Isidore of Seville on, 157; Martianus Capella on, 144–45, 194, 197, 199; monastic education and, 155; in twelfth-century schools, 205, 206; Varro on, 137
library of Alexandria, 72, 96, 375n7
light: Alhacen on, 317; Aristotle on, 313; Roger Bacon on, 320; Grosseteste on, 255, 288; Strato on, 75. *See also* optics; vision
Lindisfarne, 325
liver, 127, 338
local motion: Aristotle on, 295, 296; defined, 295, 401n17; dynamics of,

306–9; as *fluxus formae*, 298; impetus and, 308; mechanical philosophy and, 365; as a quality, 301. *See also* motion

logic: Aristotle's predecessors and, 32, 33, 34; early church's use of, 149; Gerbert of Aurillac and, 199; Ḥunayn's translations of works on, 172; invention of writing and, 10; Islamic culture and, 174, 176; Roman uses of, 136, 137; in schools, 205; translations from Arabic to Latin, 216; in university education, 223. *See also* Aristotelian logic

logographic writing, 10, 11

Lombard, Peter, 262

love and strife: Empedocles' principles of, 31, 33

Lucan, 206

Lucretius, 77, 137, 139, 157, 364

lunar month: Babylonian calculations of, 16–17, 371n29; Hipparchus's calculation of, 98, 99

Lupitus, 203

Lyceum, 45, 70, 72, 73–76, 223

Macrobius, 143–44, 194, 197, 206, 209

macrocosm-microcosm analogy: astrology and, 214, 272; of Cicero, 139; in Grosseteste's cosmology, 255; in twelfth-century natural philosophy, 213–14, 255

madrasas, 175, 191

magic: healing arts and, 8–9, 20, 185; in preliterate cultures, 6

magnetism, 271, 364

Mahdī, al-, 171

Maier, Anneliese, 359

Ma'mūn, al-, 171, 172, 177, 178

Manicheism, 163

Manṣūr, al-, 168, 170, 171, 185

mappaemundi, 280

maps, 278, 279–81. *See also* geographical knowledge

Maragha observatory, 179–80, 183, 190, 191, 267, 277, 363, 386n30

Marcus Aurelius, 72, 125, 146

Mark of Toledo, 216

Martianus Capella, 144–45; on astronomy, 145, 262; Eriugena and, 199; Isidore of Seville and, 157; *Marriage of Philology and Mercury*, 144–45, 151, 197, 199, 209; twelfth-century schools and, 206, 209

material cause: Aristotle on, 51, 52, 62; mechanical philosophy and, 365; in medieval cosmology, 259; of stellar and planetary motions, 92

materialism: of early Greek philosophers, 29–31; of Stoics and Epicureans, 79

mathematical sciences: defined, 174

mathematics: applicability to nature, 82–83; of Archimedes, 85–86, 110, 173, 217; Aristotle's natural philosophy and, 82–83, 95, 105, 360, 361; Babylonian, 12, 13–15, 371n29; definition of science and, 2; Egyptian, 12–13, 15; Greek, 12, 83–86; Hipparchus's astronomical applications of, 98–99; Islamic, 177–78, 189, 199, 201, 202–3, 215; Martianus Capella on, 144–45; in monasteries, 155; of motion, 299–306, 309–13; natural philosophy and, 3, 82; Plato and, 37, 39–41, 53, 70, 82, 83; Ptolemy's planetary models and, 99, 104–5, 262; Pythagorean philosophy and, 31, 41; Romans' limited interest in, 135, 151; schism between physics and, 360–62; translations from Arabic to Latin, 216; translations from Greek to Latin, 217; translations to Arabic or Syriac, 171, 172, 173; twelfth-century metaphysical uses of, 214–15; in university education, 223. *See also* geometry; numbers

matter: Aristotle's biology and, 62,
63, 64; Aristotle's fundamental
metaphysics of, 47, 49, 51, 52, 365;
and Aristotle's theory of elements,
287–88, 289; medieval analysis of
motion and, 307; and medieval
cosmology, 259
Maximus Planudes, 160
mean-speed theorem (Merton rule),
304–6, 366
Mecca, 166
mechanical philosophy, 365, 366;
alchemical antecedents of, 295
mechanistic worldview: of Epicureans,
77, 78, 79, 81; of Greek atomists, 30
mechanization: in physiology, 129
medical education: dissection in,
345–47; in Salerno apprenticeship, 329; of surgeons, 343, 345; in
universities, 218, 219, 221, 223, 327,
330, 331, 333–35, 345–47; in urban
schools, 205, 329
medicine: Aristotelian foundations of,
251; astrology and, 265, 275, 334,
339, 341–42; Byzantine contributions to, 162; Celsus's encyclopedia of, 139; early Greek, 111–19;
Egyptian, 18, 111; Hellenistic, 111,
122, 124–31, 322; Hippocratic (*see*
Hippocratic writings); institutionalization of, 175–76, 188, 189 (*see
also* hospitals); Islamic (*see* Islamic
medicine); local healers, 185, 321,
331, 336, 337; medieval, early period, 321–28; medieval, transitional
period, 329–30; medieval healing
activity, 335–42; medieval practitioners, 322, 330–33; Mesopotamian,
18, 20; Nestorian physicians, 164,
165, 168, 171, 172, 185; Persian,
164; preliterate healing arts and,
8–9; Roman, 135; translation from
Arabic to Latin, 215–16, 217, 328,
329–30, 333, 343, 345; translation

from Greek to Latin, 217, 321–22,
345; translation into Arabic or Syriac, 171, 172, 173; translation into
vernacular, 335–36; urbanization
and, 329; women as practitioners
of, 329, 332, 333. *See also* disease;
pharmacological therapy; surgery
medieval period: defined, 193; early vs.
high or late, 193
mendicant orders, 232, 233
menstrual blood: Aristotle's theory of,
63
Merton College, 222; scholars of,
300–301, 304–6, 311, 366
Merton rule, 304–6, 366
Mesopotamia. *See* Babylonians
metals, 290, 291–92
metaphysics: of Aristotle, 46–48, 63, 70,
249, 364, 365; of early Greek philosophers, 32; scientific revolution
and, 364–66; in university education,
223. *See also* underlying reality
meteorology: Aristotle on, 66, 137, 160,
216, 234, 272; in medieval cosmology, 277; *Meteorologica* attributed to
Aristotle, 292; Seneca on, 139
methodist physicians, 124
Meton, 86
Metonic cycle, 86
Michael Servetus, 387n43
microcosm. *See* macrocosm-microcosm
analogy
Middle Ages. *See* medieval period
midwifery, 118, 321, 322, 331
Milesians, 27–31
minima naturalia, 290
miracles: Albert the Great on flood as,
241, 243–44; Aristotle's cosmology and, 230; fourteenth-century
theology and, 252; healing and,
324, 325–26, 327; twelfth-century
scholars on, 212–13
mirrors, 107–8, 160, 183
mixed bodies, 55

mixture (*mixtum*), 287, 289, 291
modern science. *See* seventeenth-century
 science
momentum, 308–9
monasteries, 152–58, 194, 382n32;
 Charlemagne and, 195; education
 and, 152, 154–55, 198–99, 204–5;
 medical practices in, 322–24, 327;
 reform movements in, 205, 329
Mondino dei Luzzi, 346
monism: of Milesian philosophers, 29
Monophysitism, 164, 383n2
monopsychism, 230, 237, 243, 244, 246
Monte Cassino, 152, 154, 215, 241,
 322, 329
Montpellier: medical faculty at, 333
moon: in Aristotle's cosmology, 53, 55;
 astrology and, 272–75; calculated
 distance from earth, 260; in early
 Greek models, 86, 88, 90, 92; as
 Greek deity, 22; Hipparchus's measurement of diameter, 98; in Islamic
 model, 180; motion on celestial
 sphere, 59, 271; Plato on, 41. *See also*
 eclipses; lunar month
motion: accelerated, 300, 303–6,
 308, 366; Aristotle on, 50, 54–55,
 56–60, 82, 295–96, 306, 309–10,
 361, 401n17; Buridan on, 260, 298,
 308–9, 403n36; and condemnation of 1277, 247, 257; critiques
 of Aristotle on, 160, 249, 310–11;
 early Greek philosophers on, 32–33,
 372n21; mathematical description
 of, 299–306, 309–13, 361; Plato on,
 42, 43; resisted, 308–12; Strato on,
 74–75; uniform, 299, 300, 303. *See
 also* falling bodies; local motion;
 uniform circular motion
mousike, 68
Muʿāwiyah, 166
Muḥammad, 166, 168
multiple universes, 247, 249, 257
mumps, 117

Mūsā, sons of, 171, 172
Museum of Alexandria, 72, 99,
 375n7
Muslims, 166, 169
myths: African, 6; Greek (*see* Greek
 gods); invention of writing and, 11;
 medieval bestiary and, 354–55; of
 origins, 7–8, 9, 12

Naples, University of, 241
natural history: Albert the Great on,
 240; Aristotle on, 62, 65, 70, 73,
 223; medieval, 351–56
Natural History (Pliny the Elder),
 140–42, 197, 203, 279
naturalism: Aristotelian, Christian
 theology and, 228–29, 231, 237,
 240–41, 246; of early Greeks, 27;
 medical, vs. Christian supernatural-
 ism, 324; in twelfth-century natural
 philosophy, 210–13
natural law: Stoic ethics and, 78;
 twelfth-century awareness of, 212.
 See also laws of nature
natural philosophy: Aristotelian, and
 Aquinas, 243; Aristotelian, in
 medieval universities, 223, 226–28,
 234; of Aristotle, 50, 60, 62, 65,
 82; astrology as, 271; Carolingian,
 199; condemnation of 1277 and,
 248–49; defined, 3; early Christian
 church and, 149–50; early medieval, 157–58; Epicurean, 77, 78,
 81; fourteenth-century, 251–53;
 Hellenistic, defined, 67; in Hellenistic medicine, 122; of Hippocratic
 physicians, 113; mathematics and,
 3, 82; medical education and, 334,
 335; monasteries and preservation
 of, 155–57; Platonic influence on,
 38; Roman lack of interest in, 136,
 151; Stoic, 78–79, 81, 139; Theophrastus's program in, 73; twelfth-
 century, 206, 209–15

nature: Aristotle on, 27, 50–51, 287, 373n9; Aristotle on motion and, 57, 59, 295, 306–7; Cicero on, 138–39; contingency of, 253; early Greek philosophers and, 26–27, 29, 44; Galen's teleology and, 130; Plato's mathematization of, 41; prehistoric attitudes toward, 3–12

Neleus, 74

Neoplatonism: Academy's refounding and, 70–71; Albert the Great and, 237; Avicenna and, 228; Bonaventure and, 236; Byzantine culture and, 159–60; in early commentaries on Aristotle, 75–76; Eriugena and, 199; Greco-Latin translations of, 217; Grosseteste and, 234; al-Kindī and, 183; Macrobius and, 143–44; optical thought in, 317, 318; of Philoponus, 307–8; pseudo-Dionysius and, 199; universe of forces in, 318, 320. *See also* Platonic tradition

nerves, 120, 121, 126, 127, 129

Nestorian Christians, 164–65, 168, 171, 172, 185, 186, 383n2

Nestorius, 164

Neugebauer, Otto, 386n30

Newman, William, 294–95, 400n9

Newton, Isaac, 29, 44, 365

Nicholas of Damascus, 75, 352

Nicolaus Bertrucius, 346–47

Nisibis, 164, 165

Noah's flood, 241, 244

nonnaturals, 336

nosebleed, 337

nous, 25

numbers: Babylonian system, 13–15, 371n29; in Byzantine education, 160; Egyptian system, 12–13; Euclid on theory of, 85; Hindu-Arabic system of, 160, 177, 203; irrational, 83; Martianus Capella on, 144–45; Pythagoreans on, 31, 82, 144;

twelfth-century metaphysical views of, 214–15

nutrition, 122, 127, 129. *See also* diet

observation: Aristotle's use of, 53, 55, 62, 65; as defining characteristic of science, 1; Plato on, 37–38. *See also* experimentation

obstetrics, 120, 333. *See also* midwifery

Ockham, William of, 251, 297–98

Ockham's razor, 298

old logic, 148

omnipotence, divine, 230, 247, 248, 251–53, 285, 298, 327

ontological proof, 207

optics, 105–9, 313–20; Roger Bacon on, 236, 239, 318, 320; early Greek, 105, 313–14, 317, 318, 361; of Euclid, 105–7, 108, 183, 217, 314, 317, 318, 361; Grosseteste on, 318; Islamic, 182–85, 226, 313, 314–18, 320; mathematics in, 361–62; of Ptolemy, 106–9, 183, 314, 317, 318, 362, 378n29; writings on, transmitted to the West, 216, 317–18. *See also* vision

oral cultures, 4–10, 11, 369n4

order: Cicero on, 138; in Galen's teleology, 131; Greek philosophers on, 39, 43; preliterate worldview and, 5–6

Oresme, Nicole: on astrology, 249, 275, 277; Bradwardine's dynamics and, 312; geometrical analysis of qualities, 301–5; graphing techniques of, 366; on infinite void, 251, 257; on possible rotating earth, 282–85; and proof of mean-speed theorem, 304–5

Oribasius, 322

Orleans: cathedral school of, 205

Osiander, Andreas, 361

Otto III, 199, 203

Otto the Great, 215

ovaries, 120

Ovid, 206
Oxford University: Aristotle's works
at, 226, 227; Bacon, Roger, at, 234;
Grosseteste at, 234; mathematicians at, 358; medical faculty of,
333; origins of, 206, 218, 219, 221;
professorial mobility and, 224. *See
also* Merton College

Padua: medical faculty of, 333, 346
paideia, 68
pantheism, 226, 228
Pantokrator hospital, 349
paper making, 169
parallax, 96, 260
Paris: condemnations of Aristotelianism
at, 244, 246, 250–51; twelfth-century
schools of, 205, 207, 208, 212, 218
Paris, University of, 219–21, 224;
Albert the Great at, 237; Aquinas
at, 241; Aristotle's works at, 226–27,
228, 251; Bacon, Roger, at, 234;
Bonaventure at, 236; condemnations
of propositions in, 247; Gerard of
Brussels and, 299–300; mathematicians at, 358; medical faculty of, 333,
341; mendicant theologians at, 233;
Oresme at, 301
Parmenides, 32, 33, 34, 37; Aristotle's
philosophy and, 49, 53; on change,
49; Roman culture and, 133
Parthenon, 72
Paul of Taranto, 292, 294, 364, 401n13
Pecham, John, 250, 319, 320
Pergamum, 27, 124, 125, 132
peripatetics, 70, 72, 74, 75; defined, 70;
Erasistratus and, 120, 121
periplus, 279
Persia: astronomy of, 161; Greek learning in, 164, 165, 168; Islamic empire
and, 166, 168, 170
Persian Empire, 169, 170
perspectiva, 313, 361–62, 363. *See also*
optics

Peter Abelard. *See* Abelard, Peter
Peter Lombard. *See* Lombard, Peter
Peter Peregrinus of Maricourt, 363–64
Peurbach, Georg, 161–62, 270
Phaedrus, 137
Phaenomena (Aratus), 98, 136, 139
pharmacological therapy: in ancient
Egypt, 18; of Galen, 129; in Hippocratic period, 117; in medieval
period, 325, 336, 337–38, 341; in
Mesopotamia, 18. *See also* herbals;
herbs
Philinus of Cos, 122
Philo of Larissa, 137
Philoponus, John, 75, 160, 162, 307–8,
310–11, 363, 403n36
philosophy: and invention of writing,
11; origins of, 25–27
philosophy of nature: defined, 3. *See
also* natural philosophy
physics: Aristotelian, and condemnation of 1277, 247–48; Aristotle's
view of mathematics in, 82–83;
schism between mathematics and,
360–62; in university education,
223. *See also* Aristotle's works:
Physics; motion
physikoi, 27, 39, 43–44
Physiologus, 162, 353–54
physiology: Aristotle on, 62; Galen's
system of, 127–29, 379n21; of Hellenistic period, 120–22, 124, 125,
128; Islamic, 188–89; Plato on, 41;
pulmonary transit, 188–89, 387n43
physis, 27, 39, 80. *See also* nature
pictographs, 10
place, Aristotle's doctrine of, 56, 295;
medieval views, 257, 258, 298;
Strato and, 75
plague, 111, 139, 147, 324, 341
planetary astronomy: Aristotle on, 53,
59–60, 66, 92–95, 105; Babylonian,
16; Carolingian, 197; Cicero on,
139; distance calculations in, 260;

planetary astronomy (*cont.*)
in early Middle Ages, 262; Eudoxus on, 59–60, 86–92, 94, 95, 99, 136; Hellenistic, 98–105; of Heraclides, 95; Islamic, 178, 179, 180; Johannes de Sacrobosco on, 266; Martianus Capella on, 145; observed motions on celestial sphere, 59; Plato on, 41, 42–43, 86–89; Pliny on, 142, 262; retrograde motion in, 88, 90, 101, 145, 197, 262, 266; *Theorica planetarum* on, 266–67. *See also* Ptolemaic astronomy
plants: Aristotle on, 63, 64, 352; preliterate cultures' understanding of, 4, 7, 8. *See also* botany; herbs
Plato, 35–43; Academy of, 35, 45, 69–71, 78, 95; Aristotle's divergence from, 45–47, 53, 65, 70; on astronomy, 37, 38, 41, 42–43, 86–89; biblical creation and, 255; Byzantine culture and, 159; on elements, 38, 40–41, 53, 82, 287; epistemology of, 37–38; Galen's debt to, 125, 129, 130; on Heraclitus, 29; mathematics and, 37, 39–41, 53, 70, 82, 83; mechanistic worldview rejected by, 30; *Phaedo*, 37, 38; Pythagoreans and, 133; *Republic*, 35–36, 38; on the soul, 36, 37, 41, 42, 105, 127, 149; *Timaeus* (see *Timaeus*); twelfth-century schools and, 206; on vision, 41, 105, 317, 318; vital heat and, 128
Platonic solids, 40–41, 53; Euclid on, 85; Martianus Capella on, 144
Platonic tradition: Albert the Great and, 237, 239, 240; animistic strain in, 42; astrology and, 274; Boethius's theology and, 194; Cicero's relationship to, 137–38, 380n9; in commentaries on Aristotle, 187, 228; early church's uses of, 149; entrenched in first millennium, 226; Grosseteste and, 234; in Hellenistic period, 76; Islamic culture and, 173; Marcus Aurelius and, 72; in medieval cosmology, 254–56; in Persian intellectual life, 164; Roman culture and, 134, 380n9; skeptical tendencies in, 138; Stoic and Epicurean alternatives to, 81; in thirteenth-century, 237, 255; *Timaeus* and, 38, 147; twelfth-century scholars and, 209–10, 212, 214. *See also* Neoplatonism
Plato of Tivoli, 216
plenum, 54
Pliny the Elder, 139–42; on astronomy, 142, 145, 262; Carolingian availability of, 197, 203; on geography, 144, 279; on medicine, 321; *Natural History*, 140–42, 197, 203, 279; Venerable Bede and, 158
pneuma: Erastistratus on, 120–22; Galen on, 127, 128, 129, 379n22; Herophilus on, 120; in Islamic physiology, 188; of Stoics, 80, 124
pneumatist physicians, 124
popularization of Greek learning, 135, 136–46, 158
Porphyry, 148, 165, 176, 199
portolan charts, 280–81
Posidonius, 136–37
potentiality and actuality, 49–50, 62, 64, 233, 295
Praxagoras of Cos, 120
prehistoric attitudes toward nature, 3–12
pre-Socratic philosophers, 35
Prime (Unmoved) Mover, 60, 64, 93, 230, 260, 272
primitive mentality, 9
primum mobile, 258
printing: of anatomical texts and drawings, 346
Profatius Judaeus, 265
projectile motion, 57, 307–9, 366
properties, 47, 48. *See also* form, Aristotelian; qualities

prophetic medicine, 185
providence, divine, 81, 149, 230, 237, 246, 327
pseudo-Apuleius, 352
pseudo-Dionysius, 199
psyche: Plato on, 39; Stoics on, 80
Ptolemaic astronomy, 99–105; Aristotelian spheres and, 267, 269–70; Hipparchus and, 98, 99; Islamic astronomy and, 178, 179, 180, 265; Martianus Capella on, 145; medieval cosmology and, 261; as observational science, 362, 363; Persian interest in, 170. *See also* Ptolemy's *Almagest*
Ptolemaic kings, 71–72
Ptolemy, Claudius, 99; on astrology, 272–73, 274–75; *Geography*, 137, 161, 281; *Handy Tables*, 161; name of, 377n22; optics of, 106–9, 183, 314, 317, 318, 362, 378n29; *Planetary Hypotheses*, 179, 269; on rotating earth, 282; *Tetrabiblos*, 272–73, 274–75
Ptolemy's *Almagest*: assimilated by the West, 226, 261; Byzantine commentators on, 161–62; in Islam, 175, 176, 178, 179; lunar calculations based on, 260; as mathematical analysis, 262; translations of, 172, 216, 217, 223, 265; university lectures on, 267. *See also* Ptolemaic astronomy
pulmonary transit, 188–89
pulse: examination of, 126, 335, 338–39; theory of, 121
purging, 116, 336; bloodletting, 116, 122, 185, 336, 342, 343
purpose. *See* teleology
Pythagoras, 12
Pythagoreans: Aristarchus's heliocentrism and, 95; of Italy, 133; number and, 31, 82, 144; Plato and, 35, 40, 41, 82

quadrant, 180, 265
quadrivium: in Byzantine education, 159, 160; defined, 137; Gerbert of Aurillac and, 199, 201; mathematical astronomy in, 262; in monastic studies, 155, 158; in university education, 223; in urban schools, 205
qualities: accidental, 247, 249; Aristotle's theory of, 53–54, 288, 289, 291, 296, 365; in Geberian alchemy, 292; geometrical analysis of, 301–6; intensities of, 300–303, 338, 402n27; motion as, 298, 301; quantities of, 301, 303–4; secondary, 34, 77, 365; without a subject, 47
quantity: vs. intensity, 301; Oresme's analysis of, 303–4
quintessence, 53, 59, 60, 92, 255, 256, 259
Quintilian, 206
Qusṭā ibn Lūqā, 183

rain, 143, 277
rainbow: Aristotle on, 66, 82; Islamic studies of, 183–85, 363, 386n34; medieval theories of, 277
rationalism: Aristotelian, and Christian theology, 206–9, 228–29, 230–31, 240, 245, 246; of Democritus, 34, 77; Epicurus's opposition to, 77. *See also* faith and reason; reason
rationalist physicians, 122, 124, 125
rationality: earliest Greek philosophy and, 25; of Plato's cosmos, 41–42, 43; of Stoics' universe, 80, 81
Raymond of Marseilles, 267
Rāzī, al- (Rhazes), 186–88, 291, 328, 330, 335, 345
reality: in Plato's world of forms, 36–37. *See also* underlying reality
reason: Aristotle on soul and, 63; Plato on, 37, 38, 39; pneuma of Stoics and, 80; twelfth-century natural philosophy and, 214. *See also* faith and reason; rationalism; rationality

reflection of light, 107–8, 183–85, 317
refraction of light: Islamic scholars
  on, 183, 184, 316, 317; Ptolemy on,
  108–9, 362, 378n29; rainbow and,
  184, 277
regimen, 41, 116, 336
Regiomontanus, Johannes, 161–62, 267
Reichenau, 322
Reims, 205
religion. *See* Christianity; God or gods;
  Islam; Jews
Renaissance: Archimedes' rediscovery
  in, 86; beginning of, 193; Burckhardt on, 358
respiration, 41, 121, 128
resurrection of the dead, 245, 354
rete mirabile, 126, 128–29
retrograde motion: Carolingian sources
  on, 197; defined, 88; early medieval knowledge of, 262; Eudoxan
  spheres and, 88, 90; Martianus
  Capella on, 145; Ptolemy's epicycles
  and, 101, 266
revolution, scientific, 359–62, 364–67.
  See also seventeenth-century science
Rhazes. *See* Rāzī, al- (Rhazes)
Rhodes, 98, 134, 137
rhumb lines, 281
Robert de Courçon, 226
Robert of Chester, 216, 267
Roger Frugard, 343
Roman Empire, 132–33; Christianity in, 148, 159, 163–64; decline
  of, 146–47, 151–52; education in,
  150–52; separation of West and
  East, 147, 159
Roman scholarship: borrowed from
  Greeks, 132–39, 146–48, 151, 196;
  commentary tradition, 142–44, 148;
  decline of medical tradition, 321;
  encyclopedic works, 137, 139–42,
  148; mathematical arts in, 144–45;
  taught in twelfth-century schools,
  206

Rome, 67; early medieval medicine
  in, 322; Galen in, 125, 132; Greek
  scholars in, 133, 134; library of
  Lyceum shipped to, 74; methodist
  physicians in, 124

Sabra, A. I., 174
Salerno: dissection of pig in, 345;
  medical activity in, 329–30, 332,
  333, 334–35
Samarqand: madrasa in, 175, 191;
  observatory in, 180, 183, 191, 363;
  sextant, 183
Sampson hospital, 349
Sapur I, 170
Sasanids, 170
schools. *See* education
science: defined, 1–3; foundational
  questions in, 44; invention of writing and, 11; technology vs., 1, 3, 4.
  See also experimentation; natural
  philosophy
scientific revolution, 359–62, 364–67.
  See also seventeenth-century science
Scot, Michael, 392n6
seasons: Aristotle on, 272; Augustine on,
  273–74; prehistoric awareness of, 4
secondary qualities, 34, 77, 365
semen: in Aristotle's theory of generation, 63
seminal causes, 210
Seneca, 139; on rotating earth, 282;
  twelfth-century schools and, 206, 209
Sennert, Daniel, 294–95
sense perception: Aristotle on, 46, 47,
  48, 53–54; early Greek philosophers
  and, 33, 34; Epicurus on, 77; Plato
  and, 36–38, 46. *See also* empiricism; experience; experimentation;
  observation
Septimius Severus, 125
seventeenth-century science: atomism or corpuscularism in, 31, 295;
  Christian theology and, 412n36;

continuity question and, 357, 359; experimentation in, 396n48; local motion in, 296, 401n17; as revolutionary, 359–62, 364–67

sexagesimal number system, 13, 15, 177, 371n29

sextant, underground, 182

Shīrāzī, al-, 183, 386n34

sidereal motion, 94

Siger of Brabant, 244, 246, 395n31

Simplicius, 75, 160

simulacrum, 105

skeletal anatomy, 126

skepticism: fourteenth-century, 251; in literate vs. preliterate cultures, 10; within the Platonic school, 138

Smyrna, 124

Snell's law, 183, 386n33

Socrates, 34–35, 69

solstices, 87, 92, 98

sophists, 69

Soranus, 322

soul: Albert the Great on, 240; Aquinas on, 243; Aristotle on, 63, 66, 160, 230, 233; Cicero on, 138, 139; condemnation of 1277 and, 247; Plato on, 36, 37, 41, 42, 105, 127, 149; Siger of Brabant on, 244; Stoics on, 80; theories of vision and, 105; twelfth-century natural philosophy and, 214. *See also* immortality of the soul

space: Aristotle's cosmology and, 56, 75; in oral traditions, 7. *See also* place, Aristotle's doctrine of

Spain: Christian reconquest of, 190, 203, 216, 263; expulsion of Jews in 1492, 191; Gerbert's pilgrimage to, 201, 215; as multireligious state, 191, 201, 216; Muslim, 166, 169, 173, 187, 189, 191, 200–202, 215; as source of astronomical knowledge, 262–63; translation from Arabic to Latin in, 215–16; Visigothic, 157

specific form of medicinal substance, 338

stadium paradox, 33

Stagira, 45

star catalogue: of al-Battānī, 179; of Hipparchus, 98

Stark, Rodney, 387n48

stars, fixed, 42, 59, 258, 259, 272

stereographic projection, 98, 264

St. Gall, 322

Stoa, 70, 71, 78, 223

Stoic philosophy, 70, 72, 78–81; astrology and, 272; Christian conflict with, 149; Galen's debt to, 125, 129; medieval cosmology and, 254, 257, 259; pneumatist physicians and, 124; Roman culture and, 134, 136–39

Strabo, 74, 160

Strato, 72, 74–75, 121

sublunar region. *See* terrestrial region

substance: Aristotle on, 47, 287, 291; divisibility of, 290

substantial form, 287, 289, 292

Sulla, Lucius Cornelius, 74

sulphur-mercury theory, 291, 401n10

sun: astrology and, 272–73, 274–75; calculated distance from earth, 260; in early Greek models, 86–88, 90, 92; as Greek deity, 22; motion on celestial sphere, 59, 271, 272; Plato on, 41; shifting of perigee of, 179. *See also* eclipses

sundial, 177, 179, 180

supernatural: Boethius of Dacia on, 245; Epicurus on fear of, 76; Greek philosophers and, 27, 39, 43, 44; oral traditions and, 6. *See also* miracles

surgery: in ancient Egypt and Mesopotamia, 18; in hospitals, 188, 349, 351; medieval, 343–45, 407n35

Swerdlow, Noel M., 386n30

swerve, Epicurus's doctrine of, 77–78

Swineshead, Richard, 300, 312

Sylvester II (pope). *See* Gerbert of Aurillac

Syria, 163, 166, 169
Syriac, 164, 165, 171, 172

technology: in eleventh-century
    Europe, 203–4; prehistoric, 4; vs.
    theoretical science, 1, 3, 4
teleology: Aristotle on, 51, 52, 61, 62,
    64, 73, 81, 364, 365; Epicurean phi-
    losophy and, 81; of Galen's anatomy
    and physiology, 129–31; Lucretius's
    denunciation of, 139; Plato and, 39,
    42; Stoic universe and, 81; Theo-
    phrastus's reservations about, 73
Tempier, Etienne, 246–47, 248, 250
tension: Stoic theory of, 80
terrestrial region, 53, 256, 277–85, 287
Tertullian, 149, 325
Thābit ibn Qurra, 173, 260, 265
Thales, 28
Themistius, 159–60
Theodoric of Freiberg, 185, 277, 363
Theodoric the Ostrogoth, 147
theology. *See* Christian theology; God
    or gods
Theon of Alexandria, 161
Theophrastus, 28, 45, 73–74; doxo-
    graphic tradition and, 73, 138; *On
    Stones*, 73
Theorica planetarum (*Theory of the
    Planets*), 266–67
theriac, 338
Thierry of Chartres, 210, 215, 255
Thomas Aquinas. *See* Aquinas, Thomas
Thorndike, Lynn, 358
tides, 271, 274
Timaeus (Plato), 38–39, 41, 105, 129;
    Calcidius's commentary on, 197,
    206, 209, 254; Ḥunayn's transla-
    tion of, 172; Latin translations of,
    134, 147, 148, 206, 225; planetary
    deities in, 214, 260, 271–72;
    Posidonius's commentary on, 137;
    twelfth-century cosmology and,
    254–55; twelfth-century natural

philosophy and, 209, 213, 214; on
    vision, 317
timekeeping, 158, 196, 223. *See also*
    calendar
Tio people of Africa, 7
Titus, 140
Toledan tables, 161, 267
Toledo, 201, 216, 330
T-O maps, 280
translation from Arabic to Latin, 162,
    215–18, 224, 225, 263, 265; of al-
    chemical works, 291; of astrological
    works, 216, 274; of medical works,
    215–16, 217, 328, 329–30, 333, 343,
    345; of optical works, 216, 317
translation from Greek to Arabic,
    169–73, 185, 189, 308, 313, 327
translation from Greek to Latin, 134,
    136, 146, 147–48, 155, 162, 215,
    217–18, 224, 225, 265; of Aristotle's
    zoological works, 353; of astrologi-
    cal works, 274; by Boethius, 148,
    194, 217; Carolingian period and,
    196, 199; of medical works, 217,
    321–22, 345; of *Physiologus*, 354; of
    Ptolemy's *Geography*, 281
translation from Greek to Syriac, 164,
    165, 171, 172
translation from Latin to vernacular
    languages, 335–36, 343
transmutation, 54, 277, 287–88;
    alchemy and, 54, 290–92, 294,
    364
transubstantiation, 249, 395n39
trephining, 345
trigonometry, 98, 102, 178
trivium, 137, 159
Trojan War, 21
Tropic of Cancer, 87, 264
Tropic of Capricorn, 87, 264
Trotula, 332, 333
Tübingen, University of, 335
Turin, University of, 335
Ṭūsī, Naṣīr al-Dīn al-, 179

Tusi-couple, 179, 385n28
two-sphere model, 86–88

Ulugh Beg, 191
Umayyads, 166, 168, 170, 189, 201
underlying reality: Aristotle on, 46–48, 65; Cicero on, 138; as foundational issue, 43, 44; mathematical analysis and, 5; Plato on, 35–36, 37, 38; pre-Socratic philosophers on, 28–31, 33, 34. *See also* elements
uniform circular motion: in Aristotle's cosmology, 59, 256, 260; in Eudoxus's astronomy, 59–60; in nested sphere planetary models, 89–95; Plato on planetary motion as, 41; in Ptolemy's astronomy, 99–104; Tusi-couple and, 179, 385n28
uniformly accelerated motion, 300, 303–6, 366
uniform motion: ancient definition of, 299; Mertonians on, 300; Oresme's representation of, 303
universe. *See* cosmology
universities, 218–24; Aristotle's works in, 226–28; astronomical text-books in, 223, 266–67; German, 221; human dissection at, 345–47; intellectual freedom in, 224; medical learning in, 327, 331, 333–35, 343; mendicant orders and, 232, 233; mingling of philosophy and theology in, 231–32, 233. *See also* Bologna, University of; Oxford University; Paris, University of
Unmoved Movers, 259–60. *See also* Prime (Unmoved) Mover
urbanization, 204–5, 218, 329
urinalysis, 126, 335, 338–39
'Uthmān, 166

vacuum. *See* void
Varro, 137, 139, 145
veins, 120, 121, 122, 127–28

velocity: Aristotle's theory of motion and, 58, 309, 310; medieval analysis of, 300–301, 303–6, 308, 311–12, 402n25
Vesalius, Andreas, 348
Vespasian, 140
Vich, 201
Virgil, 206
vision, 105–9; Alhacen on, 183, 314–17, 320; Aristotle on, 66, 73, 105, 106, 182–83, 313, 315, 317–18; Euclid on, 105–7, 183, 314; Galen on, 314, 316, 317; Greek atomists on, 105, 313, 315; Ḥunayn ibn Isḥaq on, 186, 318; Kepler on, 320, 366; Plato on, 41, 105, 317, 318; Theophrastus on, 73; writings transmitted to the West, 317–18. *See also* eye: anatomy of; optics
vital heat: Aristotle on, 61, 63; Galen on, 128; Theophrastus on, 73
Vitruvius, 139
Vivarium, monastery of, 155, 322
vivisection of prisoners, 120
void: Aristotle's denial of, 53, 54–55, 309–10; condemnation of 1277 and, 247, 248, 249, 251, 257; denial of, by Stoics, 79; of early Greek philosophers, 29, 34; in Epicurean philosophy, 77, 79, 139, 365; Strato's corpuscularism and, 75, 121
Voltaire, 358, 359

water: as the underlying reality, 79
Wearmouth, monastery of, 157
weight: in Epicurean philosophy, 77, 79; of falling body, 308, 309, 310. *See also* heaviness and lightness
weights, science of, 109–10, 361
Westman, Robert S., 360–61
wheel, 4
White, Andrew Dickson, 358, 359
William of Auxerre, 227
William of Conches, 210, 212–13, 231

William of Moerbeke, 217
Witelo, 320
world soul, 42, 130, 213–14, 255
writing, invention of, 10–11

Yaḥyā ibn Abī Manṣūr, 178
Yaḥyā ibn Barmak, 168
York, 196

Zahrāwī, Abū al-Qāsim al- (Abulcasis), 187, 328

Zeno of Citium, 70, 78
Zeno of Elea, 32–33, 34, 133, 372n21
Zeno's paradoxes, 32–33
Zeus, 22, 31
zig-zag functions, 16–17
zodiac, 87, 145, 197, 262; signs of, 275
zoology: of Albert the Great, 239, 353; of Aristotle, 60–65, 119; Byzantine writings on, 162; medieval, 353–56
Zoroastrianism, 163, 170